Lecture Notes in Mathematics

1468

J. Noguchi T. Ohsawa (Eds.)

# Prospects in Complex Geometry

Proceedings of the 25th Taniguchi International Symposium held in Katata, and the Conference held in Kyoto, July 31–August 9, 1989

Springer-Verlag
Berlin Heidelberg New York
London Paris Tokyo
Hong Kong Barcelona
Budapest

Editors

Junjiro Noguchi
Department of Mathematics
Tokyo Institute of Technology
Ohokayama Meguro, Tokyo 152, Japan

Takeo Ohsawa
Department of Mathematics
Faculty of Science
Nagoya University
Furocho, Chigusaku, Nagoya 464-01. Japan

Mathematics Subject Classification (1985): 32C10, 32G13, 32G15, 58E20, 14D20

ISBN 3-540-54053-9 Springer-Verlag Berlin Heidelberg New York
ISBN 0-387-54053-9 Springer-Verlag New York Berlin Heidelberg


Printed in Germany

Printing and binding: Druckhaus Beltz, Hemsbach/Bergstr.
2146/3140-543210 - Printed on acid-free paper

# Preface

The 25th Taniguchi International Symposium, Division of Mathematics, titled with "Prospects in Complex Geometry" was held at Kyuzeso, Katata, July 31 through August 5, 1989, and an international conference with the same title was held successively at Research Institute for Mathematical Sciences (RIMS), Kyoto University, Kyoto, August 7 through August 9. The present volume consists of papers based on talks given at the two meetings. The central subject was complex structure and the emphasis was put on geometric aspects. The topics of the papers range therefore over various materials from complex function theory in one variable to differential geometry and algebraic geometry; e.g., the Teichmüller theory, the deformation theory of special complex manifolds, the moduli theory of holomorphic and harmonic mappings, and the cohomology theory on algebraic varieties.

The International Symposium at Kyuzeso, Katata, was fully and generously supported by the Taniguchi Foundation. The international conference at RIMS, Kyoto University was jointly supported by RIMS and the Taniguchi Foundation. The organizers wish to express their deepest gratitude to Mr. Taniguchi and the Taniguchi Foundation for supporting the two meetings and their warm hospitality, and to RIMS for supporting the second meeting. They are also very grateful to Professor S. Murakami for serving the Taniguchi International Symposium as coordinator, and last but not least to all the participants, speakers and the contributors of this volume.

All papers of this volume are in final form and no similar version will be published elsewhere.

July, 1990

Organizers
J. Noguchi (Tokyo Inst. of Tech.)
T. Ohsawa (RIMS, Kyoto Univ.)

TABLE OF CONTENTS

# HYPERKAHLER STRUCTURE ON THE MODULI SPACE OF FLAT BUNDLES

Akira Fujiki
Institute of Mathematics, Yoshida College
Kyoto University, Kyoto 606, Japan

## Introduction

Let $X$ be a compact Riemann surface. Then by the results of Hitchin [H1] and Donaldson [D] there exists a natural bijective correspondence between the set $\mathfrak{M}$ of isomorphism classes of stable Higgs bundles on $X$ with vanishing chern classes and the set $\mathfrak{R}$ of equivalence classes of irreducible complex representations of the fundamental group of $X$. It has turned out that such a bijective correspondence still exists for any compact Kähler manifold of higher dimension by the works of Simpson [S1,S2] and Corlette [C] (cf. also [JY]). Moreover, Simpson [S3] has proved that $\mathfrak{M}$ has a natural structure of a quasi-projective scheme when $X$ is projective, and that with respect to this and the well-known similar structure on $\mathfrak{R}$ the above correspondence is homeomorphic.

On the other hand, in the case of a compact Riemann surface Hitchin [H1] had shown that $\mathfrak{M}$ has moreover a natural structure of a hyperkähler manifold such that if $f: Z \to \mathbb{P}$ is the corresponding Calabi family (cf.(2.2) below), the fibers $Z_t$ are complex analytically isomorphic to $\mathfrak{R}$ for all $t \neq 0, \infty$, and $Z_0$ and $Z_\infty$ are isomorphic to $\mathfrak{M}$ and its complex conjugate respectively, where $\mathbb{P}$ is the complex projective line. In fact, it was shown that the natural $\mathbb{C}^*$-action on $U := \mathbb{P}-\{0,\infty\}$ lifts to a holomorphic $\mathbb{C}^*$-action on $Z$. The main purpose of this paper is then to show that the same result is also true in the higher dimensional cases at least when we consider only the set of non-singular points. (See Theorems (1.4.1) and (8.3.1) below.) In fact, the basic idea for the proof remains just the same as in [H1] in view of the above mentioned results of Corlette and Simpson.

In Section 1 we shall formulate and state the results in the case of vector bundles though we work more generally in the framework of principal bundles, following the formulation by Ramanathan (cf.[R][RS]); see [S2] for another approach to the principal case. In Sections 2 and 3 the basic definitions and results concerning hyperkähler manifolds and hyperkähler moment maps are summarized and the method of hyperkähler quotients will be explained; here the emphasis is laid on the case where the manifold admits a special $S^1$-action. In Section 4 we discuss the simplest but basic example of a hyperkähler vector space which is already in [H1]. In relation with the description there, in (4.7) we shall see how the hyperkähler structure on the moduli space looks like in the simplest case of line bundles, a glance at which could be helpful in understanding the general situation.

In the one dimensional case the hyperkähler quotient construction leads directly to the moduli space in question, while in the higher dimensional case the same construction leads only to an infinite dimensional hyperkähler manifold. The description of the latter is the purpose of Sections 5 and 6. The result is stated in Theorem (6.6.1) (cf. also Theorem (1.5.2)). The finite dimensional moduli space in question turns out to be a hyperkähler subspace of the above infinite dimensional manifold; this will be verified in Sections 7 and 8 using the identification $\mathfrak{M} = \mathfrak{N}$ mentioned above. In Section 7 we discuss the non-flat case also. We prove in Section 9 the existence of the moduli space of stable principal Higgs bundles in general as an analytic space along the line of Ramanathan [R]. In the Appendix we have included the proofs of certain results of Hithcin [H1] and Simpson [S1] for the convenience of the reader.

This article was written without knowing the details of the content of [S2]; as a result some part of the paper (e.g. the first part of Section 9) could have been directly quoted from [S2], but we leave it in the original form for lack of time.

During the conference Tsuji pointed out that in the projective case the result could also be obtained by reducing to the one dimensional case by considering the general curve section of $X$. In fact, this would be possible by using the Higgs version of the theorem of Mehta-Ramanathan due to Simpson [S2], though we hope that our direct construction has its own interest.

## 1. Statement of Results

(1.1) Let $X$ be a connected compact Kähler manifold with a fixed Kähler metric $g$. A *Higgs (vector) bundle* over $X$ is a pair $(E,\theta)$ consisting of a holomorphic vector bundle $E$ on $X$ and an End $E$-valued holomorphic 1-form $\theta$ on $X$ such that $\theta$ satisfies the integrability condition $[\theta,\theta] = 0$. A Higgs bundle $(E,\theta)$ is called *stable if* we have $\mu(\mathcal{F}) < \mu(E)$ for any $\theta$-invariant torsion free coherent analytic subsheaf $\mathcal{F}$ of $\mathcal{O}(E)$. Here for any torsion free coherent analytic sheaf $\mathcal{G}$ on $X$, $\mu(\mathcal{G})$ denotes the rational number $\mu(\mathcal{G}) = \deg \mathcal{G} / \operatorname{rank} \mathcal{G}$ where $\deg \mathcal{G}$ is the degree of $\mathcal{G}$ with respect to the Kähler class $\gamma$ of $g$ (cf.(7.4.2) below).

Let $(E,\theta)$ be a Higgs bundle. Let $h$ be a hermitian metric on $E$ and $D_h$ the associated hermitian connection with the curvature form $F_h$. Using the metric $h$ we form the conjugate $\theta^*$ of $\theta$ which is an End $E$-valued (0,1)-form (cf.(7.1)). Set $\psi = \theta + \theta^*$. Then define another (affine) connection $D = D(E,\theta,h)$ on $E$ by $D = D_h + \psi$. We quote a special case of a basic result of Simpson [S1][S2] on Higgs bundles as a lemma.

(1.1.1) **Lemma.** Let $(E,\theta)$ be a Higgs bundle of rank r with the vanishing first and second chern classes. Then $(E,\theta)$ is stable if and only if there exists a hermitian metric $h$ on $E$ such that the associated connection $D$ is flat and irreducible. Moreover, in this case such a metric is unique.

(1.2) The above objects are also related to harmonic metrics. In general let $V \to X$ be a $C^{\infty}$ vector bundle on $X$ and $D$ a connection on $V$. Given a hermitian metric $h$ on $V$ we may decompose $D$ uniquely into: $D = D_h + \psi$, where $D_h$ is a metric connection and $\psi$ is a $C^{\infty}$ 1-form with values in self-adjoint endomorphisms of $V$ with respect to $h$. In this notation we call $h$ *harmonic* (with respect to $D$) if $D_h\psi = D_h^*\psi = 0$, where $D_h^*$ denotes the formal adjoint of $D_h$. Note that the first condition is automatic if $D$ is flat (cf.(7.1.1)). The next result is due also to Simpson [S2]. (See (7.7.1) below.)

(1.2.1) **Lemma.** Let $(E,\theta)$ be a Higgs bundle and $h$ a hermitian metric such that $D = D(E,\theta,h)$ is flat. Then $h$ is harmonic with respect to $D$ on $V = E$. Conversely, if $V$ is a $C^{\infty}$ vector bundle with a flat connection $D$, and a harmonic metric $h$ with respect to $D$, then the following hold true: 1) the curvature $F_h$ of $D_h$ is of type (1,1) (so that the (0,1)-part $D_h''$ of $D_h$ defines a holomorphic structure on $V$), 2) (1,0)-part $\theta$ of $\psi$ is holomorphic with respect to this holomorphic structure, and 3) $\theta$ is integrable; $[\theta,\theta] = 0$.

We must also recall another basic result due to Corlette [C]:

(1.2.2) **Lemma.** Let $V$ be a $C^{\infty}$ vector bundle and $D$ a flat and irreducible connection on $V$. Then there exists a unique harmonic metric $h$ with respect to $D$.

(1.3) We next consider the moduli space of the objects considered above. Let $\pi$ be the fundamental group $X$. Fix a positive integer r and set $G = GL(r,\mathbb{C})$. Then the following is well-known (cf. [JM]):

(1.3.1) **Lemma.** The set $\mathfrak{R}$ of equivalence classes of irreducible representations of $\pi$ into $G$ has a natural structure of a quasi-projective scheme, and hence of a complex analytic space.

*Proof.* Let $R$ be the set of all the representations $\rho: \pi \to G$. Then $R$ has the natural structure of a complex affine algebraic scheme such that the action of $G$ on $R$ by conjugations is algebraic. An element $\rho \in R$ is irreducible if and only if the image of $\rho$ is not contained in any parabolic subgroup of $G$. Hence by [JM;Th.1] $\rho$ is irreducible if and only if the G-orbit of $\rho$ is closed in $R$ and the identity component of the stabilizer of $\rho$ coincides with the center of $G$. A general result of geometric invariant theory [M; Chap.1,§2] yields the desired result.

On the other hand, it is not difficult to show the following:

(1.3.2) **Proposition.** The coarse moduli space of stable Higgs bundles of a fixed rank r exists as a (Hausdorff) complex analytic space.

See Section 9 for the more detail. We are interested in those connected components of the above moduli space corresponding to Higgs bundles with $c_1 = c_2 = 0$. Let $\mathfrak{M}$ be the union of all such components. From the preceeding three lemmas we get:

(1.3.3) **Lemma.** There is a natural bijection between the two moduli spaces $\mathfrak{M}$ and $\mathfrak{R}$.

However, the complex structure of $\mathfrak{R}$ depends only on the underlying topological space of $X$, while that of $\mathfrak{M}$ depends on the complex structure of $X$.

(1.4) The main purpose of this paper is to show that these two complex structures $\mathfrak{M}$ and $\mathfrak{N}$ (on the same set) appears as generic and special members of a Calabi family of a certain hyperkähler space (as far as the nonsingular points are concerned for the moment). Denote by $\bar{\mathfrak{M}}$ the complex space which is complex conjugate to $\mathfrak{M}$, and by $\mathbb{P}$ the complex projective line. Let $\mathfrak{M}_0$ and $\mathfrak{N}_0$ be the set of nonsingular points of the underlying reduced subspace of $\mathfrak{M}$ and $\mathfrak{N}$ respectively.

(1.4.1) **Theorem.** The notations and assumptions being as above, there exists a hyperkähler manifold $M$ with a special $S^1$ action such that in the associated Calabi family $\{Y_t\}_{t\in\mathbb{P}}$ the fiber $Y_0$ (resp. $Y_\infty$) over (resp. $\infty$) is isomorphic to $\mathfrak{M}_0$ (resp. $\bar{\mathfrak{M}}_0$) and the other members $Y_t$ are all isomorphic to $\mathfrak{N}_0$.

The result is due to Hitchin (cf.[H1;§§5,6,9]) in case $\dim X = 1$. For the definitions of the terminologies used above see (2.2)-(2.4) below.

(1.5) The proof proceeds roughly as follows. Let $V \to X$ be a fixed $C^\infty$ complex vector bundle of rank, say $r$. Fix an integer $k \geq \dim X + 2$. Let $\mathcal{A} = \mathcal{A}_k$ be the space of connections on $V$ which are of Sobolev class $H_k$. $\mathcal{A}$ is a complex affine space with traslation group $A_k^1 = A_k^1(X, \mathrm{End}\ E)$ of End E-valued 1-forms of class $H_k$. The group $\mathcal{G} = \mathcal{G}_{k+1}$ of complex gauge transformations of $V$ (or of the associated principal bundle) of class $H_{k+1}$ acts on $\mathcal{A}$ by complex affine transformations. A connection $D$ of $\mathcal{A}$ with curvature $F$ is said to be *Einstein*, if

(1.5.1) $\sqrt{-1}\Lambda F = \lambda I, \qquad \lambda = 2\pi n\ \mu(E)/\int_X \omega^n$ ,

where $\omega$ is the Kähler form associated to the given Kähler metric $g$ and $\Lambda$ is the trace operator with respect to $g$ (cf.[W;p.21]). Then we consider the subset $\mathcal{E} = \mathcal{E}_k$ of $\mathcal{A}$ consisting of irreducible Einstein connections $D$ of class $H_k$ which admits a weakly harmonic metric, where we call a metric h on V *weakly harmonic* if $D_K^*\psi = 0$ in the previous notation. The action of $\mathcal{G}$ on $\mathcal{A}$ preserves $\mathcal{E}$ and we form the quotients $\mathfrak{E} = \mathcal{E}/\mathcal{G}$. The main point then is to prove the following:

(1.5.2) **Theorem.** $\mathfrak{E}$ has a natural structure of a Kähler symplectic Hilbert manifold. Moreover, there exists a hyperkähler Hilbert manifold $\mathfrak{J}$ with a special $S^1$ action such that in the associated Calabi family $\{\mathfrak{J}_t\}_{t\in P}$, $\mathfrak{J}_j$ is naturally identified with $\mathfrak{E}$ as a Kähler symplectic manifold.

Here, a Kähler symplectic manifold is a Kähler manifold with a fixed nondegenerate holomorphic 2-form which is parallel with respect to the Levi-Civita connection. $\mathfrak{J}$ is obtained as a hyperkähler quotient associated to a certain hyperkähler moment map in the sense of [HKLR]. (The proof will be given at the end of Section 6.) When dim X = 1, $\mathfrak{J}$ is of finite dimension and already gives the hyperkähler manifold of Theorem (1.4.1) (cf. Hithcin [H1]). In higher dimensional case it is necessarily of infinite dimension and we check that $\mathfrak{M}_0$ is naturally a hyperkähler submanifold of $\mathfrak{J}$ above.

## §2. Hyperkähler Moment Map

(2.1) Let H be the algebra of real quaternions with $\mathbb{R}$-basis denoted by 1, $i$, $j$, $k$ as usual. Set

$$C = \{q \in H;\ q^2 = -1\}.$$

Then we have a natural identification $C = H^*/\mathbb{C}^* = P$, where $H^* := H - 0$ acts on C by inner automorphisms, $\mathbb{C}^*$ is identified with the stabilizer $H_i^*$ of $i$, and P denotes the complex projective line. Let

$Sp(1)$ be the group of unit quaternions and $\mathfrak{s} = \mathfrak{sp}(1)$ its Lie algebra, identified with the oriented space of pure quaternions, where $i$, $j$ and $k$ form an oriented basis. For any element $q \in C \subset \mathfrak{s}$ take an element $r$ and $s$ of $C$ such that $q$, $r$ and $s$ form an oriented orthonormal basis of $\mathfrak{s}$. (Denote by $A_q$ the set of pairs $(r,s)$ of such elements in $\mathfrak{s}$.) Then the complex line

$$\ell_q := \mathbb{C}(r + \sqrt{-1}s)$$

in the complexification $\mathfrak{s}^{\mathbb{C}}$ is independent of the choices of $(r,s) \in A_q$ and $L := \bigcup_{q \in C} \ell_q$ becomes a holomorphic line bundle of degree $-2$ on $C = P$ as can be checked easily (cf. e.g.[F3;p.111,Lemma 1.3]). If we identify $\mathfrak{s}^{\mathbb{C}} \simeq \mathfrak{sl}(2,\mathbb{C})$ with its dual by using the Killing form, any element of $\mathfrak{s}^{\mathbb{C}}$ defines a linear form on each fiber $\ell_q$ of $L$ depending holomorphically on $q$; in this way we may identify $\mathfrak{s}^{\mathbb{C}}$ further with the space $\Gamma(C,L^*)$ of holomorphic sections of the dual bundle $L^*$.

(2.2) We recall the notions of a hyperkähler manifold and the associated Calabi family (cf.[HKLR][F3;p.125]). A *hyperkähler manifold* is a Riemannian manifold $(M,g)$ endowed with almost complex structures $I$ and $J$ such that $IJ = -JI$ and that $(M,I;g)$ and $(M,J;g)$ are Kähler manifolds. In this case we denote the hyperkähler manifold by a quadruple $M = (M,g;I,J)$. Each element

$$(2.2.1) \qquad q = ai + bj + ck \in C, \quad a^2 + b^2 + c^2 = 1,$$

defines an almost complex structure $J_q = aI + bJ + cIJ$ on $M$, which induces a family of Kähler structures $(M_q,g)$ parametrized by $C = P$, where $M_q = (M,J_q)$ is the underlying complex manifold. The corresponding Kähler form $\omega_q$ on $M_q$ is then given by

$$\omega_q(x,y) = g(J_q x,y).$$

Furthermore, for any $u := e(r+\sqrt{-1}s) \in \ell_q$ with a unique $(r,s) \in A_q$ and a unique positive number $e$,

$$\varphi_u := e(\omega_r + \sqrt{-1}\omega_s)$$

is a holomorphic symplectic 2-form on $M_q$.

Moreover, the complex manifolds $M_q$ fit well into a holomorphic family in the following sense: There is a unique complex structure $\mathcal{J}$ on $M \times C$ which restricts to $J_q$ on $M = M \times q$ and to the standard complex structure on each $x \times C = \mathbb{P}$. Let $Z = (M \times C, \mathcal{J})$ be the resulting complex manifold. By construction *a*) the natural projection $f\colon Z \to C$ is holomorphic with fiber over $q$ identical to $M_q$, and *b*) $x \times C$ is a complex submanifold of $Z$ for any $x \in M$. We call this holomorphic family

$$\{(M_q, g)\}_{q \in C} \quad \text{or} \quad f\colon Z \to C$$

of Kähler manifolds the *Calabi family* associated to the hyperkähler manifold $M$. We also note that the association $u(\in \ell_q) \to \varphi_u$ allows us to identify the line bundle $f^*L$ as a subbundle of the bundle $\Lambda^2 T^*_{Z/C}$ of relative holomorphic 2-froms on $Z$ over $C$.

The notions of a hyperkähler manifold and a Calabi family as well as all those introduced in what follows can naturally be extended to the category of V-manifolds. We leave it to the reader to take care of the relevant details though we shall use freely the corresponding terminology for V-manifolds.

(2.3) We write $i = 0$ and $-i = \infty$ on $C = \mathbb{P}$, and accordingly, $M_0 = M_i$ and $M_\infty = M_{-i}$. The circle group $S^1 \simeq H_i^* \cap Sp(1) \subset H^*$ acts naturally on $C$, and its complexification $\mathbb{C}^*$ in $SL(2,\mathbb{C})$ with respect to the inclusion $Sp(1) \subset SL(2,\mathbb{C})$ acts transitively on $U := C - \{0,\infty\}$. Then a hyperkähler manifold $M$ as above is said to admit a *special* $S^1$-*action* if there exists a $C^\infty$ $S^1$-action on the Riemannian manifold $(M,g)$ with the following properties: 1) the above $\mathbb{C}^*$-action on $C$ lifts to a holomorphic $\mathbb{C}^*$-action on $Z$ via $f$ such that it induces the product $S^1$-action on $Z = M \times C$, and 2) for any $q \in C$, $t \in S^1$ induces a Kähler isometry $(M_q, g) \to (M_{t(q)}, g)$. We speak also of a *special* $\mathbb{C}^*$-*action* on $Z$ in this case. Note that in this case all the $(M_q, g)$, $q \in U$, are isomorphic as Kähler manifolds; hence $M$ cannot be compact; otherwise by Prop. 13 of "A. Fujiki, Publ. RIMS, Kyoto Univ., 20 (1984)"

$M_q$ would all be isomorphic, which is absurd as is well-known.

We get a holomorphic involution on $Z$ induced by $i \in S^1$; in particular it induces an anti-holomorphic involution $\sigma_j$ on $M_j$ (since $i$ sends $j$ into $-j$ on $C$ and $J_j = -J_{-j}$), and a holomorphic involution $\sigma_0$ on $M_0$. The respective fixed point sets $F_0$ and $F$ on $M_0$ and $M_j$ are of course identical with respect to the natural identification $M_0 = M_j = M$. However, $F_0$ is a complex submanifold of $M_0$, while $F$ is a 'real part' of $M_j$ in a neighborhood of $F$.

(2.4) Let $M = (M,g;I,J)$ be a hyperkähler manifold as above. Let $\dim M = 4n$. Then the structure group of the tangent bundle of $M$ is naturally reduced to the unitary symplectic group $Sp(n)$, and hence the tangent bundle admits a natural quaternion inner product $\langle\ ,\ \rangle$ whose real part is precisely the given Riemannian metric $g$. The $\mathfrak{s}$-part $\langle\ ,\ \rangle_{\mathfrak{s}}$ of $\langle\ ,\ \rangle$ with respect to the decomposition $\mathbf{H} = \mathbf{R} + \mathfrak{s}$ is an $\mathfrak{s}$-valued 2-form on $M$, which we shall denote by $\omega = \omega_{HK}$ and call the *hyperkähler form* on $M$. We may write $\omega = i \times \omega_i + j \times \omega_j + k \times \omega_k$.

Let $G$ be a connected complex linear reductive Lie group acting holomorphically on the complex manifold $(M,J)$ and $K$ a maximal compact subgroup of $G$ whose induced action on $M$ preserves the hyperkähler structure, i.e. preserves also $I$ and $g$. We then recall the notion of the hyperkähler moment map associated to the action of $K$ on $M$ as described in [HKLR].

We fix an (ad G)-invariant nondegenerate symmetric bilinear form $(\ ,\ )_{\mathfrak{g}}$ on $\mathfrak{g}$ which restricts to a (positive definite) inner product $(\ ,\ )_{\mathfrak{k}}$ on $\mathfrak{k}$, and then using this we shall identify $\mathfrak{g}$ and $\mathfrak{k}$ with their duals respectively. Here we also assume that the semisimple part $\mathfrak{g}'$ and the center $\mathfrak{z}$ are orthogonal in $\mathfrak{g}$ with respect to $(\ ,\ )_{\mathfrak{g}}$. Now a *hyperkähler moment map* associated to the action of $K$ on $M$ is a K-equivariant map $\mu\colon M \to \mathfrak{s} \times_{\mathbf{R}} \mathfrak{k}$ such that for any $a \in \mathfrak{k}$, and any smooth vector field $u$ on $M$ we have

$$(d\mu(u),a)_{\mathfrak{k}} = \omega(\underline{a},u)$$

as an element of $\mathfrak{s}$, where $\underline{a}$ is the vector field on M defined by a. Then, since $d\mu$ is uniquely determined by the above condition, $\mu$ is unique up to the additions of constant maps $M \to \mathfrak{s} \otimes_{\mathbb{R}} \mathfrak{z}_{\mathfrak{k}}$, where $\mathfrak{z}_{\mathfrak{k}}$ is the center of $\mathfrak{k}$.

(2.5) For any $q = ai + bj + ck \in C$, by evaluating $\mu(x) \in \mathfrak{s} \otimes_{\mathbb{R}} \mathfrak{k} \simeq \mathfrak{s}^* \otimes_{\mathbb{R}} \mathfrak{k}$ on $q \in \mathfrak{s}$, we get a K-equivariant map $\mu_q : M \to \mathfrak{k}$ which is a moment map for the induced action of K on the Kähler manifold $(M_q, g)$; namely, for any element $a \in \mathfrak{k}$ and any smooth vector field u on M we have $(d\mu_q(u), a) = \omega_q(\underline{a}, u)$ as a function on M. Then we may write

$$\mu = i \otimes \mu_i + j \otimes \mu_j + k \otimes \mu_k . \tag{2.5.1}$$

Conversely, given moment maps $\mu_q$, $q = i, j, k$, for the action of K on the Kähler manifolds $M_q$ the last expression defines a hyperkähler moment map.

Suppose now that we are given holomorphic and $C^\infty$ moment maps

$$\nu_j : M \to \mathfrak{g} \quad \text{and} \quad \mu_j : M \to \mathfrak{k}$$

associated respectively to the holomorphic action of G on the holomorphic symplectic manifold $(M_j, \hat{\varphi}_j)$ (cf.(2.6.1) below) and to the $C^\infty$ action of K on the Kähler manifold $(M, \omega_j)$, where we define

$$\hat{\varphi}_j = \varphi_{k+\sqrt{-1}i} = \omega_k + \sqrt{-1}\omega_i$$

(cf.(2.2)). Write

$$\nu_j = \mu_k + \sqrt{-1}\mu_i$$

with respect to the canonical decomposition $\mathfrak{g} = \mathfrak{k} \oplus \sqrt{-1}\mathfrak{k}$. Then $\mu_k$ and $\mu_i$ themselves are moment maps for the corresponding Kähler actions of K, and therefore we have a hyperkähler moment map $\mu: M \to \mathfrak{s} \otimes_{\mathbb{R}} \mathfrak{k}$ by the formula (2.5.1).

(2.6) If we let K act on C trivially, then in the induced K-action on $Z = M \times C$ each element acts biholomorphically on Z by the property b) of the manifold Z. Suppose further that this action of K on $Z$ extends to a holomorphic action of $G$ (again inducing the identity on $C$). On each fiber $M_q$ the extended action necessarily preserves the holomorphic symplectic form $\varphi_u$ for any $u = e(r+\sqrt{-1}s) \in \ell_q$ and coincides with the original one for $q = j$. Consider $\mu$ as a $C^\infty$ section of the trivial bundle $E \to M$ with fiber $s^{\mathbb{C}}\otimes_{\mathbb{C}}\mathfrak{g} = \Gamma(C,L^*\otimes_{\mathbb{C}}\mathfrak{g})$ with respect to the natural inclusion $s\otimes_{\mathbb{R}}\mathfrak{k} \subset s^{\mathbb{C}}\otimes_{\mathbb{C}}\mathfrak{g}$ (cf.(2.1)). Let $\pi\colon Z \to M$ be the natural projection, and $\nu$ the $C^\infty$ section over $Z$ of $f^*L^*\otimes_{\mathbb{C}}\mathfrak{g}$ defined to be the image of $\pi^*\mu$ with respect to the natural homomorphism

$$v\colon \pi^*E \to f^*L^*\otimes_{\mathbb{C}}\mathfrak{g} = \mathrm{Hom}(f^*L,\mathfrak{g});$$

explicitly, for any $u \in \ell_q$ as above

$$\nu_q(z,u) = e(\mu_r(x)+\sqrt{-1}\,\mu_s(x)) \in \mathfrak{g},$$

where $\nu_q = \nu|_{M_q}$, and $z = (x,q)$, $x \in M$, $q \in C$ as follows readily from the definitions.

(2.6.1) **Lemma.** For fixed $q \in C$ and $u \in \ell_q$, $\nu_{q,u}(x) := \nu_q(x,u)\colon M \to \mathfrak{g}$ is a holomorphic moment map for the holomorphic symplectic manifold $(M_q,\varphi_u)$; namely, it is G-equivariant and satisfies the equality $(d\nu_{q,u}(v),a)_{\mathfrak{g}} = \varphi(\underline{a},v)$ for any element $a \in \mathfrak{g}$ and any vector field $v$ on $M$, where $\underline{a} = \underline{a}^q$ denotes the holomorphic vector field on $M_q$ defined by $a$.

*Proof.* We have

$$(d\nu_{q,u}(v),a) = e\{(d\mu_r(v),a) + \sqrt{-1}(d\mu_s(v),a)\}$$
$$= e\{\omega_r(\underline{a},v) + \sqrt{-1}\omega_s(\underline{a},v)\} = \varphi_u(\underline{a},v).$$

This shows that $\nu_{q,u}$ is holomorphic on $M_q$ and satisfies the second condition for a moment map. The the equivariancy follows by the usual argument as follows: For a fixed $x \in M$ consider a holomorphic map $\alpha\colon G \to \mathfrak{g}$, $\alpha(g) = g(\nu_{q,u}(x)) - \nu_{q,u}(gx)$. Then $\alpha$ vanishes on $K$ and hence on its complexfication $G$.

Since $\nu$ is holomorphic on each $x \times C$ by the definitions, we conclude that $\nu$ is actually a holomorphic section of $f^*L^*\otimes_{\mathbb{C}}\mathfrak{g}$. In view of the lemma we call $\nu$ the *moment section* on $Z$.

(2.7) Suppose now that we are given a special $S^1$-action on $M$ which commutes with the action of $K$. We say that $\mu$ is *compatible with* the special $S^1$-action if, further, $\mu$ is $S^1$-equivariant when $S^1$ acts on the factor $\mathfrak{s}$ by the adjoint representation (cf.(2.3)) and on the factor $\mathfrak{k}$ trivially. In this case, the zero $N := \mu^{-1}(0)$ of the moment map $\mu$ is preserved by the $S^1$-action as well as the K-action on M. From the definition of the $S^1$-action on $\mathfrak{s}$ we immediately get the following:

(2.7.1) **Lemma.** $\mu$ is compatible with the special $S^1$-action if $\mu_i$: $M \to \mathbb{R}$ is invariant under the $S^1$-action and $\nu_i: M \to \mathbb{C}$ is $S^1$-equivariant, where $\nu_i(z) = \nu_i(z, j+\sqrt{-1}k)$ and $t \in S^1 \subset \mathbb{C}^*$ acts on $\mathbb{C}$ by the multiplication by $t^2$.

Finally assume that the special $S^1$-action on $M$ commutes with the action of $G$ on $Z$. The adjoint $\mathbb{C}^*$-action on $\mathfrak{s}^{\mathbb{C}} = \mathfrak{sl}(2,\mathbb{C})$ gives a natural lift of the $\mathbb{C}^*$-action on $C$ to one on the vector bundle $f^*L^*\otimes_{\mathbb{C}}\mathfrak{g}$ such that the natural homomorphism $v: \pi^*E \to f^*L^*\otimes_{\mathbb{C}}\mathfrak{g}$ is $\mathbb{C}^*$-equivariant, where $\mathbb{C}^*$ acts on the $\mathfrak{g}$-factor of $\pi^*E = (\mathfrak{s}^c\otimes_{\mathbb{C}}\mathfrak{g}) \times C$ trivially.

(2.7.2) **Lemma.** If the hyperkähler moment map $\mu$ is compatible with the special $S^1$-action, then the moment section $\nu: Z \to f^*L^*\otimes_{\mathbb{C}}\mathfrak{g}$ is $\mathbb{C}^*$-equivariant with respect to the associated $\mathbb{C}^*$-action on $Z$.

*Proof.* Since $\mu$ and $v$ are $S^1$-equivariant, $\nu$ also is $S^1$-equivariant. Then $\nu(z) - t^{-1}\nu(tz)$ is locally a holomorphic function of $(z,t) \in Z \times \mathbb{C}^*$ which vanishes on $Z \times S^1$. Then it must also vanish on the whole $Z \times \mathbb{C}^*$.

(2.8) We compute the differentials of the moment maps $\nu_j$ and $\mu_j$. Fix any point $x \in M$. The action of $G$ induces the $\mathbb{C}$-linear map $d = d_x: \mathfrak{g} \to T_xM$, where $T_xM$ is the (real) tangent space of $M = M_j$ at $x$ with complex structure $J$. (Hence $d(a) = \underline{a}$ in the previous notation; cf.(2.4).) Denote by $d_K$ the restriction of $d$ to $\mathfrak{k}$ and by $d_K^*: T_xM \to \mathfrak{k}$ its adjoint with respect to $(\ ,\ )_{\mathfrak{k}}$ and the Riemannian metric underlying the given Kähler metric. Similarly, denote by $d^*$ the formal adjoint of $d$ with respect to the hermitian inner products $h(\ ,\ )_{\mathfrak{g}}$ on $\mathfrak{g}$ and $h(\ ,\ )$ on $M_j$ respectively, where $h(u,v)_{\mathfrak{g}} = (u,\bar{v})_{\mathfrak{g}}$ and $h(\ ,\ )$ is the Kähler metric of $(M_j,g)$.

(2.8.1) **Lemma.** The differentials $d\mu_j: T_xM \to \mathfrak{k}$ and $d\nu_j: T_xM \to \mathfrak{g}$ at $x$ are given respectively by

$$d\mu_j = -d_K^*J \quad \text{and} \quad d\nu_j = -\sqrt{-1}\iota d^*I,$$

where $\iota$ is the complex conjugation of $\mathfrak{g}$ with respect to $\mathfrak{k}$.

*Proof.* By definition we have $(d\mu_j(v),a) = -g(Jv,\underline{a}) = -(d_K^*Jv,\underline{a})_{\mathfrak{k}}$ and hence $d\mu_j = -d_K^*J$. By using the similar results for $IJ$ and $I$ instead of for $J$, we get $d\nu_j = d\mu_{ij} + \sqrt{-1}d\mu_i = -d_K^*IJ - \sqrt{-1}\,d_K^*I$. Denoting by $p_{\mathfrak{k}}$ and $p_{\mathfrak{p}}$ the projections onto the respective factors of $\mathfrak{g} = \mathfrak{k} \oplus \mathfrak{p}$, $\mathfrak{p} = \sqrt{-1}\mathfrak{k}$, and noting that $p_{\mathfrak{k}} - p_{\mathfrak{p}} = \iota$, we get $d_K^*IJ + \sqrt{-1}\,d_K^*I = p_{\mathfrak{k}}d^*IJ + \sqrt{-1}\,p_{\mathfrak{p}}d^*I = \iota d^*IJ = \sqrt{-1}\iota d^*I$, and hence the desired result.

(2.9) Let $B$ be the kernel of the homomorphism $G \to \mathrm{Aut}\,M_j$ induced by the action, where $\mathrm{Aut}\,M_j$ denotes the group of biholomorphic automorphisms of $M_j$. Then it is easy to see that $B$ is a complexification of $B_K := B \cap K$ (cf. (3.2.7) below). The action of $G$ factors through the effective action on the reductive Lie group $\bar{G} := G/B$. We assume that $B$ is contained in the center $Z$ of $G$. Let $\mathfrak{b}$ and $\mathfrak{b}_K$ be the Lie algebras of $B$ and $B_K$ respectively, and denote by ${}^{\perp}\mathfrak{g}$ and ${}^{\perp}\mathfrak{k}$ the orthogonal complements of $\mathfrak{b}$ and $\mathfrak{b}_K$ in $\mathfrak{g}$ and $\mathfrak{k}$ with respect to $( \, , \, )_{\mathfrak{g}}$ and $( \, , \, )_{\mathfrak{f}}$ respectively; $\mathfrak{g} = \mathfrak{b} \oplus {}^{\perp}\mathfrak{g}$, and $\mathfrak{f} = \mathfrak{b}_{\mathfrak{f}} \oplus {}^{\perp}\mathfrak{f}$. Then by the invariance property of these bilinear forms we see that ${}^{\perp}\mathfrak{g}$ and ${}^{\perp}\mathfrak{f}$ are Lie subalgebras.

It is now immediate to see that the $\mathfrak{b}$-component $\nu_{\mathfrak{b}}$ of the moment map $\nu_{J}$ is constant. Then by subtracting this constant from $\nu_{J}$ we can always assume that this constant vanishes. In this case we call the moment map *normalized*, and consider $\nu_{J}$ as a map $\nu_{J}: M \to {}^{\perp}\mathfrak{g}$. Simialar remark and definition are also valid for $\mu$ and $\mu_{q}$; for instance $\mu$ is then considered as a map into $\mathfrak{s} \otimes_{\mathbb{R}} {}^{\perp}\mathfrak{f}$. We assume in what follows that moment maps are normalized in the above sense.

## 3. Hyperkähler Quotient

(3.1) Retaining the notations of the previous section we shall explain the construction of the hyperkähler quotient associated to a hyperkähler moment map (cf.[HKLR]). The result will be summarized in (3.6). First, denote by $M'_{q}$ the open subset of $M_{q}$ defined by

$$M'_{q} = \{x \in M_{q};\ \mathrm{Ker}[d\colon \mathfrak{g} \to T_{x}M_{q}] = \mathfrak{b}\}.$$

(3.1.1) **Lemma.** 1) On $M'_{J}$ the zero of a (normalized) moment map $\nu_{J}$ is a complex submanifold. 2) The zero $N$ of a (normalized) moment map $\mu_{J}$ is a $C^{\infty}$ submanifold at any point of $M'_{J} \subset M_{J} = M$.

*Proof.* 1) It suffices to show that the differential $d\nu\colon T_{x}M \to {}^{\perp}\mathfrak{g}$ is surjective, or equivalently, $\mathrm{Im}\, d^{*} = {}^{\perp}\mathfrak{g}$ by Lemma (2.8.1), where Im denotes the image. In fact, we have

$$\mathrm{Im}\, d^{*} = (\mathrm{Ker}\, d)^{\perp} = \mathfrak{b}^{\perp} = {}^{\perp}\mathfrak{g}$$

on $M'_{J}$. 2) $N$ is defined by the equations $\nu_{J} = \mu_{J} = 0$. In view of 1) it suffices to show that $d\mu_{J}$ maps $\mathrm{Ker}\, \nu_{J}$ surjectively onto ${}^{\perp}\mathfrak{f}$

at any point of $N \cap M'_j$. Now we have by Lemma (2.8.1)

$$\operatorname{Im} d\mu_j = \operatorname{Im} d_K^* J = \operatorname{Im} d_K^* = (\operatorname{Ker} d_K)^\perp = b_K^\perp = {}^\perp \mathfrak{k}$$

on $M'_j$. On the other hand,

$$\operatorname{Im} d_K^* = \operatorname{Im} d_K^* d_K = \operatorname{Im} d_K J \cdot J d_K \subset \operatorname{Im} d\mu_j(\operatorname{Ker} d\nu_j);$$

in fact, $d\nu_j \cdot J d_K = J d\nu_j \cdot d_K = 0$ at any point of $N$. 2) follows.

The above proof shows that the differential $d\mu\colon T_x M \to s \otimes_{\mathbb{R}} {}^\perp \mathfrak{k}$ of the hyperkähler moment map $\mu\colon M \to s \otimes_{\mathbb{R}} {}^\perp \mathfrak{k}$ is surjective at any point of $N \cap M'_j$. It will turn out (cf.(3.2.1) below) that $N \cap M'_j$ coincides with the open subset $N'$ of $N$ defined by

$$N' = \{x \in N;\ \operatorname{Ker}(d_{K,x}\colon \mathfrak{k} \to T_x N) = b_K\}.$$

Consider $N' \times C$ as a subset of $Z = M \times C$. It is then contained in the zero $Y$ of the moment section $\nu$. Let $Y' = G \cdot (N' \times C)$ be the G-orbit of $N' \times C$ in $Y$.

(3.1.2) **Lemma**. $Y'$ is a locally closed complex submanifold of $Z$ which is smooth over $C$.

*Proof*. It suffices to show that for any $q \in C$, $z \in N' \times q$, and $u \in \ell_q - 0$, $d\nu_u\colon T_x M \to \mathfrak{g}$ is surjective. We observe that $d\nu_u$ is decomposed into a series of maps as follows;

$$T_x M \xrightarrow{d\mu} s \times_{\mathbb{R}} {}^\perp \mathfrak{k} \subset s \otimes_{\mathbb{C}} {}^\perp \mathfrak{g} \to s \otimes_{\mathbb{C}} {}^\perp \mathfrak{g} \to \ell_q \otimes_{\mathbb{C}} {}^\perp \mathfrak{g} \simeq {}^\perp \mathfrak{g},$$

where the last isomorphism is induced by $\ell_q^* \simeq \mathbb{C}$ determined by $u \in \ell_q$. Then since $d\mu$ is surjective by the above remark, the surjectivity of $d\nu_u$ follows from that of the projection $s \to \ell_q^*$.

(3.2) For $x \in M_q$ we denote by $G_x^q$ and $K_x$ the stabilizers at $x$ in $G$ and $K$ respectively, and set $\bar{G}_x^q = G_x^q/B$ and $\bar{K}_x = K_x/B_K$. We recall the following result due to [GS;Th.4.5,(4.6)] and [K;Th.7.4].

(3.2.1) **Lemma.** 1) For $x \in N' = N' \times q \subset M_q$, $\bar{G}^q_x$ coincides with $\bar{K}_x$. In particular, it is a finite group and $N' = N \cap M'_j$. 2) Any G-orbit in Y' intersect with a unique K-orbit. 3) Y' is open in Y.

*Proof.* We briefly recall the proof. (See [GS] and [K] for the more detail.) Fix $q \in C$ and take any point $x \in N' = N' \times q$ as in 1). Then the main observation is that

(3.2.2) $T_x M = J_q(T_x(Kx)) \oplus T_x N_q$;

here Kx denotes the K-orbit of x and note that $N_q := \mu_q^{-1}(0)$ is smooth at x. (3.2.2) follows from the two relations;

(3.2.3) $J_q(T_x(Kx)) = (T_x(Kx)^{\perp_q})^{\perp}$

and

(3.2.4) $(T_x(Kx))^{\perp_q} = T_x N_q$

together with the obvious decomposition

(3.2.5) $T_x M = T_x N_q \oplus (T_x N_q)^{\perp}$

where $\perp_q$ denotes the orthogonal space with respect to the Kähler form $\omega_q$. These in turn follow readily from the relation

$$(d\mu_q(v), a)_{\mathfrak{k}} = \omega_q(\underline{a}, v) = g(v, J_q \underline{a}),$$

where $a \in \mathfrak{k}$ and v is a vector field. Now since $Gx \subset Y_q := Y \cap f^{-1}(q)$ and $Y_q \cap N_q = N$, we may restrict (3.2.2) to $Y_q$ and get

(3.2.6) $T_x Y_q = J_q(T_x(Kx)) \oplus T_x N$.

From this 3) follows immediately (since the decomposition depends smoothly on q).

We show the assertions 1) and 2). In view of the Cartan decomposition $G = P \cdot K$, $P = \exp(\sqrt{-1}\mathfrak{k})$, we may write any element g of G uniquely in the form $g = \exp(\sqrt{-1}a) \cdot k$, $a \in \mathfrak{k}$ and $k \in K$. It follows that in order to show 1) and 2) it suffices to verify that $(\exp\sqrt{-1}a)x \notin N_q$ for any nonzero $a \in \mathfrak{k}$, and $x \in N'$. Then the essential point is that $J_q\underline{a}$, which is tangent to the orbit $(\exp\sqrt{-1}ta)x$, $0\leq t<\infty$, is a radient vector field for the function $\psi_q^a(x) := (\mu_q(x), a)_I$; $g(J_q u, \underline{a}) = \omega_q(u, \underline{a}) = (d\mu_q(u), a)_I = d\psi_q^a(u)$ for any vector field $u$. In fact the assertions follow from this since $\psi_q^a = 0$ on $N$ and $J_q\underline{a}$ is orthogonal to $N_q$ at $x$ by (3.2.2).

(3.2.7) *Remark*. From (3.2.6) we see that for any point $x \in M$ the Lie algebra of $G_x^q$ is a complexification of that of $K_x$ and hence in particular it is reductive.

(3.3) Let $x \in N' \subset N$ be as in the previous lemma. Let $\mathcal{H}_q = \mathrm{Ker}\, d^* J_q \cap \mathrm{Ker}\, d^*$. Then $\mathcal{H}_q = T_x N_q \cap J_q T_x N_q$, and hence is a maximal $J_q$-invariant subspace in $T_x N_q$; moreover we have the orthogonal decomposition

(3.3.1) $\quad T_x N_q = \mathrm{Ker}\, d^* J_q = T_x(Kx) \oplus \mathcal{H}_q.$

The corresponding orthogonal decomposition takes the following form in the hyperkähler case:

(3.3.2) $\quad T_x N = T_x(Kx) \oplus \mathcal{H},$

where

$$\mathcal{H} = \mathrm{Ker}\, d^* \cap T_x N = \mathrm{Ker}\, d^* \cap \mathrm{Ker}\, d^* I \cap \mathrm{Ker}\, d^* J \cap \mathrm{Ker}\, d^* IJ.$$

In particular, $\mathcal{H}$ has the natural hyperkähler structure. We also note that $\mathcal{H}$ also fits into the orthogonal decomposition

$$T_x Y_q = T_x(Gx) \oplus \mathcal{H},$$

noting that we may also write $\mathcal{H} = \mathrm{Ker}\, d^* \cap T_x Y$. The formulae (3.3.1) and (3.3.2) are immediate consequeces of the obvious decomposition $T_x M = \mathrm{Im}\, d \oplus \mathrm{Ker}\, d^*$.

In order to make the above argument work also in the infinite dimensional case (cf.(3.7) below) we show how (3.2.4) implies the '$\supset$' part of (3.2.3), by using the decomposition (3.3.1). In fact, we have

$\mathfrak{K}_q \cap (T_xN_q)^{\perp_q} = 0$ since any element of $v \in \mathfrak{K}_q$ is written as $Jv'$ for some $v' \in \mathfrak{K}_q$ so that $g(v,v) = \omega(v,Jv) = -\omega(v,v')$; so if $v \in (T_xN_q)^{\perp_q}$, $g(v,v) = 0$ and then $v = 0$. By (3.3.1) it follows that $(T_xN_q)^{\perp_q} = T_x(Kx)$ since $T_x(Kx) \subset (T_xN_q)^{\perp_q} \subset T_xN_q$ by (3.2.4). The assertion follows from this.

(3.4) Now we pass to the quotient (cf.[HKLR]). Let $x$ be any point of $N$ and $K_x$ the stabilizer at $x$ in $K$. Then a *slice* for the action of $K$ on $N$ at $x \in N$ is a $K_x$-invariant closed submanifold $S$ of a K-invariant open subset $\mathcal{O}$ of $N$ passing through $x$ such that the action restricted to $S$ gives a K-equivariant isomorphism $K \times_{K_x} S \simeq \mathcal{O}$, where $K_x$ acts on $K$ by $(g,f) \to gf^{-1}$, $f \in K_x$, $g \in K$, and $K$ acts on $(K \times_{K_x} S)$ by the left translations on the K-factor. In fact, any $K_x$-invariant analytic germ $S$ at $x$ will suffice if $x \in N'$ (cf. the proof of Lemma (3.5.1) below). Using such slices we can put a natural structure of a $C^\infty$ V-manifold on the quotient $\bar{N} := N'/K$ by taking all $S \to S/K_x$ as above as charts of the V-structure. (Note that for $x \in N'$, $\bar{K}_x := K_x/B_K$ is a finite group.) We now take, as we may, a slice $S$ in such a way that it is orthogonal to the orbit $Kx$ at $x$. Then $T_xS$, which is identified with the 'V-tangent space' of the quotient V-manifold at the image of $x$ in $\bar{N}$, is isomorphic to $\mathfrak{K}$ (cf. (3.3.2))) and hence admits a natural hyperkähler structure. This then defines a natural hyperkähler (V-)structure on $\bar{N}$. If we restrict to the K-invariant open subset $N'' := \{ x \in N;\ K_x = B_K\}$ of $N'$, then $N''/K$ is an (ordinary) hyperkähler manifold.) The smoothness of the hyperkähler structure can be checked, e.g., by noting that the hyperkähler form $\omega = \omega_{HK}$ descends to a smooth d-closed hyperkähler V-form $\bar{\omega}$ on $\bar{N}'$ such that $\omega|_S = (\pi|_S)^*\bar{\omega}$, where $\pi\colon N' \to \bar{N}$ is the natural projection. We denote the resulting hyperkähler (V-)manifold by $\bar{N} = (\bar{N},\bar{g};\bar{I},\bar{J})$. $N$ is called the *hyperkähler quotient* of the action of $K$ on the hyperkähler manifold $M$.

(3.5) We shall next give a direct construction of the Calabi family of the hyperkähler Kähler manifold $\bar{N}$ as the quotient of the complex manifold $Y'$ (cf.[HKLR;p.560]).

(3.5.1) **Lemma.** For any point $y = (x,q) \in N' \subset Y' \subset Z = M \times C$ there exists a holomorphic slice $S$ for the action of $G$ on $Y'$ which passes through $y$, is orthogonal to the orbit $Gy$ at $y$, and is smooth over $C$.

Here, the definition of a *holomorphic slice* is obtained by replacing $K$ by $G$ and $N$ by $Y'$ in the above definition of the slice and requiring $S$ to be a complex submanifold. As a standard consequence of the above lemma we have:

(3.5.2) **Corollary.** $\bar{Y} := Y'/G$ is a complex V-manifold and $f: Y \to C$ induces a holomorphic map $\bar{f}: \bar{Y} \to C$ which is locally trivial at each point of $\bar{Y}$.

*Proof of Lemma (3.5.1).* *a)* Suppose that we have constructed a local $G_y$-invariant submanifold $S$ of $Y'$ at $y$ such that it is smooth over $C$ and the tangent space of $S_q := S \cap Y_q$ is the orthogonal complement in $T_yY$ $(= T_yY')$ of that of the orbit $Gy$; $T_yY = T_y(Gy) \oplus T_yS_q$. Then the $\bar{G}$-equivariant morphism $\sigma: \bar{G} \times S_q \to Y$ defined by the induced action has an isomorphic differential $T_e\bar{G} \oplus T_yS \to T_yY$ in view of the smoothness of $S$ over $C$, where $e$ is the unit element of $\bar{G} = G/B$. Hence, if we restrict $S_q$ around $y$, $\sigma$ is an open map onto a G-invariant open subset, say $\mathcal{O}$, of $Y$. In order to show that the natural map $\bar{G} \times_{\bar{G}_x} S_q \to \mathcal{O}$ induced by $\sigma$ is indeed isomorphic, it suffices to show that the map is injective, i.e., any two G-equivalent points, say $y_1$ and $y_2$, in $S$ are actually $G_x$-equivalent.

Let $S_0$ be a slice for the action of $K$ on $N'$ at $x$ as described in (3.4) and set $S_1 = S_0 \times C \subset N' \times C \subset Y'$. By the choices of $S$ and $S_0$ the tangent spaces at $y$ of $S_q$ and $S_{1,q} := S_0 \times s = S_0$ coincide. Hence with respect to the local isomorphism $(Y,y) \cong (\bar{G} \times S,(e,y))$ followed by the projection $(G \times S,(e,y)) \to (S,y)$, $(S_1,y)$ $(\subset (Y,y))$ is mapped diffeomorphically onto $(S,y)$. Let $\tilde{y}_1$ and $\tilde{y}_2$ be the points of $S_1$ corresponding to $y_1$ and $y_2$ respectively. Then we may write $\tilde{y}_i = g_i y_i$ for some $g_i \in \bar{G}$ which is close to the identity. Then $\tilde{y}_1$ and $\tilde{y}_2$ are also G- equivalent; so by Lemma (3.2.1) they are in the same K-orbit. Since $S_0$ is a slice, we may write $\tilde{y}_1 = k\tilde{y}_2$ for some $k \in K_x$, or $y_1 = (g_1^{-1}kg_2)y_2$. We shall show that we can conclude that $g_1^{-1}kg_2 \in \bar{G}_y$ if we choose $S$ sufficiently small, by repeating the argument in the compact case.

Supposing the contrary we shall derive a contradiction. So we may assume that there exist sequences $\{y_{1\nu}\}$ and $\{y_{2\nu}\}$ of points of $S$ with $y_{1\nu} \neq y_{2\nu}$, sequences $\{g_{1\nu}\}$ and $\{g_{2\nu}\}$ of elements of $\bar{G}$ and a sequence $\{k_\nu\}$ of elements of $\bar{K}_x$ such that $y_{1\nu} = g_{1\nu}^{-1}k_\nu g_{2\nu}y_{2\nu}$ as above and $y_{i\nu}$ converge to $y$ as $\nu$ tend to $\infty$. Then we can also assume that $g_{i\nu}$ converge to $e$, and $k_\nu$ converge to some element of $\bar{K}_x$. Then if we replace $y_{1\nu}$ by $k^{-1}y_{1\nu}$ and $g_{1\nu}$ by $k^{-1}g_{1\nu}$, we can assume that $g_{1\nu}k_\nu g_{2\nu}^{-1}$ are close to the identity. Then, because of the local isomorphism $(G \times S,(e,y)) \to (Y,y)$, we conclude that $y_{1\nu} = y_{2\nu}$. This is a contradiction. Thus $S$ is a desired slice.

*b)* It remains to construct a local submanifold $S$ at $y$ as stated at the begining of *a)*. In fact, by the usual linearlization technique we have a $G_y$-equivariant isomorphism $\beta$: $(Y,y) \to (T_yY,o)$, where $\bar{G}_y = \bar{G}_y^q$ acts on $T_yY$ by the tangential representation (cf. [Ca;p.97]). Since the orbit $Gy$ is $\bar{G}_y$-invariant, the same is true for the tangent space $T_y(Gy)$ in $T_yY$. Let $F$ be the orthogonal complement of $T_y(Gy)$ in $T_yY$, which is $\bar{G}_y$-invariant. Then we set $S = \beta^{-1}(F)$. Since the action of $G$ is along the fibers of $f$, $S$ is necessarily smooth over $C$ and has all the required properties.

By construction each fiber $\bar{Y}_q$ of $\bar{Y} \to C$ is the holomorphic symplectic quotient $Y'_q/G$ for the holomorphic moment map $\nu_u$ for the action of $G = G^q$ on $(M'_q, \varphi_u)$, where $u \in \ell_q$ is any nonzero element. The natural $C^\infty$ maps $\bar{N} = N'/K \to \bar{Y}_q := Y'_q/G$ are diffeomorphic for all $q$ by Lemma (3.2.1) and (3.3), and give rise to a $C^\infty$ trivialization $\bar{Y} \to \bar{N} \times C$; furthermore, with respect to the latter isomorphism we can identify $\bar{f}: \bar{Y} \to C$ with the Calabi family of the hyperkähler manifold N. Moreover, if $\mu_i$ and $\nu_i$ satisfy the condition of Lemma (2.7.1), then by Lemmas (2.7.1) and (2.7.2) we have the induced special $S^1$-action on $\bar{N}$.

(3.6) We summarize the above arguments. Let $M = (M,g;I,J)$ be a hyperkähler manifold with a special $S^1$-action and $f: Z \to C$ the associated Calabi family. Let $G$ be a connected reductive linear complex Lie group with a maximal compact group $K$ acting biholomorphically on the complex manifold $M_j := (M,J)$ such that the induced action of $K$ preserves the hyperkähler structure. We assume that the resulting action of $K$ on the Calabi family $Z \to C$ extends to a holomorphic action of $G$ commuting with the special $S^1$-action. Let $B$ be the kernel of the induced homomorphism $G \to \mathrm{Aut}\, M_j$ and set $B_K = B \cap K$. Suppose that we are given moment maps $\nu_j$ and $\mu_j$ as in (2.5) with the associated hyperkähler moment map $\mu: M \to \mathfrak{s} \otimes_{\mathbb{R}} \mathfrak{l}$ and the moment section $\nu \in \Gamma(Z, f^*L^* \otimes_{\mathbb{C}} \mathfrak{g})$ (cf.(2.5)(2.6)). We assume that $\mu$ is compatible with the special $S^1$-action. Let $N$ and $Y$ be the zero set of $\mu$ and $\nu$ respectively. Consider the K-invariant open subset $N' := \{x \in N;\ \bar{K}_x = K_x/B_K$ is finite$\}$ of $N$, and a G-invariant open subset $Y' = G\cdot(N' \times C)$ of $Y \subset Z$ (cf.(3.2.1)). Then we have the following:

(3.6.1) **Theorem.** The notations and assumptions being as above, there exists a natural structure of a hyperkähler V-manifold $\bar{N} = (\bar{N}, \bar{g}; \bar{I}, \bar{J})$ on the quotient space $\bar{N} = N'/K$ such that the associated Calabi family

$\bar{f}: \bar{Z} \to C$ is obtained as a quotient $Y'/G \to C$ of complex manifolds. Moreover, the special $S^1$-action on $M$ descends to one on $N$.

(3.7) The notions of hyperkähler manifolds, hyperkähler moment maps, and hyperkähler quotients all make sense also in the category of infinite dimensional (real or complex) Hilbert manifolds; also, most of the above arguments are valid in this category. The main difference is that in the infinite dimensional case the Riemannian metrics $g$, and hence the associated Kähler forms $\omega_q$ and holomorphic symplectic forms $\varphi_u$ are only assumed to be weakly nondegenerate on each tangent space $T_x$ i.e., the associated linear map into the cotangent space is injective but in general not surjective. The similar remark also applies to the inner products on $\mathfrak{g}$ and $\mathfrak{k}$. Thus for instance, for a closed subspace $E$ of $T_x$ we have in general no orthogonal decomposition $T_x = E + E^{\perp}$. We have to see therefore that this is in fact the case in our concrete situations below. Another difficulty is the lack of the Newlander-Nirenberg theorem in infinite dimension so that we need a direct construction of the complex structure on the Calabi family.

## 4. Hyperkähler Vector Space

(4.1) Let $W$ be a real vector space of even dimension $2m$, and $C$ a complex structure of $W$. We assume that there exists a $C$-invariant inner product $( \, , \, )$ on $W$. Denote by $w = w(x,y) := (Cx,y)$ the associated $C$-invariant symplectic form on $W$. Then the complexification $V := W^{\mathbb{C}}$ decomposes into a direct sum $V = V^+ + V^-$, $V^- = \bar{V}^+$, where $V^+$ is the eigenspace of $C$ with eigenvalue $\sqrt{-1}$, $C$ denoting also its $\mathbb{C}$-linear extention to $V$. Denote by $( \, , \, )_{\mathbb{C}}$ and $w_{\mathbb{C}}$ the $\mathbb{C}$-linear extensions of $( \, , \, )$ and $w$ to $V$ respectively. Let $h_j$

be the hermitian metric on $V$ determined by $(\ ,\ )_{\mathbb{C}}$; $h_j(x,y) = (x,\bar{y})_{\mathbb{C}}$ and $g = g(\ ,\ )$ the real part of $h_j$, considered as a flat Riemannian metric on $V$, where $\bar{\ }$ denotes the complex conjugation. Then $(V,g)$ admits a natural hyperkähler structure (cf. [H1;p.109]) if we define

$I$ = the unique antilinear extention of $C$ to $V$,

and

$J$ = the multiplication by $\sqrt{-1}$ = the complex structure of $V$.

Thus we have $I(x) = C(\bar{x})$. We note that $\hat{\varphi}_j$ of (2.5) is given by

(4.1.1) $\hat{\varphi}_j := \omega_k + \sqrt{-1}\omega_i = -\sqrt{-1}\, w_{\mathbb{C}}$,

and hence $\hat{\varphi}_j(x,y) = -\sqrt{-1}h_j(x,\bar{y})$ for any $x, y \in V$.

Actually, this example is isomorphic to the following standard model. Namely, write $\mathbb{H} = \mathbb{C} \oplus \mathbb{C}j$, and accordingly, $\mathbb{H}^m = \mathbb{C}^m \oplus \mathbb{C}^m j$. We then set $W = \mathbb{C}^m$ and consider $V := \mathbb{H}^m$ as the complexification of $W$ with the complex structure given by the *left* multiplication by $j$, which is thus $J$ by definition. The map $C$ is then given by the right multiplication by $i$. Furthermore, the metric is defined by the real part of the standard $\mathbb{H}$-valued inner product $(\ ,\ )_{\mathbb{H}}$ on $V$ and if we write $(\ ,\ )_{\mathbb{H}} = (\ ,\ )_1 + (\ ,\ )_2 j$ according to the decomposition $\mathbb{H} = \mathbb{C} \oplus \mathbb{C}j$ above, then we have $h_j = (\ ,\ )_1$ and $\hat{\varphi}_j = (\ ,\ )_2$. $J_q$ is just the left multiplication by $q$.

(4.2) Let $f\colon \mathcal{V} \to C$ be the Calabi family associated to the hyperkähler manifold $V = (V,g;I,J)$ in (4.1). $\mathcal{V}$ is then a holomorphic vector bundle on $C$, and for any $v \in V$, the associated 'constant' section $s_v\colon C \to \mathcal{V}$, $x \to (v,x) \in V \times C = \mathcal{V}$, is holomorphic. We now define a canonical special $S^1$-action on $\mathcal{V}$.

(4.2.1) **Lemma.** There exists a unique $\mathbb{C}^*$-action on $\mathcal{V}$ which lifts the given $\mathbb{C}^*$-action on $C$ (cf.(2.3)) such that 1) for any $v \in W$ the associated holomorphic section $s_v$ is $\mathbb{C}^*$-equivariant, and 2) on $V_0$ (cf.(2.3)) the action is trivial on $W$ and is given by the multiplication by $t^2$ on $JW$.

*Proof.* For any $x \in U = C - \{0,\infty\}$, $s_w(x)$ for all $w \in W$ span the whole vector space $(V_q, J_q)$ over $\mathbb{C}$. Here, the condition 1) defines already a unique holomorphic lift of the action on $U$ to $\mathcal{V}_U := f^{-1}(U)$. It suffices to show that this action extends holomorphically to the whole $\mathcal{V}$ and has the property stated in 2). To verify these we may work with the standard model $H^m$ above. Moreover, from the definition of the hyperkähler structure, we easily reduce the proof to the case $m = 1$, which we assume from now on. Let $u$, $v$, $w$ denote the constant sections $s_1$, $s_j$, $s_i$ of $f$ respectively.

**Sublemma.** In a neighborhood, say $D$, of $0 \in C$ we may write $w = f(z)u + g(z)v$, where $z$ is any local parameter of $D$ at $0$ and $f$ and $g$ are holomorphic functions in $D$ with $f(0) \neq 0$, $f'(0) = 0$, $g(0) = 0$, and $g'(0) \neq 0$, where ' denotes the derivative with respect to $z$.

*Proof.* In fact, using the parametrization by $a$, $b$, $c$ of (2.2.1) we compute easily that

$$f(z) = f(a,b,c) = (-bc + \sqrt{-1}a)/(a^2+c^2),$$

and

$$g(z) = (-ab+ \sqrt{-1}c)/(a^2+c^2),$$

where $0$ corresponds to the point $(1,0,0)$. From this the assertion follows immediately.

Returning to the proof of the lemma, we deduce, from the fact that $u$ and $w$ are invariant by the $\mathbb{C}^*$-action on $\mathcal{V}_U$, that any $t \in \mathbb{C}^*$ sends $v(z)$ to $1/g(z)\,\{f(t^2z) - f(z))u + g(t^2z)v\}$ with respect to a suitable coordinate $z$. But the latter clearly extends holomorphically across $M_0$; in fact, when $z$ tends to $0$, the right hand side specializes to $(t^2-1)f'(0)/g'(0)\,u + t^2v = t^2v$. The last formula also proves the property 2) of the resulting $\mathbb{C}^*$-action.

(4.2.2) **Corollary.** The $\mathbb{C}^*$-action of the above lemma defines a special $S^1$-action on $V$.

*Proof.* One computes easily that $t \in S^1$ induces the isometry $(q;x,y) \to (t(q);x,t^2y)$, $q \in C$, $x \in W$, $y \in JW$. This checks the condition 2) for the special action in (2.3). The first condition is obvious.

(4.2.3) *Remark.* The $S^1$-action on $V$ is thus characterized by the condition that it is trivial on $W$ and via the multiplication by $t^2$ for $t \in S^1 \subset \mathbb{C}^*$ on $JW$ with respect to the complex structure $I$. (For $V = H^m$ the $S^1$-action is induced by the adjoint action of $S^1 \subset Sp(1)$.)

(4.3) Let $G$ be a connected, reductive, complex linear Lie group and $K$ a maximal compact subgroup of $G$. Suppose that the $G$ acts on $V$ J-linearly by and that the induced action of $K$ preserves $W$ as well as $C$ and $( \, , \, )$ so that $K$ acts on $V$ as a group of hyperkähler isometries and then also on the holomorphic vector bundle $\mathcal{V} \to C$ biholomorphically.

(4.3.1) **Lemma.** The above K-action extends uniquely to a G-action.

*Proof.* In general the group $A$ of automorphisms of a holomorphic vector bundle $E$ on a compact complex manifold has a natural structure of a complex linear algebraic group and any homomorphism from the compact group $K$ to such a group extends uniquely to a homomorphism from its complexification $G$. (Indeed, the algebraic structure of $A$ is induced by the Zariski open embedding $A \subset \text{End } E$.)

On the other hand, every element $w$ of $W$ acts holomorphically on $\mathcal{V}$ by the translations by the constant section $s_w(C)$. Extend this action of $W$ on $\mathcal{V} \to C$ $\mathbb{C}$-linearly to $V$ by the rule; $\sqrt{-1}w \in V$ acts by the translation by $\sqrt{-1}s_w(C)$ (instead of $s_{\sqrt{-1}w}(C)$). Thus the action of $V$ is trivial on the complex subspace $JW$ in $V_0$. Since $W$ is preserved by $K$, we can form the semidirect product $K \cdot W$

with its complexification canonically identified with the semidirect product $G\cdot V$. The action of $G$ on $\mathcal{V}$ extends to an action of $G\cdot V$ on $\mathcal{V} \to C$ by fiberwise affine transformations, and the induced action of $K\cdot W$ comes from the (constant) action on the hyperkähler manifold $V$.

(4.3.2) **Lemma.** On $\mathcal{V}$ the special $\mathbb{C}^*$-action and the action of $G\cdot V$ commute.

*Proof.* Let $t \in \mathbb{C}^*$ be an arbitrary element and $\sigma_t$ the corresponding action on $\mathcal{V}$. Since $G\cdot V$ is a complexification of $K\cdot W$ it suffices to show that $\sigma_t$ commutes with any elements of $K\cdot W$. This may be checked on the dense open subset $\mathcal{V}_U = f^{-1}(U)$. Let $e_\alpha$, $1\le\alpha\le m$, be any real basis of $W$ and $s_\alpha = s_{e_\alpha}$ the corresponding constant sections. Then $s_\alpha(q)$ form a $\mathbb{C}$-basis of $V_q = (V,J_q)$ for every $q \in U$, and with respect to these bases $\sigma_t\colon V_q \to V_{t(q)}$ is given by the identity matrix. Thus it suffices to show that the expression of the elements of $K$ and $W$ is identical in $V_q$ and $V_{t(q)}$ with respect to these bases. This in turn follows from the fact that the action of $K\cdot W$ comes from a 'constant'(= independent of $q$) automorphism of the hyperkähler manifold $V$.

(4.4) From the viewpoint of $I$ the real and imaginary parts decomposition $V = W \oplus JW$ with respect to $J$ is the eigenspace decomposition for the I-linear map $C$ with respect to the complex structure $I$; moreover, $JW$ is identified with the complex conjugate $\bar{W}$ of $W$ with respect to the real part $V'_{\mathbb{R}} = V^+$. The corresponding holomorphic symplectic forms $\hat{\varphi}_i$ is related to the hermitian metric $h_i :=$ $g + \sqrt{-1}\,\omega_i$ on $V = V_i$ by $\hat{\varphi}_i(u,v) = -\sqrt{-1}h(u,\bar{v})$ (cf.(4.1)). We also note that $G\cdot V$ preserves both $W$ and $JW$ and the induced action on $JW$ factors through the linear part $G\cdot V \to G$.

It is convenient to introduce the following K-equivariant real transformation $\beta$ of $V = V^+ \oplus V^-$;

$$\beta(a,b) = 1/2\ (a-\bar{b},\ \bar{a}+b),$$

satisfying $\beta I = -J\beta$, $\beta J = I\beta$ and inducing $(J,I)$-linear isomorphisms $W \simeq V^-$ and $JW \simeq V^+$, which is in the restrictions of the natural projections to $V^{\pm}$.

Now let $\pi$: $T^*W \to W$ be the holomorphic cotangent bundle. If we identify $T^*W$ with $W \times W^*$ and then identify the latter also with the tangent space of $T^*W$ at each point, the natural pairing $W \times W^* \to \mathbb{C}$ is considered as a holomorphic 1-form, say $\alpha$, on $T^*W$. Then $\Phi = d\alpha$ is the canonical holomorphic symplectic form on $T^*W$.

(4.4.1) **Lemma.** There is a natual identification of the bundle $T^*W$ with the projection $V \to W$ with fiber $JW$ such that with respect to this identification $\hat{\varphi}_i = \Phi$, and moreover, that the action of $G\cdot V$ on $V$ coincides with the natural lift of the induced action of $G\cdot V$ on $W$ to $T^*W$.

*Proof.* By $-\sqrt{-1}\ h_i$ we identify $\bar{W} = JW$ with $W^*$; send $v \in \bar{W}$ to $f_v$, where $f_v(u) = -\sqrt{-1}\ h_i(u,\bar{v})$ for $u \in W$. Then the equality $\hat{\varphi}_i = \Phi$ follows from the relation $\hat{\varphi}_i(u,v) = -\sqrt{-1}\ h_i(u,\bar{v})$. On the other hand, for any $g \in G$ by the G-invariance of $\hat{\varphi}_i$ we have

$$g(f_v)(u) = f_v(g^{-1}(u)) = h_i(g^{-1}u,\bar{v}) = \hat{\varphi}_i(u,gv) = h_i(u,\overline{gv}) = f_{gv}(u).$$

The last assertion follows from this.

(4.5) Let $G_1$ be a connected reductive subgroup of $G\cdot V$ such that $K_1 := K \cap G_1$ is a maximal compact subgroup of $G_1$. Let $W_0$ be a $G_1$-invariant open subset of $W$ such that $S := W_0/G$ exists as a complex manifold. Suppose that we are given a hyperkähler moment map and the associated moment section as in (2.5) and (2.6) for the hyperkähler manifold $V$ and the action of $K_1$. Then by Lemma (4.4.1) and [H2; 3.4] if we set $Y_{i0} = Y' \cap (W_0 \times JW)$ $(\subset Y' \cap V_i)$ in the notation of (3.1) (for $G = G_1$ there), $Y_{i0}$ is also G-invariant and we have a canonical isomorphism

(4.5.1) $\quad Y_{i0}/G_1 \simeq T^*S$

of holomorphic symplectic manifolds, where $T^*S$ is the holomorphic cotangent bundle with its canonical symplectic structure. Identifying $S$ with the zero section of $T^*S$ we have the canonical direct sum decomposition $T(T^*S)|S = TS \oplus T^*S$, where $T$ denotes the holomorphic tangent bundle. Then if $J$ again denotes the almost complex structure on $T^*S$ induced from $J$ on $V$ via the isomorphism (4.5.1) and the $C^\infty$ inclusion $Y_{i0}/G_1 \subset N'/K_1$, then $J$ induces an anti-linear isomorphism $J$: $TS \to T^*S$ over $S \subset T^*S$, as follows from the descriptions in (4.4) and (3.3).

(4.5.2) **Lemma.** For any complex submanifold $N$ of $S$, if $J(TN)$ is a holomorphic subbundle of $T^*S|_N$, then it is isomorphic to the cotangent bundle $T^*N$. In particular $T^*N$ is naturally realized as a closed holomorphic symplectic submanifold of $Y_{i0}/G_1 \subset \bar{Z}_i$.

*Proof.* Since the restriction of the induced holomorphic symplectic form $\varphi_i$ on $TN \times JTN$ is nondegenerate, $JTN$ is isomorphic to the cotangent bundle of $N$.

Even when $S$ is only assumed to be a complex V-manifold, after obvious modifications (4.5.2) is still true for the complex V-manfold $N$ and its cotangent V-bundle. Moreover, even if $N$ has singularities, the proof show that its linear tangent fiber space is mapped isomorphically onto a linear fiber subspace $T^*N$ of $T^*S$ which is isomorphic to the cotangent bundle on the nonsingular locus. Denote by $C^*(N) \subset T^*N$ the image of the tangent cone of $N$, which we shall call the cotangent cone of $N$.

(4.6) By the definition $C$ commutes with every complex structure $J_q$ for $q \in C$; hence $C$ defines a holomorphic vector bundle endomorphism

$f_C$ of $\mathcal{V} \to C = \mathbb{P}$ with $C^2 = - \mathrm{Id}$. Hence, we have the holomorphic eigenspace decomposition $\mathcal{V} = \mathcal{V}^+ \oplus \mathcal{V}^-$ of $\mathcal{V}$ with respect to $f_C$. Denote by $p^\pm : \mathcal{V} \to \mathcal{V}^\pm$ the natural projection. Note that over $i$ (resp. $j$) $p^+$ for example gives the projection to $\mathcal{V}_i = W$ (resp. $\mathcal{V}_j = V^+$). The action of C on V commutes obviously with the special $S^1$-action and then $f_C$ also commutes with the special $\mathbb{C}^*$-action on $\mathcal{V}$ so that $\mathcal{V}^\pm$ are also $\mathbb{C}^*$-invariant. Similarly, since C commutes with the action of K, $f_C$ commutes with the action of G on $\mathcal{V}$ by the same argument as in (4.3.2). Therefore, G preserves the subspaces $\mathcal{V}^\pm$, and $p^\pm$ are G-equivariant. For any element $v \in V$, $t_v^\pm := p^\pm \cdot t_v$ is a holomorphic section of $\mathcal{V}^\pm$. Let V act on $\mathcal{V}^\pm$ via the translation by $t_v^\pm$. Then it is clear that $G \cdot V$ acts naturally also on $\mathcal{V}^\pm$ making $p^\pm$ equivariant.

(4.7) Choose now any lattice $\Gamma$ in W and set $T = (W/\Gamma) \times JW$. Then T has the induced hyperkähler structure and if we set $Z = \mathcal{V}/\Gamma = (V/\Gamma) \times C$, the natural projection $Z \to C$ is considered as the Calabi family of T. Moreover, the special $S^1$-action on V descends to the one on T. Note that the general fiber $Z_t$, $t \neq 0, \infty$, is isomorphic to $\mathbb{C}^{*2m}$, and for $t = 0, \infty$, we have $Z_t \cong A \times \mathbb{C}^m$, where A is a complex torus $(W, \pm I)/\Gamma$.

The above hyperkähler manifold in fact gives us a model for the group $\Xi$ of charactors of the fundamental group $\pi$ of a compact Kähler manifold. Namely, we have the line bundle version of Theorem (1.4.1). We denote by $\mathrm{Pic}\, X = H^1(X, \mathcal{O}_X)/H^1(X, \mathbb{Z})$ the (identity component of) the Picard variety of X, where $\mathcal{O}_X$ denotes the sheaf of germs of holomorphic functions on X. Also, $H^0(X, \Omega_X^1)$ denotes the vector space of holomorphic 1-forms on X.

(4.7.1) **Proposition.** The identity component $\Xi_0$ of $\Xi$ has the natural structure of a hyperkähler manifold (or group) with a special $S^1$-action. In the associated Calabi family $\{Z_t\}$ the general fiber

$Z_j$ is naturally identified with $H^1(X,\mathbb{C})/H^1(X,\mathbb{Z}) \simeq \mathbb{C}^{*2m}$ and the special fiber $Z_i$ with the product $\operatorname{Pic} X \times H^0(X,\Omega^1_X)$, where $2m$ is the first Betti number of $X$. The latter is identified naturally with the moduli space of topologically trivial holomorphic Higgs line bundles on $X$.

*Proof.* $\Xi$ is naturally identified with the cohomology group $H^1(X,\mathbb{C}^*)$ and the structure of the latter is described by the following commutative diagram of standard cohomology exact sequences

$$\begin{array}{ccccccccc}
 & & & & 0 & & & & \\
 & & & & \downarrow & & & & \\
 & & & & H^0(X,d\mathcal{O}_X) & = & H^0(X,\Omega^1_X) & & \\
 & & & & \downarrow & & & & \\
0 & \to & H^1(X,\mathbb{Z}) & \to & H^1(X,\mathbb{C}) & \to & H^1(X,\mathbb{C}^*) & \to & \operatorname{Tor} H^1(X,\mathbb{Z}) & \to & 0 \\
 & & \| & & \downarrow \alpha & & \downarrow & & \downarrow & & \\
0 & \to & H^1(X,\mathbb{Z}) & \to & H^1(X,\mathcal{O}_X) & \to & H^1(X,\mathcal{O}^*_X) & \to & H^2(X,\mathbb{Z}) & \to & 0 \\
 & & & & \downarrow & & & & \\
 & & & & 0 & & & &
\end{array}$$

where Tor denotes the torsion part, and the vertical sequence comes from the sheaf exact sequence $0 \to \mathbb{C} \to \mathcal{O}_X \to d\mathcal{O}_X \to 0$. Thus $\Xi_0$ is identified with the group $H^1(X,\mathbb{C})/H^1(X,\mathbb{Z})$ which is also naturally a (nontrivial) extension of $\operatorname{Pic} X$ by $H^0(X,\Omega^1_X)$. The associated "graded space" $\operatorname{Pic} X \times H^0(X,\Omega^1_X)$ is then considered as the moduli space of topologically trivial Higgs line bundles.

Now we shall see how this fits into the hyperkähler description mentioned so far. We set $W = H^1(X,\mathbb{R})$ so that $V = H^1(X,\mathbb{C})$. On $W$ we have a natural complex structure $C$ by the real isomorphism induced by $\alpha$. The Kähler structure on $X$ and the cup product defines a canonical $C$-invariant symplectic form $w$ by $w(\alpha,\beta) = \alpha \cup \beta \cup \gamma \in H^{2n}(X,\mathbb{R}) = \mathbb{R}$, $\alpha$, $\beta \in H^1(X,\mathbb{R})$, where $\gamma$ is the given Kähler class. Then we can apply the preceeding construction to this $(W,C,w)$; in particular if we set $\Gamma = H^1(X,\mathbb{Z})$, we have $\Xi_0 = V/\Gamma$ and the desired assertion follows. The identifications $W/\Gamma = \operatorname{Pic} X$, and $JW = H^0(X,\Omega^1_X)$ also follow from the description in (4.3).

Returning to the abstract setting we note the following:

(4.7.2) Lemma. Let $\Gamma_q^{\pm}$ be the image of $\Gamma$ in $V_q$ by $p^{\pm}: V \to V_q$. Then $\Gamma_q$ is a lattice in $V_q$ if $q \neq + i$ and $= \{0\}$ if $q = + i$; moreover, $V_q^{\pm}/\Gamma_q$ fit into a commutative complex Lie group $Z^{\pm} := V^{\pm}/\Gamma$ over $C$ such that the natural projection $\bar{p}^{\pm}: Z \to Z^{\pm}$ is a homomorphism of complex Lie groups over $C$, and $\bar{p}^{+} \times \bar{p}^{-}: Z \to Z^{+} \times_C Z^{-}$ is locally biholomorphic.

*Proof.* We may work in the standard model $V = \mathbb{H}$. Since $\Gamma$ is contained in the $\sqrt{-1}$-eignespace $W$ of $I$, it suffices to show that $W \cap V_q^{\pm} \neq \{0\}$ if and only if $q = \mp i$, and that in the latter case $W = V_q^{\pm}$. In fact, if $w$ is a nonzero element of $W \cap V_q^{\pm}$, then we have $iw = wi = \mp qw$, or $i = \mp q$. This proves the first assertion, and then the second one is clear.

Applying the lemma in the line bundle case above we see that the projection $p^{+}$ gives the projections: Kähler metric $g$ and with the

$$(\mathrm{Pic}\ X)^{-} \overset{p_j^{-}}{\leftarrow} Z_j = \Xi_0 \overset{p_j^{+}}{\rightarrow} \mathrm{Pic}\ X$$

$$H^0(X,\Omega_X^1) \overset{p_i^{-}}{\leftarrow} Z_i \overset{p_i^{+}}{\rightarrow} \mathrm{Pic}\ X\ ,$$

where $(\mathrm{Pic}\ X)^{-}$ is the group of isomorphism class of antiholomorphic line bundles on $X$, and the projections $p_j^{\pm}$ are induced by $a$ and its complex conjugate.

(4.8) In our case of interest the vector spaces $W$ and $V$ appear in the following form. For $G$ and $K$ as above let $\mathfrak{g}$ and $\mathfrak{k}$ be the corresponding Lie algebras with the symmetric bilinear forms $(\ ,\ )_{\mathfrak{g}}$ and $(\ ,\ )_{\mathfrak{k}}$ as in (2.4). Let $T$ be a real vector space of even dimension with a complex structure $C_T$ and with a $C_T$-invariant inner product $(\ ,\ )_T$. Then we set $W = T \otimes_{\mathbb{R}} \mathfrak{k}$, and $V = T \otimes_{\mathbb{R}} \mathfrak{g} = T_{\mathbb{C}} \otimes_{\mathbb{C}} \mathfrak{g}$.

Define a complex structure $C$ on $W$ by $C_T \otimes_{\mathbb{R}} id_{\mathfrak{k}}$ and the C-invariant inner product $( \ , \ )$ by $( \ , \ ) = ( \ , \ )_T \otimes_{\mathbb{R}} ( \ , \ )_{\mathfrak{k}}$. Then, starting from the data $(W, C, ( \ , \ ))$ we get a hyperkähler structure $V = (V, g; I, J)$ on $V$ as explained in (4.1). The action of the group $G$ on $V$ is via $id_T \otimes_{\mathbb{R}} ad$, where $ad: G \to GL(\mathfrak{g})$ is the adjoint representation. Thus all the previous results apply to this case.

## 5. Hyperkähler Vector Bundles

(5.1) Let $G$, $K$, $\mathfrak{g}$, and $\mathfrak{k}$ together with $( \ , \ )_{\mathfrak{g}}$ and $( \ , \ )_{\mathfrak{k}}$ be as in (4.8). We fix a compact connected Kähler manifold $X$ with a Kähler metric $g$ and with the corresponding Kähler form $\omega$. We also fix a $C^\infty$ principal G-bundle $P$ with a fixed reduction $P_K \to X$ of the structure group of $P$ to $K$. Let $\mathrm{Ad}\, P \to X$ (resp. $\mathrm{ad}\, P \to X$) be the associated fiber bundle with fiber $G$ (resp. $\mathfrak{g}$) on which the structure group $G$ acts via the adjoint representations. This admits a similarly defined real subbundles $\mathrm{Ad}\, P_K \subset \mathrm{Ad}\, P$ (resp. $\mathrm{ad}\, P_K \subset \mathrm{ad}\, P$) with fiber $K$ (resp. $\mathfrak{k}$) and with structure group $K$. In particular, we have a direct sum decomposition

(5.1.1) $\quad \mathrm{ad}\, P = \mathrm{ad}\, P_K \oplus \sqrt{-1}\mathrm{ad}\, P_K$

as a real vector bundle. On the other hand, the (real) cotangent bundle $T \to X$ has a natural complex structure $C_T$ induced from the complex structure of $X$. (By a *complex structure* on a vector bundle we mean a $C^\infty$ family of complex structures on each fiber; thus in this case it becomes canonically a complex vector bundles.) Form the real and complex vector bundles $W = T \otimes_{\mathbb{R}} \mathrm{ad}\, P_{\mathfrak{k}}$ and $V = T^{\mathbb{C}} \otimes_{\mathbb{C}} \mathrm{ad}\, P$ respectively. We get a complex structure $C$ on $W$ by $C = C_T \otimes_{\mathbb{R}} id$ and a natural C-invariant inner product $( \ , \ )$ on the fibers of $W$ defined by the Kähler metric on $X$ and the invariant inner product

$( \ , \ )_I$. We denote the $\mathbb{C}$-linear extension of $C$ to $V$ also by $C$; it is the multiplication by $\sqrt{-1}$ (resp. $-\sqrt{-1}$) on (1,0)- (resp. (0,1)-) forms. Then on each fiber of $W$ the situation is exactly the one which was described in (4.8). Since the constructions in Section 4 depend smoothly on any smooth parameter, we get the corresponding objects on the vector bundles $W$ and $V$; in particular we have a family of the Calabi families of the fibers of $V$ parametrized by $X$. But we shall first introduce some general definitions to describe the resulting structure more conveniently.

(5.2) Let $f: Y \to B$ be any $C^\infty$ fiber bundle on a $C^\infty$ manifold $B$. Then we call $f$ *of type (H)* if its typical fiber $F$ is a complex manifold and the structure group acts on $F$ by biholomorphic transformations. Let $S$ be a complex manifold and $\mathfrak{Q} \to X \times S$ be a $C^\infty$ fiber bundle of type (H). For any point $x \in X$ (resp. $s \in S$) we identify $S_x := x \times S$ (resp. $X_s := X \times s$) with $S$ (resp. $X$). We say that $\mathfrak{Q}$ is *holomorphic with respect to* $S$ if for any $x \in X$ the restriction $\mathfrak{Q}|_{S_x}$ is a holomorphic fiber bundle. This means that there exists a system of $C^\infty$ transition functions $\{f_{ij}\}$ for $\mathfrak{Q}$ with respect to a suitable open covering $\{U_i\}$ of $X \times S$ such that the restriction of $f_{ij}$ to each $U_i \cap U_j \cap S_x$ is holomorphic.

A real $C^\infty$ vector bundle $V$ on $X$ is said to have a *hyperkähler structure* if it admits *a)* two $C^\infty$ complex structures $I$ and $J$ satisfying $IJ = -JI$, and *b)* an $I$- and $J$-invariant inner product $( \ , \ )_V$. In this case for any $q = ai + bj + ck \in C$, $J_q := aI+bJ+cIJ$ defines a complex structure on $V$, and then also on the vector bundle $V \times C \to X \times C$ such that $V_q := V \times q$ is given a complex structure $J_q$ on $X_q$. We call the resulting complex vector bundle $\mathcal{V} \to X \times C$ the *Calabi family* of $V$. In fact, the restriction of $\mathcal{V}$ to $x \times C$ for any $x \in X$ is clearly the Calabi family for the hyperkähler vector space $V_x$; in particular $\mathcal{V}$ is holomorphic with respect to $C$. We say that $V$ admits a *special* $S^1$*-action* if the natural $\mathbb{C}^*$-action on $X \times C$ which is trivial on the X-factor lifts to a $S^1$-action on $\mathcal{V} \to X \times C$ such that for each $x \in x$, this gives rise to a special $\mathbb{C}^*$-action of the hyperkähler vector space $V_x$. Specializing to our situation we get:

(5.2.1) **Lemma.** The vector bundle $V = T^{\mathbb{C}} \otimes_{\mathbb{C}} \mathrm{ad}\, P$ defined above admits a natural hyperkähler structure (as a real bundle) $\mathbf{V} = (V, (\ ,\ ); I, J)$, where $V_j$ coincides with the standard complex structure on $V$. In particular we have the associated Calabi family $\mathcal{V} \to X \times C$ with a fixed $C^\infty$ trivialization $\mathcal{V} \simeq V \times C$. Moreover, $\mathbf{V}$ admits a special $S^1$-action.

(5.3) Before passing to the space of sections of $V$ we first give the merit of the general definitions given above.

a) Let $f: Y \to X$ be any $C^\infty$ fiber bundle of type (H). Fix a positive integer $k \geq n + 1$, where $n = \dim X$, and denote by $H_k(X,Y)$ the space of sections of the bundle $f$ of (Sobolev) class $H_k$. Here, in general, a section of a fiber bundle is said to be *of class* $H_k$ if with respect to any local coordinates its coefficients are distributions having $L^2$-integrable derivatives up to order k. (The terminology will be applied also for similar geometric objects in what follows.) By the Sobolov's lemma every section of $H_k(X,Y)$ is continuous. We have the following variation of the results of Palais [P]:

(5.3.1) **Lemma.** 1) $H_k(X,Y)$ has a natural structure of a complex Hilbert manifold. 2) If $f': Y' \to X$ is another fiber bundle of type (H) and $h: Y \to Y'$ is a $C^\infty$ bundle map of type (H), i.e., holomorphic on each fiber, then the induced map $H(h): H_k(X,Y) \to H_k(X,Y')$ is holomorphic.

*b)* Next, we consider a fiber bundle $\mathcal{Q} \to X \times S$ of type (H) which is holomorphic with respect to $S$, where $S$ is a connected complex manifold. Since $X$ is compact and $S$ is connected, the restriction $Q_s := \mathcal{Q}|_{X_s}$ of $\mathcal{Q}$ to $X_s$ are isomorphic to one and the same $C^\infty$ fiber bundle, say, $Q \to X$. For any open subet $U$ of $S$ we consider a $C^\infty$ isomorphism $\psi: \mathfrak{Q}|X{\times}U \simeq Q \times U$ as bundles over $X \times U$. We say that $\psi$ is an *admissible* trivialization of $\mathfrak{Q}$ over $U$ if for any $x \in X$, the restriction of $\psi_x$ of $\psi$ to $S_x \cap U$ is a biholomorphic map onto $Q_x \times (S_x \cap U)$. The following is then immediate to see:

(5.3.2) **Lemma.** For any point $s \in S$ there exists a neighborhood $U$ of $s$ in $S$ and an admissible $C^\infty$ trivialization of $\mathfrak{Q}$ over $U$.

Regard $H := \cup_{s \in S} H_k(X, Q_s)$ as a fiber space over $S$. Then as another variation of Palais [P] we can prove the following:

(5.3.3) **Lemma.** 1) There exists a natural structure of a holomorphic fiber bundle on the fiber space $H \to S$ with typical fiber $H_k(X,Q)$ (cf. Lemma (5.3.1)) such that for any admissible trivialization $\psi$: $\mathfrak{Q}|_{X \times U} \simeq Q \times U$ as above the induced map $\psi_*: H|_U \simeq H_k(X,Q) \times U$ is a biholomorphic isomorphism. Here, if $\mathfrak{Q} \to X$ is a vector bundle, $H$ also is a vector bundle. 2) If $\mathfrak{Q}' \to X \times S$ is another fiber bundle of type (H) which is holomorphic with respect to $S$ and with the associated fiber bundle $H' \to S$, and if $g: \mathfrak{Q} \to \mathfrak{Q}'$ is a bundle map of type (H) which is holomorphic over each $S_x$, then the induced map $H \to H'$ is a holomorphic bundle map.

Unfortunately, the author knows no good reference for Lemmas (5.3.1) and (5.3.3) though the former for example is known for long time. However, we shall not give a proof here, which is actually not very difficult along the line of [P], but requires more preparations which has little relevance to the subject of this paper.

(5.4) We apply the above general consideration to the situation of (5.1). We set $\omega^{[k]} = \omega^k/k!$ for any positive integer k. Then $\omega^{[n]}$ is the volume form of our Kähler metric $g$. Together with the given Riemannian metric on $V$ and the integration over $X$ this makes $H:=$

$H_k(X,V)$ into a real Hilbert space with the inner product denoted by $g_H$. Moreover, the complex structures $I$ and $J$ on $V$ give rise to a natural hyperkähler (vector space) structure $\mathbf{H} = (H, g_H; I, J)$ on $H$. Note that in our case $H = A^1_k(X, \mathrm{ad}\, P)$ is the space of ad $P$-valued 1-forms of class $H^k$. $\mathbf{H}$ has the real structure with respect to $J$ with real part $H_{\mathbb{R}} := H_k(X,W) = A^1_k(X, \mathrm{ad}\, P_K)$. For any $q \in C$, $J_q = aI + bJ + cIJ$ defines a (continuous) complex structure on $H$ leaving $g_H$-invariant. We denote by $H_q$ the resulting complex Hilbert space.

(5.4.1) **Proposition.** 1) The projection $\mathcal{H} := \bigcup_{q \in C} H_q \to C$ has the natural structure of a hermitian holomorphic vector bundle with fiber a complex Hilbert space. 2) There exists a special $S^1$-action on $\mathbf{H}$ inducing a $\mathbb{C}^*$-action on $\mathcal{H}$.

*Proof.* 1) We may identify $H_q$ with the complex Hilbert space $H_k(X, V_q)$, where $V_q = \mathcal{V} | X \times q$. Then $J_q$ on $H = H_q$ is induced by the corresponding complex structure on $V$. Hence by Lemma (5.3.3) $\mathcal{H}$ admits a natural holomorphic vector bundle structure. Furthermore, the hermitian metric on $\mathcal{V}$ and the integration with respect to $\omega^{[n]}$ gives rise to a $C^\infty$ hermitian metric on $\mathcal{H}$.

2) By the above identification $H_q = H_k(X, V_q)$ we see that the speicial $\mathbb{C}^*$-action $\sigma_{\mathcal{V}}\colon \mathbb{C}^* \times \mathcal{V} \to \mathcal{V}$ on the vector bundle $\mathcal{V}$ induces a $\mathbb{C}^*$-action on the vector bundle $\mathcal{H}$. In order to show that this action is holomorphic, we consider $\mathbb{C}^* \times \mathcal{V}$ as a fiber bundle with typical fiber $\mathbb{C}^* \times V$ over $C$ via the map

$$\mathbb{C}^* \times \mathcal{V} \xrightarrow{\ \mathrm{id}_{\mathbb{C}^*} \times \pi\ } \mathbb{C}^* \times C \times X \xrightarrow{\ \sigma_C \times \mathrm{id}_X\ } C \times X ,$$

where $\sigma_C$ denotes the action of $\mathbb{C}^*$ on $C$. It is then immdeiate to see that this bundle is of type (H) and is holomorphic with respect to $C$, and moreover that $\sigma_{\mathcal{V}}\colon \mathbb{C}^* \times \mathcal{V} \to \mathcal{V}$ is a bundle map over $C$ which is holomorphic on each $C_x$, $x \in X$. Then by Lemma (5.3.3) we see that the action $\mathbb{C}^* \times \mathcal{H} \to \mathcal{H}$ itself is holomorphic.

(5.5) As is well-known (cf.[FU;Prop.A2]) the gauge transformation group $\mathcal{K} = \mathcal{K}_k := H_k(X, \mathrm{Ad}\, P_K)$ is naturally a (real) Hilbert Lie group, while by Lemma (5.3.3) the complex gauge transformation group $\mathcal{G} = \mathcal{G}_k := H_k(X, \mathrm{Ad}\, P)$ has the natural structure of a complex Hilbert Lie group. The action $\mathrm{Ad}\, P \times_X V \to V$ of the relative Lie group $\mathrm{Ad}\, P \to X$ on the complex vector bundle $V \to X$ over $X$ induces a $\mathbb{C}$-linear action of $\mathcal{G}$ on $H = H_j$ such that the induced action of $\mathcal{K}$ preserves $H_{\mathbb{R}}$ and the hyperkähler structure on $H$ (cf.(4.8)). Then the semidirect product $\mathcal{G}\cdot H$ acts on $H$ by affine transformations and induces a similar action of $\mathcal{K}\cdot H_{\mathbb{R}}$, the latter being one of hyperkähler structure. In particular $\mathcal{K}\cdot H_{\mathbb{R}}$ acts on each (flat) Kähler manifold $(H_q, g_H)$.

(5.5.1) **Lemma.** The above action $\mathcal{G}\cdot H \times H \to H$ of $\mathcal{G}\cdot H$ on $H = H_j$ is holomorphic. Moreover, this action extends to a holomorphic action of the vector bundle $\mathcal{H} \to C$ over $C$ which induces the above 'constant' action of $\mathcal{K}\cdot H_{\mathbb{R}}$ on each fiber $H_q$.

*Proof.* The first assertion follows easily from Lemma (5.3.3). Denote by $(\mathrm{Ad}\, P)\cdot V$ the semidirect product $\mathrm{Ad}\, P$ and $V$ considered as Lie groups over $X$. Then by Lemma (4.3.1) (cf.(5.1)) we have the relative action of the relative Lie group $(\mathrm{Ad}\, P)\cdot V \times C \to X \times C$ on $\mathcal{V}$ which is holomorphic on each $C_x$. If we apply Lemma (5.3.3) to this map we get that the action $\mathcal{G}\cdot H \times \mathcal{H}$ $(= H_k(X, (\mathrm{Ad}\, P)\cdot V) \times \mathcal{H}) \to \mathcal{H}$ over $C$ is holomorphic. The last assertion is clear.

(5.6) We also add one remark on the structure of the gauge group when $G$ is not semisimple. The center $Z$ of $G$, considered naturally as a subgroup of $\mathcal{G}_k$, acts trivially on $H_j$ and the induced action of the quotient group $\bar{\mathcal{G}}_k := \mathcal{G}_k\ Z$ is effective. Similarly, $\bar{\mathcal{K}}_k := \mathcal{K}_k/Z_K$ acts effectively on $H_j$, where $Z_K$ is the center of $K$. Thus $Z$ and $Z_K$ correspond respectively to $B$ and $B_K$ in the setting of (2.9).

(5.6.1) **Lemma.** $\bar{\mathcal{G}}_k$ (resp. $\bar{\mathcal{K}}_k$) has a natural structure of a complex (resp. real) Hilbert Lie group whose Lie algebra is naturally identified with the orthogonal complement $\mathfrak{z}^\perp$ of the center $\mathfrak{z}$ of $\mathfrak{g}$ (resp. $\mathfrak{z}_K^\perp$ of $\mathfrak{z}_K$ of $\mathfrak{k}$) in $H_k(X, \mathrm{ad}\, P)$. Moreover, the holomorphic actions of $\mathcal{G}_k$ on $H_j$ and on $\mathcal{H} \to C$ descend to those of $\bar{\mathcal{G}}_k$.

*Proof.* We consider only the case of $\mathcal{G} = \mathcal{G}_k$. (The proof in the other case is similar.) First, we consider the case where $G = T = \mathbb{C}^{*r}$ is an algebraic torus. In this case the exponential exact sequence $0 \to \Gamma \to \mathfrak{t} \to T \to 0$ yields an exact sequence

$$0 \to \Gamma \to H_k(X, \mathrm{ad}\, P_T) \to \mathcal{T} := H_k(X, \mathrm{Ad}\, P_T) \to H^1(X, \Gamma) \to 0.$$

(Note that $\mathrm{ad}\, P_T$ and $\mathrm{Ad}\, P_T$ are both trivial in this case.) In particular, the identity component $\mathcal{T}_o$ of the gauge group is written as $\mathcal{T}_o = H_k(X, \mathrm{ad}\, P_T)/\Gamma$. On the other hand, we have the orthogonal decomposition $H_k(X, \mathrm{ad}\, P_T) = \mathfrak{t} \oplus \mathfrak{t}^\perp$, where

$$\mathfrak{t}^\perp = \{f \in H_k(X, \mathrm{ad}\, P_T);\ \int_X f\omega^{[n]} = 0 \text{ as an element of } \mathfrak{t}\}.$$

Thus we get $\mathcal{T}_o = T \times \mathfrak{t}^\perp$ as a Hilbert Lie group.

In the general case, let $G'$ be the commutator subgroup of $G$ and set $T = G/G'$. The natural bundle map $\mathrm{Ad}\, P \to \mathrm{Ad}\, P_T$ induces a homomorphism of complex Hilbert groups $\mathcal{G}_o \to \mathcal{T}_o$. Composed with the natural projection $\mathcal{T}_o \to T$ this yields a surjective homomorphism $\mathcal{G}_o \to T$. Let $\mathcal{G}'_o$ be the kernel of this homomorphism. This is a Hilbert subgroup of $\mathcal{G}_o$ and then its quotient $\mathcal{G}''_o = \mathcal{G}'/Z \cap \mathcal{G}'_o$ by the finite group $Z \cap \mathcal{G}'_o$ is again a Hilbert Lie group. We may write $\mathcal{G}_o = (\mathcal{G}'_o \times Z)/(\mathcal{G}'_o \cap Z)$. Thus the quotient group $\bar{\mathcal{G}}$, whose identity component is identified with $\mathcal{G}_o/(Z \cap \mathcal{G}_o) = \mathcal{G}'_o\ /(Z \cap \mathcal{G}'_o)$, has the natural structure of a complex Hilbert group. The assertion about the Lie algebra follows easily from the above construction together with our choice of the inner product on $\mathfrak{g}$ for which $\mathfrak{g}'$ and $\mathfrak{z}$ are orthogonal. The last assertion also is clear from the above construction.

## 6. Moduli of Einstein Connections with Weakly Harmonic Metric

(6.1) With all the above preparations we now turn to our moduli problem itself. We fix the situation as follows: $X$ is a compact connected complex manifold with a fixed Kähler metric $g$ and with the associated Kähler form $\omega$. $G$ is a connected complex reductive linear Lie group with a maximal compact group $K$. Fix also a principal $G$-bundle $P \to X$ and a reduction $P_K \to X$ of the structure group of $P$ to $K$. We refer to (5.1) for the other notations. We fix any integer $k$ with $k \geq \dim X + 2$.

Let $\mathcal{A} = \mathcal{A}_k$ (resp. $\mathcal{A}_R = \mathcal{A}_{k,R}$) be the space of connections of class $H_k$ on the principal bundle $P$ (resp. $P_K$) (cf.(5.3). This is the affine space with the corresponding vector space $H = H_k = A_k^1(X, \mathrm{ad}\, P)$ (resp. $H_R = H_{k,R} = A_k^1(X, \mathrm{ad}\, P_K)$). In particular, it inherits a natural hyperkähler structure from $H$. In what follows, however, we fix a connection $D_0$ in $\mathcal{A}_R \subset \mathcal{A}$ and identify $\mathcal{A}$ with $H$ (resp. $\mathcal{A}_R$ with $H_R$) once and for all. (All the constructions below are actually independent of the choice of such a $D_0$ up to canonical isomorphisms.) Accordingly, we identify the Calabi family of $\mathcal{A}$ with the vector bundle $\mathcal{H} \to C$ (cf. (5.4.1)). Write $\mathcal{A} = \mathcal{A}_j$ when we identify $\mathcal{A}$ with the *complex* vector space $H_j$. Then the complex gauge transformation group $\mathcal{G}_{k+1}$ of class $H_{k+1}$ acts naturally on $\mathcal{A}_j$ by complex affine transformations and preserves the flat holomorphic symplectic structure $\hat{\varphi}_j$ determined by the hyperkähler structure as before.

(6.1.1) **Lemma.** The above action $\mathcal{G}_{k+1} \times \mathcal{A}_j \to \mathcal{A}_j$ is holomorphic and extends to a holomorphic action of $\mathcal{G}_{k+1}$ on the Calabi family $\mathcal{H} \to C$ with respect to the identification $\mathcal{A}_j = H_j$. The action descends to an effective holomorphic action of $\bar{\mathcal{G}}_{k+1}$ on $\mathcal{A}_j$.

*Proof.* The above action factors through a homomorphism $a$: $\mathcal{G}_{k+1} \to \mathcal{G}_k \cdot H_k$ and the natural action of the latter on $H_j$, where $a$

composed with the projection $\mathcal{G}_k \cdot H_k \to \mathcal{G}_k$ is just the inclusion $\iota$: $\mathcal{G}_{k+1} \subset \mathcal{G}_k$, and $a - \iota$: $\mathcal{G}_{k+1} \to H_k$ is well-known to be holomorphic. (If $G = GL(r,\mathbb{C})$ for instance, then $(a - \iota)(g) = D_0 g \cdot g^{-1}$ with $g$ considered naturally as an element of $A^0_{k+1}(X, \text{ad } P)$). The lemma follows from this and Lemma (5.5.1). For the last assertion see (5.6).

Note that the restriction of the above action to $\mathcal{K}_{k+1}$ induces the automorphisms of the hyperkähler manifold $\mathcal{A}$.

(6.2) We next consider the moment maps 1) for the action of $\mathcal{G} = \mathcal{G}_{k+1}$ on the holomorphic symplectic manifold $(\mathcal{A}_J, \varphi_J)$ and 2) for the action of $\mathcal{K} = \mathcal{K}_{k+1}$ on the Kähler manifold $(H_J, g_H)$. We begin with the case 1). For simplicity we write $A^i_\ell = A^i_\ell(X, \text{ad } P)$. In analogy with the finite dimensional case we consider a moment map as a $\mathcal{G}$-equivariant holomorphic map $\nu_J$: $\mathcal{A}_J \to A^0_{k-1} := A^0_{k-1}(\text{ad } P)$ with respect to the natural $\mathcal{G}$-action on $A^0_{k-1}$ satisfying the equality $(d\nu_J(u), a)_A = \hat{\varphi}_J(\underline{a}, u)$, where $(\ ,\ )_A$ is the hermitian inner product in $A^0_{k-1}$, or in view of (4.1.1) and [W;p.23,Th.2] more explicitly,

$$\int_X \langle d\nu_{JD}(u), a\rangle \omega^{[n]} = \sqrt{-1} \int_X u \wedge Da \wedge \omega^{[n-1]}, \quad a \in A^0_{k+1}, \quad u \in A^1_k,$$

for any $D \in \mathcal{A}_J$, where $\langle\ ,\ \rangle$ denotes the symmetric bilinear form on the fibers of ad P defined by $(\ ,\ )_{\mathfrak{g}}$ and $\wedge$: $A^p_\ell \times A^q_\ell \to A^{p+q}_\ell(X)$ is defined by the exterior product and $\langle\ ,\ \rangle$.

(6.2.1) **Proposition.** A normalized moment map $\nu_J$ for the action of $\mathcal{G} = \mathcal{G}_{k+1}$ on the holomorphic symplectic manifold $(\mathcal{A}_J, \varphi_J)$ is given by $\nu_J(D) = \sqrt{-1}\Lambda F_D - c \in A^0_{k-1}$, $D \in \mathcal{A}_J$, where $F_D$ is the curvature form of D, $\Lambda$ is the adjoint of the multiplication operator L by the Kähler form $\omega$ (cf.[W]) and c is an element of the center $\mathfrak{z}$ of $\mathfrak{g}$, considered as a constant section of ad P, which is uniquely determined by the topological condition on the bundle P.

*Proof*. The proof is the same as in the real case (cf.[AB;p.587]). Set $f = \Lambda F_D$. Then $f$ is holomorphic in $D$ and we have

$$((df)_D(u),a)_A = \int_X \langle a, \Lambda Du\rangle \omega^{[n]}$$

$$= \int_X a \wedge Du \wedge \omega^{[n-1]} = \int_X u \wedge Da \wedge \omega^{[n-1]}.$$

Hence, $\sqrt{-1}\, f - c$ is a moment map for any element $c$ in $\mathfrak{z}$. Let $f_{\mathfrak{z}}$ and $F_{\mathfrak{z}}$ be the $\mathfrak{z}$-components of $f$ and $F_D$ respectively. Then we have

$$\int_X f_{\mathfrak{z}}\omega^{[n]} = \int_X F_{\mathfrak{z}} \wedge \omega^{[n-1]} = 2\pi\sqrt{-1}\, c_1\cdot\gamma^{n-1} \in H^{2n}(X,\mathbb{C}) \otimes_{\mathbb{C}} \mathfrak{z} \simeq \mathfrak{z},$$

where $c_1$ denotes the first chern class of $P$ (cf.(8.1) below). Thus $\sqrt{-1}\, f - c$ is normalized if and only if $c = 2\pi\sqrt{-1}\, c_1\cdot\gamma^{n-1}$.

We say that a connection $D$ in $\mathcal{A}$ is *Einstein* if $\Lambda F_D$ is a constant section contained in the center of $\mathfrak{g}$ (cf.[RS]), which then must coincide with the above constant $c$. Thus the zero set of the moment map $\nu_j$ is exactly the set $\mathcal{E}_0$ of Einstein connections on $P$ of class $H_k$.

(6.3) Next, we consider the case 2). First we note that every connection $D \in \mathcal{A}$ is decomposed uniquely into a sum $D = D_K + \psi$, where $D_K$ is a connection which is reducible to $P_K$ and $\psi$ is an element of $A^1_k(\sqrt{-1}\text{ad } P_K)$ (cf.(5.1.1)). Denote by $D_K^*$ the formal adjoint of the associated covariant derivative $D_K\colon A^0_{k+1}(\text{ad } P_K) \to A^1_k(\text{ad } P)$ with respect to the natural real inner products of these spaces. In this case we consider a moment map as a $\mathcal{K}$-equivariant smooth map $\mu_j\colon \mathcal{A} \to A^0_{k-1}(\text{ad } P_K)$ with respect to the adjoint action on the latter which satisfies the equality: $(d\mu_j(u),a) = \omega_j(\underline{a},u)$ for any $u \in A^1_k$ and $a \in A^0_{k-1}(\text{ad } P_K)$. The following is then due to Corlette [C;Prop.2.1]:

(6.3.1) **Proposition.** A normalized moment map $\mu_j$ for the action of $\mathcal{K} = \mathcal{K}_{k+1}$ on $(\mathcal{A}_j, \omega_j)$ is given by $\mu_j(D) = \sqrt{-1}D_K^*\psi = \sqrt{-1}\Lambda D_K^c\psi$, $D \in \mathcal{A}_j$, where $D = D_K + \psi$ and $D_K^*$ are as above, and $D_K^c = C^{-1}D_K C$ in the notation of [W].

*Proof.* Set $f = f(D_K, \psi) = D_K^*\psi = \Lambda D_K^c\psi$. (The first order Kähler identities hold for possibly non-integrable connections also.) For any $g \in \mathcal{K}$ we have

$$f\cdot g = \Lambda(g\cdot D_K\cdot g^{-1})^c(g\psi g^{-1}) = g\cdot\Lambda D_K^c\cdot g^{-1}(g\psi g^{-1}) = g\cdot\Lambda D_K^c\psi = g\cdot f.$$

This shows the $\mathcal{K}$-equivariancy of $f\colon A_k^1(\mathrm{ad}\ P) \to A_{k-1}^0(\mathrm{ad}\ P_K)$. The differential of $f$ is computed as follows. Take a smooth 1-parameter family $(D_{K,t}, \psi_t)$ of elements of $A_k^1$, $t \in (-\varepsilon, \varepsilon)$, $\varepsilon > 0$. Set $(\alpha, \varphi) = (\dot{D}_K, \dot{\psi})$, where $(\dot{\ }) = \frac{d}{dt}|_{t=0}$. Then we get

(6.3.2) $\dot{f} = \Lambda(\dot{D}_K^c\psi + D_K^c\dot{\psi}) = \Lambda(D_K^c\varphi + [\psi, \alpha^c]) = D_K^*\varphi - \Lambda[\psi^c, \alpha]$,

where $\alpha^c = C(\alpha)$ etc. On the other hand, the differential of the orbit map $\mathcal{K} \to \mathcal{A}$, $g \to g(D_K, \psi)$, is given by $\beta \to D_K\beta + [\psi, \beta]$, and its adjoint by $D_K^*\alpha + r_\psi^*\varphi$, where $r_\psi = [\psi,\ ]$. Then, noting that $r_\psi^* = \Lambda[\psi^c,\ ]$, we compute

$$D_K^*J(\alpha, \varphi) = -\sqrt{-1}D_K^*(\varphi) + \sqrt{-1}\Lambda[\psi^c, \alpha].$$

Comparing this with (6.3.2) we conclude by Lemma (2.8.2) that $\sqrt{-1}f$ is a moment map. It is clearly normalized.

We call a connection $D \in \mathcal{A}$ divergence-free, or *weakly harmonic* if $D_K^*\psi = 0$ (with respect to the given reduction $P_K$). Then the set $\mathcal{D}_0$ of the zeroes of $\mu_j$ is nothing but the set of weakly harmonic connections in $\mathcal{A}$. The differentials of the above moment mappings $\mu_j$ and $\nu_j$ at $D \in \mathcal{A}$ are given respectively by

$$d\nu_j(\beta) = \sqrt{-1}\Lambda D\beta, \qquad \beta \in A_k^1(\mathrm{ad}\ P),$$
$$d\mu_j(\alpha, \varphi) = D_K^*\varphi + \Lambda[\psi^c, \alpha], \qquad \alpha \in A_k^1(\mathrm{ad}\ P_K),\ \psi \in A_k^1(\sqrt{-1}\mathrm{ad}\ P_K)$$

where $\psi^c = C(\psi)$ (cf.(5.1)). Note that $\sqrt{-1}\Lambda D = -\sqrt{-1}D^{*c} = -\sqrt{-1}\iota D^* I$ in accordence with Lemma (2.8.1), where $D^{*c} = C^{-1}D^*C$.

(6.4) If we apply the transformation $\beta$ in (4.4) to our hyperkähler vector space $V = H$, the complex vector space $V_i = (H,I)$ is viewed as a direct sum $A_k^{0,1}(\text{ad } P) \oplus A_k^{1,0}(\text{ad } P)$, where $A^{p,q}$ denote the space of (p,q)-forms. In respect to our identification $\mathcal{A} = H$ in (6.1) the first factor $A_k^{0,1}$ should be considered as the space $\mathcal{A}^{0,1}$ of (0,1)-connections, i.e., the (0,1)-parts of the connections in $P_K$. Thus, if we write $D = D_K + \psi$ according to the direct sum decomposition $V = W \oplus JW = A_k^1(\text{ad } P_K) \oplus A_k^1(\sqrt{-1}\text{ad } P_K)$, or $\mathcal{A} = \mathcal{A}_K \oplus A_k^1(\sqrt{-1}\text{ad } P_K)$, then $\beta(D)$ is represented by the pair $(D''_K, \theta)$, where $D''_K$ is the (0,1)-part of $D_K$ and $\theta$ is the (1,0)-part of $\psi$. The actions of the gauge transformation group are then the usual one on $\mathcal{A}^{0,1}$ and $A_k^{1,0}(\text{ad } P)$. Furthermore, the moment maps $\mu_i$ and $\nu_i$ in this expression (corresponding to considering $\mu_i \cdot \beta^{-1}$ and $\mu_j \cdot \beta^{-1}$ respectively) take the following form:

$$\mu_i(D''_K, \theta) = \Lambda(F_{D_K} + [\theta, \theta^*]),$$

$$\nu_i(D''_K, \theta) = -2\,\Lambda D''_K\,\theta = 2\sqrt{-1}\,D'^*_K\theta$$

where $F_{D_K}$ denotes the curvature form of $D_K$ (cf.[H1;pp.79,90]). It follows that $\mu_i$ is $S^1$-invariant, and $\nu_i$ is $S^1$-equivariant if we let $t \in S^1$ act on $\mathbb{C}$ by the multiplication by $t^2$ (cf. Remark (4.2.3)).

(6.5) We use the abbreviations $A_\ell^i = A_\ell^i(\text{ad } P)$ and $A_{K,\ell}^i = A_\ell^i(\text{ad } P_K)$. The same arguments as in (2.5) and (2.6) shows that $\nu_j$ and $\mu_j$ give rise to a (normalized) hyperkähler moment map $\mu\colon\ \mathfrak{s} \to \mathfrak{s} \otimes_{\mathbb{R}} {}^{\perp}A_{k-1}^0$ and a holomorphic moment section $\nu$ of the holomorphic vector bundle $f^*L^* \otimes_{\mathbb{C}} {}^{\perp}A_{k-1}^0$ on $\mathcal{R}$. Together with the remark at the end of (6.4) and Lemmas (2.7.1) and (2.7.2) we conclude that $\mu$ is compatible with the special $S^1$-action and $\nu$ is $\mathbb{C}^*$-equivariant. We denote by $\mathcal{F}_0 = \mathcal{E}_0 \cap \mathcal{D}_0$ $(\subset \mathcal{A})$ and $\mathcal{G}_0$ $(\subset \mathcal{R})$ the zero sets of $\mu$ and $\nu$ respectively.

In order to be able to apply the results of Section 3 to our infinite dimensional situation we have only to note, as indicated in (3.7), the orthogonal decomposition property of the tangent spaces of $\mathcal{A}$ relative to the $\mathcal{G}$-orbits. We identify any connection $D$ in $\mathcal{A}$ with the associated covariant derivative $D: A^0_{k+1} \to A^1_k = H$. Let $D^*: A^1_k \to A^0_{k-1}$ be the adjoint of $D$. Similarly, we consider $D_K: A^0_{K,k+1} \to A^1_k$ and its (real) adjoint $D^*_K: A^1_k \to A^0_{K,k-1}$. Since $D^*D$ and $D_K D^*_K$ are self-adjoint strongly elliptic operators, we deduce the $L^2$-orthogonal decompositions

$$(6.5.1)\quad \begin{cases} A^0_{k-1} = \operatorname{Ker} D \oplus \operatorname{Im} D^* \\ A^0_{K,k-1} = \operatorname{Ker} D_K \oplus \operatorname{Im} D^*_K \end{cases}$$

and

$$(6.5.2)\qquad H = A^1_k = \operatorname{Im} D \oplus \operatorname{Ker} D^* = \operatorname{Im} D_K \oplus \operatorname{Ker} D^*_K.$$

Note that from (6.5.1) we also have $\operatorname{Im} D^* = \operatorname{Im} D^*D$ and (6.5.2) implies that we also have $A^1_k = \operatorname{Im} J_q d_K \oplus \operatorname{Ker} D^* J_q$ for all $q$. We also note that for any connection $D \in \mathcal{A}$ the differential of the orbit map $\mathcal{G} \to \mathcal{A}$ (resp. $\mathcal{K} \to \mathcal{A}$), $g \to g(D)$, is given by $D: A^0_{k+1} \to A^1_k$ (resp. $D_K: A^0_{K,k+1} \to A^1_k$).

For the proof of the assertions corresponding to 2) and 3) of Lemma (3.2.1) it is sufficient to note the Cartan decomposition of the gauge transformation group $\mathcal{G}_k$ in general. First, the exponential mapping $\mathfrak{g} \to G$ induces a bundle map $\operatorname{Exp}: \operatorname{ad} P \to \operatorname{Ad} P$ and if we set $R = \operatorname{Exp}(\sqrt{-1}\operatorname{ad} P_K)$ (cf.(5.1.1)), $\operatorname{Ad} P$ decomposes into a product $\operatorname{Ad} P_K \times_X R$ which gives the Cartan decomposition of each fiber. This in turn gives rise to the desired decomposition $\mathcal{G}_k = \mathcal{K}_k \times \mathcal{R}_k$, where $\mathcal{R}_k = H_k(X,R)$ and it is a cell because Exp induces a diffeomorphism $JA^0_{K,k} = H_k(X,\sqrt{-1}\operatorname{ad} P_K) \simeq \mathcal{R}_k$.

(6.6) Recall that a connection $D \in \mathcal{A}$ is said to be *irreducible* if $\operatorname{Ker} D = \mathfrak{z}$ in $A^0_{k+1}$, where $\mathfrak{z}$ denotes the center of $\mathfrak{g}$ identified with a constant section of $\mathfrak{z} \times X \subset \operatorname{ad} P$. Let $\mathcal{E}$ be the set of irreducible connections in $\mathcal{E}_0$ which is equivalent to a weakly harmonic one, and set $\mathcal{F} = \mathcal{F}_0 \cap \mathcal{E} = \mathcal{D}_0 \cap \mathcal{E}$. The latter corresponds to the set N' in (3.1). Moreover, we denote by $\mathcal{Y}$ the $\mathcal{G}$-orbit $\mathcal{G}(\mathcal{F} \times C)$ of $\mathcal{F} \times C$ in $\mathcal{H} = \mathcal{A} \times C$, where $\mathcal{G} = \mathcal{G}_{k+1}$. Denote by $\mathfrak{E} = \mathcal{E}/\mathcal{G}$ (resp. $\mathfrak{F} = \mathcal{F}/\mathcal{K}$) be the moduli space of irreducible Einstein connections on P which is equivalent to a weakly harmonic one (resp. the moduli space of irreducible, weakly harmonic, Einstein connections) of class $H^k$ considered up to isomorphisms of class $H^{k+1}$, where $\mathcal{K} = \mathcal{K}_{k+1}$. Then with all the above remarks together with the construction technique of the quotient by $\mathcal{K}$ in a real situation as in [AHS;§6][FU;§3] we can see that the arguments in Sect. 3 can be applied word for word also to our infinite dimensinal situation and we get the following:

(6.6.1) **Theorem.** $\mathfrak{E}$ has a natural structure of a Hausdorff Kähler symplectic Hilbert V-manifold. Furthermore, $\mathfrak{F}$ has a natural structure of a hyperkähler Hilbert V-manifold with a special $S^1$-action such that in the associated Calabi family $\{\mathfrak{Z}_t\}_{t\in C}$, $\mathfrak{Z}_j$ is naturally identified with $\mathfrak{E}$ as a Kähler symplectic Hilbert V-manifold.

The associated Calabi family $\bar{f}\colon \bar{\mathfrak{Z}} \to C$ is obtained as a quotient $\mathcal{Y}/\mathcal{G} \to C$ of the holomorphic map $\mathcal{Y} \to C$ by $\mathcal{G}$.

Let Aut(P,D) be the group of D-preserving automorphisms of P. Let $\mathcal{E}'$ be the subsets of $\mathcal{E}$ consisting of all the connections D such that Aut(P,D) = Z. $\mathcal{E}'$ is a $\mathcal{G}$-invariant open subset of $\mathcal{E}$ and we have the open inclusion $\mathfrak{E}' := \mathcal{E}'/\mathcal{G} \subset \mathfrak{E}$. Since the stabilizer of $\bar{\mathcal{G}} = \bar{\mathcal{G}}_{k+1}$ at each point of $\mathcal{E}'$ is trivial, $\mathfrak{E}'$ is an (ordinary) Hilbert manifold. Similarly if we set $\mathcal{F}' = \mathcal{E}' \cap \mathcal{F}$, the quotient $\mathfrak{F}' =: \mathcal{F}'/\mathcal{K} \subset \mathfrak{F}$ is an (ordinary) hyperkähler manifold. Moreover, as follows from Lemma (3.2.1) $\mathcal{E}' = \mathcal{G}(\mathcal{F}')$ in $\mathcal{A}_j$, and therefore, $\mathfrak{E}'$ is identified with $\mathfrak{Z}'_j$ in the Calabi family $\{\mathfrak{Z}'_t\}_{t\in C}$ of $\mathfrak{F}'$. When $G = GL(r,\mathbb{C})$, by Schur's lemma we have $\mathcal{E} = \mathcal{E}'$ and $\mathcal{F} = \mathcal{F}'$. Therefore, Theorem (1.5.2) follows easily from Theorem (6.6.1) above.

# 7. Finite Dimensional Moduli

(7.1) Suppose that we are in the situation of Section 6 (cf.(6.1)). Recall that any connection $D$ in $\mathcal{A}$ is decomposed into $D = D_K + \psi$. Then the curvature form $F = F_D$ of $D$ and the curvature form $F_K = F_{D_K}$ are related by $F = F_K + D_K\psi + 1/2\,[\psi,\psi]$. Let $\theta = \psi^{1,0} \in A^{1,0}(X,\mathrm{ad}\,P)$ be the (1,0)-part of $\psi$ and $\theta^* \in A^{0,1}(X,\mathrm{ad}\,P)$ the conjugate of $\theta$ defined by the complex conjugations of 1-forms and the *minus* of the conjugation of the vector bundle ad P with respect to $\mathrm{ad}\,P_K$. Then we have $\psi = \theta + \theta^*$, and according to the types of forms and the decom- position (5.1.1) of ad P we have the following 6 components of the curvature form:

(7.1.1) $\quad F_K^{1,1} + [\theta,\theta^*], \qquad D''_K\theta + D'_K\theta^*$

(7.1.2) $\quad F_K^{2,0} + [\theta,\theta], \qquad D'_K\theta$

$\qquad\qquad F_K^{0,2} + [\theta^*,\theta^*], \qquad D''_K\theta^*$

where the superscripts denote the types of forms.

We want to obtain finite dimensional moduli spaces as a subspace of $\mathfrak{F}$ by imposing additional restrictions on th curvature form so that the resulting moduli problem is locally described by a suitable elliptic graded Lie algebra complex. There seem two natural ways for doing this, one from the J-point of view, and the other from the I-point of view. In general, these lead to a complex subspace of $\mathfrak{C} = \mathfrak{Z}_j$ and $\mathfrak{Z}_i$ respectively, but not of both in general.

In what follows any connected principal bundle, i.e., a principal bundle with a connection, or any (anti-)holomorphic principal bundle is assumed *to have the given* $P$ *as the underlying* $C^\infty$ *bundle*. The moduli space which appear below is in general not reduced. In order to treat such structures we would have to rely on the functorial formulation of the problem, which however will be omitted in the following descriptions.

(7.2) From the J-point of view it is natural to require the curvature form F is of type (1,1) so that each term of (7.1.2) vanishes. In this case the (1,0)-part D' and the (0,1)-part D" of D give rise to anti-holomorphic and holomorphic principal bundles P' and P" respectively. For simplicity we call D *bi-integrable* if F is of type (1,1). We formulate the corresponding moduli problem as follows. We denote by $\mathfrak{N}$ the set of isomorphism classes of $C^\infty$ connected principal bundles (P,D) such that D is irreducible, Einstein and bi-integrable and admits a weakly harmonic metric, i.e., a metric $P_K$ for which $D_K^*\psi = 0$. Then we have the following:

(7.2.1) **Theorem.** The set $\mathfrak{N}$ has a natural structure of a (finite dimensional) complex space; moreover, it is realized as a locally closed analytic subspace of $\mathfrak{C}$.

(7.2.2) *Remark.* In particular, $\mathfrak{N}$ has the natural Kähler V-metric induced from the Kähler V-manifold structure of $\mathfrak{C}$ (though it can be defined intrinsically). However, $\mathfrak{N}$ is in general not a hyperkähler subspace of $\mathfrak{J} = \mathfrak{C}$; in other words, it is not a complex subspace of $\mathfrak{J}_i = \mathfrak{J}$, or equivalently, it is not a symplectic subspace of $\mathfrak{C}$.

Let $\mathfrak{S}$ (resp. $\mathfrak{S}'$) be the moduli space of stable holomorphic (resp. antiholomorphic) principal bundles on X (cf.(7.6.1) below). $\mathfrak{S}$ and $\mathfrak{S}'$ are complex spaces which are canonically conjugate to each other, the antiholomorphic isomorphism $\iota: \mathfrak{S} \to \mathfrak{S}'$ being geometrically realized by the map $Q \to \bar{Q}$, where $\bar{Q}$ is the conjugate bundle of Q. In particular, the product $\mathfrak{S} \times \mathfrak{S}'$ contains a distinguished real analytic subset, namely the graph of $\iota$. On the other hand, $\mathfrak{N}$ contains a distinguished Zariski open subset $\mathfrak{U}$ corresponding to the pairs (P,D) whose associated anti-holomorphic and holomorphic principal bundles P' and P" are both stable (cf.(9.6.1) below).

(7.2.3) **Proposition.** The natural map $b: \mathfrak{U} \to \mathfrak{G} \times \mathfrak{G}'$ is locally biholomorphic.

(7.3) One of the ingredients of the proofs of (7.2.1) and (7.2.3) is the construction of the local moduli space, or the *Kuranishi space*, for a given connected principal bundle $(P,D)$ in $\mathfrak{A}$. This can be explained as follows. Namely, the Kuranishi family is constructed, according to the standard method (cf.[Gr1]), by using the following elliptic graded Lie algebra complex $(B^{\cdot},d^{\cdot})$, where $B^i$ are subpaces of $A^i(\text{ad } P) = A^i(X,\text{ad } P)$ defined by

$$B^0 = A^0(\text{ad } P),\ B^i = A^{i,0}(\text{ad } P) \oplus A^{0,i}(\text{ad } P),\ i \neq 0,\ 2,$$

and

$$B^2 = A^{2,0}(\text{ad } P) \oplus \mathbb{C}\omega \otimes A^0(\text{ad } P) \oplus A^{0,2}(\text{ad } P);$$

here in $B^2$ the middle term is considered as a subspace of $A^{1,1}(\text{ad } P)$, and the differentials are defined by the composition of those in $A^{\cdot}(\text{ad } P)$ and the natural projections $r_i: A^i(\text{ad } P) \to B^i$. Note that the bi-integrable, Einstein condition for $(P,D)$ is precisely the condition that this in fact forms a complex, i.e., $d_B d_B = 0$. (cf. (6.2.1).) Similarly the structure of a graded Lie algebra complex (cf.(9.1) below) is introduced by defining the bracket on $B^{\cdot}$ as the one induced by that of $A^{\cdot}(\text{ad } P)$ followed by the projections $r_i$. It turns out that this is an elliptic complex and its harmonic space, which describes the tangent space of the Kuranishi space, is isomorphic to $H^1(X,\text{ad } P')$ $\oplus$ $H^1(X,\text{ad } P'')$; this would then prove (7.2.3). The detail, however, will be given elsewhere since the construction of the above complex is a special case of a more general construction and is better to be discussed in that framework.

(7.4) From the I-point of view, we require that 1) the curvature form $F_K$ of $D_K$ is of type (1,1), 2) $\theta$ is holomorphic with respect to the resulting holomorphic structure on $P$ defined by $D''_K$ and finally, 3) $\theta$ is integrable; namely we assume that

(7.4.1) $F_K^{0,2} = D_K''\theta = [\theta,\theta] = 0.$

It is then immediate to see that these conditions determine a $\mathcal{G}$-invariant analytic subset of $\mathcal{D}_i$ and the moduli space is given by the quotient of it by $\mathcal{G}$. In fact, this amounts to considering the moduli space of *(principal holomorphic) Higgs (G-)bundles* $(R,\theta)$ consisting by definition of a holomorphic principal G-bundle $R$ and a holomorphic 1-form $\theta$ with values in the associated holomorphic vector bundle ad $R$ satisfying $[\theta,\theta] = 0$.

So we first recall some basic definitions related to Higgs bundles. We call any reduction $R_K$ of the structure group of $R$ to $K$ a (hermitian) *metric* on $R$. Given a metric $R_K$ we have a unique connection $D_K$ on $R$ which is reducible to $R_K$ and whose connection form is of type (1,0) (cf.[KN;p.185,Remark]). $D_K$ is called the *hermitian connection* on $R$ with respect to $R_K$. A metric on $(R,\theta)$ is called *irreducible* if any covariant constant section $s$ of ad $R$ (with respect to $D_K$) which is annihilated by $\theta$ (i.e. $[\theta,s] = 0$) is a constant section coming from an element of the center of $\mathfrak{g}$. A metric on $(R,\theta)$ is called *Hermitian-Einstein* (resp. *flat*) if the associated connection $D := D_K + \psi$ is Einstein (resp. flat), where $\psi = \theta + \theta^*$.

On the other hand, the notion of stability is defined as follows (cf.[RS;p.22]): First, let $Q$ be a complex Lie subgroup of $G$. Then a holomorphic reduction $R_Q$ of the structure group of $R$ to $Q$ on an open subset $U$ of $X$ is said to be *$\theta$-admissible* if the restriction of $\theta$ to $U$ is a section of the subbundle ad $R_Q \subset$ ad $R|U$. Let $\gamma$ be the Kähler class of our Kähler metric $g$. A Higgs bundle $(R,\theta)$ is said to be *stable* (with respect to $\gamma$) if for any Zariski open subset $U$ of $X$ with $\operatorname{codim}(X-U) \geq 2$, any maximal parabolic subgroup $Q$ of $G$, and any $\theta$-admissible holomorphic reduction $R_Q$ of the structure group of $R$ to $Q$ over $U$, the line bundle $L$ on $X$ associated to $R_Q$ and to any character of $Q$ which is dominant with respect to a Borel subgroup of $G$ contained in $Q$ (cf.[R;p.131]) has always a negative degree with respect to $\gamma$. Here, L is defined at first on U, but turns out to extend uniquely to the whole X, and the degree of L is the real number defined by

(7.4.2) $\deg L = c_1(L)\cdot\gamma^{n-1} \in H^{2n}(X,\mathbb{R}) \simeq \mathbb{R}$, where $n = \dim X$ (cf.[RS;§1]).

(7.5) The basic relation between the above two notions of stability and Hermitian-Einstein connection is given by the following:

(7.5.1) **Theorem.** A Higgs G-bundle $(R,\theta)$ is stable if and only if it admits an irreducible Hermitian-Einstein metric.

In the basic case where $G = GL(r,\mathbb{C})$, the result is due to Hitchin [H1;Th.2.18,Th.4.3] in case of dimension one, and to Simpson [S1] in the general case, the case $\theta = 0$ being due to Donaldson, and Uhlenbeck-Yau. (See (A.3.1) for the proof of the easier implication.) In the general case the proof is done by the redution to the case of $GL(r,\mathbb{C})$ (cf.[F4]) just as in [RS][D], where this reduction is made in the case of an ordinary principal bundle, i.e., the case $\theta = 0$. In fact, the line of proof is exactly the same as in Ramanathan-Subramanian [RS] with necessary modification in the Higgs case. One point worth mentioning is as follows. In [RS] the result of Ramanan-Ramanathan [RR] was used to deduce that any quasistable (see below) principal bundle admits a Hermitian-Einstein metric. We thus need an analogue of the results of [RR]: The situation is as follows. Let $\rho: G \to G'$ be a homomorphism of connected complex reductive linear algebraic groups and $d\rho: \mathfrak{g} \to \mathfrak{g}'$ the associated Lie algebra homomorphism. Let $(R,\theta)$ be a principal Higgs G-bundle and $(R',\theta')$ the induced Higgs G'-bundle with the associated bundle homomorphism $f: R \to R'$, where $\theta'$ is the image of $\theta$ by the map $A^1(\mathrm{ad}\, R) \to A^1(\mathrm{ad}\, R')$ induced by $d\rho$. In the case $\theta = \theta' = 0$, the main result of [RR] states that under the condition that X is projective, if R is quasi-stable, so is R'.

The obvious analogue in our case is the following result, where the generalization of nonalgebraic case is also relevant (cf.[F4]).

(7.5.2) **Proposition.** Let $X$ be any compact Kähler manifold. Then if $(R,\theta)$ is quasistable, then so is $(R',\theta')$ in the above notations.

Here, a Higgs bundle $(R,\theta)$ is called *quasi-stable* if there exist a parabolic subgroup $Q$ of $G$, a Levi factor $M$ of $Q$, and a reduction $(R_M,\theta_M)$ of the group of $(R,\theta)$ to $M$ such that a) $(R_M,\theta_M)$ is stable, and b) for every character of $M$ which is trivial on the center of $G$ the associated line bundle on $X$ has degree zero with respect to the Kähler class $\gamma$.

(7.6) Denote by $\mathfrak{M}$ the set of isomorphism classes of stable Higgs bundles $(R,\theta)$ on $X$. By (7.5.1) we shall identify $\mathfrak{M}$ also with the set of isomorphism classes of Higgs bundles which admits an irreducible Hermitian-Einstein metric. Then the following holds true:

(7.6.1) **Theorem.** The set $\mathfrak{M}$ has a natural structure of a finite dimensional complex space; further, $\mathfrak{M}$ can also be realized naturally as a locally closed analytic subspace of $\mathfrak{Z}_i$ of Theorem (6.6.1).

(7.6.2) *Remark.* 1) $\mathfrak{M}$ thus inherits a natural Kähler V-metric induced from that of $\mathfrak{Z}_i$, which can also be defined intrinsically. But it is in general not a symplectic subspace of $\mathfrak{Z}_i$ (cf. Remark (7.2.2)).

2) When $X$ is projective and $G = GL(r,\mathbb{C})$, Simpson [S3] has proved more precisely that $\mathfrak{M}$ has the structure of a quasiprojective scheme.

The proof of (7.6.1) will be sketched in Section 9. Note that $\mathfrak{M}$ contains a distinguished Zariski open subset $\mathfrak{M}_1 := \{(R,\theta) \in \mathfrak{M};\ R$ is stable$\}$. (Here, and in what follows, we often identify an isomorphism class with one of its representatives.) This open set is described as follows. Let $\mathfrak{S}$ be the moduli space of stable holomorphic principal bundles on $X$ as in (7.2). Then it is not difficult to construct a fiber space $u: \mathfrak{B} \to \mathfrak{S}$ whose fiber over a point $R \in \mathfrak{S}$ is isomorphic to the quotient $C_R/\mathrm{Aut}\ R$ of the cone $C_R := \{\theta \in H^0(X, \mathrm{ad}\ R); [\theta,\theta] = 0\}$ of Higgs fields by the finite group $\mathrm{Aut}\ R$ of automorphism of $R$. Then we have a natural map $\mathfrak{B} \to \mathfrak{M}$, which actually gives an isomorphism of $\mathfrak{B}$ and $\mathfrak{M}_1$, the inverse map being easily specified.

(7.7) We shall give a result which relates the two classes of objects considered in (7.2) and (7.4), and which is due essentially to Simpson [S2].

(7.7.1) **Proposition.** Let $(P,D)$ be a connected principal bundle which is bi-integrable and Einstein. Let $P_K$ be a metric on $P$ and write $D = D_K + \psi$ with $\psi = \theta + \theta^*$ as before. Then the metric is harmonic with respect to $D$ if and only if the condition (7.4.1) is satisfied. In this case, if $R$ denotes the resulting holomorphic principal bundle $(P, D_K'')$, $(R,\theta)$ becomes a Higgs bundle with a Hermitian-Einstein metric $P_K$; moreover, $(R,\theta)$ is irreducible if and only if so is the connection $D$.

Here, in analogy with the case of a vector bundle, $P_K$ is said to be *harmonic* with respect to $D$ if $D_K\psi = D_K^*\psi = 0$. The proposition thus gives a bijection of the following two subsets of the moduli spaces $\mathfrak{X}$ and $\mathfrak{M}$ respectively:

$$\mathfrak{N} = \{(P,D) \in \mathfrak{X};\ (P,D) \text{ admits a harmonic metric}\}$$

and

$$\mathfrak{M}' = \{(R,\theta) \in \mathfrak{M};\ (R,\theta) \text{ admits a H-E metric such that } D := D_K + \psi \text{ is bi-integrable}\}.$$

In fact, these sets may also be considered as subsets of $\mathfrak{Z}_j$ and $\mathfrak{Z}_i$ of Theorem (6.6.1) respectively, and hence of the same space $\mathfrak{T}$; as such they can also be identified. Note however that these spaces are in general not complex subspaces.

*Proof of (7.7.1).* Since $D$ is Einstein, by applying the operator $\Lambda$ to the second term of (7.1.1) and by using the standard Kähler identities we get

(7.7.2) $D_K'^{*}\theta = D_K''^{*}\theta^{*}$.

(The Kähler identities holds true also when the connection is not integrable.) Now suppose first that the condition (7.4.1) is satisfied. Then since $D_K''\theta = 0$, and hence $D_K'^{*}\theta = 0$, the weak harmonicity condition

(7.7.3) $D_K'^{*}\theta + D_K''^{*}\theta^{*} = 0$

is also satisfied. We shall see also that $D_K\psi = 0$. The vanishing of (2,0)- and (0,2)-parts of $D_K\psi$ follow from the bi-integrability condition. For the (1,1)-part we have

$$(D_K\psi)^{1,1} = D_K'\theta^{*} + D_K''\theta \quad = (D_K''\theta\ )^{*} + D_K''\theta \ = 0,$$

as desired.

Conversely, suppose that the metric $P_K$ is harmonic. First we note that from (7.7.2) and (7.7.3) we actually get

(7.7.4) $D_K'^{*}\theta = D_K''^{*}\theta^{*} = 0$.

From (7.1.2) it is clear that the vanishing of $F_K^{0,2}$ and the integrability of $\theta$ are equivalent. So it suffices to show that $\theta$ is holomorphic and integrable. The proof depends on the variation of the Siu-Bochner technique due to Corlette [C]: First writing $\partial = D_K'$, and $\bar\partial = D_K''$ for simplicity one notes the following formula of Siu- Bochner type:

**Sublemma.** $\partial\bar\partial(\theta^{*}\wedge\theta) = -\ \partial\theta^{*}\wedge\ \partial\theta\ -\ [\theta^{*},\theta^{*}]\wedge[\theta,\theta]$, where $\wedge$: $A^{i}(X,\mathrm{ad}\ P)\times A^{j}(X,\mathrm{ad}\ P)\to A^{i+j}(X)$ is induced by the exterior product and the Hermitian inner product of ad P.

*Proof.* Using the conditons $\bar\partial\theta^{*} = 0$, $\partial\theta^{*} = -\ \bar\partial\theta$, and $\partial\partial = -[\theta,\theta]$

succesively, we get:

$$\partial\bar{\partial}(\theta^* \wedge \theta) = -\partial(\theta^* \wedge \bar{\partial}\theta) = -\partial\theta^* \wedge \bar{\partial}\theta + \theta^* \wedge \partial\bar{\partial}\,\theta,$$

with

$$\theta^* \wedge \partial\bar{\partial}\,\theta = -\theta^* \wedge \partial\partial\,\theta^* = \theta^* \wedge [[\theta,\theta],\theta^*] = -[\theta^*,\theta^*] \wedge [\theta,\theta]$$

as desired.

Continuing the proof of the lemma we first note:

$$-\partial\theta^* \wedge \bar{\partial}\theta \wedge \omega^{[n-2]} = |\partial\theta^*|^2_\omega \,\omega^{[n]} \geq 0,$$

since $\Lambda\partial\theta^* = \sqrt{-1}\bar{\partial}^*\theta^* = 0$ by (7.7.4) so that $\partial\theta^*$ is primitive (cf. [W;p.23;Th.2]). Here, $\omega^{[k]} = \omega^k/k!$ in general. On the other hand, one gets

$$([\theta^*,\theta^*] \wedge [\theta,\theta]) \wedge \omega^{[n-2]} = -([\theta^*,\theta^*] \wedge [\theta^*,\theta^*]^*) \wedge \omega^{[n-2]}$$

$$= -\,|[\theta^*,\theta^*]|^2\,\omega^{[n]}.$$

Hence,

$$\partial\bar{\partial}(\theta^* \wedge \theta) = |\partial\theta^*|^2_\omega\,\omega^{[n]} + |[\theta^*,\theta^*]|^2\,\omega^{[n]}.$$

Integrating this over $X$ and taking complex conjugates we have

$$\bar{\partial}\theta = [\theta,\theta] = 0.$$

Finally, the last assertion follows from Corollary (A.2.4) in the Appendix below.

## 8. Case of Flat Bundles

(8.1) We restrict ourselves to the flat connections. So we assume in this section that the underlying $C^\infty$ principal bundle $P$ has the vanishing first and second chern classes $c_1$ and $c_2$, which we define as follows: Let $\bar{G} = G/G'$, where $G'$ is the semisimple part of $G$. Let $\bar{P} \to X$ be the principal $\bar{G}$- bundle associated to $P$ and to the natural homomorphism $G \to \bar{G}$. Since $\bar{G}$ is an algebraic torus, we have its characterstic class in $H^1(X,\Gamma)$, which we call the *first chern class* $c_1$ of P, where $\Gamma$ is the fundamental group of $\bar{G}$. On the other hand, $c_2$ is just the second chern class of ad P; $c_2 = c_2(\mathrm{ad}\ P)$.

(8.1.1) **Lemma.** If $c_1 = c_2 = 0$ in the sense defined above, then we have $\mathfrak{M} = \mathfrak{M}'$, and they are also identified with the set of $(R,\theta) \in \mathfrak{M}$ which admits a flat metric; moreover $\mathfrak{R}$ is identified with the set of $(P,D) \in \mathfrak{R}$ such that D is flat and (P,D) admits a harmonic metric.

*Proof.* By Corollary (A.1.3) in the Appendix any Hermitian-Einstein metric on $(R,\theta) \in \mathfrak{M}$ is flat; in particular the associated connection $D = D_K + \psi$ is bi-integrable. Hence, the assertion concerning $\mathfrak{M}$ are proved. In view of the bijective correspondnce of the sets $\mathfrak{R}$ and $\mathfrak{M}_0$ we also see that the connections D in $\mathfrak{R}$ are flat. (One may also deduce this fact directly by the same argument as in (A.1.1).)

In view of the lemma we have an identification

$$\mathfrak{R} = \{(P,D);\ D \text{ irreducible, flat and admits a harmonic metric}\}.$$

It is well-known that $\mathfrak{R}$ has a natural structure of a complex space whose local structure is described by the elliptic graded Lie algebra complex $(A^{\bullet}(X,\mathrm{ad}\ P),D)$ (cf.e.g. [G]). The result corresponding to the second assertion of (7.2.1) also is standard:

(8.1.2) **Lemma.** The complex space $\mathfrak{R}$ is realized naturally as a locally closed analytic subspace of $\mathfrak{C}$.

*Proof.* Let $\mathcal{N}$ be the set of irreducible flat connections on our fixed $C^{\infty}$ G-bundle P on X which admits a weakly harmonic metric. This is a $\mathcal{G}$-invariant closed analytic subspace of $\mathcal{C}$ because the curvature $F = F_D$ defines a holomorphic map $\mathcal{A} = A^1_k(\mathrm{ad}\ P) \to A^2_{k-1}(\mathrm{ad}\ P)$ (cf.[FU;p198,Prop.A4] for the smooth case) and the last condition is an open condition for irreducible connections by the inverse mapping theorem (cf. Lemma (2.8.1) and Proposition (6.3.1)). Let $\mathcal{A}_\infty$ be the subset of $\mathcal{A}$ consisting of $C^\infty$ connections, and $\mathcal{G}_\infty$ the group of $C^\infty$ gauge transformations of P. Then $\mathfrak{M}$ is naturally identified with $\mathcal{A}_\infty/\mathcal{G}_\infty$ and we have a natural map $b: \mathfrak{M} \to \mathfrak{M}_k := \mathcal{A}/\mathcal{G}$. We show that this is an open embedding.

First, we note that b is injective. In fact, any isomorphism $u:(P,D) \to (P,D_1)$ of class $H^k$ induces an biholomorphic isomorphism of the associated holomorphic bundles $P''_D$ and $P''_{D_1}$, and hence in particular of class $C^\infty$. Next, we fix any $C^\infty$ connection $D \in \mathcal{A}_\infty$. Then the intersection of $\mathcal{A}_\infty$ with the canonical slice

$$\tilde{S} = \{D + \alpha \in \mathcal{E};\ \alpha \in A^1_k(X,\mathrm{ad}\ P),\ D^*\alpha = 0\},$$

is given by

$$S = \{D \in \mathcal{A};\ D\alpha + [\alpha,\alpha] = D^*\alpha = 0\},$$

which in turn is a canonical slice for the action of $\mathcal{G}_\infty$ on $\mathcal{A}_\infty$. The equation is clearly a quasi-linear elliptic equation with $C^\infty$ coefficients. Hence, by the elliptic regularity we conclude that the connections in S are all in $\mathcal{A}_\infty$ and therefore b is an open embedding. This proves the lemma.

(8.2) The space $\mathfrak{M}$ can also be identified with the space of stable homomorphisms of the fundamental group $\pi$ of X into G via the basic result of Corlette [C] (cf. also [JY]). We shall explain this.

Let $\mathcal{A}_0$ be the set of $C^\infty$ flat connections D on P. Fix a base point $x_0 \in X$ and then fix a point $u \in P_{x_0}$ of the fiber over $x_0$. Then the holonomy representation at u yields a natural map $\alpha: \mathcal{A}_0 \to R$, where $R = R(\pi,G)$ is the set of homomorphisms of $\pi = \pi_1(X,x_0)$ into G. We may identify G with the group of automorphisms of the fiber $P_{x_0}$ which commutes with the right actions by G. Then by restrictions we get a natural homomorphism $\mathcal{G}_\infty \to G$ of the group $\mathcal{G}_\infty$ of $C^\infty$ complex gauge transformations onto G. Moreover, if we let G act on R by conjugations, the map $\alpha$ is $(\mathcal{G}_\infty, G)$-equivariant and induces a bijection $\mathcal{N}_0/\mathcal{G}_\infty \to R/G$ which is independent of the choices of the base points $x_0$ and $u$. We shall identify the image of the subset $\mathfrak{N} \subset \mathcal{N}_0/\mathcal{G}_\infty$ in $R/G$ as follows. First, we recall a theorem of Johnson-Millson [JM;Th.1.1]:

(8.2.1) **Lemma.** For an element $\rho \in R$ the following conditions a) and b) are equivalent; a) the G-orbit of $\rho$ in $R$ is closed and the identity component of the stabilizer of $\rho$ in $G$ coincides with the center of $G$, and b) the image $\operatorname{Im} \rho$ of $\pi$ is not contained in any parabolic subgroup of $G$.

In [JM] the case where $G$ is (real) semisimple was treated. The proof however works also in the general reductive case. Now an element $\rho$ of $R$ is called *stable* if the above equivalent conditions are satisfied. Note that when $G = GL(r,\mathbb{C})$, the notion coincides with the usual notion of the irreducibility of the representation of a group. We also recall the result of Corlette [C] mentioned above (cf. due to Donaldson [D] in the case of dimension one; see also [JY]):

(8.2.2) **Lemma.** If an irreducible connection $D \in \mathcal{N}$ determines a homomorphism $\rho \in R$ via $\alpha$, then $D$ admits a harmonic metric if and only if $\rho$ is stable.

As is well-known (cf.e.g.[G][JM]), $R$ has a natural structure of an affine algebraic subscheme of a product space $G^N$, where $N$ can be taken for instance to be a number of minimal generators of $\pi$. (The structure is independent of such an embedding $R \to G^N$.) The action of $G$ on $R$ is algebraic. Let $R_s$ be the G-invariant set of stable elements in $R$. Then by a general result in the geometric invariant theory [M;p.38,Th.1.10] we know that the quotient $R_s/G$ has a natural structure of a quasi-projective scheme and the natural projection $p\colon R_s \to R_s/G$ is a morphism of algebraic schemes. By Lemma (8.2.2) we have the identification: $\mathfrak{N} = R_s/G$. We omit the proof of the following more or less standard fact:

(8.2.3) **Lemma.** The above identification is compatible with the analytic structures of the two spaces.

(8.3) Let $\mathfrak{R}$ be the moduli space of stable homomorphisms of the fundamental group of $X$ into $G$ with the analytic structure defined as above and $\mathfrak{M}$ the moduli space of stable principal Higgs G-bundles with vanishing first and second chern classes (cf.(8.1)). Then the following is our main theorem, which is the principal bundle version of Theorem (1.4.1). Let $\mathfrak{M}_0$ (resp. $\mathfrak{R}_0$) be the set of nonsingular points of the underlying reduced subspace of $\mathfrak{M}$ (resp. $\mathfrak{R}$), and $\bar{\mathfrak{M}}_0$ the complex conjugate of $\mathfrak{M}_0$.

(8.3.1) **Theorem.** There exists a hyperkähler manifold $M$ with a special $S^1$-action such that in the associated Calabi family $\{Y_t\}_{t \in C}$ the fiber $Y_0$ (resp. $Y_\infty$) is isomorphic to $\mathfrak{M}_0$ (resp. $\bar{\mathfrak{M}}_0$) and the other members $Y_t$ are all isomorphic to $\mathfrak{R}_0$ (cf.(2.2)-(2.4)).

*Proof.* Identify $\mathfrak{M}$ (resp. $\mathfrak{R}$) naturally with an analytic subspace of $\mathfrak{Z}_i$ (resp. $\mathfrak{Z}_j = \mathfrak{C}$) by Theorem (7.6.1) (resp. Lemma (8.1.2)). Then with respect to the natural identification $\mathfrak{Z}_i = \mathfrak{Z}_j = \mathfrak{U}$ in Theorem (6.6.1) we have $\mathfrak{M} = \mathfrak{R}$ by Proposition (7.7.1) and Lemmas (8.1.1), (8.2.2) and (8.2.3), and hence also $\mathfrak{M}_0 = \mathfrak{R}_0$. Call the latter set $U$. Then for any point $p$ of $U$ the real tangent space $T_p = T_pU$ is invariant by both by $I$ and $J$ since $\mathfrak{Z}_i = (\mathfrak{U}, I)$ and $\mathfrak{Z}_j = (\mathfrak{U}, J)$, where $I$ and $J$ are the complex structures on $\mathfrak{U}$ defining the hyperkähler structure of $\mathfrak{U}$. This shows that $U$ is a (finite dimensional) hyperkähler submanifold of $\mathfrak{U}$, and $Z_0 := U \times C$ is a complex submanifold of the total space $\mathfrak{Z} = \mathfrak{U} \times C$ of the Calabi family of $\mathfrak{U}$, giving the Calabi family of $U$. Moreover, the $\mathbb{C}^*$-action on $\mathfrak{Z}_i$ obviously preserves the subspace $\mathfrak{M}$. It follows that the special action of $S^1$ on $\mathfrak{Z}$, and hence also of $\mathbb{C}^*$, preserves $Z_0$; thus $U$ admits a special

$S^1$-action. The rest of the assertion is immediately seen.

Even at *any* point of the moduli space $\mathfrak{M} = \mathfrak{N}$ considered as a subspace of $\mathfrak{J}$ as above, its Zariski tangent space also is invariant by both I and J. This will be checked in the next two subsections.

(8.4) Let D be any flat connection on P and $\mathcal{H}^1(X;P,D)$ the space of harmonic 1-forms associated to the elliptic complex $(A^{\cdot}(\mathrm{ad}\, P),D)$ with respect to some metric $P_K$. Then this harmonic space is naturally considered as the Zariski tangent (V-)space of the moduli space $\mathfrak{N}$ at the corresponding point. First, we give a general description of the harmonic space. We start with an arbitrary smooth connection D.

We first recall some formulas for the Laplacians of the associated covariant derivatives. Following Simpson [S1] we set

$$(8.4.1)\quad \begin{cases} d' := D_K' + \theta^* \\ d'' := D_K'' + \theta \,. \end{cases}$$

Then $D = d' + d''$. The usual Kähler identities still hold for these operators

$$(8.4.2)\quad \begin{cases} [\Lambda,d'] = i d''^* \\ [\Lambda,d''] = -i d'^* \end{cases}$$

where $^*$ denotes the formal adjoint as before (cf.[S1;p876,Lemma 3.1]). Denote by $\Delta$, $\square'$, and $\square''$ the associated Laplacians $DD^* + DD^*$, $d'd'^* + d'^*d'$, and $d''d''^* + d''^*d''$ respectively. We then define the operators $d^c$, $d^{*c}$ and $\Delta^c$ by the formulae

$$i d^c = d' - d'', \quad i d^{*c} = d'^* - d''^*, \text{ and } \Delta^c = d^c d^{*c} + d^{*c} d^c.$$

Clearly, $d^{c*}$ is the formal adjoint of $d^c$. Assume for simplicity that $F = F_D$ is of type (1,1). Then just as in [W;II,n°5,6] one can prove succesively the following formulae:

$$[\Lambda,d\ ] = -d^{*c}$$
$$[\Lambda,d^c] = d^*$$

$[\Delta, L] = 0$, $\Delta = \Delta^c$, and finally

$$2\Box' = \Delta - i/2\ ([L, i(F)] + [e(F), \Lambda]$$
$$2\Box'' = \Delta + i/2\ ([L, i(F)] + [e(F), \Lambda]$$

Especially as operators on the space of sections we have

(8.4.3) $2\,\Box' = \Delta + i\Lambda F$, $2\Box'' = \Delta - i\Lambda F$, and $\Box' - \Box'' = i\Lambda F$.

Specializing to the flat case we get:

(8.4.4) **Proposition.** Let $(P,D)$ be a principal bundle with a flat connection. Then the following hold true: 1) $\Delta = 2\Box' = 2\Box''$, and 2) for an element $\alpha \in A^i(X, \mathrm{ad}\, P)$ the following three conditions are equivalent: *i*) $\Delta\alpha = 0$, *ii*) $\Box'\alpha = \Box''\alpha = 0$, and *iii*) $d'\alpha = d''\alpha = d'^*\alpha = d''^*\alpha = 0$.

(8.4.5) **Corollary.** A 1-form $\alpha = \alpha' + \alpha''$ in $A^1(X, \mathrm{ad}\, P)$ is harmonic if and only if it satisfies the following conditions:

$$D'_K\alpha'' + [\theta^*, \alpha'] = 0\ , \qquad D''_K\alpha' + [\theta, \alpha''] = 0,$$
$$D'_K\alpha' = [\theta^*, \alpha''] = 0\ , \qquad D''_K\alpha'' = [\theta, \alpha'] = 0\ .$$

(8.5) Let $S$ and $\tilde{S}$ be the natural slices at $D$ for the action of $\mathcal{G}$ on $\mathcal{N}_\infty$ and $\mathcal{E}$ respectively as defined in the course of the proof of Lemma (8.1.2). The Zariski tangent space $T_D S = \mathcal{H}^1 := \mathcal{H}^1(X, \mathrm{ad}\, P; D)$ at $D$ is defined in the tangent space

$$T_D\tilde{S} = \{\alpha;\ \alpha \in A^1(X, \mathrm{ad}\, P),\ \Lambda D\alpha = 0,\ D^*\alpha = 0\}$$

of $S$ by

$$\mathcal{H}^1 = \{\alpha;\ D\alpha = 0,\ D^*\alpha = 0\}.$$

(8.5.1) **Proposition.** $\mathcal{H}^1$ is both I- and J-invariant.

*Proof.* Since $D$ and $D^*$ are J-linear, J-invariance of $\mathcal{H}^1$ is clear. Since we know that $T_D\check{S}$ is I-invariant by (6.2.1) and (3.3) for the I-invariance of $\mathscr{H}^1$ it suffices to show that for any element $\alpha \in \mathscr{H}^1$ we have $DI\alpha = 0$. Write $\alpha = \xi + \eta$, where $\xi \in A^1(\mathrm{ad}\, P_K)$ and $\eta \in A^1(\sqrt{-1}\mathrm{ad}\, P_K)$. Then the equations $D\alpha = 0$ and $DI\alpha = 0$ are divided respectively into two equations:

$$(8.5.2) \quad \begin{cases} D_K\xi + [\theta,\eta] = 0 \\ D_K\eta + [\theta,\xi] = 0 \end{cases}$$

$$(8.5.3) \quad \begin{cases} D_K^c\xi + [\theta^c,\eta] = 0 \\ D_K^c\eta + [\theta^c,\xi] = 0 \end{cases}$$

where $\theta^c = C(\theta)$. By (the proof of) Proposition 7.7.1 we have $D_K^c\theta = 0$. From this, by taking the linear part we get the second equation of (8.5.3). Let $\gamma := D_K^c\xi + [\theta^c,\eta]$ be the right hand side of the first equation. Then we have

$$\gamma^{2,0} = D_K'\xi' + [\theta,\eta'] = 0, \quad \gamma^{0,2} = D_K''\xi'' + [\theta^*,\eta''] = 0.$$

where ' and " denote the (1,0)- and (0,1)-parts respectively. But this is the same equation that comes from the first one of (8.5.2); hence it holds true. Finally, $\gamma^{1,1} = 0$ if and only if

$$(8.5.4) \quad D_K''\xi' + [\theta,\eta''] = D_K'\xi'' + [\theta^*,\eta']$$

The (1,1)-part of the second equation of (8.5.3) yields

$$(8.5.5) \quad D_K''\eta' + [\theta,\xi''] = D_K'\eta'' + [\theta^*,\xi']$$

Hence, we have the implications (8.5.4) $\leftrightarrow$ (8.5.4) +(8.5.5) $\leftrightarrow$

$$D_K''\alpha' + [\theta,\alpha''] = D_K'\alpha'' + [\theta^*,\alpha'].$$

But (8.4.5) shows that the both hand sides are zero since $\alpha \in \mathscr{H}^1$.

(8.5.6) *Remark*. In particular, every Zariski tangent V-space of $\mathfrak{M}$ has always of even complex dimension. This is clearer at the "real points". Namely, if we assume that $D = D_K$, ad $P_K$ defines a flat real structure on ad $P$ and we have the usual Hodge decomposition

(8.5.7) $\mathcal{H}^1(\mathrm{ad}\ P) = \mathcal{H}^{1,0}(\mathrm{ad}\ P) \oplus \mathcal{H}^{0,1}(\mathrm{ad}\ P), \quad \mathcal{H}^{1,0} = \bar{\mathcal{H}}^{0,1}$

of the harmonic space.

(8.6) Let $\mathfrak{S}$ be the moduli space of stable principal bundles, which is also considered the moduli space of equivalence classes of irreducible unitary representations of the fundamental group under our assumption of the vanishing chern classes [MR][UY]. Then as in the case of dimension 1 [H2], the moduli space $\mathfrak{M}$ contains the total space of the "V-cotangent cone" of $\mathfrak{S}$ (cf.(4.5)) as a Zariski open subset. But in the higher dimensional case the role of the symplectic structure is more apparent. For simplicity of the statement, however, we restrict ourselves to the Zariski open subset $\mathfrak{S}_o$ consisting of V-smooth points. Let $\mathfrak{D}$ be the inverse image of $\mathfrak{S}_o$ for the natural projection $\mathfrak{M}_1 \to \mathfrak{S}$ of (7.6).

(8.6.1) **Proposition.** There is a natural isomorphism of the total space $T^*\mathfrak{S}_o$ of the cotangent V-bundle of $\mathfrak{S}_o$ and the Zariski open subset $\mathfrak{D}$ above as a holomorphic symplectic V-manifold.

*Proof.* If we restrict to a suitable $\mathcal{G}$-invariant open subset $W_0$ of $W = A^1(X, \mathrm{ad}\ P_K)$, we can realize the situation of (4.5) so that $\mathfrak{S}_o$ is realized as a closed analytic subspace of $W_0/\mathcal{G}$. Then by using $J$ as in Lemma (4.5.2) there exists a natural symplectic Zariski open embedding of the total space of $T^*\mathfrak{S}_o$ into $\mathfrak{M}$. In fact, in view of (4.5), (4.1), (8.5.7) and the natural identification of the (V-)tangent space of $\mathfrak{S}$ at $R$ with $\mathcal{H}^{0,1}$ we see that on each fiber, say over $R \in \mathfrak{S}_o$, this embedding is realized by identifying the (V-)cotangent space of $\mathfrak{S}$ at $R$ with $\mathcal{H}^{1,0} = H^0(X, \Omega^1(\mathrm{ad}\ R))$ by Lemma (4.5.2) and then considering the elements of $\mathcal{H}^{1,0}$ as Higgs fields on $R$. Thus the image coin cides with $\mathfrak{D}$. (Under our assumption of V-smoothness, the bracket on $\mathcal{H}^{1,0}$ vanishes.)

(8.6.2) *Remark*. Even at a singular point of $\mathfrak{G}$ the result of Goldman-Millson [GM] implies that the total space of the V-cotangent cone coincides with the fiber of $\mathfrak{M}_1 \to \mathfrak{G}$.

## 9. The Moduli Space of Principal Higgs Bundles

(9.1) In this section we shall prove Theorem (7.6.1). In [R] Ramanathan has constructed the moduli space of stable principal bundles on a compact Riemann surface as a normal analytic space. We follow mainly the line of the proof in [R] and give modifications which are necessary in the higher dimensional Higgs case. Some standard parts of the arguments will, however, only be indicated without detailed proof. Throughout this section we fix a compact Kähler manifold X with a Kähler class $\gamma$, and a connected complex linear reductive Lie group G as in the prvious sections.

We begin with recalling some definitions. A *graded Lie algebra* is a pair $(A,[\ ,\ ])$ consisting of a graded vector space $A = \oplus_{p\geq 0} A^p$ and bilinear maps $[\ ,\ ]\colon A^p \times A^q \to A^{p+q}$ satisfying the following conditions:

$$(9.1.1) \qquad (-1)^{pr}[\varphi,[\psi,\eta]] + (-1)^{qr}[\eta,[\varphi,\psi]] + (-1)^{pq}[\psi,[\eta,\varphi]] = 0$$

$$\varphi \in A^p, \quad \psi \in A^q, \quad \eta \in A^r$$

$$(9.1.2) \qquad (-1)^{pq+1}[\varphi,\psi] = [\psi,\varphi], \qquad \varphi \in A^p, \quad \psi \in A^q$$

A *graded Lie algebra complex* is a triple $(A,[\ ,\ ],d)$, where $(A,[\ ,\ ])$ is a graded Lie algebra and $(A,d)$, $d\colon A^q \to A^{q+1}$, is a complex such that the following condition is satisfied:

$$(9.1.3) \qquad d\,[\varphi,\psi] = [d\varphi,\psi] + (-1)^p[\varphi,d\psi] \qquad \varphi \in A^p,\ \psi \in A.$$

In this case an element $\varphi \in A^1$ is called *integrable* if the following condition is fulfilled:

(9.1.4) $d\varphi + 1/2\ [\varphi,\varphi] = 0$.

(9.2) Let $(R,\theta)$ be a principal Higgs (G-)bundle on X. We consider the double complex $(A^{\cdot\cdot}, \partial, \bar\partial)$, where $A^{s,t} = A^{s,t}(X,\mathrm{ad}\ R)$, and the two differentials are defined respectively by

$$\partial := [\theta,\ ]: A^{s,t} \to A^{s+1,t}$$

and by the usual $\bar\partial$-operator

$$\bar\partial: A^{s,t} \to A^{s,t+1}.$$

In fact, the fact that $\partial\partial = 0$ is verified by setting $\varphi = \psi = \theta$ $(p = q = 1)$ in (9.1.1) and getting $[\theta,[\eta,\theta]] = 0$. Denote the associated double complex by $(A^{\cdot},d)$, where

$$A^{p} = A^{p}(X,\mathrm{ad}\ P) = \underset{s+t=p}{+} A^{s,t}(X,\mathrm{ad}\ P) \quad \text{and} \quad d = \partial + \bar\partial = \bar\partial + [\theta,\ ].$$

Indeed, the vanishing of $d^{2} = \partial\bar\partial + \bar\partial\partial$ is checked by setting $\varphi = \theta$ in (9.1.3) for the graded Lie algebra complex $(A^{0,\cdot},\bar\partial)$ and getting $\bar\partial\ [\theta,\varphi] = -\ [\theta,\bar\partial\varphi]$ since $\theta$ is holomorphic.

(9.2.1) **Lemma.** $(A^{\cdot}(X,\mathrm{ad}\ R),[\ ,\ ],d)$ is a graded Lie algebra complex.

*Proof.* We have to show (9.1.3). It is enough to verify this for both of $\partial$ and $\bar\partial$ instead of for $d$. For $\bar\partial$ this is well-known. For $\partial$ this can be checked easily by putting $\eta = \theta$ in (9.1.1).

We shall call the above complex the *Dolbeault-Higgs complex*. Now, according to (9.1.4) an element $\varphi \in A^{1,0}$ (resp. $\psi \in A^{0,1}$) is integrable in the graded Lie algebra complex $(A^{\cdot,0},\partial)$ (resp. $(A^{0,\cdot},\bar\partial)$) if and only if $[\varphi,\varphi] = 0$ (resp. $\bar\partial\psi + 1/2\ [\psi,\psi] = 0$) in accordence with the preceedng (resp. usual) definition. For any $\psi \in A^{0,1}$ we set $\bar\partial_{\psi} = \bar\partial + [\psi,\ ]$. We say that an element $\nu$ of $A^{1,0}$ is holomorphic with respect to $\psi$ if $\bar\partial_{\psi}\nu = 0$.

(9.2.2) **Lemma**. An element $\Phi = (\varphi,\psi) \in A^1$, $\varphi \in A^{1,0}$, $\psi \in A^{0,1}$ is integrable if and only if $\psi$ is integrable, $\theta + \varphi$ is holomorphic with respect to $\psi$, and $\theta + \varphi$ is integrable in the sense defined above.

*Proof*. By the definitions we have $d\Phi = ([\theta,\varphi],\bar{\partial}\varphi + [\theta,\psi],\bar{\partial}\psi)$, and $[\Phi,\Phi] = ([\varphi,\varphi],\ 2\,[\varphi,\psi],\ [\psi,\psi])$. Hence, $\Phi$ is integrable if and only if

$$[\theta,\varphi] + 1/2\,[\varphi,\varphi] = 0, \quad \bar{\partial}\varphi + [\varphi,\psi] + [\theta,\psi] = 0, \quad \bar{\partial}\psi + 1/2\,[\psi,\psi] = 0,$$

where the first and second equations can be rewritten as

$$[\theta + \varphi,\ \theta + \varphi] = 0, \quad \text{and} \quad \bar{\partial}_{\psi}(\theta + \varphi) = 0,$$

respectively. The lemma follows.

(9.2.3) *Remark*. The differential of the above complex is nothing but $d''$ in (8.4.1), and the condition $d^2 = 0$ is exactly the vanishing of the curvature $F'' := d''^2 = F^{0,2} + D''_K\theta + [\theta,\theta]$ of $d''$. The integrability of $\Phi$ is exactly the vanishing of the curvature of the new Higgs differential $d_\Phi = (d'' + \psi) + (\theta + \varphi)$.

(9.3) We next give an interpretation of the cohomology groups of the Dolbeault-Higgs complex in terms of sheaf cohomology groups. We denote by $\Omega^{\bullet}(\mathrm{ad}\,R;\theta)$ the complex $(\Omega^{\bullet}(\mathrm{ad}\,R),d)$ of sheaves on $X$, where $d = [\theta,\ ]$ and $\Omega^p(\mathrm{ad}\,R)$ is the sheaf of germs of ad $R$-valued holomorphic $p$-forms. Then the composition of the natural inclusions of the complexes

$$\Omega^{\bullet}(\mathrm{ad}\,R;\theta) \subset \mathcal{A}^{\bullet,0}(\mathrm{ad}\,R;\theta) \subset \mathcal{A}^{\bullet}(\mathrm{ad}\,R).$$

are quasi-isomorphic since for each $p$, $\mathcal{A}^{p,\bullet}(\mathrm{ad}\,R)$ is a (fine) resolution of $\Omega^p(\mathrm{ad}\,R)$, where the second and the third complexes of sheaves are the sheaf versions of the complexes $(A^{\bullet,0},\partial)$ and $(A^{\bullet},d)$ considered above. Hence we have

(9.3.1) **Proposition.** The cohomology group $H^i(A^{\cdot}(X,\mathrm{ad}\,R))$ for each $i$ is naturally isomorphic to the hypercohomology group $\mathbb{H}^i(X,\Omega^{\cdot}(\mathrm{ad}\,R;\theta))$ of the complex $\Omega^{\cdot}(\mathrm{ad}\,R;\theta)$.

We have the standard spectral sequence

$$E_1^{p,q} = H^q(X,\Omega^p(\mathrm{ad}\,R)) \qquad \mathbb{H}^{p+q}(X,\Omega^{\cdot}(\mathrm{ad}\,R;\theta)),$$

from which we obtain an exact sequence

$$0 \to H^1(\Gamma(X,\Omega^{\cdot}(\mathrm{ad}\,R))) \to \mathbb{H}^1(X,\Omega^{\cdot}(\mathrm{ad}\,R;\theta)) \to H^0(H^1(X,\Omega^{\cdot}(\mathrm{ad}\,R)) \to$$

$$H^2(\Gamma(X,\Omega^{\cdot}(\mathrm{ad}\,R))) \to \mathbb{H}^2(X,\Omega^{\cdot}(\mathrm{ad}\,R;\theta)),$$

where the differentials of the complexes $\Gamma(X,\Omega^{\cdot}(\mathrm{ad}\,R;\theta))$ and $H^1(X,\Omega^{\cdot}(\mathrm{ad}\,R;\theta))$ are induced by $[\theta,\ ]$. We also note that $\mathbb{H}^0(X,\Omega^{\cdot}(\mathrm{ad}\,R;\theta))$ is the space of holomorphic sections of $\mathrm{ad}\,R$ which are annihilated by $\theta$.

(9.4) Clearly, the Dolbeault-Higgs complex $(A^{\cdot},d)$ is an elliptic complex because its principal symbols coincide with those of $\bar\partial$. So the associated harmonic theory is available, and we can apply the Griffiths' result [Gr1] on the extremal slice. Let $d^*: A^p \to A^{p-1}$ be the formal adjoint of $d$, and $\mathcal{H}^i$ the space of harmonic forms of degree $i$ (with respect to $d$). We take suitable Sobolev completions of all the spaces under consideration and denote them again by the same letters. For any closed subspace $F$ of $A^p$, $\pi_F$ denotes the orthogonal projection to $F$, and $N(*)$ a neighborhood of the origin of a vector space $*$. Then by a general result of Griffiths [Gr1;Lemma 2.3] we obtain the following:

(9.4.1) **Lemma.** The integrable elements in $\mathrm{Ker}\, d^* \subset A^1$ which have sufficiently small norms are parametrized by the analytic subspace $V := h^{-1}(0)$ of $N(\mathcal{H}^1)$, where $h: \mathcal{H}^1 \to \mathcal{H}^2$ is the holomorphic map defined by $h(\varphi) = \pi_{\mathcal{H}}\,[p(\varphi),p(\varphi)]$.

In fact, there exists a $C^\infty$ map $p: N(\mathcal{K}^1) \to N(\pi_d*(A^1))$ such that $\{p(\varphi) := \varphi + p(\varphi)\}_{\varphi \in V}$ exhausts the set of integrable elements in $\operatorname{Ker} d^*$. Now we call a Higgs bundle $(R,\theta)$ *simple* if any holomorphic section of ad $R$ annihilated by $\theta$ is a constant section coming from an element of the center of $\mathfrak{g}$.

(9.4.2) **Theorem.** Let $(R,\theta)$ be any Higgs G-bundle. Then there exists a Kuranishi family $(\mathcal{R},\tilde{\theta}) \to T \times X$ of deformations of $(R,\theta) = (\mathcal{R},\tilde{\theta})_o$, where $T = (T,o)$ is a germ of complex spaces; moreover it is even universal if $(R,\theta)$ is simple. The Zariski tangent space of $T$ at $o$ is naturally isomorphic to the hypercohomology group $\mathbb{H}^1(X,\Omega^\cdot(\mathrm{ad}\ R;\theta))$.

The first assertion of the theorem means that the Higgs bundle $(\mathcal{R},\tilde{\theta})$ on $T \times X$ has the following versal property: Let $t \in T$ be any point (which is sufficiently close to $o$) and $(\mathcal{R}',\tilde{\theta}') \to T' \times X$ any deformation of $(R_t,\theta_t) = (\mathcal{R}',\tilde{\theta}')_{o'}$. Then there exists a morphism $\tau$: $(T',o') \to (T,t)$ of complex spaces such that $(\mathcal{R}',\tilde{\theta}')$ is isomorphic to the Higgs bundle obtained by pulling back $(\mathcal{R},\tilde{\theta})$ by $\tau$ defined in the obvious way; moreover, when $t = o$, the differential $\tau_*$ of $\tau$ is determined uniquely by $(\mathcal{R}',\theta')$. The family is called universal, if $\tau$ itself is uniquely determined by $(\mathcal{R}',\theta')$.

(9.5) We give a sketch of the proof of Theorem (9.4.2). We follow the argument in [Gr1;3.2], where the case of ordinary principal bundles are treated. Let $P$ be the underlying $C^\infty$ principal bundle of $R$. An almost complex structure $J$ on $P$ is said to be *admissible* if a) it is invariant under the right action of $G$ on $P$, b) each fiber of $P \to X$ is an (almost) complex submanifold of $(P,J)$, and it is naturally isomorphic to $G$ as a complex manifold, and finally c) the induced almost complex structure on the base manifold $X$ coincides with the given complex structure of $X$. Denote by $\mathcal{C}$ the set of all the pairs $(J,\theta)$ consiting of an admissible integrable almost complex structure

$J$ on $P$ and an ad $P$-valued $(1,0)$-form $\theta$ on $X$ such that $\theta$ is integrable and is holomorphic with respect to $J$. Any pair in $\mathcal{E}$ clearly determines a Higgs bundle with underlying smooth structure $P$. Let $(J_0,\theta_0)$ correspond to $(R,\theta)$. Then conversely, in view of Lemma (9.2.2) (cf. Remark (9.2.3)) we can prove the following:

(9.5.1) **Lemma**. The elements in $\mathcal{E}$ which are sufficiently close to $(J_0,\theta)$ are canonically parametrized by a neighborhood of the origin of the set of integrable elements in $A^1(X,\text{ad } R)$, where $(J_0,\theta_0)$ corresponds to the origin.

By using this lemma we can constructs a desired Kuranishi family parametrized by $V$ in Lemma (9.4.1) along the same line with [Gr1] in the case of ordinary principal bundles. We only note that with respect to the natural action of the gauge transformation group $\mathcal{G}$ on $\mathcal{E}$, the differential at the identity of the orbit map $\mathcal{G} \to \mathcal{E}$, $g \to g((J_0,\theta_0))$, is expressed just as the differential $d\colon A^0(X,\text{ad } R) \to A^1(X,\text{ad } R)$ of the Dolbeault-Higgs complex.

Finally, when $(R,\theta)$ is simple, the identity component $\text{Aut}_0(X,\text{ad } R)$ of the automorphism group of $(R,\theta)$ reduces to the center $Z$ of $G$ acting by right multiplications on $R$. Hence, any element of $\text{Aut}_0(X,\text{ad } R)$ extends to an automorphism of the Kuranishi family inducing the identity on the base space. From this the universality of the family follows by a formal argument. In view of the upper semicontinuity of the dimensions of harmonic spaces, all the nearby fibers are also simple so that the universality hodlds true at any point of $T$. The assertion of the tangent space is standard.

(9.6) The following result is due to Ramanathan [R;Prop.4.1] in the case of a compact Riemann surface and (ordinary) principal bundles.

(9.6.1) **Proposition**. Let $T$ be a complex space and $(R,\theta)$ a Higgs $G$-bundle on $X \times T$. Then the set of points $t$ of $T$ for which

$(R,\theta)_t$ is a stable Higgs bundle on $X$ forms a Zariski open subset of $T$.

Fix any maximal parabolic subgroup $Q$ of $G$ and consider the associated fiber bundle $u: Y := R/Q \to X \times T$ with typical fiber the flag manifold $G/Q$. Consider $\theta$ naturally as a $\mathfrak{g}$-valued holomorphic 1-form on $R$. Let $\theta'$ be the induced $\mathfrak{g}/\mathfrak{q}$-valued form, where $\mathfrak{q}$ is the Lie algebra of $Q$. Let $S'$ be the set of zeroes of $\theta'$ and $S$ the image of $S'$ in $Y = R/Q$. Then $S$ is an analytic subset of $Y$; indeed, if $\theta$ is $\mathfrak{q}$-valued at some point $r \in R$, then by the tensorial property of $\theta$ it is still $\mathfrak{q}$-valued at any point of the 'coset' $rQ$, and then $S = \{y \in Y;\ \dim \nu^{-1}(y) \geq \dim Q\}$, where $\nu: R \to Y$ is the natural morphism. Let $f: S \to T$ be the natural morphism. Denote by $\alpha$: $D_f \to T$ the relative Douady space of the subspaces of the fibers of $f$ (cf.[F1]). For $d \in D_f$, $A_d$ denotes the subspace of a fiber of $f$ corresponding to $d$. Then the set

$$B := \{d \in D_f;\ A_d \text{ reduced and } A_d \to X \text{ bimeromorphic}\}$$

is Zariski open in $D_f$ (cf.[F2;p.484,Lemma 15]). Note that for $d \in B$ with $\alpha(d) = t$, $A_d$ is a graph of a meromorphic section $\sigma: X \to Y_t$, which is automatically defined on a Zariski open subset $U$ with $\operatorname{codim}(X-U) \geq 2$, and hence $A_d$ defines a reduction of the structure group of $R_t$ to $Q$ over $U$, which is $\theta_t$-admissible (7.4) since $\sigma(U) \subset S$. Conversely, any $\theta_t$-admissible reduction is obtained as above from a unique element $d$ of $B$.

Let $D_0 = \{D_\alpha\}$ be the set of those irreducible components of $D_{f,red}$ which intersect with $B$, where $_{red}$ denotes the underlying reduced subspace. Let $L \to Y$ be the line bundle defined by a fixed dominant character $\chi$ of $Q$ so that $L$ is negative on the fibers. We consider the class $\beta := c_1(L)\cdot u^*\gamma^{n-1} \in H^n(Y,\mathbb{R})$, where $n = \dim X$. For any n-dimensional analytic subspace $A$ of a fiber of $f$ we can evaluate $\beta$ on the n-cycle associated to $A$; we shall denote the resulting real number by $\deg A$, which is a constant, say $d_\alpha$, on each irreducible component $D_\alpha$ of $D_f$.

Now note that for any $d \in D_\alpha$ there exists a unique reduced and irreducible component $C_d$ of $A_d$ which is mapped bimeromophically onto $X$.

(9.6.2) **Lemma.** For any $d \in D_\alpha$ we have $\deg C_d \geq d_\alpha$.

*Proof.* We have

$$d_\alpha = \deg A_d = \sum_i \deg A_{d,i}$$

where $A_{d,i}$ are irreducible components of $A_d$. It suffices thus to check that if $A_{d,i} \neq C_d$, we have $\deg A_{d,i} \leq 0$. In fact, if $A_{d,i}$ is mapped to an analytic subset of codimension $\geq 2$, then the degree is clearly zero, while if the image is of codimension one, then the degree is negative because a general fiber of $A_{d,i} \to X$ is of dimension one and $L$ is negative on the fibers.

(9.7) We now set

$$\mathfrak{D} = \mathfrak{D}(Q) = \{D_\alpha \in \mathfrak{D}_0;\ d_\alpha \geq 0\} = \bigcup_{\alpha\in\mathfrak{A}} D_\alpha.$$

By the above lemma for any $d \in D_\alpha$ in $\mathfrak{D}$ we necessarily have

(9.7.1) $\deg C_d \geq 0$.

(9.7.2) **Lemma.** The natural image of $\bigcup_{\alpha\in\mathfrak{A}} D_\alpha$ in $T$ is an analytic subset of $T$.

*Proof.* Fix a maximal compact subgroup $K$ of $G$. Then there exists a $K$-invariant hermitian metric on the line bundle on $G/Q$ defined by the dominant character $\chi$ whose first chern form is negative definite. Then by reducing the structure group of $R$ to $K$ we obtain a hermitian metric on $L$ whose chern form $\eta$ is negative definite on each fiber $Y_t$. Fix also a Kähler form $\omega$ in the Kähler class $\gamma$

and set $\tilde{\omega} = u^*\omega$. Then for any relatively compact subdomain N of T there exists a positive integer m such that $\psi_m := \tilde{\omega}_m - \eta$ is a Kähler form on $f^{-1}(N) \subset Y$, where $\tilde{\omega}_m = m\,\tilde{\omega}$. In particular, $\varphi_m := \psi_m \wedge \tilde{\omega}^{n-1}$ is a closed positive n-form on $f^{-1}(N)$ (cf.[F1;p.6]). Now we have for any point $d \in B_N := a^{-1}(N)$

$$\deg A_d = \int_{A_d} \eta\wedge\tilde{\omega}^{n-1} = \int_{A_d} \tilde{\omega}_m\wedge\tilde{\omega}^{n-1} - \int_{A_d} \varphi_m = m\int_X \omega^n - \int_{A_d} \varphi_m$$

The condition $\deg A_d \geq 0$ then implies that the function $a(d) := \int_{A_d} \varphi_m$ is bounded from above on $B_N$. Hence, by [F1;p.32,Th.4.3] $B_N$ is relatively compact in $D_f$ and its closure $D = \cup_{\alpha\in\mathfrak{A}} D_\alpha$ is proper over N. Thus the lemma follows by a theorem of Remmert.

(9.7.3) *Proof of Proposition (9.6.1).* Fix a set $Q = \{Q_j\}$ of representatives of the conjugacy classes of maximal parabolic subgroups of G. Let $D_0$ be the union of the irreducible components in at least one of $D(Q_i)$, and S the image of $D_0$ in T. S is an analytic subset of T by the previous lemma. On the other hand, by (9.7.1) for any point $t \in S$ we can find a meromorphic section $\sigma: X \to Y_t$ which violates the condition of stability, and vice versa. Hence the points of $T - S$ precisely correspond to stable Higgs bundles.

(9.8) When we try to construct local charts of the moduli space of principal bundles by using the Kuranishi families the main difficulty is the description of the isomorphisms of two bundles. Here, we follow the method of Ramanathan [R] of embedding the set of such isomorphisms into that of isomorphisms of suitable vector bundles.

Let V be a finite dimensional complex vector space and $\rho: G \to GL(V)$ a representation. Let $V = V_1 \oplus \cdots \oplus V_r$ be the irreducible decomposition of V so that $\rho = \rho_1 \times \cdots \times \rho_r$ with $\rho_i: G \to GL(V_i)$. Then we define $C = \{(t_1\rho_1(g),\cdots,t_r\rho_r(g) \in GL(V);\ t_i \in \mathbb{C}^*,\ g \in G\}$. Then $G_\rho := \rho(G)$ and C are Zariski closed in $GL(V)$. Let $\bar{C}$ be the

(Zariski) closure of $C$ in $\operatorname{End} V$.

Now suppose that we are given two Higgs G-bundles $(R,\theta)$ and $(R',\theta')$ on $X$. We then consider the vector bundles $E$ and $E_i$ (resp. $E'$ and $E'_i$) on $X$ associated to $\rho$ and $\rho_i$ and to $R$ (resp. $R'$). These are naturally Higgs vector bundles with Higgs fields in- duced by $\theta$ and $\theta'$. We may consider the vector bundle $W := \operatorname{Hom}(E,E')$ as the vector bundle associated to the princial $G\times G$-bundle $F := R \times_X R'$ and to the natural action of $G \times G$ on $\operatorname{End} V$. Then, since $G_\rho$, $C$ and $\bar{C}$ are invariant by the right and left multiplications of $G$ on $\operatorname{End} V$, we can consider the fiber bundles $F(G_\rho)$, $F(C)$, and $F(\bar{C})$ associated to $F$ as a subbundle of $W$. If we set $W_i = \operatorname{Hom}(E_i,E'_i)$, we also have the natural inclusion $W_1 \oplus\cdots\oplus W_r \subset W$. We then define

$$A = \{s = (s_1,\cdots,s_r):\ s_i \in H^0(X,W_i)^\theta \text{ and } s(X) \subset F(\bar{C})\},$$

and $A^* = \{s \in A;\ s_i \neq 0\}$, where $H^0(X,W_i)^\theta = \{s_i \in H^0(X,W_i);\ \theta'_i s_i = s_i\theta_i \text{ in } H^0(X,\Omega^1(W_i))\}$, $\theta_i$ and $\theta'_i$ denoting the Higgs fields on $E_i$ and $E'_i$ respectively. Then we have the natural inclusion $\mathbb{C}^{*r}\cdot\Gamma(X,F(G_\rho))^\theta \subset A^*$, where $\Gamma(X,F(G_\rho))^\theta = \Gamma(X,F(G_\rho)) \cap (\bigoplus_{i=1}^{r} H^0(X,W_i)^\theta)$ in $H^0(X,W)$. The next result corresponds to [R]; Prop. 3.1, and a part of the proof of Prop.3.2.

(9.8.1) **Proposition.** We have actually $A^* = \mathbb{C}^{*r}\cdot\Gamma(X,F(G_\rho))^\theta$. Moreover, $A$ is an r-dimensional algebraic group whose identity component is isomorphic to $\mathbb{C}^{*r}$.

*Proof.* The proof is the same as [R]. So we shall only give an outline. In order to show that $A^* \subset \mathbb{C}^{*r}\cdot\Gamma(X,F(G_\rho))^\theta$ we show the two inclusions

$$(9.8.2)\qquad A^* \subset \Gamma(X,F(C))^\theta := \Gamma(X,F(C)) \cap \Gamma(X,W)^\theta \subset \mathbb{C}^{*r}\cdot\Gamma(X,F(G_\rho))^\theta.$$

For the first one it suffices to show that each $s = (s_1,\cdots,s_r) \in A^*$ is isomorphic since $GL(V) \cap \bar{C} = C$. The proof of this is then reduced to the case $r = 1$. Let $\mathcal{F}$ be the $\theta'$-invariant torsion free subsheaf of the vector bundle $E'$ generated by the image of some $s \in A^* \subset \Gamma(X,W)^\theta$. $\mathcal{F}$ is locally free over some Zariski open subset $U$ of $X$ with $\mathrm{codim}(X - U) \geq 2$. First one shows that $\mathrm{rk}\,\mathcal{F} = \mathrm{rk}\,E'$, by deriving a contradiction by assuming that $\mathrm{rk}\,\mathcal{F} < \mathrm{rk}\,E'$, where rk = rank. This is done in two steps: a) One shows that the restricted bundle $\mathcal{F}|U \to U$ is of type $\Phi$ in the sense of [R;p.135, Def.3.1], whose proof in the general case is completely the same, and then b) by using the obvious analogue of Lemma 3.3 of [R] in the Higgs case we arrive at a contradiction exactly as in p.136 of [R], where we use our assumption on the (semi)stability of the Higgs bundles. Thus we may assume that $\mathrm{rk}\,E' = \mathrm{rk}\,\mathcal{F}$, which we denote by $k$. Then $s$ defines a nonzero section (on the whole $X$) of the line bundle $\mathrm{Hom}(\Lambda^k E, \Lambda^k E')$ which has the vanishing chern classes. It then follows that $s$ has no zeroes since $X$ is Kähler. Hence $s$ is isomorphic as desired.

The second inclusion of (9.8.2) is seen as follows. Let $s$ be any element of $\Gamma(X,F(\mathbb{C}))^\theta$. Then for each $x \in X$, we can find an element $t(x) \in \mathbb{C}^{*r}$ such that $t(x)s(x) \in F(G_\rho)_x$. The point then is to show that $t(x)$ can actually be taken to be independent of $x$, by using the characters defining the subgroup $G_\rho \cap \mathbb{C}^{*r}$ in $\mathbb{C}^{*r}$; see [R; p.136] for the detail.

It remains to show the last assertion. The determinant of a linear transformation induces a polynomial map $\det: H^0(X,W) \to H^0(X,\mathbb{C}) \simeq \mathbb{C}$. By the first part, the intersection of the zero of det and $A$ consists only of the origin. It follows that every irreducible component $A'$ of $A$ is of dimension one since $A'$ passes through the origin, $A$ being a $\mathbb{C}^{*r}$-invariant analytic space in $H^0(X,W)$. The desired conclusion follows from this easily.

(9.9) Let $(R,\theta)$ be any Higgs G-bundle on $X$. The center $Z$ of $G$ is contained naturally in the group $\mathrm{Aut}(R,\theta)$ of the automorphisms of $(R,\theta)$ as a normal subgroup. As a corollary of the above proposition we obtain the following result corresponding to [R;Prop.3.2]:

(9.9.1) **Proposition.** If $(R,\theta)$ is stable, then $(R,\theta)$ is simple and $\mathrm{Aut}(R,\theta)/Z$ is finite.

*Proof.* Take a faithful representation $\rho: G \to GL(V)$. Then by Proposition (9.8.1) we have the natural inclusion $\mathrm{Aut}(R,\theta) \subset A^*$, and hence, $\mathrm{Aut}(R,\theta)/Z = \mathrm{Aut}(R,\theta)/(\mathrm{Aut}(R,\theta)\cap\mathbb{C}^{*r}) \subset A^*/\mathbb{C}^{*r}$, where the latter is a finite group. The proposition follows.

In order that the above description of the isomorphisms of the Higgs bundles actually lead to a proper equivalence relation on the Kuranishi spaces one uses also the following general lemma of Simha [R;Lemma 4.1]:

(9.9.2) **Lemma.** Let $T$ be any complex space and $(E_i,\theta_i)$, $1\le i\le r$, Higgs vector bundles on $T\times X$. Let $C$ be an analytic subspace of (the total space of) $E := \bigoplus_{1\le i\le r} E_i$. Then: 1) the set $\Sigma := \bigcup_{t\in T} \{\sigma \in \oplus H^0(X,\mathrm{End}(E_i,\theta_i)_t);\ \sigma(X) \subset C\}$ has a natural structure of an analytic space over $T$, and 2) if $C$ is invariant by the natural $\mathbb{C}^{*r}$-action on $E$, then the induced $\mathbb{C}^{*r}$-action on the open subset $\Sigma' := \{\sigma = (\sigma_i) \in \Sigma;\ \sigma_i \neq 0\}$ is proper, and the induced morphism of complex spaces $\Sigma'/\mathbb{C}^{*r} \to T$ is proper.

*Proof of Theorem (7.6.1).* For the construction of the moduli space in our higher dimensional Higgs case the proof of Theorems 4.1 and 4.3 of [R;pp139-141] is also applicable word for word if we use (9.4.2), (9.6.1), (9.8.1), (9.9.1) and (9.9.2) above instead respectively of Theorem 4.2, Proposition 4.1, Proposition 3.1, and Proposition 3.2 and Lemma 4.1 there. The second assertion follows easily from the fact that by construction the Kuranishi space is realized as a subspace of $A_k^{1,0}(\mathrm{ad}\,P) \times \mathcal{A}_k^{0,1}$ (cf.(6.4)).

**Appendix**

(A.1) We shall prove the generalization by Simpson [S1;Prop.3.4] of the inequality of Lübke-Kobayashi to the Higgs case.

(A.1.1) **Proposition.** Let $(E,\theta)$ be a Higgs vector bundle on $X$ of rank r admitting a Hermitian-Einstein metric $h$. Then we have the inequality of chern classes: $((r-1)\, c_1^2 - 2r\, c_2)\cdot\gamma^{n-2} \geq 0$. Moreover, if the equality holds here, then the trace-free part $F^0$ of the curvature $F$ of the associated affine connection vanishes, and the associated projective bundle $\mathbb{P}(E)$ is flat. If $c_1 = c_2 = 0$, then $E$ is flat.

*Proof.* If we set $\rho = 1/r \operatorname{Tr} F \otimes I$, then $F^0$ is given by $F^0 = F - \rho$, where $I$ is the identity endomorphism of $E$ and $\operatorname{Tr}$ denotes the trace. Then the usual computation shows that $\tau := (r-1)c_1^2 - 2rc_2$ is represented by the (ordinary) 4-form $(i/2\pi)^2\, r \operatorname{Tr}(F^0 \wedge F^0)$. Then we shall check in general the formula

$$\text{(A.1.2)} \qquad \operatorname{Tr}(\eta\wedge\eta)\wedge\omega^{[n-2]} = (|\eta|^2 - |\Lambda\eta|^2)\cdot\omega^{[n]}, \quad \omega^{[n]} = \omega^n/n!$$

for any End E-valued 2-form $\eta$ satisfying the two conditions 1) (1,1)-part $\eta^{1,1}$ is skew-hermitian, and 2) $\eta^{2,0} + \eta^{0,2}$ hermitian. In fact, we compute

$$\operatorname{Tr}(\eta\wedge\eta)\wedge\omega^{n-2} = \operatorname{Tr}(\eta^{1,1}\wedge\eta^{1,1})\wedge\omega^{n-2} + 2\operatorname{Tr}(\eta^{2,0}\wedge\eta^{0,2})\wedge\omega^{n-2}$$

for any End E-valued 2-form $\eta$. With respect to a unitary frame we have the formula for an End E-valued 2-form $\varphi$;

$$\operatorname{Tr}(\varphi\wedge\varphi)\wedge\omega^{n-2} = \sum_{\alpha,\beta} \varphi_{\alpha\beta}\wedge\varphi_{\alpha\beta}\wedge\omega^{n-2} = \pm \sum_{\alpha,\beta} \varphi_{\alpha\beta}\wedge\bar{\varphi}_{\alpha\beta}\wedge\omega^{n-2},$$

where $\pm$ is taken according as $\varphi$ is hermitian or skew-hermitian. (A.1.2) then follows if we combine this formula with the following one deduced easily from [W;p.23, Th.2]:

$$\pm(1/n(n-1))\; v\wedge\bar{v}\wedge\omega^{n-2} = (|v|^2 - |\Lambda v|^2)\wedge\omega^n ,$$

where $v$ is an ordinary 2-form and the + sign is taken if $v$ is of type (2,0) or of type (0,2) and the - sign is relevant if $v$ is of type (1,1).

Hence, we conclude that if $F^0$ satisfies the two conditions 1) and 2) mentioned above, then we get

$$((r-1)c_1^2 - 2r\, c_2)\wedge\omega^{[n-2]} = (i/2\pi)^2 r(|F^0|^2 - |\Lambda F^0|^2)\wedge\omega^{[n]}.$$

Thus, if, further, the metric is Einstein so that $\Lambda F^0 = 0$, we have the desired conclusion. It remains to check the above two conditions. In view of (7.1.1) and (7.1.2) these are equivalent to the system of equations;

$$D_h''\theta + D_h'\theta^* = 0,$$

$$F_h^{2,0} + [\theta,\theta] = F_h^{0,2} + [\theta^*,\theta^*] = 0,$$

which is satisfied for any Higgs bundle. The rest of the assertion is immediate.

(A.1.3) **Corollary**. Let $(R,\theta)$ be a principal Higgs G-bundle as in (7.4) with a Hermitian-Einstein metric $R_K$. If $c_1 = c_2 = 0$ (cf. (8.1)), then the metric is actually flat in the sense of (7.4).

*Proof*. Since $c_1 = 0$, the induced Hermitian-Einstein metric on the Higgs $\bar{G}$-bundle $(\bar{P},\bar{\theta})$ associated to the projection $G \to \bar{G} = G/G'$ is flat (cf.(1.5.1)). On the other hand, by the above proposition the induced Hermitian-Einstein metric on the Higgs vector bundle (ad P, ad $\theta$) is flat. Hence, the induced metric on the Higgs G/Z-bundle $(R',\theta')$ associated to the projection $G \to G/Z$ also is flat, where $Z$ is the center of $G$. Since $G \to G/Z \times G/G'$ is isogenous, we deduce that the original metric also is flat (cf.[RS; §2]).

(A.2) We formulate and prove two vanishing theorems which are essentially in [H1] (cf. p.67 and p.110,111 of [H1]).

(A.2.1) **Proposition.** Let $(E,\theta)$ be a Higgs vector bundle on $X$ admitting a Hermitian-Einstein metric $h$. 1) If $\mu(E) < 0$, there exist no $\theta$-invariant holomorphic sections of $E$. 2) If $\mu(E) = 0$, any such section is covariant constant; moreover in this case if $\theta s = \lambda s$ for some holomorphic 1-form $\lambda$ on $X$, then we have $\theta^* s = \bar{\lambda} s$. In particular, if $s$ is annihilated by $\theta$, then it is covariant constant also with respect to the Einstein connection $D = D_h + \psi$.

*Proof.* First, one notes the following: For a linear transformation $A$ of a unitary vector space $V$ in general if we have $Av = cv$ for some $v \in V$ and a constant $c$, then we get $\langle [A,A^*]v,v\rangle \geq 0$, and the equality holds true if and only if $A^* v = \bar{c} v$ where $\langle\ ,\ \rangle$ is the inner product and $A^*$ is the adjoint of $A$ (cf.[H1;p.68]). Hence, we see that $i\langle\Lambda[\theta,\theta^*]s,s\rangle \geq 0$ and the equality holds if and only if $\theta^* s = \bar{\lambda} s$, where $\langle\ ,\ \rangle$ is the point-wise inner product defined by the given Hermitian-Einstein metric $h$. Now we use the integrated form of the Weitzenbock formula [K;p53,(1.8)]

$$0 \leq \|D_h' s\|^2 = i\int_X \langle\Lambda F_h s,s\rangle\ \omega^{[n]} = -i\int_X \langle\Lambda[\theta,\theta^*]s,s\rangle\ \omega^{[n]} + \lambda\int_X \langle s,s\rangle\ \omega^{[n]}$$

where we have used the equality $\Lambda(F_h + [\theta,\theta^*]) = \lambda I$, $I$ denoting the identity endomosphism (cf.(7.1.1)) and (1.5.1)). Since $\lambda = c\mu(E)$ with $c > 0$, the result now follows from this and the remark at the beginning of the proof.

(A.2.2) *Remark.* The proposition also follows from the formula (8.4.3) for the Laplacians noting that the following conditions for a section is equivalent: 1) $\square'' s = 0$, 2) $d'' s = 0$, and 3) $\bar{\partial} s = \theta s = 0$.

(A.2.3) **Proposition.** Let $D$ be a bi-integrable connection on a smooth vector bundle $E$ with a harmonic metric $h$. Then any covariant constant section $s$ of $E$ is also covariant constant with repect to the Hermitian connection $D_h$; moreover it is also annihilated by both the (1,0)- and (0,1)-parts $\theta$ and $\theta^*$ of $\psi$.

*Proof.* Let $D_+$ and $D_-$ be the Hermitian connections with respect to the metric $h$ and the holomorphic structures $D'' = D''_h + \theta^*$ and $D''_h - \theta^*$ respectively. In other words, $D_\pm = D_h \pm \hat{\psi}$, where $\hat{\psi} = -\theta + \theta^*$. Their curvatures $F_\pm$ are given by $F_\pm = F_h + \hat{\psi} \wedge \hat{\psi} \pm D_h\hat{\psi}$. Since the metric is weakly harmonic, we have $\Lambda D_h\hat{\psi} = \sqrt{-1}\, D_h^*\psi = 0$, and therefore $\Lambda F_+ = \Lambda F_-$. Now, for our section $s$ we have $D''_+s = D''s = 0$, and $D'_-s = D's = 0$. Furthermore, the Weitzenbock formula for $D_\pm$ yields

$$0 \leq \|D'_+s\|^2 = i\int\langle\Lambda F_+s,s\rangle\, \omega^{[n]} = i\int\langle\Lambda F_-s,s\rangle\, \omega^{[n]} = -\|D''_-s\|^2 \leq 0.$$

Hence, we also have $D'_+s = D''_-s = 0$. The proposition follows from this in view of the formulae: $2\,D_h = D_+ + D_-$, $2\,\psi = D - D_h$ and $2\,\hat{\psi} = D_+ - D_-$.

(A.2.4) **Corollary.** Let $(P,\theta)$ be a principal Higgs G-bundle with a Hermitian-Einstein metric $P_K$. Then the metric $P_K$ is irreducible in the sense of (7.4) if and only if the associated affine connection $D = D_K + \psi$ is irreducible.

*Proof.* We apply the above propositions to the vector bundle $E = \text{ad}\, P$ of degree zero; indeed, the necessity follows from the second one, while the sufficiency from the first one.

(A.3) The following result is due to Hitchin [H;p.67,Th.2.1] in the one dimensional case and to Simpson [S1;p.878,Prop.3.3] in the general case.

(A.3.1) **Proposition.** Let $(E,\theta)$ be a Higgs bundle which admits an Hermitian-Einstein metric $h$. Then for any torsion free $\theta$-invariant coherent analytic subsheaf $\mathcal{F}$ of $E$, we have the inequality $\mu(\mathcal{F}) \leq$

$\mu(E)$; moreover if the equality holds, then $\mathcal{F}$ is a subbundle, and E is a holomorphic direct sum of $\mathcal{F}$ and its orthgonal complement $\mathcal{F}^{\perp}$, which is also preserved by $\theta$.

(A.3.2) **Corollary.** If the Hermitian-Einstein metric h is irreducible, then $(E,\theta)$ is stable.

*Proof.* Let k be the rank of $\mathcal{F}$. Then $\mathcal{F}$ defines a natural inclusion $\det \mathcal{F} := (\Lambda^k \mathcal{F})^{**} \subset \Lambda^k E$, or equivalently a holomorphic section s of the vector bundle $N := \Lambda^k E \otimes (\det \mathcal{F})^*$, where $(\ )^{**}$ denotes taking the double dual. Note that $\mu_N = k(\mu_E - \mu_{\mathcal{F}})$. We can make N naturally into a Higgs bundle $(N,\theta_N)$ with a Hermitian-Einstein metric such that s is annihilated by $\theta_N$. Then by Proposition (A.2.1) we have $\mu_E \geq \mu_{\mathcal{F}}$ and if $\mu_E = \mu_{\mathcal{F}}$, s is parallel and so $\mathcal{F} = F$ is a subbundle. In this case let $\varphi$ be the section of the bundle $M := F^* \otimes E$ corresponding to the inclusion $F \subset E$. Since $\mu(M) = 0$, with respect to the natural Higgs bundle structure $\theta_M = \theta_F{}^* \otimes I_E + I_F{}^* \otimes \theta_E$ on M with an Einstein-Hermitian metric $h_M = h_F{}^* \otimes h_E$ $\varphi$ is covariant constant. Then by Proposition (A.2.1) $\varphi$ is also covariant constant with respect to $D_{h_M}$ and is annihilated by $\theta_M^*$. In terms of the bundles this implies that the restriction of $D_h$ to the subbundle F is the metric connection $D_{h|F}$, and $F^{\perp}$ is also preserved by $\theta$. The first implies that the second fundamental form of F in E vanishes, which means that $F^{\perp}$ is again a holomorphic subbundle of E (cf.[Gr2; VI,3]).

## References

[AB] Atiyah, M.F., and Bott, R: The Yang-Mills equations over Riemann surfaces, Phil. Trans. R. Soc. London A, 308 (1982), 523-615.

[AHS] Atiyah, M.F., Hitchin, N.J., and Singer, I.M.: Self-duality in four dimensional Riemannian geometry, Proc. R. Soc. London A,

362 (1978), 425-461.

[Ca] Cartan, H.: Quotient d'un espace analytique par un groupe d'automorphismes, In: Algebraic geometry and topology in honour of Lefschetz, 90-102, Princeton Univ. Press, Princeton, 1957.

[C] Corlette, K.: Flat G-bundles with canonical metrics, J. Diff. Geom. 28 (1988), 361-382.

[Do] Doi, H.: Stable principal bundles on a projective plane, Sûri-kenkôkyûroku, (1988), 132-147. (in Japanese)

[D] Donaldson, D.K.: Twisted harmonic maps and the self-duality equations, Proc. London Math. Soc., 55 (1987), 127-131.

[F1] Fujiki, A.: Closedness of the Douady spaces of compact Kähler spaces, Publ. RIMS, Kyoto Univ., 14 (1978), 1-52.

[F2] ——— : On the Douady space of a compact complex space in the category $\mathscr{C}$, II, Publ. RIMS, Kyoto Univ., 20 (1984), 461-189.

[F3] ——— : On the de Rham cohomology group of a compact symplectic Kähler manifold, Adcanced Studies in Pure Math., 10 (1987), 105-165.

[F4] ——— : Stability and Hermitian-Einstein connection on a principal Higgs bundle on a compact Kähler manifold, in preparation.

[FU] Freed, D.S., and Uhlenbeck, K.K.: Instantons and four-manifolds, MSRI Publ. Springer, 1984.

[G] Goldman, W.M.: The symplectic nature of fundamental group of surfaces, Advance in Math., 54 (1984), 200-225.

[GM] Goldman, W.M., and Millson, J.J.: The deformation theory of representations of fundamental groups of compact Kähler manifolds, Publ. Math. IHES, 67 (1988), 43-78.

[Gr1] Griffiths, P.A.: Extension problem in complex analysis I, In: Proc. Conf. on Complex Analysis, 113-142, Minneapolis 1964, Springer, 1965.

[Gr2] Griffiths, P.A.: Extension problem in complex analysis II, Amer. J. Math., 88 (1966), 306-446.

[GS] Guillemin, V., and Sternberg, S.: Geometric quantization and multiplicities of group representations, Invent. math., 67

(1982), 515-538.

[H1] Hitchin, N.J.: The self-duality equations on a surface, Proc. London Math. Soc., 55 (1987), 59-126.

[H2] ——— : Stable bundles and integrable systems, Duke Math. J., 54 (1987), 91-114.

[HKLR] Hithcin, N.J.,Karlhede, A., Lindström, U. and Rocek, M.: Hyper-kähler metrics and supersymmetry, Commun. Math. Phys., 108 (1987), 535-589.

[JM] Johnson, D., and Millson, J.J.: Deformation spaces associated to compact hyperbolic manifolds, 48-105, In: discrete groups in geometry and analysis, Papers in Honor of G.D. Mostow, Progress in Math., 67, 1986, Birkhäuser.

[JY] Jost, J., and Yau, S.-T.: Harmonic maps and group representations in complex geometry, Lecture at Conf. in Kyoto, 1989.

[K] Kirwan, F.C.: Cohomology of quotients in symplectic and algebraic geometry, Math. Notes 31, Princeton Univ. Press, 1984.

[Ko] Kobayashi, S.: Differential geometry of complex vector bundles, Publ. Math. Soc. Japan 15, Iwanami and Princeton Univ. Press, 1987.

[KN] Kobayashi, S., and Nomizu, K.: Foundations of differential geometry, vol. 2, Wiley-Interscience, 1969.

[MR] Mehta, V.B., and Ramanathan, A.: Restriction of stable sheaves and representations of the fundamental group, Invent. math., 77 (1984), 163-172.

[M] Mumford, D.: Geometric invariant theory, Springer Verlag, Berlin-Heidelberg-New York, 1965.

[P] Palais, R.S.: Foundations in global nonlinear analysis, W.A. Benjamin, Inc., 1968.

[R] Ramanathan, A., Stable principal bundles on a compact Riemann surfaces, Math. Ann., 213 (1975), 129-152.

[RR] Ramanan, S., and Ramanathan, A.: Some remarks on the instbility flag, Tohoku Math. J., 36 (1984), 269-281.

[RS] Ramanathan, A., and Subramanian, S.: Einstein-Hermitian connections on principal bundles and stability, J. reine angew. Math., 390 (1988), 21- 31.

[S1] Simpson, C.T.: Constructing variations of Hodge structure using Yang-Mills theory and applications to uniformization, J. of AMS, 1 (1988), 867-918.

[S2] —— : Higgs bundles and local systems, preprint.

[S3] —— : Moduli of representations of the fundamental group of a smooth projective variety, preprint.

[UY] Uhlenbeck, K., and Yau, S.-T.: On the existence of Hermitian-Yang-Mills connections on stable vector bundles, Comm. Pure and Appl. Math., 39 (1986), 257-293.

[W] Weil, A.: Variétés kählériennes, nouvelle édition, Hermann, Paris, 1971.

# Hardy spaces and $BMO$ on Riemann surfaces

HIROSHIGE SHIGA

Department of Mathematics
Tokyo Institute of Technology

Dedicated to Professor Tatsuo Fuji'i'e on his 60th birthday

## INTRODUCTION

Functions of bounded mean oscillation called BMO-functions, are related to various fields of analysis. For example, they have come from the singular integrals and have been joined with the theory of quasiconformal mappings. As for these topics, see [Ga], [Ge], [RR]. Among them, the duality theorem between Hardy space $H^1$ and $BMO$, the function space of bounded mean oscillation, on $\mathbf{R}^n$ which was proved by Fefferman-Stein[FS], is one of the most important results about $BMO$. When $n = 1$, both spaces of functions are regarded as boundary values of spaces of certain kinds of analytic functions in the unit disk. Thus, we have the duality theorem between $H^1$ and $BMOA$(analytic functions of bounded mean oscillation) on *the unit disk*. About this theorem, Metzger[M] proposed the following question.

*Let $R$ be a compact bordered Riemann surface(a Riemann surface of finite genus with a finite number of boundary curves). Then, is the dual space of $H^1(R)$, $(H^1(R))^*$, equal to $BMOA(R)$ ?*

However, this problem has been already solved affirmatively by Gotoh[G](his master thesis, in Japanese) and the author(unpublished) several years ago, whose methods are slightly different to each other.

In this paper, first we will give an argument under more general setting. Secondly, we will show a decomposition theorem on compact bordered Riemann surfaces of $BMOA$ and prove the duality theorem(the answer to Metzger's question) there by means of the author's method mentioned above. Finally, we shall show Read's theorem on parabolic ends and discuss related topics. The author is grateful to the referee for his helpful comments.

## 1. $H^p$ AND $BMO$ ON RIEMANN SURFACES

In this section, we shall define Hardy space $H^p$ and $BMO$ on general Riemann surfaces and give fundamental properties of them. For the details, see [Ga], [H], [RR].

Let $R$ be an open Riemann surface whose universal covering surface is conformally equivalent to the unit disk $\Delta$. Then, we can take a Fuchsian group $\Gamma$ acting on $\Delta$ so that $\Delta/\Gamma = R$.

**Definition 1-1.** Let $f$ be a holomorphic function on $\Delta$ and let $p \geq 1$. For $1 \leq p < \infty$, the function $f$ is called $H^p(\Delta)$ *function* if $|f|^p$ has a harmonic majorante on $\Delta$. For $p = \infty$, it is called $H^\infty(\Delta)$ *function* if it is bounded in $\Delta$. Furthermore, a function $f$ on a Riemann surface $R$ is called $H^p(R)$ *function* if the lift of $f$ to the unit disk $\Delta$, say $\tilde{f}$ belongs to $H^p(\Delta)$ and $\Gamma$-automorphic, namely, $\tilde{f}$ belongs to $H^p(\Delta)$ and for all $\gamma \in \Gamma$ and for all $z \in \Delta$ equalities

$$\tilde{f}(\gamma(z)) = \tilde{f}(z)$$

hold. And we denote by $H^p(\Delta)$ and $H^p(R)$, the sets of all $H^p(\Delta)$ functions and $H^p(R)$ functions, respectively.

Since $|f|^p$ is a subharmonic function, we have easily the followings;

PROPOSITION 1-1. *Let $f$ be a holomorphic function on $\Delta$. Then, the followings are equivalent.*

(1) *$f$ belongs to $H^p(\Delta)$ $(1 \le p < \infty)$,*

(2) *$\sup_{0 \le r < 1} \frac{1}{2\pi} \int_0^{2\pi} |f(re^{i\theta})|^p d\theta < \infty$ $(1 \le p < \infty)$.*

PROPOSITION 1-1'. *Let $f$ be a holomorphic function on $R$ and let $\{R_n\}_{n=1}^{\infty}$ is a regular exhaustion of $R$. We denote by $g^{R_n}(\cdot; q)$ Green's function on $R_n$ with the pole at $q \in R_1$. Then, the followings are equivalent.*

(1) *$f$ belongs to $H^p(R)$ $(1 \le p < \infty)$,*

(2) *$\sup_{1 \le n < \infty} \frac{1}{2\pi} \int_{\partial R_n} |f(z)|^p (-{}^*dg^{R_n}(z; q)) < \infty$ for all $q \in R_n$ $(1 \le p < \infty)$.*

For a fixed $q_0 \in R$, we set

$$\|f\|_p^R = \inf_{1 \le n < \infty} \{ \int_{\partial R_n} |f(z)|^p (-{}^*dg^{R_n}(z; q_0)) \}^{1/p}$$

$$\|f\|_\infty^R = ess.\sup_{z \in R} |f(z)|.$$

Then, it is easily seen that $\|f\|_p^R = \{L.H.M.|f|^p(q_0)\}^{1/p}$ for $1 \le p < \infty$, where $L.H.M.|f|^p$ is the least harmonic majorante of $|f|^p$ on $R$. Furthermore, we can show that for every $z_0 \in \Delta$ with $\pi(z_0) = q_0$ $\|f\|_p^R = \{L.H.M.|\tilde{f}|^p(z_0)\}^{1/p} = \|\tilde{f}\|_p^\Delta$.

A holomorphic function $f$ which belongs to $H^p(\Delta)$ has non-tangential limit $f^*(e^{i\theta})$ for almost all $e^{i\theta} \in \Delta$ and $f^* \in L^p(\partial\Delta)$. Conversely, the boundary function $f^*$ determines $f \in H^p(\Delta)$ uniquely. Thus, in this paper we will identify an $H^p$-function on $\Delta$ and the boundary function.

**Definition 1-2.** Let $\tilde{h}$ be a measurable function on $\partial\Delta$. The function $\tilde{h}$ is called $BMO$-function if there exists a constant $M$ such that

$$\frac{\int_I |\tilde{h}(e^{i\theta}) - \tilde{h}_I| d\theta}{|I|} < M,$$

for every interval $I \subset \partial\Delta$, where $\tilde{h}_I = \frac{1}{|I|} \int_I \tilde{h}(e^{i\theta}) d\theta$. A harmonic(resp. holomorphic) function $\tilde{h}$ is called $BMOH(\Delta)$(resp. $BMOA(\Delta)$) function if $\tilde{h}$ is represented by Poisson integral of a $BMO$-function on $\partial\Delta$.

**Definition 1-2'.** Let $h$ be a harmonic(resp. holomorphic) function on a Riemann surface $R$ is called $BMOH(R)$(resp. $BMOA(R)$) function if the lift $\tilde{h}$ of $h$ on $\Delta$ is $BMOH(\Delta)$.

For each $BMO$-function $\tilde{h}$ on $\partial\Delta$, we set

$$\|\tilde{h}\|_* = \sup_I \frac{\int_I |\tilde{h}(e^{i\theta}) - \tilde{h}_I| d\theta}{|I|} + |h(0)|.$$

$BMOH$ and $BMOA$ on $\Delta$ are closely related to bounded analytic functions and $H^p$-functions on $\Delta$. The set of all $BMOH(R)$ (resp. $BMOA(R)$) functions is denoted by $BMOH(R)$ (resp. $BMOA(R)$).

PROPOSITION 1-2. *Let $h$ be a harmonic function on $\Delta$. Then the followings are equivalent.*

(1) *$h$ belongs to $BMOH(\Delta)$.*
(2) *$h = u + {}^*v$, where both $u$ and $v$ are bounded harmonic functions on $\Delta$, and ${}^*v$ means the conjugate harmonic function of $v$.*

The decomposition of $BMOH$-function in (2) is called *Fefferman-Stein* decomposition for $BMO$.

The group of all conformal automorphisms of $\Delta$ is denoted by $M$ and for a holomorphic function $f$ on $\Delta$ we set

$$M(f) = \{g : g(z) = f(S(z)) - f(S(0)), S \in M\}.$$

Then,

PROPOSITION 1-3. *Let $f$ be a holomorphic function on $\Delta$. Then the followings are equivalent.*

(1) *$f$ belongs to $BMOA(\Delta)$*
(2) *the set $M(f)$ is bounded in $H^p(\Delta)$ for every $p$ ($1 \leq p < \infty$)*
(3) *the set $M(f)$ is bounded in $H^p(\Delta)$ for some $p$ ($1 \leq p < \infty$).*

*Furthermore, there exists a constant $C(p)$ depending only on $p$ such that*

$$\sup_{g \in M(f)} \|g\|_p \leq C(p)\|f\|_*.$$

Fefferman-Stein established an important theorem about the relation between $BMOA$ and $H^1$. Namely,

PROPOSITION 1-4. $H^1(\Delta)^* = BMOA(\Delta)$.

## 2. Duality Theorem on Compact Bordered Riemann Surfaces

Let $R$ be a Riemann surface which does not belongs to class $O_G$, namely the Riemann surface $R$ has Green's function. $R$ has non-constant positive superharmonic function. So,the universal covering surface of $R$ is (conformally equivalent to) the unit disk $\Delta$. Therefore, $R$ is represented by a Fuchsian group $\Gamma$ as $\Delta/\Gamma$.

Now, we shall consider about the duality between $H^1(R)$ and $BMOA(R)$ on a Riemann surface $R$. But, unfortunately;

PROPOSITION 2-1. *There exists a Riemann surface $R$ such that $H^1(R)^* \neq BMOA(R)$.*

*Proof.* First, we remark that $BMOA(R) \subset H^p(R)$ for all $p$ with $1 \leq p < \infty$. In fact, for each $f = u + iv \in BMOA(R)$ there exists the lift $\tilde{f}$ of $f$ on $\Delta$. Thus, it suffices to show that $|\tilde{f}|^p$, has a harmonic majorante on $\Delta$ for p ($1 < p < \infty$). Put $\tilde{f} = \tilde{u} + i\tilde{v}$. For a bounded harmonic function $h$, $|h|^p$ has a harmonic majorante, of course. Thus, by Riesz' theorem(cf. [H ]) and Proposition 2 we verify that $|\tilde{u}|^p$ and $|\tilde{v}|^p$ have harmonc majorantes on $\Delta$ for $p$ ($1 < p < \infty$). Thus, $|\tilde{f}|^p$ does so.

Heins[H] constructed a Riemann surface $R$ such that $R$ admits non-constant $H^1$-function but it has no non-constant $H^p$-function for every $p > 1$. This Riemann surface $R$ is our desired one. Because, as vector spaces over $\mathbf{C}$, $\dim H^1(R) > \dim BMOA(R) = \dim H^p(R) = 1$ ($p > 1$). Therefore, $H^1(R)^* \neq BMOA(R)$. q. e. d.

But, generally we can show the following;

PROPOSITION 2-2. *Let $R$ be a Riemann surface not in $O_G$ as before. There exists an isomorphism $\iota$ of $BMOA(R)$ into $H^1(R)^*$. Thus, $BMOA(R)$ is regarded as a subset of $H^1(R)^*$ via $\iota$.*

*Proof.* We take a Fuchsian group $\Gamma$ on the unit disk $\Delta$ as above. For each $f \in BMOA(R)$, we take the lift $\tilde{f} \in BMOA(\Delta)$. From Fefferman-Stein duality theorem mentioned at Proposition 1-4, $\tilde{f}$ induces an element of $H^1(\Delta)^*$. Since $H^1(R)$ is regarded as a subset of $H^1(\Delta)$, it also induces an element of $H^1(R)^*$, say $l_f$. We define $\iota(f)$ as $l_f$. Obviously, $\iota$ is a linear mapping of $BMOA(R)$ to $H^1(R)^*$. It suffices to show that it is injective. Take distinct functions $f$ and $g$ in $BMOA(R)$ and denote by $\tilde{f}$ and $\tilde{g}$ these lifts on $\Delta$. Since $BMOH(R) \subset H^2(R) \subset H^1(R)$ and $H^2(R)$ is a Hilbert space, $f$ and $g$ are regarded as distinct elements in $(H^2(R))^* = H^2(R)$. Thus, they are distinct as elements of $(H^1(R))^*$. q.e.d.

Here, we assume that $R$ is a compact bordered Riemann surface of genus $g$ and that the relative boundary $\partial R$ consists of a finite number, say $n$, of analytic Jordan curves. Then we can take the *double* $\hat{R}$ of $R$. $\hat{R}$ is a compact Riemann surface of genus $2g+n-1$. We take a point $a_0 \in R$ and fix it. The Riemann surface $R$ has Green's function $g^R(\cdot, a_0)$ with the pole at $a_0$, and $dm^R_{a_0} = -{}^*dg^R(\cdot, a_0)/2\pi$ is a probabilty measure on $\partial R$ which is called *harmonic measure with respect to* $a_0$ on $R$. We denote by $L^p(\partial R)$ $(1 \leq p \leq \infty)$ the set of all measurable $L^p$ functions with respect to $dm^R_{a_0}$. Function spaces $H^p(R)$ $(1 \leq p \leq \infty)$, $BMOH(R)$ and $BMOA(R)$ are regarded as subspaces of $L^p(R)$. It is easily seen that the space $L^p(\partial R)$ does not depend on the choice of $a_0$. Since Green's function $g^R(\cdot, a_0)$ is harmonic on $R$ except at $a_0$ and vanishes identically on $\partial R$, $dg^R(\cdot, a_0) + i^*dg^R(\cdot, a_0)$ is extended to an Abelian differential $\omega_{a_0}$ on $\hat{R}$ with simple poles at $a_0$ and $\hat{a}_0$, where $\hat{a}_0$ is the mirror image of $a_0$ with respect to $\partial R$. We denote by $\delta_{a_0}$ and $\hat{\delta}_{a_0}$ the devisors of $\omega_{a_0}$ in $R$ and in $\hat{R} - R$, respectively. Then, we can see from classical Riemann-Roch's theorem that

$$\deg \delta_{a_0} = \deg \hat{\delta}_{a_0} = 2g + n - 2.$$

THEOREM 2-1. *Let $R$ be a compact bordered Riemann surface as above. Then, as a subspace of $L^1(\partial R)$*

$$BMOH(R) = BMOA(R) \oplus \overline{BMOA_0(R)} \oplus M({\delta_{a_0}}^{-1}\hat{\delta}_{a_0}^{-1})|_{\partial R}, \tag{2.1}$$

*where $M({\delta_{a_0}}^{-1}\hat{\delta}_{a_0}^{-1})$ is the space of all meromorphic functions on $\hat{R}$ whose divisors are multiple of ${\delta_{a_0}}^{-1}\hat{\delta}_{a_0}^{-1}$ on $\hat{R}$ and $\overline{BMOA_0(R)}$ is the set of holomorphic functions in $BMOA(R)$ which vanish at $a_0$.*

*Proof.* To show this theorem, we need some lemmas. Since $R$ is a compact bordered Riemann surface of genus $g$ with $n$ borderes, we can take $(2g+n-1)$ simple closed curves $\{C_1, C_2, \ldots, C_{2g+n-1}\}$ on $R$ and $(2g+n-1)$ harmonic functions $\{u_1, u_2, \ldots, u_{2g+n-1}\}$ on $R \cup \partial R$ so that $\{C_1, C_2, \ldots, C_{2g+n-1}\}$ is a basis of the homology group of $R$ and

$$\int_{C_j} {}^*du_i = \delta_{ij}.$$

We denote by $N$ a vector space over $\mathbf{C}$ spanned by $\{u_1, u_2, \ldots, u_{2g+n-1}\}$. Obviously, $\dim N = 2g + n - 1$. Then, we have

LEMMA 2-1.

$$BMOH(R) = BMOA_0(R) \oplus \overline{BMOA(R)} \oplus N. \tag{2.2}$$

*Proof of Lemma 2-1.* First, we suppose that $u$ is a real valued harmonic function in $BMOH(R)$. Set $a_j = \int_{C_j} {}^*du$ and $U = u - \sum_{j=1}^{2g+n-1} a_j u_j$. By the definition, $F = U - U(a_0) + i{}^*U$ is a holomorphic function on $R$, where ${}^*U$ is a conjugate harmonic function of $U$ with ${}^*U(a_0) = 0$. Furthermore, from Proposition 2-1 we see that and ${}^*U$ belongs to $BMOH(R)$ as well as $U$. Thus, $F \in BMOA(R)$. And we have a decomposition

$$u = \frac{F + \overline{F}}{2} + U(a_0) + \sum_{j=1}^{2g+n-1} a_j u_j. \tag{2.3}$$

It is also true for a complex-valued harmonic function in $BMOH(R)$. We have the desired result. q. e. d.

We proceed to prove Theorem 2. 1. From Riemann-Roch's theorem, we have

$$\dim M({\delta_{a_0}}^{-1}\hat{\delta}_{a_0}^{-1}) = \dim N = 2g + n - 1.$$

Take a basis $m_1, m_2, \ldots, m_{2g+n-1}$ of $M({\delta_{a_0}}^{-1}\hat{\delta}_{a_0}^{-1})$. Then each $m_j|_{\partial R}$ $(j = 1, 2, \ldots, 2g + n-1)$ is bounded continuous on $\partial R$. Hence, from Lemma 2-1 there exist $f_j \in BMOA_0(R)$, $g_j \in BMOA(R)$ and $v_j \in N$ for each $j$ such that

$$m_j|_{\partial R} = f_j + \overline{g}_j + v_j, \text{on } \partial R.$$

We show;

LEMMA 2-2. *The assignment $m_j \mapsto v_j$ gives an isomorphism of $M({\delta_{a_0}}^{-1}\hat{\delta}_{a_0}^{-1})$ to $N$.*

*Proof of Lemma 2-2.* It is seen that the assignment $m_j \mapsto v_j$ gives a homomorphism $\ell$ of $M({\delta_{a_0}}^{-1}\hat{\delta}_{a_0}^{-1})$ to $N$. We show that $\ell$ is injective. Suppose that there exists a meromorphic function $m \in M({\delta_{a_0}}^{-1}\hat{\delta}_{a_0}^{-1})$ such that the image of $m|_{\partial R}$ via the homomorphism is zero. Then, we have

$$m = f + \overline{g} \quad (f \in BMOA_0(R),\, g \in BMOA(R)). \tag{2.4}$$

Hence, $f = m - \overline{g}$ on $\partial R$. This contradicts with Theorem in Heins[H, Chap. VI Theorem 7] which is an application of Read's theorem (see also Corollary 3-1).

This lemma implies that $\{v_1, v_2, \ldots, v_{2g+n-1}\}$ is a basis of $N$ and

$$v_j = -f_j - \overline{g}_j + m_j.$$

So, from Lemma 2-1 we have the decomposition (2.1). q. e. d.

COROLLARY 2. 1. *For a compact bordered Riemann surface $R$, $H^1(R)^* = BMOA(R)$.*

*Proof.* From Theorem 2-1, $BMOA(R) \subset H^1(R)^*$. By the Hahn-Banach theorem, we see that for each $\ell \in H^1(R)^*$, there exists $h \in L^\infty(R)$ so that for all $F \in H^1(R)$

$$\ell(F) = \int_{\partial R} F(z)\overline{h(z)}(-{}^*dg^R(z, a_0)/2\pi),$$

because $H^1(R) \subset L^1(R)$ and $L^1(R)^* = L^\infty(R)$. The function $\overline{h}$ is regarded as a boundary function of some $BMOH$-function. From Lemma 2-2, we have the decomposition

$$\overline{h} = f + \overline{g} + m,$$

where $f \in BMOA_0(R)$, $g \in BMOA(R)$ and $m \in M(\delta_{a_0}{}^{-1}\hat{\delta}_{a_0}^{-1})$. Since $g^R(z, a_0)$ vanishes identically on $\partial R$,

$${}^*dg^R(z, a_0)/2\pi = \omega_{a_0}/2\pi i \qquad (z \in \partial R).$$

Hence,

$$\begin{aligned}\int_{\partial R} F(z)\overline{h(z)}(-{}^*dg^R(z, a_0)/2\pi &= \int_{\partial R} F(z)\overline{h(z)}(-\omega_{a_0}/2\pi i)\\ &= \int_{\partial R} F(z)\{f(z) + \overline{g(z)} + m(z)\}(-\omega_{a_0}/2\pi i).\end{aligned}$$

Differentials $f\omega_{a_0}$ and $m\omega_{a_0}$ are holomorphic on $R$. Therefore,

$$\begin{aligned}\ell(F) &= \int_{\partial R} F(z)\overline{h(z)}(-{}^*dg^R(z, a_0)/2\pi\\ &= \int_{\partial R} F(z)\overline{g(z)}(-\omega_{a_0}/2\pi i).\end{aligned}$$

Namely, $g \in BMOA(R)$ induces $\ell \in H^1(R)^*$. q. e. d.

**Remark.** The above method is a slight modification of that of Shiga[Sh1, Sh2]. Gotoh[G] takes another method similar to that of Heins[H].

## 3. READ'S THEOREM ON PARABOLIC ENDS

First, we shall recall the definition of parabolic ends.

**Definition 3-1.** Let $R$ be a subregion of an open Riemann surface $R' \in O_G$. It is called *parabolic end* (or *$SO_{HB}$-end*) if it is not relatively compact and the relative boundary $\partial R$ is compact.

In this section, we assume that the relative boundary $\partial R$ consists of a finite number of analytic Jordan curves. Since $R$ has Green's function $g^R(z, a_0)$ with the pole at $a_0 \in R$, we can also define $L^p(\partial R)$ $(1 \leq p \leq \infty)$ as in Sec. 2. Then, we extend Read's theorem for $H^2(R)$.

**THEOREM 3-1.** *Let $h$ be a function in $L^p(\partial R)$ $(1 < p \leq \infty)$. Then, $h$ is a boundary function of some $H^p$-function if and only if for every holomorphic differential $\omega$ on $R \cup \partial R$ with finite Dirichlet integral,*

$$\int_{\partial R} h\omega = 0. \tag{3.1}$$

*Proof.* We set $\Gamma_h(R)$ and $\Gamma_a(R)$ the sets of all harmonic and analytic differentials whose Dirichlet integrals are finite, respectively. For any simple closed curve $C$ in $R$, there exist period reproducing differentials $\sigma_h(C) \in \Gamma_h(R)$ and $\sigma_a(C) \in \Gamma_a(R)$ for $C$ in $\Gamma_h(R)$ and $\Gamma_a(R)$, respectively. Namely, for every $\omega_h \in \Gamma_h(R)$ and $\omega_a \in \Gamma_a(R)$,

$$\int_C \omega_h = (\omega_h, \sigma_h(C)),$$
$$\int_C \omega_a = (\omega_a, \sigma_a(C))$$

hold, where $(\omega_1, \omega_2) = \iint_R \omega_1 \wedge {}^*\overline{\omega}_2$. In fact, $\sigma_h(C)$ and $\sigma_a(C)$ have the following representations.

$$\sigma_h(C) = \frac{1}{2\pi}(dP_{I\sigma_C} + {}^*dP_{I\tau_C})$$
$$\sigma_a(C) = \frac{1}{4\pi}\{dP_{I\sigma_C} + {}^*dP_{I\tau_C} + i({}^*dP_{I\sigma_C} - dP_{I\tau_C})\},$$

where $P_{I\sigma_C}$ and $P_{I\tau_C}$ are $(\mathbf{I})L_1$ principal functions corresponding to $C$(cf. Sario-Rodin[SR] and Sario-Oikawa[SO]). Since $P_{I\sigma_C}$ is a harmonic function with finite Dirichlet integral on $R \cup \partial R$ and $P_{I\sigma_C}|\partial R$ is a constant, $P_{I\sigma_C}$ is the constant on $R$ by the maximal principle. Thus,

$$\sigma_h(C) = \frac{1}{2\pi}{}^*dP_{I\tau_C}$$
$$\sigma_a(C) = \frac{1}{4\pi}({}^*dP_{I\tau_C} - i dP_{I\tau_C}).$$

Furthermore,

$$\sigma_a(C) = \frac{1}{4\pi}{}^*dP_{I\tau_C} = \frac{1}{2}\sigma_h(C) \quad \text{on } \partial R,$$

because $P_{I\tau_C}$ is a constant on $\partial R$. Therefore, we have

$$0 = \int_{\partial R} h\sigma_a(C) = \frac{1}{2}\int_{\partial R} h\sigma_h(C) \tag{3.2}$$

Here, we note that

$$\int_{\partial R} h\sigma_h(C) = \int_C {}^*dH_h^{\partial R}, \tag{3.3}$$

where $H_h^{\partial R}$ is the Dirichlet solution on $R$ with the boundary value $h$. To show this, we take a sequence $\{h_k\}_{k=1}^{\infty}$ of harmonic functions on a neighbourhood of $\partial R$ so that

$$\lim_{k\to\infty}\int_{\partial R}|h_k-h|(-{}^*dg^R(;a_0))=0.$$

Then, the Dirichlet solutions $\{H_{h_k}^{\partial R}\}_{k=1}^{\infty}$ converges to $H_h^{\partial R}$ uniformly on every compact subset of $R$. Hence, we have

$$\lim_{k\to\infty}\int_C {}^*dH_{h_k}^{\partial R}=\int_C {}^*dH_h^{\partial R}.$$

Since each $H_{h_k}^{\partial R}$ is a harmonic function on $R\cup\partial R$ and the Dirichlet integral is finte, we have

$$\int_C {}^*dH_{h_k}^{\partial R}=(H_{h_k}^{\partial R},\sigma_h(C))$$

$$=\iint_R {}^*dH_{h_k}^{\partial R}\wedge\overline{{}^*\sigma_h(C)}=\iint_R dH_{h_k}^{\partial R}\wedge\sigma_h(C).$$

Noting that $R$ is a parabolic end, we conclude that

$$\int_C {}^*dH_h^{\partial R}=\lim_{k\to\infty}\int_C {}^*dH_{h_k}^{\partial R}=\lim_{k\to\infty}\iint_R dH_{h_k}^{\partial R}\wedge\sigma_h(C)$$
$$=\lim_{k\to\infty}\int_{\partial R}h_k\sigma_h(C)=\int_{\partial R}h\sigma_h(C).$$

This completes the proof of (3.3). Thus, from (3.2) and (3.3) we see that $H_h$ has a conjugate harmonic function ${}^*H_h$ on $R$. By the same argument as in the proof of Lemma 2. 1, we verify that there exist $f,g\in H^p(R)$ such that $g(a_0)=0$ and

$$H_h^{\partial R}=f+\overline{g},\qquad \text{in } R.$$

Thus, it suffices to show that $g$ is a constant.

Let $\hat{R}$ be the *double* of $R$ with respect to the relative boundary $\partial R$. Then, there exists an anti-conformal automorphism $J$ of $\hat{R}$ satisfying

(1) $J(R)=\hat{R}-R\cup\partial R$,
(2) $J^2=id$.
(3) $J|_{\partial R}=id$.

And it is seen that $\overline{g\circ J}$ belongs to $H^p(\hat{R}-R\cup\partial R)$. For any $a_1,a_2\in\hat{R}-R\cup\partial R$ we can take an Abelian differential $\omega_{a_1a_2}$ on $\hat{R}$ with finite Dirichlet integral near the ideal boundary so that

(1) $\omega_{a_1a_2}$ is holomorphic except at $a_1$ and $a_2$,
(2) $\omega_{a_1a_2}$ has simple poles at $a_1$ and $a_2$, and
(3) the residue of $\omega_{a_1a_2}$ at $a_1$ is 1 and the residue at $a_2$ is $(-1)$.

The existence of such a differential is well known(cf. Sario-Oikawa[SO]). Therefore, $\omega_{a_1a_2}|_{R\cup\partial R}$ is a holomorphic differential with finite Dirichlet integral on $R$. From (3. 1), we have

$$0=\int_{\partial R} h\omega_{a_1a_2}=\int_{\partial R}\bar{g}\omega_{a_1a_2}.$$

On the other hand, $\overline{g(z)}=\overline{g(J(z))}$ on $\partial R$. Thus, we have

$$\int_{\partial R}\bar{g}\omega_{a_1a_2}=\int_{\partial R}\overline{g\circ J}\omega_{a_1a_2}=2\pi i\{\overline{g(J(a_1))}-\overline{g(J(a_2))}\}.$$

Since $a_1,a_2\in\hat{R}-R\cup\partial R$ are arbitrary points, $g$ must be a constant. This implies that $H_h^{\partial R}=f$ and we complete the proof. q. e. d.

**Remark.** This theorem does not hold for *any* ends $R$. For example, if $R=\{z\in\mathbf{C}\colon 1<|z|<r\}$ and $\partial R=\{|z|=1\}$, then $h(e^{i\theta})=1$ is, of course, a boundary function of $H^\infty(R)$ and $\omega=\frac{1}{z}dz$ is a holomorphic differential on $R$ with finite Dirichlet integral. But,

$$\int_{|z|=1} h\omega=2\pi i\neq 0.$$

Let $\tilde{M}(\delta_{a_0})$ be the set of all anti-analytic meromorphic functions $g$ on $R$ such that $g|_{\partial R}\in L^2(\partial R)$ and that for $\omega_{a_0}=dg(\cdot,a_0)+i^*dg(\cdot,a_0)$, $g\overline{\omega_{a_0}}\in\Gamma_a(R\cup\partial R)$.

COROLLARY 3-1. *Let $R$ be a parabolic end. Then, an orthogonal decomposition of $L^2(\partial R)$*

$$L^2(\partial R)=H^2(R)\dot{+}C\ell_2\tilde{M}(\delta_{a_0})|_{\partial R}\tag{3.4}$$

*holds, where $C\ell_2\tilde{M}(\delta_{a_0})|_{\partial R}$ means the closure of $\tilde{M}(\delta_{a_0})|_{\partial R}$ in $L^2(\partial R)$.*

*Proof.* Take $g\in\tilde{M}(\delta_{a_0})|_{\partial R}$. Then, $\omega=\bar{g}\omega_{a_0}\in\Gamma_a(R\cup\partial R)$. Thus we have

$$\int_{\partial R} f\bar{g}\omega_{a_0}=\int_{\partial R} f\omega=0\quad\text{for every } f\in H^2(R).$$

This implies that $C\ell_2\tilde{M}(\delta_{a_0})|_{\partial R}$ is a closed subspace of the orthogonal complement of $H^2(R)$ in $L^2(\partial R)$. Hence, the orthogonal complement of $C\ell_2\tilde{M}(\delta_{a_0})|_{\partial R}$ contains $H^2(R)$. Conversely, suppose that $h\in L^2(\partial R)$ is in the orthogonal complement of $C\ell_2\tilde{M}(\delta_{a_0})|_{\partial R}$. By the definition, $\frac{\omega}{\omega_a}\in\tilde{M}(\delta_{a_0})$ for any $\omega\in\Gamma_a(R\cup\partial R)$. Therefore, we have

$$\int_{\partial R} h\left(\frac{\omega}{\omega_{a_0}}\right)\omega_{a_0}=\int_{\partial R} h\omega=0.$$

So, we conclude that $h$ is in $H^2(R)$ from Theorem 3-1. q. e. d.

Now, we shall consider about $H^p(R)^*$ $(1<p)$ on a parabolic end $R$. Take an arbitrary $\ell\in H^p(R)^*$. Since $H^p(R)\subset L^p(\partial R)$, by the Hahn-Banach theorem, there exists a measurable function $h\in L^q(\partial R)$ $(p^{-1}+q^{-1}=1)$ such that

$$\ell(f)=\int_{\partial R} f(z)\overline{h(z)}(-{}^*dg^R(z,a_0)/2\pi),\qquad\text{for all } f\in L^p(\partial R).$$

From Theorem 3-1, we verify that $\ker\ell$ is $C\ell_q(\tilde{M}(\delta_{a_0})|_{\partial R}$, the closure of $\tilde{M}(\delta_{a_0})|_{\partial R}$ in $L^q(\partial R)$. Thus, we have

COROLLARY 3-2. *The dual space $H^p(R)$ $(p>1)$ is identified with $L^q(\partial R)/C\ell_q(\tilde{M}(\delta_{a_0})|_{\partial R}$.*

## REFERENCES


[EM] Earle, C. J. and Marden, A., *On Poincare series with application to $H^p$ spaces on bordered Riemann surfaces*, Illinois J. Math. **13** (1969), 202-219.

[F] Fisher, S., "Function Theory on Planar Domains," John Wiley & Sons, New York - Chichester - Brisbane - Tronto - Singapore, 1983.

[FS] Fefferman, C. and Stein, E. M., *$H^p$ spaces of several variables*, Acts Math. **129** (1972), 137-193.

[Ga] Garnett, J, "Bounded Analytic Functions," Academic Press, New York, 1981.

[Ge] Gehring, F. W., "Characteristic properties of quasidisks," Séminaire de Mathématiques Supérieures, Les Presses de l'Université de Montréal, 1982.

[G] Gotoh, Y., *On BMO spaces on Riemann surfaces (in Japanese)*, Master thesis at Kyoto University (1985).

[H] Heins, M, "Hardy classes on Riemann surfaces.," Lecture Notes in Math. No 98, Springer, Berlin, 1969.

[M] Metzger, T. A., *Bounded mean oscillation and Riemann surfaces*, in "Bounded Mean Oscillation in Complex Analysis," University of Joensuu Publications in Science vol. **14**, 1989, pp. 79-100.

[RR] Reimann H. M. and Rychener, T., "Functionen beschränkter mittelerer Oszillation," Lecture Notes in Math. 487, Springer-Verlag, Berlin - Heidelberg - New York, 1975.

[SR] Sario, L. and Rodin, B., "Principal Functions," Van Nostrand Co., Inc., Princeton, N. J., 1968.

[SO] Sario, L. and Oikawa, K., "Capacity Functions," Springer, Berlin, 1969.

[Sh1] Shiga, H., *On the boundary of $H^p$ classes (in Japanese)*, RIMS Kōkyuroku **366** (1979), 30-47.

[Sh2] Shiga, H., *On the boundary values of analytic functions on bordered Riemann surfaces*, Acta Human. Sci. Univ. Sangyo Kyotien. Natur. Sci. Ser. **9** (1980), 11-28.


Oh-okayama Meguro-ku Tokyo 152 Japan

# Application of a Certain Integral Formula to Complex Analysis

By

Kensho Takegoshi

Department of Mathematics
College of General Education
Osaka University , Toyonaka Osaka 560 Japan

## Introduction

In [D-X] , Donnelly and Xavier induced an integral formula for differential forms with compact supports on a complete Riemannian manifold $( M , g_M )$ , which involves the gradient and real Hessian of a certain real valued function on M . Combining this formula with a certain curvature condition for $( M , g_M )$ , they induced an integral inequality to show a vanishing theorem for $L^2$- harmonic forms and $L^2$-closedness of the range of Laplace-Beltrami operator relative to $g_M$ . Here it should be noted that Bochner technique can not be applied to their case to show these properties . This is a crucial reason why they had to induce such a new technique .

After receiving an indication from their work , T.Ohsawa and the author have investigated its application to complex Analysis by reformulating and modifying their formula in the category of complete Kähler manifolds ( cf. $[O_1]$,$[O_2]$,$[O\text{-}T_1]$,$[O\text{-}T_2]$,$[T_1]$,$[T_2]$ ) .

On the other hand one can also reformulate their formula for a certain vector bundle-valued differential forms on a bounded domain with smooth boundary over a non-compact Kähler manifold . Necessarily this case raises some boundary integrals as a consequence of Stokes theorem and those boundary integrals represent something like a boundary condition for differential forms which should be estimated . From this observation , we can get a quite new aspect to analyze several problems on con-compact complex manifolds i.e. Liouville type theorem , the $\bar{\partial}$ - Neumann problem on pseudoconvex domains etc . The main purpose of this article is to introduce a new approach by the integral formula to those problems .

In the first section we induce two integral formulae which are reformulations of Donnelly and Xavier's integral formula on bounded domains with smooth boundary over non-compact Kähler manifolds .

In the next sections we introduce their applications to several problems in complex analysis i.e. 1) Liouville theorem for harmonic functions on a Stein manifold satisfying a certain slow volume growth condition 2) the complex analyticity of harmonic maps satisfying the tangential Cauchy-Riemann equation on the boundary of a bounded domain and 3) a priori estimates for the $\bar{\partial}$ - Neumann problem on pseudoconvex domains with smooth boundary and its application to a global regularity theorem for the Neumann operator .

## 1. Integral formulae on bounded domains with smooth boundary

Let $( M , ds_M^2 )$ be an $m$ dimensional Kähler manifold with the metric tensor

$$ds_M^2 = 2\mathrm{Re} \sum g_{i\bar{j}} dz^i dz^{\bar{j}}$$

Let $( \mathbb{E} , h )$ be a differentiable complex vector bundle on $M$ with the fibre metric

$$h = \{ h_{\alpha\bar{\beta}} \}$$

For any domain $D$ of $M$ , we denote by $C^{p,q}(D,\mathbb{E})$ ( resp. $C^r(D,\mathbb{E})$ ) the space of $\mathbb{E}$ - valued $C^\infty$- differential forms of type $(p,q)$ ( resp. $r$ ) and denote by $C^{p,q}(\bar{D},\mathbb{E})$ the image of the homomorphism : $C^{p,q}(M,\mathbb{E}) \longrightarrow C^{p,q}(D,\mathbb{E})$ and so on . We denote $C^{p,q}(M,\mathbb{E})$ by $C^{p,q}(M)$ whenever $\mathbb{E}$ is the trivial vector bundle .

The star operator $*$ : $C^{p,q}(M,\mathbb{E}) \longrightarrow C^{m-q,m-p}(M,\mathbb{E})$ is defined by

$$* u = C(m,p,q) \sum_{A_q,B_p} \mathrm{sign} \begin{pmatrix} 1,\dots,m \\ A_q, A_{m-q} \end{pmatrix} \mathrm{sign} \begin{pmatrix} 1,\dots,m \\ B_p, B_{m-p} \end{pmatrix}$$
$$\times \det(g_{i\bar{j}}) \, u^{\bar{B}_p A_q} dz^{A_{m-q}} \wedge d\bar{z}^{B_{m-p}}$$

for $C(m,p,q) = (\sqrt{-1})^m (-1)^{(1/2)m(m-1)+mp_2p+q-m}$ and $u \in C^{p,q}(M,\mathbb{E})$ . The pointwise inner product $< , >_{\mathbb{E}}$ on $C^{p,q}(M,\mathbb{E})$ is defined by the equality

$$<u,v>_{\mathbb{E}} \, dv_M = u \wedge *(\overline{h(v)})$$

for $u , v \in C^{p,q}(M,\mathbb{E})$ . Here $dv_M$ is the volume form of $M$ relative to $ds_M^2$ defined by

$$dv_M = \frac{\overset{m}{\wedge} \omega_M}{m!} \quad \text{for} \quad \omega_M = \frac{\sqrt{-1}}{2} \sum g_{i\bar{j}} \, dz^i \wedge d\bar{z}^j$$

Then we have

$$\langle u,v\rangle_{\mathbb{E}} = \frac{2^{p+q}}{p!q!} \sum h_{\nu\bar{\mu}}\, u^{\nu}_{A_p\bar{B}_q}\, \overline{v^{\mu \bar{A}_p B_q}}$$

The inner product $(\ ,\ )_{\mathbb{E}}$ on $C^{p,q}(D,\mathbb{E})$ is defined by

$$(u,v)_{\mathbb{E}} = \int_D \langle u,v\rangle_{\mathbb{E}}\, dv_M \quad \text{for } u \text{ and } v \in C^{p,q}(D,\mathbb{E})$$

For any $\xi \in C^{s,t}(M)$, we denote by $e(\xi)$ the left exterior multiplication by $\xi$ on $C^{p,q}(M,\mathbb{E})$ and by $e(\xi)^*$ its adjoint operator relative to $(\ ,\ )_{\mathbb{E}}$ i.e. $e(\xi)^* = (-1)^{(p+q)(s+t+1)} * e(\xi) *$ on $C^{p,q}(M,\mathbb{E})$

Let $D$ be a bounded domain with smooth boundary $\partial D$ over $M$ and let $r$ be a defining function of $D$ i.e. $D = \{ r < 0 \}$ and $dr \neq 0$ on $\partial D$. The volume element $dS$ of $\partial D$ is defined by

$$dv_M = \frac{dr}{|dr|_M} \wedge dS$$

We set

$$\lambda = \frac{dS}{|dr|_M}$$

For any $u \in C^1(M)$ we have by Stokes theorem

$$(1.1) \qquad \int_D d*u = \int_{\partial D} e(dr)^* u\, \lambda$$

When we write $u = \sum_{i=1}^m u_i dz^i + u_{\bar{i}} d\bar{z}^i$, setting $v^i = g^{\bar{j}i} u_{\bar{j}}$ and $v^{\bar{i}} = g^{\bar{i}j} u_j$, we have

$$(1.2) \qquad d*u = 2\Big( \sum_{i=1}^m \nabla_i v^i + \bar{\nabla}_i v^{\bar{i}} \Big)\, dv_M$$

Here $\nabla$ ( resp. $\bar{\nabla}$ ) is the covariant differentiation of type $(1,0)$ ( resp. $(0,1)$ ) relative to $ds^2_M$. From (1.1) and (1.2), we have

$$(1.3) \qquad \int_D 2\Big( \sum_{i=1}^m \nabla_i v^i + \bar{\nabla}_i v^{\bar{i}} \Big)\, dv_M = \int_{\partial D} e(dr)^* u\, \lambda$$

We set

$$[u,v] := \int_{\partial D} \langle u,v\rangle_{\mathbb{E}}\, \lambda \quad \text{for } u, v \in C^{p,q}(\bar{D},\mathbb{E})$$

Let $\Phi$ be a real valued $C^\infty$ function on $M$. We prepare the following notation :

(1.4) $\langle \mathcal{L}(\Phi)(u),u\rangle_{\mathbb{E}} = \frac{2^{p+q+1}}{p!(q-1)!} \sum h_{\mu\bar{\nu}}\, g^{\bar{j}i} \Phi_{i\bar{k}}\, u^{\mu}{}_{A_p \bar{j} \bar{B}_{q-1}}\, \overline{u^{\nu \bar{A}_p k B_{q-1}}}$

for $u \in C^{p,q}(M,\mathbb{E})$ and $\Phi_{i\bar{j}} = \partial^2\Phi/\partial z^i \partial \bar{z}^j$ .

Let $(\mathbb{E}, h)$ be a holomorphic vector bundle over $(M, ds^2_M)$ . We denote by $\vartheta_{\mathbb{E}}$ the formal adjoint operator of the $\bar{\partial}$ operator relative to the inner product $(\ ,\ )_{\mathbb{E}}$ .

First we show the following integral formula .

Proposition 1

$$(1.5)\quad (\mathcal{L}(\Phi)(u),u)_{\mathbb{E}} - (\vartheta_{\mathbb{E}}u, e(\bar{\partial}\Phi)^* u)_{\mathbb{E}} - (u, e(\bar{\partial}\Phi)^* \bar{\partial}u)_{\mathbb{E}} + (u, e(\bar{\partial}\Phi)^* \bar{\nabla}u)_{\mathbb{E}}$$

$$= [e(\bar{\partial}r)^* u, e(\bar{\partial}\Phi)^* u]$$

for any $u \in C^{p,q}(\bar{D},\mathbb{E})$ and $\Phi \in C^\infty_{\mathbb{R}}(M)$ .

Proof. By the Kählerity of $ds^2_M$ , we obtain the following formulae ( cf. [M-K] Chapter 3 ) :

$$(1.6)\qquad (\bar{\partial}u)^{\nu}{}_{A_p \bar{B}_{q+1}} = \sum (-1)^{p+k+1} \nabla_{\beta_k} u^{\nu}{}_{A_p \bar{\beta}_1,\ldots,\hat{\bar{\beta}}_k,\ldots,\bar{\beta}_{q+1}}$$

$$(1.7)\qquad (\vartheta_{\mathbb{E}}u)^{\nu}{}_{A_p \bar{B}_{q-1}} = -\,2 \sum \nabla^{\mathbb{E}}_i\, u^{\nu i}{}_{A_p \bar{B}_{q-1}}$$

for $u \in C^{p,q}(M,\mathbb{E})$ .

For $u \in C^{p,q}(M,\mathbb{E})$ , we construct the following differential 1-form :

$$w = C_{p,q} \sum h_{\mu\bar{\nu}}\, u^{\mu}{}_{A_p \bar{i} \bar{B}_{q-1}}\, \Phi_{\bar{j}}\, \overline{u^{\nu \bar{A}_p j B_{q-1}}}\, d\bar{z}^i$$

for $C_{p,q} = 2^{p+q}/p!(q-1)!$ . From $w$ , we construct a tangent vector $\{v^i\}$ . Then we have

$$\nabla_i v^i = C_{p,q}\{ h_{\mu\bar{\nu}} \Phi_{i\bar{j}}\, u^{\mu\ i}_{A_p\ \bar{B}_{q-1}}\, \overline{u^{\nu \bar{A}_p j B_{q-1}}} + h_{\mu\bar{\nu}} \nabla^{\mathbb{E}}_i\, u^{\mu\ i}_{A_p\ \bar{B}_{q-1}}\, \Phi_{\bar{j}}\, \overline{u^{\nu \bar{A}_p j B_{q-1}}}$$

$$+ h_{\mu\bar{\nu}} \Phi_{\bar{j}}\, u^{\mu}{}_{A_p \bar{k} \bar{B}_{q-1}}\, \overline{\nabla^k u^{\nu \bar{A}_p j B_{q-1}}} \}$$

Using (1.6) and the fact that u is alternating , we have

$$(-1)^p h_{\mu\bar{\nu}}\, u^{\mu}{}_{A_p\bar{k}\bar{B}_{q-1}}\, \overline{(\bar{\partial}u)^{\nu\bar{A}_p kjB_{q-1}}}$$

$$= q\, h_{\mu\bar{\nu}}\, u^{\mu}{}_{A_p\bar{k}\bar{B}_{q-1}}\, \overline{\nabla^k u^{\nu\bar{A}_p jB_{q-1}}} - h_{\mu\bar{\nu}}\, u^{\mu}{}_{A_p\bar{k}\bar{B}_{q-1}}\, \overline{\nabla^j u^{\nu\bar{A}_p kB_{q-1}}}$$

By substituting this into the above equality and by (1.7) , we have

$$2\sum \nabla_i v^i$$

$$= \langle \mathcal{L}(\Phi)(u),u\rangle_E - \langle \vartheta_E u, e(\bar{\partial}\Phi)^* u\rangle_E - \langle u, e(\bar{\partial}\Phi)^*\, \bar{\partial}u\rangle_E + \langle u, e(\bar{\partial}\Phi)^*\, \bar{\nabla}u\rangle_E$$

Hence the formula follows from (1.3) . q.e.d.

Next we consider a differentiable map $f : ( M , ds_M^2 ) \longrightarrow ( N , ds_N^2 )$ into a Kähler manifold $( N , ds_N^2 )$ with the metric tensor

$$ds_N^2 = 2 \operatorname{Re} \sum h_{\alpha\bar{\beta}}\, dw^{\alpha} d\bar{w}^{\beta}$$

Then the induced bundle $f^*TN$ is a differentiable bundle over M . The exterior differentiation $D_{1,0} : C^{p,q}(M,f^*TN) \longrightarrow C^{p+1,q}(M,f^*TN)$ ( resp. $D_{0,1} : C^{p,q}(M,f^*TN) \longrightarrow C^{p,q+1}(M,f^*TN)$ ) is defined by the covariant differentiation of type (1,0) ( resp. (0,1) ) induced from the connection on $T^*M \otimes f^*TN$ relative to $ds_M^2$ and $f^*ds_N^2$ . We denote by $D_{1,0}^*$ ( resp. $D_{0,1}^*$ ) the formal adjoint operator of $D_{1,0}$ ( resp. $D_{0,1}$ ) relative to the inner product $( \, , \, )_{T^*M \otimes f^*TN}$ .

Since the differential df of f is an $f^*TN$- valued differential 1 form , we obtain an $f^*TN^{1,0}$ - valued differential (1,0) form $\partial f$ and an $f^*TN^{1,0}$ - valued differential (0,1) form $\bar{\partial}f$ by composing the map $\Pi^{1,0}\circ df : TM \longrightarrow TN^{1,0}$ , $\Pi^{1,0} : TN \longrightarrow TN^{1,0}$ being the projection , with the inclusions $TM^{1,0}$ into TM and $TM^{0,1}$ into TM respectively . Then $\partial f$ ( resp. $\bar{\partial}f$ ) is represented by $(f_i^{\alpha})$ ( resp. $(f_{\bar{i}}^{\alpha})$ ) locally where $f_i^{\alpha} = \partial f^{\alpha}/\partial z^i$ and $f_{\bar{i}}^{\alpha} = \partial f^{\alpha}/\partial \bar{z}^i$ ( cf. [E-L] ).

Next we show the following integral formula .

Proposition 2

$$(1.8) \qquad ((-\Box\Phi)\bar\partial f, \bar\partial f)_{f^*TN} - (\mathcal{L}(\Phi)(\bar\partial f), \bar\partial f)_{f^*TN}$$

$$+ (D_{0,1}^{*}\bar\partial f, e(\bar\partial\Phi)^{*}\bar\partial f)_{f^*TN} + (e(\partial\Phi)^{*}D_{1,0}\bar\partial f, \bar\partial f)_{f^*TN}$$

$$= [\langle\partial r, \partial\Phi\rangle\bar\partial f, \bar\partial f]_{f^*TN} - [e(\bar\partial r)^{*}\bar\partial f, e(\bar\partial\Phi)^{*}\bar\partial f]_{f^*TN}$$

where $-\Box\Phi = 2\mathrm{Trace}_{ds_M^2}\, \partial\bar\partial\Phi$ and $\Phi \in C^\infty_{\mathbb{R}}(M)$

$$(1.9) \qquad 2\langle(D_{0,1}D_{0,1}^{*} - D_{1,0}^{*}D_{1,0})\bar\partial f, \bar\partial f\rangle_{f^*TN}$$

$$= - R^N_{\alpha\bar\beta\gamma\bar\delta}(f)(f^\alpha_{\bar i}\overline{f^\beta_{j}} - f^\alpha_{\bar j}\overline{f^\beta_{i}})\overline{(f^{\delta,i}\overline{f^{\gamma,\bar j}} - f^{\delta,j}\overline{f^{\gamma,\bar i}})}$$

where $\langle\ ,\ \rangle_{f^*TN}$ is the pointwise inner product on $C^{p,q}(M, f^*TN)$ relative to $ds_M^2$ and $f^*ds_N^2$ and $R^N_{\alpha\bar\beta\delta\bar\gamma}$ is the Riemannian curvature tensor of $ds_N^2$ .

Proof. We consider the following differential 1-forms :

$$u_1 = h_{\alpha\bar\beta}(f)\, g^{\bar j i} f^\alpha_{\bar j}\overline{f^\beta_{\bar i}}\, \Phi_{\bar k}\, d\bar z^k$$

$$u_2 = h_{\alpha\bar\beta}(f)\, g^{\bar j i} f^\alpha_{\bar k}\overline{f^\beta_{\bar i}}\, \Phi_{\bar j}\, d\bar z^k$$

From $u_i$ , we construct tangent vectors $v^i_{(j)}$ ( $j = 1 , 2$ ) . The integrand of the left hand-side of (1.8) can be obtained by calculating $2(\sum_i \nabla_i(v^i_{(1)} - v^i_{(2)}))$ ( cf. [$T_1$] p.26. (1.21) ) and so we obtain (1.8) by (1.3) .

To show the formula (1.9) , we choose holomorphic normal coordinate systems $(z^i)$ around $x \in M$ and $(w^\alpha)$ around $y = f(x) \in N$ i.e. $g_{i\bar j}(x) = \delta_{ij}$ , $dg_{i\bar j}(x) = 0$ and $h_{\alpha\bar\beta}(y) = \delta_{\alpha\beta}$ , $dh_{\alpha\bar\beta}(y) = 0$ respectively . Then all the Christofell symbols ${}^M\Gamma^k_{ij}$ and ${}^N\Gamma^\gamma_{\alpha\beta}$ of $ds_M^2$ and $ds_N^2$ vanish at $x$ and $y$ respectively and moreover it holds

$R^N_{\alpha\bar\beta\gamma\bar\delta} = -\partial_\gamma \bar\partial_\delta h_{\alpha\bar\beta}$ and $\partial^N_\sigma \Gamma^\gamma_{\alpha\beta} = \partial^N_\beta \Gamma^\gamma_{\alpha\sigma}$ at $y$ respectively . Using these properties and the formulæ (1.5) , (1.7) and (1.12) in $[T_1]$ , the formula follows from a routine calculation . q.e.d.

## 2. Liouville theorem for harmonic functions on a Stein manifold

Let $(A, ds^2_A) \overset{\iota}{\hookrightarrow} (\mathbb{C}^n, ds^2_e)$ be an $m$ dimensional closed submanifold of $\mathbb{C}^n$ provided with the induced metric $ds^2_A = \iota^* ds^2_e$ and let $\Phi$ be the restriction of the norm function $\| z \|$ , $z \in \mathbb{C}^n$ onto $A$ . Then it is known that the function $n(A,r) := \mathrm{Vol}(A(r))/r^{2m}$ ( $A(r) = \{ \Phi < r \}$ ) is a non-decreasing function of $r$ .

As an application of the integral formula (1.8) , we show the following theorem :

<u>Theorem</u> Let $(A, ds^2_A, \Phi)$ be as above and suppose

$$\int_\delta^\infty \frac{dt}{t n(A,t)} = \infty$$

Then $(A, ds^2_A)$ admits no non-constant bounded harmonic functions .

<u>Remark 1</u>. Since the Ricci curvature of $ds^2_A$ is non-positive , in a sense the above theorem is a nice counterpart of the following theorem obtained by Yau [Y] :

Theorem Any complete Kähler manifold whose Ricci curvature is non-negative admits no non-constant bounded harmonic functions .

In this case , $\mathrm{Vol}(B(r))/r^{2m}$ is a non-increasing function of $r$ and so bounded from above where $B(r)$ is a geodesic ball of radius $r$ centered at some fixed point and $m$ is the complex dimension of that manifold .

The class of closed submanifolds satisfying the above volume growth condition contains smooth affine algebraic varieties properly because $A$ is affine algebraic if and only if $n(A,r)$ is bounded ( cf. [St] ). However it is not so clear whether the above condition is optimal for the non-existence of non-constant bounded harmonic functions .

<u>Remark 2</u>. Any holomorphic function on $A$ is harmonic . However with

respect to the Liouville property for holomorphic functions , we can show a stronger result i.e. Casorati-Weierstrass theorem under the assumption of Theorem ( cf. [$T_2$] ).

Proof of Theorem . Since we may assume that $0 \notin A$ and $\Phi$ has only non-degenerate critical points on $M$ , Stokes theorem can be applied to every sublevel set $A(r) = \{ \Phi < r \}$ . For any real-valued $C^\infty$ function $f : ( A , ds_A^2 ) \to ( \mathbb{R} , dx^2 ) ( \hookrightarrow ( \mathbb{C} , dz d\bar{z} ) , z = x + \sqrt{-1}\, y )$ , substituting $\Phi$ and $\Psi = \Phi^2$ instead of $r$ and $\Phi$ into (1.8) respectively , we have

$$(2m-2)\| df \|_r^2 + (*)_r = r \{ [ |d\Phi| df ]_r^2 - 4[e(\bar\partial\Phi)^* \bar\partial f]_r^2 \}$$

and

$$(*)_r = 2[(\vartheta\bar\partial f, e(\bar\partial\Psi)^* \bar\partial f)_r + (e(\partial\Psi)^* \partial\bar\partial f, \bar\partial f)_r]$$

for any $A(r)$ , $r > r_0 := \inf \Phi$ .

Here we have used the fact $g_{i\bar j} = \Psi_{i\bar j}$ . We assert that the integral $(*)_r$ is non-negative whenever $f$ is harmonic i.e. $\Delta f = 2\vartheta\bar\partial f = 0$ ( by the Kählerity of $ds_A^2$ ) . In fact if $f$ is harmonic , then we have

$$(*)_r = 4\int_{r_0}^{r} t\, [e(\partial\Phi)^* \partial\bar\partial f, \bar\partial f]_t \, dt$$

$$= 4\int_{r_0}^{r} t\, \| \partial\bar\partial f \|_t^2 \, dt \geq 0$$

since $\mathfrak{D}\partial\bar\partial f = 0$ by the harmonicity of $f$ ( cf. [M-K] p.113 ) . Hence setting $E(f,r) = \| df \|_r^2$ and $B(f,r) = 4[e(\bar\partial\Phi)^* \bar\partial f]_r^2$ , we have

$$r \frac{\partial}{\partial r} E(f,r) - (2m-2)E(f,r) \geq r\, B(f,r) \tag{2.1}$$

for any non-critical value $r$ of $\Phi$ whenever $f$ is harmonic . Here we have used the fact $|d\Phi| \leq 1$ on $A$ ( since $\partial\bar\partial \log \Phi \geq 0$ ). From (2.1) , setting $H(f,r) = E(f,r)/r^{2m-2}$ , we have

$$\frac{\partial}{\partial r} H(f,r) \geq \frac{B(f,r)}{r^{2m-2}} \tag{2.2}$$

for any non-critical value $r$ of $\Phi$ .

From now on , we assume that $f$ is a non-constant bounded harmonic function . First we have by integration by parts

$$E(f,r) = 2[f,e(\bar{\partial}\Phi)^{*}\bar{\partial}f]_r$$

By Schwarz inequality and the boundedness of $f$ , we have

$$E(f,r)^2 \leq C_1\frac{\partial}{\partial r}V(r)\ B(f,r)$$

for $\frac{\partial}{\partial r}V(r) = \int_{\partial A(r)}\lambda_r$ and $C_1 > 0$ .

By dividing the above inequality by $r^{4m-4}$ and by (2.2) , we have

$$H(f,r)^2 \leq C_1\ \frac{\frac{\partial}{\partial r}V(r)}{r^{2m-2}}\ \frac{\partial}{\partial r}H(f,r)$$

Hence by Schwarz inequality we have for $r_2 > r_1 > r_0$ and $C_2 > 0$

$$(\ r_2 - r_1\ )^2 \leq C_2\ r_2^2 n(A,r_2)\ \left(\ \frac{1}{H(r_1)} - \frac{1}{H(r_2)}\ \right)$$

Since $n(A,r)$ is non-decreasing , taking a sequence $\{r_n\}_{n\geq 1}$ with $r_{n+1} = 2r_n$ , we have

$$\int_{r_2}^{\infty}\frac{dt}{tn(A,t)} \leq \sum_{i=2}^{+\infty}\frac{C_3}{n(A,r_i)} \leq \frac{C_4}{H(r_1)} < \infty$$

This is a contradiction . q.e.d.

## 3. Analyticity of harmonic maps

Let $f : (\ M\ ,\ ds_M^2\ ) \longrightarrow (\ N\ ,\ ds_N^2\ )$ be a differentiable map of Kähler manifolds . $f$ is called <u>harmonic</u> if

$$\mathrm{Trace}_{ds_M^2}\ \nabla^{1,0}_{f^{*}TN}\bar{\partial}f = 0$$

and $f$ is called <u>pluriharmonic</u> if

$$\nabla^{1,0}_{f^{*}TN}\bar{\partial}f = 0$$

As another application of the integral formula (1.8), we show the following theorem :

<u>Theorem</u> Let $D = \{ r < 0 \}$ be a bounded ( connected ) domain with smooth boundary $\partial D$ over an $m \geq 2$ dimensional Kähler manifold $( M , ds_M^2 )$ and let $f : \bar{D} \longrightarrow ( N , ds_N^2 )$ be a differentiable map into a Kähler manifold $( N , ds_N^2 )$ which is smooth up to the boundary $\partial D$. Suppose

1) $f$ is harmonic on $D$ and satisfies the tangential Cauchy-Riemann equation on $\partial D$ i.e. $\bar{\partial}_b f = 0$ on $\partial D$.
2) $\partial D$ is hyper $m - 1$ convex and strongly hyper $m - 1$ convex at some point of $\partial D$ i.e. the sum of any $m - 1$ eigen-values of $\partial\bar{\partial}r$ relative to $ds_M^2$ is non-negative on $\partial D$ and positive at some point of $\partial D$.
3) The Riemannian curvature of $ds_N^2$ is semi-negative in the sense of Siu i.e.

$$R^N_{\alpha\bar{\beta}\gamma\bar{\delta}}(w)(A^\alpha\overline{B^\beta} - C^\alpha\overline{D^\beta})\overline{(A^\delta\overline{B^\gamma} - C^\delta\overline{D^\gamma})}$$

is non-positive for any complex numbers $A^\alpha$, $B^\beta$, $C^\gamma$, $D^\delta$ and $w \in N$. Then $f$ is holomorphic on $D$.

<u>Remark</u>. The above theorem has been already proved by Siu [Si], p.93 5.14 Theorem, (a) as a by-product of showing the complex analyticity of harmonic maps of compact Kähler manifolds by the $\partial\bar{\partial}$ - Bochner - Kodaira technique which was suggested by Bedford and Tayler's method [B-T]. Siu induced this theorem from an integral formula obtained by Morrey's trick which uses the first Neumann condition i.e. the dual condition of the tangential Cauchy-Riemann equation by the star operator ( cf. [Si] 5.7 Proposition ). However substantially we can induce the same integral formula from (1.8) and our method is quite different from their works [B-T] and [Si].

<u>Proof of Theorem</u>. For a differentiable map $f : ( M , ds_M^2 ) \longrightarrow ( N , ds_N^2 )$ of Kähler manifolds and a bounded domain $D = \{ r < 0 \}$ with smooth boundary $\partial D$ over $( M , ds_M^2 )$, we set

$$[\bar\partial_b f]_t^2 := \int_{\partial D_t} |\bar\partial_b f|^2 \lambda_t$$

for $|\bar\partial_b f|^2 := 2 \sum_{i<j} h_{\alpha\bar\beta}(f)(r_{\bar i} f^{\alpha,j} - r_{\bar j} f^{\alpha,i})\overline{(r_{\bar i} f^{\beta,j} - r_{\bar j} f^{\beta,i})}$

and $\partial D_t = \{ r = t \}$ $( 0 \leq |t| << 1 )$.

Since $|\bar\partial_b f|^2 = |\partial r|^2 \langle \bar\partial f, \bar\partial f \rangle_{f^* TN} - |e(\bar\partial r)^* \bar\partial f|^2_{f^* TN}$ by Lagrange equality, differentiating (1.8), we have

$$(3.1) \quad \frac{\partial}{\partial r}[\bar\partial_b f]_r^2\Big|_{r=t} = [(-\Box\, r)\bar\partial f, \bar\partial f]_t - [\mathcal{L}(r)(\bar\partial f), \bar\partial f]_t$$

$$+ [D_{0,1}^* \bar\partial f, e(\bar\partial r)^* \bar\partial f]_t + [e(\partial r)^* D_{1,0} \bar\partial f, \bar\partial f]_t$$

By integration by parts, the sum of the last two terms of the right hand-side of (3.1) is equal to

$$\| D_{1,0} \bar\partial f \|_t^2 - \| D_{0,1}^* \bar\partial f \|_t^2 + ((D_{0,1} D_{0,1}^* - D_{1,0}^* D_{1,0})\bar\partial f, \bar\partial f)_t$$

Hence we have from (1.9)

(3.2)

$$\frac{\partial}{\partial r}[\bar\partial_b f]_r^2\Big|_{r=t} = [(-\Box\, r)\bar\partial f, \bar\partial f]_t - [\mathcal{L}(r)(\bar\partial f), \bar\partial f]_t$$

$$+ \| D_{1,0} \bar\partial f \|_t^2 - \| D_{0,1}^* \bar\partial f \|_t^2$$

$$- \frac{1}{2}\int_{D_t} \sum R^N_{\alpha\bar\beta\gamma\bar\delta}(f)(f^\alpha_{\bar i}\overline{f^\beta_{\bar j}} - f^\alpha_{\bar j}\overline{f^\beta_{\bar i}})\overline{(f^{\delta,i}\overline{f^{\gamma,\bar j}} - f^{\delta,j}\overline{f^{\gamma,\bar i}})}\, dv_M$$

for any $0 \leq |t| << 1$.

If $f$ is harmonic on $D$ i.e. $D_{0,1}^* \bar\partial f = 0$ on $D$ and $\bar\partial_b f = 0$ on $\partial D$, then setting $r = 0$ in (3.2) and by the condition (3), we have

$$(3.3) \quad [(-\Box r)\bar\partial f, \bar\partial f]_{\partial D} - [\mathcal{L}(r)(\bar\partial f), \bar\partial f]_{\partial D} + \| D_{1,0} \bar\partial f \|_D^2 \leq 0$$

Since the left hand-side of (3.3) is non-negative by the condition (2), we have $D_{1,0}\bar\partial f = 0$ on $D$ i.e. $f$ is pluriharmonic on $D$ and $\bar\partial f$ vanishes on an open subset of $\partial D$. Therefore $\bar\partial f$ vanishes identically on $D$ ( cf. [Si] 5.11 Lemma ) i.e. $f$ is holomorphic on $D$.

## 4. A priori estimates for the $\bar\partial$ - Neumann problem on pseudoconvex domains

Since Morrey has introduced the Morrey trick to handle boundary integrals when $\bar\nabla$ Bochner-Kodaira technique is applied to a bounded domain with smooth boundary over an arbitrary Kähler manifold , several a priori estimates for the $\bar\partial$ - Neumann problem on pseudoconvex domains ( cf. [C],[H],[K] ) have been induced from this technique .

In this section we introduce a new method to induce a priori estimates and apply them to show a compactness estimate for the $\bar\partial$ - Neumann problem on a pseudoconvex domain whose boundary satisfies somewhat weaker non-degenerate condition than that known up to now .

We hope that our method brings us a new insight to solve the global regularity for the Neumann operator on ( general ) pseudoconvex domains.

We begin to show the following integral formula .

<u>Propositon 1</u> Let $D = \{ r < 0 \}$ be a bounded domain with smooth boundary on an $m$ dimensional Kähler manifold $( M , ds^2_M )$ . Then for any domain $D_t = \{ r < t \}$ $( 0 \leq |t| << 1 )$ , it holds that

(4.1)

$$\frac{\partial}{\partial r}[e(\bar\partial r)^* u]^2_r\Big|_{r=t} = [\mathscr{L}(r)(u),u]_t + \| \bar\nabla u \|^2_t + (\mathscr{K}(u),u)_t$$

$$— \| \bar\partial u \|^2_t — \| \vartheta_E u \|^2_t — 2\ \mathrm{Re}[\vartheta_E u, e(\bar\partial r)\ u]_t$$

for any $u \in C^{p,q}(M,E)$

where $\mathscr{K} : C^{p,q}(M,E) \longrightarrow C^{p,q}(M,E)$ is a linear map defined by

$$(\mathscr{K}u)^{\nu}_{A_p\bar B_q} = \sum_{h=1}^{q} R^M{}_{\bar\beta_h}{}^{\bar\gamma}\ u^{\nu}{}_{A_p,\bar\beta_1,\ldots,(\gamma)_h,\ldots,\bar\beta_q}$$

$$— \sum_{i=1}^{p}\sum_{j=1}^{q} R^{M\tau}{}_{\alpha_i\bar\beta_j}{}^{\bar\gamma}\ u^{\nu}{}_{\alpha_1,\ldots,(\tau)_i,\ldots,\alpha_p,\bar\beta_1,\ldots,(\gamma)_j,\ldots,\bar\beta_q}$$

$$+ \sum_{k=1}^{q} \Theta^{E}{}_{\mu}{}^{\nu\bar\sigma}{}_{\bar\beta_k}\ u\ {}_{A_p,\bar\beta_1,\ldots,(\sigma)_k,\ldots,\bar\beta_q}$$

for the curvature tensor $\Theta^E{}_{\mu\bar\nu i\bar j}$ of $( E, h_E )$ and the Riemannian curvature tensor $R^M{}_{\alpha\bar\beta\gamma\bar\delta}$ relative to $ds^2_M$ .

Proof. From (1.1), we have

$$(u,\bar\partial\vartheta_E u)_t - \| \vartheta_E u \|_t^2 = [e(\bar\partial r)^* u, \vartheta_E u]_t$$

$$\| \bar\partial u \|_t^2 - (u, \vartheta_E \bar\partial u)_t = [u, e(\bar\partial r)^* \bar\partial u]_t$$

$$\| \bar\nabla u \|_t^2 - (u, \bar\nabla^* \bar\nabla u)_t = [u, e(\bar\partial r)^* \bar\nabla u]_t$$

$$(u, \square_E u)_t = (u, \bar\nabla^* \bar\nabla u)_t + (u, \mathcal{K}(u))_t$$

for any $u \in C^{p,q}(M,E)$ and $\square_E = \bar\partial\vartheta_E + \vartheta_E\bar\partial$ ( cf. [Si] p.63 (1.3.2) One should be careful to the sign of curvatures ) .
Setting $\Phi = r$ and differentiating (1.5) , we have

$$\frac{\partial}{\partial r}[e(\bar\partial r)^* u]_r^2 \Big|_{r=t} = [\mathcal{L}(r)(u), u]_t - (**)_t$$

$$(**)_t = [\vartheta_E u, e(\bar\partial r)^* u]_t + [u, e(\bar\partial r)^* \bar\partial u]_t - [u, e(\bar\partial r)^* \bar\nabla u]_t$$

From these equalities , we obtain (4.1) immediately . q.e.d.

Moreover from (1.5) , we obtain the following integral inequality by Schwarz inequality .

Proposition 2 Let $D = \{ r < 0 \}$ be a bounded domain with smooth boundary $\partial D$ over a Kähler manifold $( M , ds_M^2 )$ . Then
(4.2)

$$|(\mathcal{L}(\Phi)(u), u)| \le \sup_{z \in \mathrm{Supp}(u)} |\partial\Phi|(z) \, \| u \| \, ( \| \bar\partial u \| + \| \vartheta_E u \| + \| \bar\nabla u \| )$$

for any $\Phi \in C^\infty_{\mathbb{R}}(M)$ and $u \in C^{p,q}(\bar D, E)$ with $\partial r \wedge * u = 0$ on $\partial D$ .

We call a bounded domain $D$ with smooth boundary $\partial D$ pseudoconvex if there exists a defining function $r$ of $D$ i.e. $D = \{ r < 0 \}$ and $dr \neq 0$ on $\partial D$ such that its Levi form $\partial\bar\partial r$ is positive semi-definite whenever restricted to the complex tangent space of $\partial D$ .

We obtain the following a priori estimate for the $\bar\partial$ - Neumann problem on pseudoconvex domains .

Theorem 3 Let $D = \{ r < 0 \}$ be a pseudoconvex domain with smooth boundary $\partial D$ over an $m$ dimensional Kähler manifold $( M , ds_M^2 )$. Then

$$(4.3) \qquad |(\mathcal{L}(\Phi)(u),u)| \leqq 2 \sup_{z \in \mathrm{Supp}(u)} |\partial\Phi|(z) \; \| u \| (\| \bar{\partial} u \| + \| \vartheta u \|)$$

for any $\Phi \in C^\infty_{\mathbb{R}}(M)$ and $u \in C^{m,q}(\bar{D})$ with $\partial r \wedge * u = 0$ on $\partial D$ ( $q \geq 1$ ).

In particular if $\Phi$ is strictly plurisubharmonic on $M$, then there exists $C_\Phi > 0$ such that

$$(4.4) \qquad \| u \| \leq C_\Phi (\| \bar{\partial} u \| + \| \vartheta u \|)$$

for any $u \in C^{m,q}(\bar{D})$ with $\partial r \wedge * u = 0$ on $\partial D$ ( $q \geq 1$ ).

Proof. We note that the map $\mathcal{K} : C^{m,q}(M) \longrightarrow C^{m,q}(M)$ is the zero map since $\Theta_{\mathbb{E}}$ vanishes identically and the Riemannian curvature terms cancel each other. Setting $r = 0$ in (4.1), if $u \in C^{m,q}(\bar{D})$ satisfies $\partial r \wedge * u = 0$ on $\partial D$, then the integral $\| \bar{\nabla} u \|^2$ is dominated by the integral $\| \bar{\partial} u \|^2 + \| \vartheta u \|^2$ by the pseudoconvexity of $D$. Combining this fact with (4.2), we can obtain (4.3).

If $\Phi$ is strictly plirusubharmonic on $M$, then there exists $C'_\Phi > 0$ such that $\langle \mathcal{L}(\Phi)(u),u \rangle \geq C'_\Phi \langle u,u \rangle$ on $\bar{D}$ for any $u \in C^{p,q}(\bar{D})$. Hence we obtain (4.4) from (4.3). q.e.d.

Let $D = \{ r < 0 \}$ be a bounded domain with smooth boundary $\partial D$ on a complex manifold $( M , ds_M^2 )$. The following definition is due to Kohn and Nirenberg [K-N].

Definition A compactness estimate is said to hold for the $\bar{\partial}$ - Neumann problem for $(p,q)$ forms on $D$ if for any $\varepsilon > 0$, there exists a positive constant $C(\varepsilon)$ such that

$$\| u \|^2 \leq \varepsilon (\| \bar{\partial} u \|^2 + \| \vartheta u \|^2) + C(\varepsilon) \| u \|^2_{-1}$$

for any $u \in C^{p,q}(\bar{D})$ with $\partial r \wedge * u = 0$ on $\partial D$ where $\| \;\; \|_{-1}$ is the Sobolev norm of order $-1$ on $D$.

If the compactness estimate holds , then the Neumann operator is defined modulo the finite dimensional space of harmonic forms and satisfies global regularity property ( cf. [K-N] ) .

When $D \Subset M$ is a pseudoconvex domain with smooth boundary $\partial D$ , the set of degeneracy of the Levi form of $\partial D$ is defined by the set of points where the Levi form restricted on the complex tangent space of $\partial D$ is not positive definite .

In the rest of this section we show the compactness estimate for a class of pseudoconvex domains with smooth boundary which contains the class defined by D.Catlin [C] i.e. weakly regular pseudoconvex domains .

Theorem 4 Let $D = \{ r < 0 \}$ be a pseudoconvex domain with smooth boundary $\partial D$ over an $m$ dimensional Kähler manifold $( M , ds_M^2 )$ and let $E$ be the set of degeneracy of the Levi form of $\partial D$ .
Suppose the following conditions :

(i) $\partial D \setminus E \neq \phi$ and $E \neq \phi$

(ii) there are positive constants $C_*'$ and $C_*''$ such that for any $\varepsilon > 0$ there exists a real-valued $C^\infty$ function $\rho_\varepsilon$ on $M$ which satisfies

$$(\alpha) \qquad <\mathcal{L}(r + \rho_\varepsilon)(a),a>(p) \geqq C_*'<a,a>(p)$$

for any $p \in E$ and $a \in T_p^{1,0}(\partial D)$ i.e. $<\partial r,a>(p) = 0$

$$(\beta) \qquad |\partial\rho_\varepsilon|_M \leqq C'' \varepsilon \quad \text{on} \quad U(\varepsilon) : = \{ x \in M : d_M(x,E) < \varepsilon \}$$

Then a compactness estimate holds for the $\bar\partial$ - Neumann problem for $(m,q)$ forms on $D$ $( q \geqq 1 )$ .

Here we note that the condition (ii) does not depend on the choice of the defining function $r$ by retaking the constants $C_*'$ and $C_*''$ .

Remark 1. Any Stein manifold is Kähler and the condition (i) is satisfied whenever $M$ is Stein ( cf. [D-F] §3, Lemma ). In particular if $( M , ds_M^2 )$ is the $m$ dimensional complex Euclidean space $( \mathbb{C}^m , ds_e^2 )$ provided with Euclidean metric $ds_e^2$ , then under the above assumption the compactness estimate holds for the $\bar\partial$ - Neumann problem for $(p,q)$ forms on $D$ for any $p \geqq 0$ and $q \geqq 1$ since the Riemannian curvature of $ds_e^2$ vanishes identically .

Remark 2. Here we recall the definition of weakly regular pseudoconvex domain by Catlin [C] :

A pseudoconvex domain $D = \{ r < 0 \}$ with smooth boundary $\partial D$ over a complex manifold $M$ is called to be weakly regular if there exists a finite number of compact subsets $S_i$ of $\partial D$ ( $0 \leqq i \leqq \ell$ ) such that

(i)′ $\phi = S_\ell \subset S_{\ell-1} \subset \cdots \subset S_1 \subset S_0 = \partial D$

(ii)′ If $p \in S_i \setminus S_{i+1}$ , then there are a neighborhood $U$ of $p$ and a $C^\infty$ submanifold $N$ of $U \cap \partial D$ with $p \in N$ such that $S_i \cap U \subset N$ , the dimension of the vector space of tangent vectors $T^{1,0}(N)$ is constant in $U$ and

$$<\mathcal{L}(r)(a),a>(p) > 0 \quad \text{for any} \quad p \in N \quad \text{and} \quad a \in T^{1,0}(N)$$

Such a non-degenerate condition for pseudoconvex domains was first introduced by Diederich and Fornaess [D-F] in a slightly stronger form and was used by Catlin [C] as an intermediate conception to understand that the compactness estimate holds for the $\bar\partial$ - Neumann problem on pseudoconvex domains of finite type over $\mathbb{C}^m$ .

Here we would like to point out that the condition (ii) of Theorem 4 follows from the above weak regularity condition .

If any point $p$ of the set of degeneracy $E$ satisfies the condition (ii)′ , then we may conclude that there exist $C^\infty$ functions $r_{\ell+1}, \ldots, r_m$ with $r_m = r$ on $U$ ( $\ell = \dim_{\mathbb{C}} T^{1,0}(N)$ ) such that $E \cap U \subset N \subset \{ z \in U : r_k(z) = 0 , \ell+1 \leqq k \leqq m \}$ and $\partial r_{\ell+1}, \ldots, \partial r_m$ are linearly independent at each point of $U$ .
Since the Levi form $<\mathcal{L}(r_m)(a),a>(q)$ is positive for any $q \in N$ and $a \in T^{1,0}_q(M)$ satisfying $<\partial r_k,a>(q) = 0$ , $\ell+1 \leqq k \leqq m$ , and the Levi form is non-negative , after shrinking $U$ if necessary , there exists a positive constant $C_*'$ such that

$$<\mathcal{L}(r_m)(a),a>(q) + \sum_{k=\ell+1}^{m} |<\partial r_k,a>(q)|^2 \geqq C_*'<a,a>(q)$$

for any $q \in U \cap E$ and $a \in T^{1,0}_q(\partial D)$ .
We set $\rho := \sum_{k=\ell+1}^m r_k^2$ . Since $E$ is compact , covering $E$ by a finite number of such neighborhoods $\{ U_i , \rho_i \}$ and taking a partition of unity $\{\lambda_i\}$ subordinate to $\{U_i\}$ , we have only to set $\rho := \sum \lambda_i \rho_i$ .

Proof of Theorem 4.

We take an arbitrary $\varepsilon > 0$ and fix $\varepsilon_1 > 0$ with $0 < \varepsilon_1 < \varepsilon$. From the condition (ii), ($\alpha$), setting $\rho_1 = \rho_{\varepsilon_1}$ it holds that there exists $A = A(\varepsilon_1) >> 1$ such that

$$<\mathcal{L}(r + Ar^2 + \rho_1)(a),a> \geqq \frac{C_*'}{2}<a,a>$$

for any $p \in E$ and $a \in T_p^{1,0}(M)$

Hence setting $r_1 = r + Ar^2$, there exists $\delta_1 > 0$ such that

$$<\mathcal{L}(r_1 + \rho_1)(u),u>(z) \geqq \frac{C_*'}{3}<u,u>(z) \tag{4.5}$$

for any $u \in C^{p,q}(\bar{D})$ and $z \in U(\delta_1) \cap \{ |r| < \delta_1 \}$.

We set

$$\delta := \min ( \varepsilon_1 , \delta_1 )/AL \quad \text{for } L \geqq 1$$

The constant $L$ is determined later. We set

$$\Lambda(\delta) := \{ u \in C^{m,q}(\bar{D}) : \operatorname{supp}(u) \subset U(\delta) \cap \{ -\delta < r \leqq 0 \}$$
$$\text{and } \partial r \wedge * u = 0 \text{ on } \partial D \}$$

We first show the following estimate :

$$\| u \| \leqq \frac{\varepsilon}{16} (\| \bar{\partial} u \|^2 + \| \vartheta u \|^2) \quad \text{for any } u \in \Lambda(\delta) \tag{4.6}$$

By setting $\Phi = r_1 + \rho_1$, we have from (1.5)

$$(\mathcal{L}(\Phi)(u),u) = \Sigma(r_1) + \Sigma(\rho_1) \quad \text{for any } u \in \Lambda(\delta) \tag{4.7}$$

where $\Sigma(\lambda) := (\vartheta u, e(\bar{\partial}\lambda)^* u) + (u, e(\bar{\partial}\lambda)^* \bar{\partial} u) - (u, e(\bar{\partial}\lambda)^* \nabla u)$ for any $\lambda \in C^\infty_{\mathbb{R}}(M)$.

By the pseudoconvexity of $D$, (4.5) and (4.7), there exists $C_1 > 0$ not depending on $\varepsilon$ such that

$$\| \nabla u \|^2 + \| u \|^2 \leqq C_1 (\| \bar{\partial} u \|^2 + \| \vartheta u \|^2) \quad \text{for any } u \in \Lambda(\delta) \tag{4.8}$$

By Schwarz inequality and the condition (ii), ($\beta$), we have

$$|\Sigma(\rho_1)| \leqq \varepsilon_1 \| u \| (\| \bar{\partial} u \| + \| \vartheta u \| + \| \bar{\nabla} u \|)$$

$$\leqq C_2 \, \varepsilon_1 \, (\| \bar{\partial} u \|^2 + \| \vartheta u \|^2) \quad \text{by} \quad (4.8)$$

On the other hand

$$|\Sigma(r_1)| = \left| \int_{-\delta}^{0} (1 + 2At)(**)_t \, dt \right| \quad (\text{ See the proof of Proposition 1 })$$

$$\leqq \delta \, (\| \bar{\partial} u \|^2 + \| \vartheta u \|^2) + \int_{-\delta}^{0} [\vartheta u]_t [e(\bar{\partial} r)^* u]_t \, dt$$

Here we need the following estimate obtained by Hardy's inequality :

There exists $C_3 > 0$ such that

$$\| \frac{v}{r} \|_D \leqq C_3 \, \| v \|_{D,1} \quad \text{for any} \quad v \in C^\infty(\bar{D}) \quad \text{with} \quad v_{|\partial D} = 0$$

( cf. [A] p. 209 , Theorem B.1 ) .

Here $\| \, , \|_{D,1}$ is the Sobolev norm of order 1 on D . Hence we have

$$\int_{-\delta}^{0} [\vartheta u]_t [e(\bar{\partial} r)^* u]_t \, dt$$

$$\leqq \{ \int_{-\delta}^{0} (-t)^2 [\vartheta u]_t^2 \, dt \}^{1/2} \, \| \frac{e(\bar{\partial} r)^* u}{r} \|$$

$$\leqq \delta \, C_4 \| \vartheta u \| \; \| e(\bar{\partial} r)^* u \|_1$$

and

$$\| e(\bar{\partial} r)^* u \|_1^2 \leqq C_5 (\| \bar{\nabla} u \|^2 + \| u \|^2)$$

$$\leqq C_6 (\| \bar{\partial} u \|^2 + \| \vartheta u \|^2) \quad \text{by} \quad (4.8)$$

and so

$$|\Sigma(r_1)| \leqq \delta \, C_7 (\| \bar{\partial} u \|^2 + \| \vartheta u \|^2)$$

Finally we have from (4.7)

$$\| u \|^2 \leqq ( C_8 \varepsilon_1 + C_9 \delta )(\| \bar{\partial} u \|^2 + \| \vartheta u \|^2)$$

Here clearly $C_8$ and $C_9$ do not depend on $\varepsilon$ . Since we may assume $C_8 \geqq 1$ , setting $\varepsilon_1 = \varepsilon/32C_8$ and $L = 32C_9$ , we obtain (4.6) .

Since $d_M(x,E)$ is a Lipschitz continuous function ( we do not assume any smoothness condition for $E$ ) , by smoothing $d_M(x,E)$ in a higher dimensional real Euclidean space , we can conclude that for any small $\eta > 0$ there exists a non-negative $C^\infty$ function $\chi_\eta$ such that $\chi_\eta \equiv 1$ on $U(\eta/2)$ , $\mathrm{supp}(\chi_\eta) \Subset U(2\eta/3)$ and $|\partial\chi_\eta|_M^2 \leqq C_{10}/\eta^2$ for some positive constant $C_{10}$ not depending on $\eta$ .

For any $u \in C^{m,q}(\bar{D})$ with $\partial r \wedge * u = 0$ on $\partial D$ , we set $u_1 := \chi_\eta u$ and $u_2 := ( 1 - \chi_\delta )u$ . Then we have

$$\| u \|^2 \leqq 2 (\| u_1 \|^2 + \| u_2 \|^2)$$

$$\leqq \frac{\varepsilon}{8}(\| \bar\partial u_1 \|^2 + \|\vartheta u_1 \|^2) + 2\| u_2 \|^2 \quad \text{by} \quad (4.6)$$

$$\leqq \frac{\varepsilon}{2} \| \bar\partial u \|^2 + \|\vartheta u \|^2) + C_1(\varepsilon)\| u_3 \|^2$$

where $C_1(\varepsilon) = \varepsilon C_{10}/\delta^2 \geqq 4$ and $u_3 := ( 1 - \chi_{(1/2)\delta} )u$ .

Since $\mathrm{supp}(u_3) \cap E = \phi$ , from (4.1) , there exists $C_\varepsilon > 0$ not depending on $u_3$ but depending on $\varepsilon$ such that

$$[u_3]^2 + \| \nabla u_3 \|^2 \leqq C_\varepsilon (\| \bar\partial u_3 \|^2 + \|\vartheta u_3 \|^2)$$

This estimate implies that the graph norm $\| \, , \| + \| \bar\partial \quad \| + \|\vartheta \quad \|$ is completely continuous ( with respect to $\| \, , \|$ ) on the space of $(m,q)$ forms whose supports are contained in a fixed compact subset of $\bar{D}$ not intersecting $E$ ( cf. [K] 6.2. Theorem and 6.16. Proposition ) . Hence setting $\varepsilon_2 = \delta^4/16C_{10}^3$ , from [K-N] p.454 Lemma 1.1 , there exists $C(\varepsilon_2) > 0$ such that

$$\| u_3 \|^2 \leqq \varepsilon_2(\| \bar\partial u_3 \|^2 + \|\vartheta u_3 \|^2) + C(\varepsilon_2) \| u_3 \|_{-1}^2$$

Hence we have

$$\| u_3 \|^2 \leqq 2\varepsilon_2 (\| \bar\partial u \|^2 + \|\vartheta u \|^2) + \varepsilon_3 \| u \|^2 + C(\varepsilon_2) \| u \|_{-1}^2$$

for $\varepsilon_3 = \delta^2/C_{10}^2$ .

Therefore we have

$$\left(1 - \frac{\varepsilon}{C_{10}}\right)\| u \|^2 \leq \frac{\varepsilon}{2}\left(1 + \frac{\delta^2}{C_{10}^2}\right)(\| \bar{\partial} u \|^2 + \| \vartheta u \|^2)$$

$$+ C_2(\varepsilon)\| u \|_{-1}^2$$

Taking $C_{10} > 0$ sufficiently large , we can obtain the compactness estimate .

## References

[A] Agmon,S., Spectral properties of Schrödinger operators and scattering theory , Ann. Scuola Norm. Sup. Pisa Cl. Sci. (4) 2 (1975) , 151-218

[B-T] Bedford,E., Taylor,B.A., Variational properties of the complex Monge - Ampère equation I , Dirichlet principle , Duke Math. J. 45 (1978) , 375-403

[C] Catlin,D., Global regularity of the $\bar{\partial}$ - Neumann problem , Proccedings of Sym. in Pure Math., 41 (1984) , 39-49

[D-F] Diederich,K., Fornaess,J.E., Pseudoconvex domains : Existence of Stein neighborhoods , Duke Math. J., 44 (1977) , 641-661

[D-X] Donnelly,H., Xavier,F., On the differential form spectrum of negatively curved Riemannian manifolds , Amer. J. Math., 106 (1984) , 169-185

[E-L] Eells,J., Lemaire,L., A report on harmonic maps , Bull. London Math. Soc., 10 (1978) , 1-68

[H] Hörmander,L., $L^2$ estimates and existence theorems for the $\bar{\partial}$ operator , Acta Math., 113 (1965) , 89-152

[K] Kohn,J.J., Harmonic integrals on strongly pseudoconvex manifolds , Ann. of Math., 78 (1963) , 112-148

[K-N] Kohn,J.J., Nirenberg,L., Non-coercive boundary value problems , Comm. Pure Appl. Math., 18 (1965) , 443-492

[M-K] Morrow,J.A., Kodaira,K., Complex Manifolds , Holt, Rinehart and Winston , New York , 1977

[$O_1$] Ohsawa,T., Hodge spectral sequence on compact Kähler spaces , Publ. RIMS, Kyoto Univ., 23 (1987) , 265-274

[$O_2$] Ohsawa,T., On the rigidity of noncompact quotient of bounded symmetric domains , Publ. RIMS, Kyoto Univ., 23 (1987) , 881-894

[O-$T_1$] Ohsawa,T., Takegoshi,K., On the extension of $L^2$ holomorphic functions , Math. Zeit., 195 (1987) , 197-204

[O-$T_2$] Ohsawa,T., Takegoshi,K., Hodge spectral sequence on pseudoconvex domains , Math. Zeit., 197 (1988) , 1-12

[Si] Siu,S.T., Complex analyticity of harmonic maps , vanishing and Lefschetz theorems , J. of Diff. Geometry , 17 (1982) , 55-138

[St] Stoll,W., The growth of area of a transcendental analytic set I , II , Math. Ann., 156 (1964) , 47-78 , 144-170

[$T_1$] Takegoshi,K., A non-existence theorem for pluriharmonic maps of finite energy , Math. Zeit., 192 (1986) , 21-27

[$T_2$] Takegoshi,K., Energy estimates and Liouville theorems for harmonic maps , to appear in Annales Scientifiques de l'Ècole Normale Superièure

[Y] Yau,S.T., Harmonic functions on complete Riemannian manifolds , Comm. Pure Appl. Math., 28 (1975) , 201-228

# On inner radii of Teichmüller spaces *

Toshihiro Nakanishi
and
John A. Velling †

1 December 1989

## 1 Statement of results

Let $\Gamma$ be a Fuchsian group acting on the upper half plane $\mathcal{H}$ and hence also on the lower half plane $\mathcal{H}^*$ in the complex plane. Let $Q(\Gamma)$ denote the complex Banach space of bounded quadratic differentials for $\Gamma$ defined in $\mathcal{H}^*$. The Teichmüller space $\mathbf{T}(\Gamma)$ of $\Gamma$ is represented by means of the Bers embedding as a bounded region in $Q(\Gamma)$. The *inner radius* $i(\Gamma)$ of $\mathbf{T}(\Gamma)$ is the supremum of radii of balls in $Q(\Gamma)$ centered at the origin which are contained in $\mathbf{T}(\Gamma)$. The most general estimate for $i(\Gamma)$ is the following ([2], [14]):

$$(1.1) \qquad i(\Gamma) \geq 2 \quad \text{whenever} \quad \mathbf{T}(\Gamma) \neq \{0\}.$$

If $\Gamma$ is a hyperbolic cyclic group, then we have equality in (1.1), [7]. On the other hand it can be shown, via a result of Gehring and Pommerenke [6], that the inequality is strict in (1.1) for a finitely generated Fuchsian group of the first kind ([8], [12]). We now state our main theorem:

THEOREM 1. *Let* $\{\Gamma_n\}_{n=1}^{\infty}$ *be a sequence of Fuchsian groups. Assume that for any* $d > 0$ *a collar of width* $d$ *about the axis of a hyperbolic element of* $\Gamma_n$ *exists if* $n$ *is sufficiently large. Then* $\inf(i(\Gamma_n)) = 2$.

---

*This paper is in final form and no version of it will be submitted for publication elsewhere.

†Part of this work was done while the second author was visiting Kyoto University, supported by a grant from the JSPS.

Some estimates for inner radii are derived from this theorem. If $\mathbf{T}(\Gamma) \neq \{0\}$, the quotient space $\mathcal{H}/\Gamma$ admits a simple closed geodesic on it with respect to the Poincaré metric. Bers employed a *pinching deformation* about a simple closed curve in [3] to show the existence of cusps in the boundary of Teichmüller spaces. By this pinching deformation and the collar lemma ([5]), we obtain

THEOREM 2. *If* $\mathbf{T}(\Gamma) \neq \{0\}$, *then*

$$\inf\{i(\Gamma') : \Gamma' \text{ is quasiconformally equivalent to } \Gamma\} = 2.$$

Let $\sigma = (g; \nu_1, \ldots, \nu_n)$ be the signature of a finitely generated Fuchsian group of the first kind, where $g \geq 0$, $2 \leq \nu_1 \leq \cdots \leq \nu_n \leq \infty$ and $(g,n) \neq (0,3)$. Since two Fuchsian groups with the same signature are quasiconformally equivalent we obtain, as a special case of the above theorem,

THEOREM 3 [12]. *For a signature* $\sigma = (g; \nu_1, \ldots, \nu_n)$ $((g,n) \neq (0,3))$ *of a finitely generated Fuchsian group of the first kind,*

$$\inf\{i(\Gamma) : \Gamma \text{ has signature } \sigma\} = 2.$$

The hypothesis in the following theorem is the same as the condition $(O_2)$ in [11].

THEOREM 4. *Let* $\Gamma$ *be a Fuchsian group and assume that for any* $d > 0$ *a collar of width* $d$ *exists about the axis of some hyperbolic element of* $\Gamma$. *Then* $i(\Gamma) = 2$.

To obtain this theorem apply Theorem 1 to $\Gamma_n = \Gamma$, $n = 1, 2, \ldots$

We would like to note that this is very similar to aspects of the discussion found in §3 of Curt McMullen's article [10], as he pointed out to the second author.

# 2 Preliminaries

Let $D$ be a subregion of the extended complex plane $\hat{\mathbf{C}} = \mathbf{C} \cup \{\infty\}$ endowed with a Poincaré metric $\rho_D(z)|dz|$. We normalize $\rho_D$ so that the metric has curvature $-4$. Let $\Gamma$ be a discontinuous group of conformal self-mappings of $D$. A holomorphic function $\phi$ in $D$ is called a bounded quadratic differential for $\Gamma$ if it satisfies

$$(2.1) \qquad \phi(z) = \phi(\gamma(z)) \cdot \gamma'(z)^2 \qquad \text{for all} \quad \gamma \in \Gamma,$$

and the norm of $\phi$ defined by

$$(2.2) \qquad \|\phi\|_D = \sup_{z \in D} \rho_D(z)^{-2} |\phi(z)|$$

is finite. (If $\infty \in D$, then $\rho_D(\infty)^{-2}|\phi(\infty)|$ means $\lim_{z\to\infty} \rho_D(z)^{-2}|\phi(z)|$). Let $Q(D,\Gamma)$ denote the complex Banach space of bounded quadratic differentials for $\Gamma$. Assume that a conformal homeomorphism $h : D \longrightarrow h(D)$ is given. Then $h\Gamma h^{-1} = \{h \circ \gamma \circ h^{-1} : \gamma \in \Gamma\}$ is also a discontinuous group acting on $h(D)$. The mapping $h^* : Q(h(D), h\Gamma h^{-1}) \longrightarrow Q(D,\Gamma)$, $(h^*\phi)(z) = \phi(h(z))h'(z)^2$ for $\phi \in Q(h(D), h\Gamma h^{-1})$ is an isometry:

$$\rho_D(z)^{-2}|(h^*\phi)(z)| = \rho_{h(D)}(h(z))^{-2}|\phi(h(z))|$$

and hence

$$\|h^*\phi\|_D = \|\phi\|_{h(D)}.$$

Let $f : D \to \hat{\mathbf{C}}$ be locally univalent. The Schwarzian derivative of $f$ is

$$S_f(z) = \left( (\frac{f''}{f'})' - \frac{1}{2}(\frac{f''}{f'})^2 \right)(z).$$

We now take $\mathcal{H}$ as $D$. Then a discontinuous group $\Gamma$ of conformal self-mappings is a Fuchsian group, that is a discrete subgroup of $\text{Möb}(\mathbf{R}) = \{z \mapsto \frac{az+b}{cz+d} : a,b,c,d \in \mathbf{R}, ad-bc = 1\}$. We shall also use the notation $\text{Möb}(\mathbf{C})$ for the full Möbius group. Let $B(\Gamma)$ be the unit ball of the $L^\infty$-space of Beltrami differentials for $\Gamma$, that is measurable functions $\mu$ in $\mathcal{H}$ satisfying

$$\text{ess. sup.} |\mu| < 1$$

and

$$\mu(z) = \mu(\gamma(z)) \frac{\overline{\gamma'(z)}}{\gamma'(z)} \qquad \text{for all } \gamma \in \Gamma.$$

Extend $\mu$ to be a function defined in $\mathbf{C}$ by setting $\mu|_{\mathbf{C}\setminus\mathcal{H}} = 0$ and consider the Beltrami equation

$$\frac{\partial f}{\partial \bar{z}} = \mu \cdot \frac{\partial f}{\partial z}. \tag{2.3}$$

This equation has a solution $f^\mu$ defined uniquely up to post-composition by elements of Möb($\mathbf{C}$). Hence the Schwarzian derivative $\mathcal{S}_{f^\mu}$ in $\mathcal{H}^*$ depends only on $\mu$ ($f^\mu$ is conformal in $\mathcal{H}^*$). By well known properties of Schwarzian derivatives one can show that $\mathcal{S}_{f^\mu}$ is an automorphic form of weight $-4$, and by Kraus' inequality we have

$$\|\mathcal{S}_{f^\mu}\|_{\mathcal{H}^*} = \sup_{z\in\mathcal{H}^*} 4(\Im z)^2 |\mathcal{S}_{f^\mu}(z)| \leq 6. \tag{2.4}$$

Hence $\mathcal{S}_{f^\mu} \in \mathrm{Q}(\Gamma) = \mathrm{Q}(\mathcal{H}^*, \Gamma)$.

DEFINITION: *The Teichmüller space* $\mathbf{T}(\Gamma)$ of $\Gamma$ is the image of $B(\Gamma)$ in $\mathrm{Q}(\Gamma)$ under the mapping $\mu \mapsto \mathcal{S}_{f^\mu}$.

The following numbers are called the outer radius and the inner radius of $\mathbf{T}(\Gamma)$, respectively:

$$\begin{aligned} o(\Gamma) &= \sup\{\|\phi\|_{\mathcal{H}^*} : \phi \in \mathbf{T}(\Gamma)\}, \\ i(\Gamma) &= \inf\{\|\phi\|_{\mathcal{H}^*} : \phi \in \mathrm{Q}(\Gamma) \setminus \mathbf{T}(\Gamma)\}. \end{aligned}$$

By (2.4) we have $o(\Gamma) \leq 6$. For refined estimates of the outer radius, see [4], [11], [13], [15], and [16]. Estimates of the inner radius other than (1.1) are given in [8] and [12].

# 3 Several lemmas

Let $\Delta = \{z : |z| < 1\}$ be the unit disc and $\Delta^* = \hat{\mathbf{C}} \setminus \mathrm{cl}(\Delta)$ ($\mathrm{cl}(X)$ means the closure of the set $X$). Let $A \in$ Möb($\mathbf{C}$) be a transformation sending $\Delta$ onto $\mathcal{H}$ and $\mu \in B(\Gamma)$, where $\Gamma$ is a Fuchsian group acting on $\mathcal{H}$. Set $\nu(z) = \mu(A(z))\frac{\overline{A'(z)}}{A'(z)}$ (and $\nu(z) = 0$ for $z \in \Delta^*$). We consider the Beltrami equation

$$\frac{\partial f}{\partial \bar{z}} = \nu \cdot \frac{\partial f}{\partial z}. \tag{3.1}$$

If $f^\nu$ is a solution of (3.1), then $f^\mu = f^\nu \circ A^{-1}$ satisfies $\frac{\partial f^\mu}{\partial \bar{z}} = \mu \cdot \frac{\partial f^\mu}{\partial z}$. For more details of the following description, see [1, ch.V]. Consider the following two operators from $L^p(\mathbf{C})$ $(p > 2)$ to itself:

$$(3.2) \qquad \begin{aligned} Ph(z) &= \frac{1}{2\pi i} \iint \frac{h(\zeta)}{\zeta - z}\, d\zeta \wedge d\bar{\zeta} \\ Th(z) &= \text{p.v.} \frac{1}{2\pi i} \iint \frac{h(\zeta)}{(\zeta - z)^2}\, d\zeta \wedge d\bar{\zeta}. \end{aligned}$$

The operator $T$ is also defined on $L^2(\mathbf{C})$, where it is an isometry. For $p > 2$, there exists a constant $C_p$ with $\lim_{p\to 2} C_p = 1$ and $\|T\|_p < C_p$. Let $k =$ ess. sup.$|\mu| =$ ess. sup.$|\nu|$ and choose $p > 2$ so that $kC_p < 1$. Let $h$ be a solution in $L^p$ of the equation $h = T(\nu(h+1))$. Then the following function $f$ is a solution of (3.1):

$$(3.3) \qquad \begin{aligned} f(z) &= z + P(\nu(h+1))(z) \\ &= z + \frac{1}{2\pi i} \iint \nu(\zeta)(h(\zeta)+1) \left[\frac{1}{\zeta - z}\right] d\zeta \wedge d\bar{\zeta}. \end{aligned}$$

*Remark:* The definition of $P$ is slightly different from that in [1]. This is because in our treatment we do not need the normalized solution such that $f(0) = 0$.

LEMMA 3.1. *Let $z_0 \in \Delta$ and $z_0^* = \bar{z}_0^{-1} \in \Delta^*$. Let $\Delta(z_0, \rho)$ denote the hyperbolic disc with center $z_0$ and radius $\rho$. For $z_0$, $\rho$ and $k$, $0 \le k < 1$, define*

$$\begin{aligned} B(z_0, \rho, k) &= \{\mu : \mu \textit{ is measurable in } \mathbf{C},\ \textit{ess. sup. } |\mu| \le k, \\ &\qquad \textit{and } \operatorname{supp}\mu \subset \operatorname{cl}(\Delta \setminus \Delta(z_0, \rho))\}. \end{aligned}$$

*Then there exists a constant $a(k, \rho)$, depending only on $k$ and $\rho$, for which*

$$\rho_{\Delta^*}(z_0^*)^{-2} |S_{f^\mu}(z_0^*)| = (|z_0^*|^2 - 1)^2 |S_{f^\mu}(z_0^*)| < a(k, \rho)$$

*for all $\mu \in B(z_0, \rho, k)$, where $f^\mu$ is a solution of the Beltrami equation $\frac{\partial f}{\partial \bar{z}} = \mu \cdot \frac{\partial f}{\partial z}$. Moreover, for each fixed $k$,*

$$a(k, \rho) \to 0 \quad \textit{as} \quad \rho \to \infty.$$

PROOF: Let $A$ be a Möbius transformation preserving $\Delta^*$ with $A(\infty) = z_0^*$ (and $A(0) = z_0$). For each $\mu \in B(z_0, \rho, k)$, set $\nu(z) = \mu(A(z))\frac{\overline{A'(z)}}{A'(z)} \in B(0, \rho, k)$. Then if $f^\nu$ is a solution of the equation $\frac{\partial f}{\partial \bar{z}} = \nu \cdot \frac{\partial f}{\partial z}$ we have that

$$\rho_{\Delta^*}(z_0^*)^{-2}|S_{f^\mu}(z_0^*)| = \rho_{\Delta^*}(\infty)^{-2}|S_{f^\nu}(\infty)|.$$

Thus we may assume that $z_0 = 0$. For $\nu \in B(0, \rho, k)$, let $f$ be the solution (3.3). The integration takes place only in $\Delta \setminus \Delta(0, \rho)$. If $f(z) = z + b_0 + b_1 z^{-1} + \cdots$ is the expansion of $f$ at $z = \infty$, then

$$b_1 = \frac{-1}{2\pi i} \int\int \nu(\zeta)(h(\zeta) + 1)\, d\zeta \wedge d\bar{\zeta}$$

and $\rho_{\Delta^*}(\infty)^{-2}|S_f(\infty)| = 6|b_1|$. Let $\sigma(\rho) = \pi(\cosh \rho)^{-2}$, the Euclidean area of $\Delta \setminus \Delta(0, \rho)$.

Then we have

$$\begin{aligned} 6|b_1| &\le 3\pi^{-1}\{k\sigma(\rho) + \|h\|_p \|\nu\|_q\} \qquad (p^{-1} + q^{-1} = 1) \\ &\le 3\pi^{-1}\{k\sigma(\rho) + C_p(1 - kC_p)^{-1}\|\nu\|_p \|\nu\|_q\} \\ &\le 3k\sigma(\rho)\pi^{-1}\{1 + C_p(1 - kC_p)^{-1}k\} \quad = a(k, \rho), \end{aligned}$$

where $p$ is chosen so that $kC_p < 1$. To derive the second inequality we used equation (8) from [1, ch.V]. Noting that $a(k, \rho)$ has the desired property completes the proof.

**A section of the Teichmüller space of hyperbolic cyclic groups** [7].

Let $\Gamma$ be the group $\langle \gamma_\lambda \rangle$ generated by a hyperbolic transformation $\gamma_\lambda(z) = \lambda z$ $(\lambda > 1)$ or an extension of $\langle \gamma_\lambda \rangle$ of index 2. Then functions $\phi_\alpha(z) = \alpha z^{-2}$ $(\alpha \in \mathbf{C})$ belong to $Q(\Gamma)$. Let $\delta$ be a value defined by $2\alpha = 1 - \delta^2$. If $|\delta - 1| < 1$, then $\mu(z) = (\delta - 1)\frac{z}{\bar{z}} \in B(\Gamma)$. In this case the Beltrami equation $\frac{\partial f}{\partial \bar{z}} = \mu \cdot \frac{\partial f}{\partial z}$ has an explicit solution $f_\alpha(z) = z\bar{z}^{\delta - 1}$ which has a conformal continuation $f_\alpha(z) = z^\delta$ to $\mathcal{H}^*$. We can easily check that $S_{f_\alpha} = \phi_\alpha$ in $\mathcal{H}^*$. Then $|\delta - 1| < 1$ if and only if

$$\alpha \in \Lambda = \{\alpha = \frac{(1 - re^{2i\theta})}{2} : r < 4\cos^2\theta, 0 \le \theta \le \pi\}.$$

We can conclude that $\{\phi_\alpha : \alpha \in \Lambda\} \subset \mathbf{T}(\Gamma)$ and observe that $o(\Gamma) = 6$ and $i(\Gamma) = 2$. We remark also that a function $f$ in $\mathcal{H}^*$ with $S_f = \phi_\alpha$ is univalent in $\mathcal{H}^*$ if and only if $\alpha \in \mathrm{cl}(\Lambda)$.

LEMMA 3.2. *For $t$, $0 < t \leq \infty$, let $\mathrm{D}(t) = \{z \in \mathcal{H} : |z - it| < t\}$ be the disc centered at $it$ and tangent to the real line if $t < \infty$, and $\mathrm{D}(\infty) = \mathcal{H}$. For $\delta \in \mathbf{C}$, $|\delta - 1| < 1$, define*

$$\mu_{t,\alpha}(z) = \begin{cases} (\delta - 1)\frac{z^2}{\bar{z}^2} & \text{for } z \in \mathrm{D}(t) \\ 0 & \text{elsewhere,} \end{cases}$$

*where $\alpha = \frac{(1-\delta^2)}{2}$. Then a solution $f$ of the equation $\frac{\partial f}{\partial \bar{z}} = \mu_{t,\alpha} \cdot \frac{\partial f}{\partial z}$ coincides with a Möbius transformation in $\mathcal{H}^*$. Therefore $S_f = 0$ in $\mathcal{H}^*$.*

PROOF: Let $A_t(z) = \frac{-iz}{(z-it)}$ for $t < \infty$. Then $\mathrm{D}(\infty) = \mathcal{H} = A_t(\mathrm{D}(t))$ and $\mu_{t,\alpha} = (\mu_{\infty,\alpha} \circ A_t)\frac{\overline{A_t'}}{A_t'}$. Thus, for a solution $f$ of $\frac{\partial f}{\partial \bar{z}} = \mu_{t,\infty} \cdot \frac{\partial f}{\partial z}$, $f \circ A_t$ has the Beltrami coefficient $\mu_{t,\alpha}$. Therefore we need only consider the case of $t = \infty$. In this case the function defined by

$$f(z) = \begin{cases} \delta|z|^2((\delta - 1)z + \bar{z})^{-1} & \text{for } z \in \hat{\mathbf{C}} \setminus \mathcal{H}^* \\ z & \text{for } z \in \mathcal{H}^* \end{cases}$$

has the Beltrami coefficient $\mu_{\infty,\alpha}$. QED

# 4 Proof of Theorem 1

Let $Hol(\mathcal{H}^*)$ be the space of all holomorphic functions on $\mathcal{H}^*$, with the topology of normal convergence: For $\phi$, $\phi_n \in Hol(\mathcal{H}^*)$, $n = 1, 2, \ldots$,

$$\phi_n \to \phi \text{ in } Hol(\mathcal{H}^*)$$

means that $\phi_n$ converge to $\phi$ uniformly in every compact subset of $\mathcal{H}^*$. The space of all univalent functions in $\mathcal{H}^*$ is a closed subspace of $Hol(\mathcal{H}^*)$. In what follows, $d_{\mathbf{H}}(a, b)$ denotes the hyperbolic distance between $a$ and $b$ in $\mathcal{H}$.

Let $\{\Gamma_n\}$ be a subsequence of Fuchsian groups satisfying the hypothesis of Theorem 1. Let $\alpha \in \Lambda$. Then there is a $\phi_n \in \mathbf{T}(\Gamma_n)$ such that $\{\phi_n\}$

contains a subsequence $\{\phi_{n_j}\}$ with $\phi_{n_j} \to \phi_\alpha(z) = \alpha z^{-2}$ in $Hol(\mathcal{H}^*)$ and $\|\phi_{n_j}\|_{\mathcal{H}^*} \to 4|\alpha| = \|\phi_\alpha\|_{\mathcal{H}^*}$ as $j \to \infty$.

First we see how Theorem 1 is derived from this proposition. Assume that $\inf i(\Gamma_n) > 2 + 2\epsilon$ for an $\epsilon > 0$. Let $\alpha = \frac{1}{4} \in \Lambda$ and apply the proposition to obtain $\phi_{n_j} \in \mathbf{T}(\Gamma_{n_j})$ as above. Then $(2+\epsilon)\phi_{n_j} \in \mathbf{T}(\Gamma_{n_j})$ for large $j$ and converges to $\frac{(2+\epsilon)}{4} z^{-2}$ in $Hol(\mathcal{H}^*)$. Let $g_j$ be a function in $\mathcal{H}^*$ satisfying $S_{g_j} = (2+\epsilon)\phi_{n_j}$. Since $(2+\epsilon)\phi_{n_j} \in \mathbf{T}(\Gamma_n)$, $g_j$ is univalent. By composing with a suitable Möbius transformation we can assume that each $g_j$ fixes the points $-i$, $-2i$, and $-3i$. Then $g_j$ forms a normal family and has an accumulation point $g$ in $Hol(\mathcal{H}^*)$. This function $g$ must be unvalent in $\mathcal{H}^*$ and satisfies $S_g(z) = \frac{(2+\epsilon)}{4} z^{-2}$ as well. However $\frac{(2+\epsilon)}{4} \notin \mathrm{cl}(\Lambda)$, which is a contradiction. Therefore $\inf i(\Gamma_n) = 2$.

PROOF OF PROPOSITION 4.1: As before assume that the axis of a hyperbolic element $\gamma_n$ of $\Gamma_n$ has the collar width $d_n$ with $d_n \to \infty$ as $n \to \infty$. Because of (2.2) we can assume, after conjugation by suitable elements of Möb($\mathbf{R}$), that $\gamma_n(z) = \lambda_n z$ with $\lambda_n > 1$. The axis of $\gamma_n$ coincides with the imaginary axis $I = \{ti : t > 0\}$ and the collar of width $d_n$ about $I$ is the set

$$C_n^* = \{z \in \mathbf{H} : \theta_n < \arg z < \pi - \theta_n\}, \quad d_n = \log\cot\frac{\theta_n}{2}.$$

Let $C_n$ denote the collar of width $\frac{d_n}{2}$ about $I$.

Let $\alpha \in \Lambda$ and $\delta = (1-2\alpha)^{\frac{1}{2}}$ with $|\delta - 1| < 1$. Define

$$\hat{\mu}_n(z) = \begin{cases} (\delta - 1)\frac{z}{\bar{z}} & \text{for } z \in C_n \\ 0 & \text{for } z \in \hat{\mathbf{C}} \setminus C_n \end{cases}.$$

Note that $\hat{\mu}_n \in B(\Gamma_{n,I})$, where $\Gamma_{n,I} = \{\gamma \in \Gamma_n : \gamma(I) = I\}$ is the stability subgroup of $I$. For $\Gamma_{n,I}$ contains only transformations of the form $\gamma(z) = \lambda z$ ($\lambda > 0$) and $\gamma(z) = \frac{\lambda}{z}$ ($\lambda > 0$). Since $\gamma(C_n) \cap \delta(C_n) \neq \emptyset$ only if $\gamma$ and $\delta$ are in the same right coset of $\Gamma_n/\Gamma_{n,I}$ and since in this case $\gamma(C_n) = \delta(C_n)$, the function

$$\mu_n(z) = \sum_{\gamma \in \Gamma_n/\Gamma_{n,I}} \hat{\mu}_n(\gamma(z)) \frac{\overline{\gamma'(z)}}{\gamma'(z)}$$

is a Beltrami differential in $B(\Gamma_n)$. Since $\theta_n \to 0$ as $n \to \infty$, $\mu_n$ converges to the function

$$\mu_\infty(z) = \begin{cases} \delta\frac{z}{\bar{z}} & \text{for } z \in \mathcal{H} \\ 0 & \text{for } z \in \hat{\mathbf{C}} \setminus \mathcal{H} \end{cases}.$$

except at most on $\mathbf{R} \cup \{\infty\}$. Let $f_n$ be the solution of the equation $\frac{\partial f_n}{\partial \bar{z}} = \mu_n \cdot \frac{\partial f_n}{\partial z}$ normalized so that $f_n(-ki) = (-ki)^\delta$ for $k = 1, 2, 3$. Passing to a subsequence, if necessary, we assume that $f_n$ converges to $f_\infty$ uniformly in $\hat{\mathbf{C}}$ (with respect to the spherical metric), where $f_\infty$ satisfies $\frac{\partial f_\infty}{\partial \bar{z}} = \mu_\infty \cdot \frac{\partial f_\infty}{\partial z}$. Then $\phi_n = \mathcal{S}_{f_n|\mathcal{H}^*}$ belongs to $\mathbf{T}(\Gamma_n)$ and converges to $\mathcal{S}_{f_\infty}(z) = \phi_\alpha(z) = \alpha z^{-2}$ in $Hol(\mathcal{H}^*)$.

*Remark:* In the above arguement we used some properties of quasiconformal mappings. These are found in [9] (in particular §5 of chapter II and §4 of chapter V).

Next we show that $\|\phi\|_{\mathcal{H}^*}$ (more precisely its subsequence) converges to $4|\alpha| = \|\phi_\alpha\|_{\mathcal{H}^*}$. First note that $\|\phi\|_{\mathcal{H}^*} = \sup_{z \in R_n} \rho_{\mathcal{H}^*}(\bar{z})^{-2}|\phi_n(\bar{z})|$ for any fundamental region $R_n$ ($\subset \mathcal{H}$) of $\Gamma_n$. We choose $R_n$ so that

$$C_n^* \cap \{z \in \mathcal{H} : 1 \leq |z| < \lambda_n\} \subset R_n.$$

Let $\epsilon$ ($> 0$) be sufficiently small. Since $(\mathrm{supp}\mu) \cap R_n \subset C_n$, for each $z \in R_n \backslash C_n^*$ it holds that $\mu_n|_{\Delta(z, \frac{d_n}{2})} \equiv 0$, where $\Delta(z, \frac{d_n}{2})$ denotes the hyperbolic disc with center $z$ and radius $\frac{d_n}{2} = d_{\mathbf{H}}(C_n, \partial C_n^*)$. Let $A$ be a Möbius transformation sending $\Delta$ onto $\mathcal{H}$ and $g_n$ be a solution of the equation $\frac{\partial g}{\partial \bar{z}} = [(\mu_n \circ A)(\frac{\bar{A}'}{A'})] \cdot \frac{\partial g}{\partial z}$. Then $\phi_n = \mathcal{S}_{f_n} = A^* \mathcal{S}_{g_n}$. We apply lemma 3.1 to $g_n$. Since $\frac{d_n}{2} \to +\infty$ as $n \to \infty$, we have by (2.2) that if $n$ is sufficiently large,

$$\rho_{\mathcal{H}^*}(\bar{z})^{-2}|\phi_n(\bar{z})| < \epsilon \leq 4|\alpha| + \epsilon \quad \text{for } z \in R_n \setminus C_n^*.$$

So next we need to estimate $\sup_{z \in R_n \cap C_n^*} \rho_{\mathcal{H}^*}(\bar{z})^{-2}|\phi_n(\bar{z})|$. Since $\mathrm{cl}(R_n \cap C_n^*)$ is compact in $\mathcal{H}$, this value is attained by some $z_n \in \mathrm{cl}(R_n \cap C_n^*)$. Before proceedings further we see that we can assume that $|z_n| = 1$. Let $B_n(z) = |z_n| z \in \text{Möb}(\mathbf{R})$. We replace $\Gamma_n$ by $B_n^{-1}\Gamma_n B_n$. Accordingly we replace $\mu_n$, $f_n$ and $\phi_n$ by $(\mu_n \circ B_n)\frac{\overline{B_n'}}{B_n'}$, $f_n \circ B_n$ and $(\phi \circ B_n)B_n^2$. Note that this new $\mu_n$ also converges to $\mu_\infty$ almost everywhere in $\hat{\mathbf{C}}$. Repeating the above arguement and passing to a subsequence, if necessary, we know that the new $\phi_n$ converges to $\phi_\alpha$ in $Hol(\mathcal{H}^*)$. Moreover, $B_n$ preserves the collar $C_n^*$.

Now assume that $|z_n| = 1$. If $d_{\mathbf{H}}(z_n, I) \to +\infty$, then conjugate $\Gamma_n$ by the Möbius transformation

$$C_n(z) = \xi_n \frac{z + \epsilon\eta_n}{z - \epsilon\eta_n} \quad \text{with } C_n(z_n) = i = \sqrt{-1},$$

where $\xi_n$, $\eta_n > 0$ and $\epsilon = 1$ if $\Re z > 0$, $= -1$ if $\Re z < 0$.

Then

$$\mu_n(B_n(z))\frac{\overline{B_n'(z)}}{B_n'(z)} = (\delta - 1)\frac{(z^2 - \eta_n^2)}{(\bar{z}^2 - \eta_n^2)} \quad \text{for } z \in B_n^{-1}(C_n).$$

Since $d_{\mathbf{H}}(z_n, I) \to +\infty$, $\eta_n \to 0$ as $n \to \infty$. Moreover $d_{\mathbf{H}}(z_n, \gamma(C_n)) > d_n \to +\infty$ for $\gamma \in \Gamma_n \setminus \Gamma_{n,I}$. Thus, if we choose a subsequence, $(\mu_n \circ B_n)(\frac{\overline{B_n'}}{B_n'})$ converges to some $\mu_{t,\alpha}$ as in lemma 3.2 or to $\mu_0 \equiv 0$ almost everywhere in $\hat{\mathbf{C}}$. In both cases $f_n \circ B_n$ converges to a Möbius transformation in $\mathcal{H}^*$. Hence,

$$\rho_{\mathcal{H}^*}(\bar{z_n})^{-2}|\phi_n(\bar{z_n})| = \rho_{\mathcal{H}^*}(-i)^{-2}|S_{f_n \circ B_n}(-i)| \to 0 \quad \text{as } n \to \infty.$$

If $\alpha = 0$ the proof is completed. If $\alpha > 0$, then $\rho_{\mathcal{H}^*}(-i)^{-2}|\phi_n(-i)| \to 4|\alpha|$ since $\phi_n \to \phi_\alpha$ in $\text{Hol}(\mathcal{H}^*)$. Therefore (4.1) contradicts the fact that $\rho_{\mathcal{H}^*}(\bar{z}_n)^{-2}|\phi_n(\bar{z}_n)| = \sup_{z \in R_n \setminus C_n^*} \rho_{\mathcal{H}^*}(\bar{z})^{-2}|\phi_n(\bar{z})|$. Hence $d_{\mathbf{H}}(z_n, I)$ is bounded. Since $|z_n| = 1$, this means that $\bar{z_n}$ belongs to a compact subset of $\mathcal{H}^*$. Therefore

$$\rho_{\mathcal{H}^*}(\bar{z}_n)^{-2}|\phi_n(\bar{z}_n)| < 4|\alpha|\frac{(\Im z_n)^2}{|z_n|^2} + \epsilon \leq 4|\alpha| + \epsilon$$

for sufficiently large $n$ (and not necessarily $z_n \to i$). Now we conclude that $\|\phi_n\|_{\mathcal{H}^*} < 4|\alpha| + \epsilon$ for sufficiently large $n$.

**Remark:** Proposition 4.1 does not mean that $\|\phi_{n_j} - \phi_\alpha\|_{\mathcal{H}^*} \to 0$ (see §4 of [11]).

# References


[1] L. V. Ahlfors, *Lectures on quasiconformal mappings,* Van Nostrand (1966).

[2] L. V. Ahlfors and G. Weill, A uniqueness theorem for Beltrami equations, *Proc. AMS* **13** (1962) 975–978.

[3] L. Bers, On boundaries of Teichmüller spaces and Kleinian groups, I., *Ann. Math.* **91** (1970) 570–600.

[4] T. Chu, On the outradius of finite dimensional Teichmüller spaces, *Ann. Math. Studies* **79** Princeton University Press, Princeton, New Jersey (1974) 75–79.

[5] F. Gehring and C. Pommerenke, On the Nehari univalence criterion and quasicircles, *Comment. Math. Helv.* **59** (1984) 226–242.

[6] N. Halperin, A proof of the collar lemma, *Bull. London Math. Soc.* **13** (1981) 141–144.

[7] C. I. Kalme, Remarks on a paper by Lipman Bers, *Ann. Math.* **91** (1970) 601–606.

[8] S. L. Kruskal' and B. D. Golovan', Approximation of analytic functions and Teichmüller spaces (russian), *reprint 3* Novosibirsk (1989).

[9] O. Lehto and K. I. Virtanen, *Quasiconformal mappings in the plane,* Springer, Berlin (1973).

[10] C. McMullen, Geometric limits of quadratic differentials, *preprint* (1988).

[11] T. Nakanishi, A theorem on the outradii of Teichmüller spaces, *J. Math. Soc. Japan* **40** (1988) 1–8.

[12] T. Nakanishi, The inner radii of finite-dimensional Teichmüller spaces, *Tôhoku Math. J.* **41** (1989) 679–688.

[13] T. Nakanishi and H. Yamamoto, On the outradius of the Teichmüller space, *Comment. Math. Helv.* **64** (1989) 288–299.

[14] Z. Nehari, The Schwarzian derivative and schlicht functions, *Bull. AMS* **55** (1949) 545–551.

[15] H. Sekigawa, The outradius of the Teichmüller space, *Tôhoku Math. J.* **30** (1978) 607–612.

[16] H. Sekigawa and H. Yamamoto, Outradii of Teichmüller spaces of the second kind, *Tôhoku Math. J.* **38** (1986) 365–370.

Toshihiro Nakanishi: Department of Mathematics, Shizuoka University, Shizuoka 422, JAPAN
and Department of Mathematics, University of Helsinki, Hallituskatu 15, 00100 FINLAND

John A. Velling: Department of Mathematics, Brooklyn College, Brooklyn, New York 11210 USA

# ON THE CAUSAL STRUCTURES OF THE SILOV BOUNDARIES OF SYMMETRIC BOUNDED DOMAINS

Soji Kaneyuki
Department of Mathematics, Sophia University
Kioicho, Chiyoda-ku, Tokyo 102

Dedicated to Professor Tadashi Nagano on his sixtieth birthday

## Introduction

Let D be a domain in $C^n$. Then it seems to be an interesting problem to study the interaction between the complex structure of D and an appropriate geometric structure on the boundary of D or on a distinguished subset of the boundary. In this direction, in the case where D is a nondegenerate Siegel domain of the second kind, Tumanov-Khenkin [28] worked out the study on the relationship between automorphisms of the CR-structure on the Silov boundary S and holomorphic automorphisms of D. In the case where D is holomorphically equivalent to a Siegel domain of the first kind, the Silov boundary is totally real, and hence the CR-structure on it becomes trivial. We have to seek another appropriate geometric structure on the Silov boundary in that case.

In this paper, we are concerned with the afore-mentioned problem exclusively for the case where D is an irreducible symmetric bounded domain which is holomorphically equivalent to a Siegel domain of the first kind. In that case, its Silov boundary S is an irreducible symmetric R-space. The main purpose of this paper is, first, to introduce the causal structure C on S derived from the self-dual cone V, which is the noncompact dual of S (Proposition 5.4), and next , to study the relationship between the causal automorphism group $G(S,\underline{C})$ and the holomorphic automorphism group G(D) of D (§6). For basic facts on causal structures, see §5 (See also Segal[20]).

§2 has an independent interest of its own. Here we obtain the orbit decomposition of a compact simple Jordan algebra A under the automorphism group of the self-dual cone V formed by the interior points of the set of the squares of all elements in A (Theorem 2.6). This decomposition is viewed as a generalization of the Sylvester's law of inertia, and V is regarded, in some sense, as the cone of positive definite elements in A. Under a more restrictive assumption, this decomposition was obtained by Satake and by the author independently ([19],[7]). We also obtain the closure structures of these orbits (Theorem 2.8). In §3 we describe the isometry group and the holomorphic automorphism group of D in terms of the Jordan algebra A (Proposition 3.4). In §6, applying the Tanaka's theory on the normal Cartan connections [23],[26], we prove the main theorem which asserts that if $\dim D > 1$, then every causal automorphism of the Silov boundary S extends to a unique holomorphic automorphism of the domain D, and conversely every holomorphic automorphism of D is obtained in this manner (Theorem 6.2). A similar result can be proved for anti-causal (=causality-reversing) automorphisms of S and anti-holomorphic automorphisms of D (Theorem 6.6). We also prove the coincidence between the affine automorphism group of the Siegel domain D(V) of the first kind and the causal automorphism group of the Jordan algebra A with causal structure coming from the cone V (Theorem 6.10). This result is a generalization of E.C. Zeeman's [31]. A.D.Alexandrov [1] has obtained a similar result under some weaker assumption.

## §1. Basic facts on Jordan algebras

Let B be a simple Jordan algebra over K, where K = R or C. Let $a \in B$ and let $L_a$ be the left multiplication by a, which is a K-linear operator on B. The trace form $t_B$ of B is defined by

$$(1.1) \qquad t_B(x,y) = \operatorname{Tr} L_{xy} , \qquad x,y \in B.$$

Note that $t_B$ is nondegenerate. Let g be a K-linear endomorphism of B. We will denote by $g^*$ the adjoint operator of g with respect to $t_B$. The quadratic operator $P(a)$, $a \in B$, is defined by $P(a) = 2L_a^2 - L_{a^2}$. Then one can consider the structure group of B:

$$(1.2) \quad \operatorname{Str} B = \left\{ g \in GL_K(B) : gP(a)g^* = P(ga),\ a \in B \right\},$$

where $GL_K(B)$ denotes the group of invertible K-linear endomorphisms of B. According to Koecher [10] and Kantor [9], to the Jordan algebra B, there corresponds a simple graded Lie algebra over K of the first kind:

$$(1.3)\qquad \underline{g}(B) = \underline{g}_{-1}(B) + \underline{g}_0(B) + \underline{g}_1(B),$$

where $\underline{g}_{-1}(B)$ and $\underline{g}_1(B)$ are two copies of the underlying vector space of B, and $\underline{g}_0(B)$ is the Lie algebra of Str B. An element $a \in B$, viewed as an element of $\underline{g}_1(B)$, will be denoted by $\bar{a}$. Let $X = a + T + \bar{b}$ and $X' = a' + T' + \bar{b}'$ be two elements of $\underline{g}(B)$, where $a,a',b,b' \in \underline{g}_{-1}(B)$ and $T,T' \in \underline{g}_0(B)$. Then the Lie bracket $[X,X']$ in $\underline{g}(B)$ is given by ([17])

$$(1.4)\qquad [X,X'] = (Ta' - T'a) + (2a' \square b + [T,T'] - 2a \square b') + \overline{(T'^{*}b - T^{*}b')}.$$

Here we put

$$(1.5)\qquad x \square y = L_{xy} + [L_x,L_y], \qquad x,y \in \underline{g}_{-1}(B).$$

Let $\tau$ be a linear endomorphism of $\underline{g}(B)$ defined by

$$(1.6)\qquad \tau(a + T + \bar{b}) = b - T^{*} + \bar{a}.$$

Then $\tau$ is a grade-reversing involutive automorphism of $\underline{g}(B)$. Let $\mathrm{Aut}\,\underline{g}(B)$ denote the automorphism group of the Lie algebra $\underline{g}(B)$ over K, and $\mathrm{Aut}_{gr}\underline{g}(B)$ denote the subgroup of $\mathrm{Aut}\,\underline{g}(B)$ consisting of grade-preserving automorphisms of $\underline{g}(B)$. For an element $g \in \mathrm{Str}\,B$ we define a grade-preserving K-linear endomorphism $\varphi_B(g)$ of $\underline{g}(B)$ by putting

$$(1.7)\qquad \varphi_B(g) = \begin{cases} g & \text{on } \underline{g}_{-1}(B), \\ \mathrm{Ad}_{\underline{g}_0(B)}g & \text{on } \underline{g}_0(B), \\ \tau(g^{*})^{-1}\tau & \text{on } \underline{g}_1(B). \end{cases}$$

Lemma 1.1 ([17]). $\varphi_B$ is a K-isomorphism of Str B onto $\mathrm{Aut}_{gr}\underline{g}(B)$, and the restriction $(\mathrm{Aut}_{gr}\underline{g}(B))|_{\underline{g}_{-1}}$ coincides with Str B.

## §2. The Sylvester's law of inertia

Let A be a Jordan algebra over R. We say that A is compact if the trace form $t_A$ is positive definite. In this section, we shall always assume that A is compact and simple. Let V be the interior of the set of the squares of all elements of A. It is known [29] that V is an irreducible homogeneous self-dual open convex cone in A. Let G(V) be the automorphism group of V, that is,the group of all invertible linear endomorphisms of A leaving V stable. Then we have [17]

$$\text{(2.1)} \qquad \operatorname{Str} A = G(V)\times\{\pm 1_A\} .$$

Let $a \in A$, and let $\mu_a(\lambda)$ and $m_a(\lambda)$ be the minimum polynomial and the generic minimum polynomial of the element a, respectively (Jacobson [5]). We can write $m_a(\lambda)$ in the form

$$\text{(2.2)} \qquad m_a(\lambda) = \lambda^r - \sigma_1(a)\lambda^{r-1} + \cdots + (-1)^r\sigma_r(a),$$

where $\sigma_i(a)$ is a homogeneous polynomial of degree i in the components of a with respect to a basis of A. The positive integer r does not depend on the choice of the element $a \in A$ and is uniquely determined by A. r is called the degree of the Jordan algebra A. Later on we assume that the degree of A is equal to r. Let e be the unit element of A. Since A is of degree r, one can choose a system of primitive orthogonal idempotents $\{e_1, \cdots, e_r\}$ of A such that $\sum_{i=1}^r e_i = e$ (Jacobson [5]). Such systems are conjugate to each other under the automorphism group Aut A of the Jordan algebra A. We choose and fix such a system $\{e_1, \cdots, e_r\}$ .

As was mentioned in §1, to the Jordan algebra A, there corresponds the real simple graded Lie algebra $\underline{g}(A) = \underline{g}_{-1}(A) + \underline{g}_0(A) + \underline{g}_1(A)$. For simplicity, we write $\underline{g}$ and $\underline{g}_\lambda$ ($\lambda = 0, \pm 1$) for $\underline{g}(A)$ and $\underline{g}_\lambda(A)$, respectively. Thus we have

$$\text{(2.3)} \qquad \underline{g} = \underline{g}_{-1} + \underline{g}_0 + \underline{g}_1 .$$

The involutive automorphism $\tau$ of g in (1.6) is a grade-reversing Cartan involution of $\underline{g}$ ([2]). Let

(2.4) $$\underline{g} = \underline{k} + \underline{p}$$

be the Cartan decomposition by $\tau$, where $\tau|_{\underline{k}} = 1$ and $\tau|_{\underline{p}} = -1$. Since $\underline{g}_0$ is stable under $\tau$, one has the Cartan decomposition of $\underline{g}_0$:

(2.5) $$\underline{g}_0 = \underline{k}_0 + \underline{p}_0,$$

where $\underline{k}_0 = \underline{k} \cap \underline{g}_0$ and $\underline{p}_0 = \underline{p} \cap \underline{g}_0$. $\underline{k}_0$ is the Lie algebra of derivations of A, and $\underline{p}_0$ is the space of left multiplications by elements in A. Consider the subspace $\underline{a} = \sum_{i=1}^{r} RL_{-e_i}$ of $\underline{p}_0$ containing $L_{-e}$. It is known [3] that $\underline{a}$ is a maximal abelian subspace of $\underline{p}$. Let $\{\beta_1, \cdots, \beta_r\}$ be the dual basis to the basis $\{L_{-e_1}, \cdots, L_{-e_r}\}$ of $\underline{a}$. Then the root systems $\Delta(\underline{g},\underline{a})$ of $\underline{g}$ and $\Delta(\underline{g}_0,\underline{a})$ of $\underline{g}_0$ with respect to $\underline{a}$ are given by

(2.6)
$$\Delta(\underline{g},\underline{a}) = \{\pm(\beta_i \pm \beta_j)/2 \ (1 \leq i < j \leq r), \ \pm\beta_i \ (1 \leq i \leq r)\},$$
$$\Delta(\underline{g}_0,\underline{a}) = \{\pm(\beta_i - \beta_j)/2 \ (1 \leq i < j \leq r)\}.$$

If we denote by $\underline{g}^{\gamma}$ the root space for the root $\gamma \in \Delta(\underline{g},\underline{a})$, then we have $\dim \underline{g}^{\beta_i} = 1 \ (1 \leq i \leq r)$. If we put

(2.7) $$H_{\beta_i} = 2L_{-e_i}, \quad E_{-\beta_i} = e_i, \quad E_{\beta_i} = -\tau e_i \quad (1 \leq i \leq r),$$

then we have $E_{\pm\beta_i} \in \underline{g}^{\pm\beta_i}$. Since (1.4) is valid in the graded Lie algebra $\underline{g}$, it follows easily that

(2.8) $$\beta_i(H_{\beta_i}) = 2, \quad \tau E_{\beta_i} = -E_{-\beta_i}, \quad [E_{\beta_i}, E_{-\beta_i}] = H_{\beta_i} \quad (1 \leq i \leq r).$$

Let $a \in A$. Then the polynomial equation $\mu_a(\lambda) = 0$ has only real roots ([4]); furthermore each irreducible factor of $m_a(\lambda)$ is a factor of $\mu_a(\lambda)$ ([5]). Hence the equation $m_a(\lambda) = 0$ also has only real roots.

Definition 2.1. By the rank of an element $a \in A$ (denoted by rk ( a)) we mean the number of nonzero roots of the equation $m_a(\lambda)$

= 0. By the signature of $a \in A$ (denoted by sgn (a)) we mean the pair of the integers $(p,q)$ such that p and q are the numbers of positive roots and negative roots of the equation $m_a(\lambda) = 0$, respectively. Here the number of a root should be counted by including its multiplicity. Specifically, if $\mathrm{sgn}(a) = (r,0)$, then we say that a is positive definite.

Let $A_h$ (resp. $A_{p,q}$) denote the set of elements $a \in A$ with $\mathrm{rk}(a) = h$ (resp. $\mathrm{sgn}(a) = (p,q)$). Then we have the decomposition:

$$(2.9) \qquad A = \coprod_{h=0}^{r} A_h , \qquad A_h = \coprod_{p+q=h} A_{p,q} ,$$

$$(2.10) \qquad A = \coprod_{p,q \leq r} A_{p,q} .$$

Note that $A_0 = A_{0,0} = (0)$. Let us consider the elements of A:

$$(2.11) \qquad o_{p,q} = \sum_{i=1}^{p} e_i - \sum_{j=p+1}^{p+q} e_j , \qquad p,q \geq 0,\ p+q \leq r;$$

here we adopting the convention that the first or the second term of the right-hand side should be zero, according as p = 0 or q = 0, respectively.

Lemma 2.2. A is expressed as the union of $G^0(V)$-orbits through the points $o_{p,q}$:

$$(2.12) \qquad A = \bigcup_{p+q \leq r} G^0(V) o_{p,q},$$

where $G^0(V)$ denotes the identity component of G(V).

Proof. Let K(V) be the isotropy subgroup of G(V) at the point e. Note that $\mathrm{Lie}\, K(V) = \underline{k}_0$. Denoting the identity component of K(V) by $K^0(V)$, we have the Cartan decomposition

$$(2.13) \qquad G^0(V) = K^0(V) \exp \underline{p}_0 .$$

Since an element of the Weyl group of the root system $\Delta(\underline{g}_0, \underline{a})$ has its representative in $K^0(V)$, it follows from a result of Satake [18] that given an element $a \in A$, there exists $k \in K^0(V)$ such that

(2.14) $$ka = \sum_{i=1}^{p+q} d_i e_i,$$

where $d_1, \cdots, d_p > 0$, $d_{p+1}, \cdots, d_{p+q} < 0$, $p+q \leqslant r$, $p, q \geqslant 0$. Since $\exp tL_{e_i}$ acts on $\underline{a}$ as a transformation $e_j \mapsto \exp(t\delta_{ij})e_j$ $(1 \leqslant j \leqslant r)$, it follows that there exists an element $g \in \exp \underline{p}_0$ such that $g(ka) = o_{p,q}$. q.e.d.

Lemma 2.3. Let h be an integer such that $0 \leqslant h \leqslant r$. For any integer p, $q \geqslant 0$ satisfying $p+q = h$, we have

(2.15) $$A_{p,q} \subset G^0(V)o_{p,q} \subset A_h.$$

Proof. Let $a \in A_{p,q}$. Then, as is seen in the proof of Lemma 2.2, there exists an element $k \in K^0(V)$ such that $ka = \sum_{i=1}^{p_1+q_1} d_i e_i$, where $d_1, \cdots, d_{p_1} > 0$, $d_{p_1+1}, \cdots, d_{p_1+q_1} < 0$, $p_1+q_1 \leqslant r$, $p_1, q_1 \geqslant 0$. Note that the automorphism group Aut A of the Jordan algebra A coincides with K(V) ([17]). Also the generic minimum polynomial is invariant under Aut A ([5]). Therefore, by direct computations, we have

$$m_a(\lambda) = m_{ka}(\lambda) = m_{\Sigma d_i e_i}(\lambda)$$
$$= (\lambda - d_1)(\lambda - d_2) \cdots (\lambda - d_{p_1+q_1})\lambda^{r-p_1-q_1}.$$

Consequently, we have $p_1=p$ and $q_1=q$. An argument in the proof of Lemma 2.2 now shows the inclusion $A_{p,q} \subset G^0(V)o_{p,q}$. To prove the second inclusion in (2.15), let us consider the subset $V_h = \bigcup_{p+q=h} (\mathrm{Str}\, A)\, o_{p,q}$. Then, in view of Lemma 1.1, (2.7) and (2.8), it follows from a result of Takeuchi [22] that

(2.16) $$A = \underline{g}_{-1} = \coprod_{h=0}^{r} V_h.$$

On the other hand, by the definition of $V_h$, one has $G^0(V)o_{p,q} \subset V_h$, $p+q = h$. So we get $A_{p,q} \subset V_h$ $(p+q = h)$, and consequently $A_h \subset V_h$. Therefore, by (2.9) and (2.16), we have $A_h = V_h$. q.e.d.

Corollary 2.4. The rank rk, viewed as an integer-valued

function on A, is invariant under Str A .

Proof. Since $A_h = V_h$ holds, one has $(\mathrm{Str}\,A)A_h = A_h$, $0 \leq h \leq r$.
q.e.d.

The following proposition was obtained by Satake [19] and Kaneyuki [7] independently.

Proposition 2.5. $A_{p,q} = G^0(V)o_{p,q}$, $p+q \leq r$.

Proof. First we shall show that $A_{p,q}$ is open in $A_h$, where $h = p+q$. Since $\sigma_i(b)$ is continuous in $b \in A$, a root of the equation $m_b(\lambda) = 0$ depends continuously on b. Now let $a \in A_{p,q}$. If we choose a sufficiently small neighborhood U of a in A, then for an arbitrary point $a' \in U$, we have $p' \geq p$ and $q' \geq q$, where we set $\mathrm{sgn}(a') = (p',q')$. Hence, if a' lies in $U \cap A_h$, then we have $p'+q' = \mathrm{rk}(a') = \mathrm{rk}(a) = p+q$, and consequently $p' = p$ and $q' = q$. This implies that $U \cap A_h \subset A_{p,q}$, and hence $A_{p,q}$ is open in $A_h$. Now suppose that $A_{p,q} \subsetneqq G^0(V)o_{p,q}$ (cf. (2.15)). Then, by (2.9), we have

$$G^0(V)o_{p,q} = A_{p,q} \amalg \left( \coprod_{\substack{0 \leq k \leq h \\ k \neq p}} (G^0(V)o_{p,q} \cap A_{k,h-k})\right).$$

By the assumption, there exists at least one k, say $k_1$ $(k_1 \neq p)$, such that $G^0(V)o_{p,q} \cap A_{k_1,h-k_1}$ is not empty. Since $A_{k_1,h-k_1}$ is open in $A_h$, this intersection is open in $G^0(V)o_{p,q}$. But this contradicts the fact that $G^0(V)o_{p,q}$ is connected. Thus we should have $A_{p,q} = G^0(V)o_{p,q}$. q.e.d.

We finally obtain the following theorem.

Theorem 2.6 (The Sylvester's law of inertia). Let A be a compact simple Jordan algebra of degree r. Let V be the self-dual cone formed by the interior points of the set of squares of all elements of A. Let G(V) be the automorphism group of V. Then the decomposition (2.10) is the G(V)-orbit decomposition of A. More precisely, each subset $A_{p,q}$ is the $G(V)$-orbit through the point $o_{p,q}$ $(p,q \geq 0,\ p+q \leq r)$.

Proof. When the group $G(V)$ is connected, the theorem is reduced to Proposition 2.5. So we assume that $G(V)$ is not connected. In this case, $G(V)$ has two connected components, and further there exists an element $\varepsilon \in G(V) - G^0(V)$ such that $\varepsilon^2 = 1$ and that for each $p,q$,

$$\varepsilon o_{p,q} \in A_{p,q}. \tag{2.17}$$

This follows from a result of Satake [17]. To prove the theorem, it suffices to show that $G(V)$ leaves $A_{p,q}$ stable. Let $a \in A_{p,q}$. Since $G^0(V)$ acts transitively on $A_{p,q}$ by Proposition 2.5, there exists an element $g \in G^0(V)$ such that $a = go_{p,q}$. Put $g' = \varepsilon g \varepsilon \in G^0(V)$. Then, by (2.17) we have $\varepsilon a = \varepsilon g o_{p,q} = g'\varepsilon o_{p,q} \in g'A_{p,q} \subset A_{p,q}$, which shows $\varepsilon A_{p,q} = A_{p,q}$. Thus $A_{p,q}$ is $G(V)$-invariant. q.e.d.

We will give a supplementary result:

Lemma 2.7. (1) $V = A_{r,0}$. (2) $A_{p,q}$ is open if and only if $p+q = r$. (3) $A_{q,p} = -A_{p,q}$.

Proof. (1) By a result of Vinberg [29], $V$ is the $G(V)$-orbit through $e = o_{r,0}$. (2) $A_h$ is open in $A$ if and only if $h = r$. On the other hand, $A_{p,q}$ is open in $A_{p+q}$, as was shown in the proof of Proposition 2.5. (3) Since the polynomials $\sigma_i(a)$ are homogeneous of degree $i$, it follows that $m_a(\lambda_1) = 0$ if and only if $m_{-a}(-\lambda_1) = 0$. q.e.d.

We want to consider the closures of the $G(V)$-orbits.

Theorem 2.8. Under the same assumptions as in Theorem 2.6, the closure $\overline{A}_{p,q}$ of the orbit $A_{p,q}$ is given by

$$\overline{A}_{p,q} = \coprod_{p_1 \leq p,\, q_1 \leq q} A_{p_1,q_1}. \tag{2.18}$$

Proof. Let $a \in \overline{A}_{p,q}$ and let $\mathrm{sgn}(a) = (p_1,q_1)$. Let $U$ be the neighborhood of a in A, which was chosen in the proof of Proposition 2.5. Then, by choosing an element a' in $U \cap A_{p,q}$, and by following the argument in that proof, we can deduce $p \geq p_1$ and $q \geq q_1$. This shows the inclusion $\subset$ in (2.18). To prove the converse inclusion take an element $a \in A_{p_1,q_1}$ $(p_1 \leq p,\ q_1 \leq q)$. Then, by Theorem 2.6, there exists $g \in G(V)$ such that $ga = o_{p_1,q_1}$. Consider the sequence in $A_{p,q}$:

$$b_n = o_{p_1,q_1} + \sum_{k=p_1+q_1+1}^{p+q_1} (1/n)e_k - \sum_{h=p+q_1+1}^{p+q} (1/n)e_h, \quad n=1,2,\cdots.$$

Then we have $\lim_{n\to\infty} b_n = o_{p_1,q_1} = ga$. Hence $\lim_{n\to\infty} g^{-1}b_n = a$. By Theorem 2.6, we have $g^{-1}b_n \in A_{p,q}$, and consequently $a \in \overline{A}_{p,q}$. q.e.d.

Corollary 2.9. The boundary $\partial A_{r,0}$ of the cone $V = A_{r,0}$ is given by

$$(2.19) \qquad \partial A_{r,0} = A_{r-1,0} \amalg A_{r-2,0} \amalg \cdots \amalg A_{1,0} \amalg (0).$$

The right-hand side is a G(V)-invariant stratification of the boundary $\partial A_{r,0}$.

$\partial A_{r,0}$ is called a generalized light cone. It can be seen that $\overline{A}_{r,0}$ coincides with the convex closure of $\partial A_{r,0}$.

## §3. Structure of the isometry group I(D(V))

3.1. Let A be a compact simple Jordan algebra of degree r, and $A^C$ be its complexification which is a complex simple Jordan algebra. Let $A^{CR}$ be the real Jordan algebra obtained from $A^C$ by the scalar restriction to R. Then the trace forms $t_{A^C}$ and $t_{A^{CR}}$ are nondegenerate. We have the relations:

$$(3.1) \qquad t_{A^{CR}}(x,y) = 2\,\mathrm{Re}\, t_{A^C}(x,y),$$

$$(3.2) \qquad t_{A^C}(x,y) = (1/2)\{t_{A^{CR}}(x,y) - i t_{A^{CR}}(Jx,y)\},$$

where $x,y \in A^C$ and J denotes the complex structure of $A^{CR}$. One can easily see from (3.2) that if $g \in \mathrm{End}\, A^{CR}$ is C-linear, then the adjoint operator of g with respect to $t_{A^{CR}}$ coincides with that of g with respect to $t_{A^C}$. Taking this into consideration and proceeding analogously as in the proof of Lemma 1.5 in Neher [15], we obtain the following lemma (cf. (1.2)).

Lemma 3.1. Let $\theta$ be the conjugation of $A^C$ with respect to A. Then we have

(3.3) $$\mathrm{Str}\, A^{CR} = \mathrm{Str}\, A^C \amalg (-\theta)\mathrm{Str}\, A^C .$$

Let V be the cone of positive definite elements in A, and consider the Siegel domain of the first kind

(3.4) $$D(V) = \{ z = x + iy \in A^C :\ x \in A,\ y \in V \} .$$

D(V) is holomorphically equivalent to an irreducible symmetric bounded domain of tube type. Conversely, every irreducible symmetric bounded domain of tube type is obtained by the above manner ([11]). By (2.1), one has the equality $\mathrm{Str}\, A = G(V) \amalg (-1_A)G(V)$. Let $a \in G(V)$, and let us put $\iota(a) = \tilde{a}$ and $\iota((-1)a) = (-\theta)\tilde{a}$, where $\tilde{a}$ denotes the C-linear extension of a to $A^C$. Since $\tilde{a}$ commutes with the conjugation $\theta$, it follows that $\iota$ is an injective homomorphism of $\mathrm{Str}\, A$ into $\mathrm{Str}\, A^{CR}$. From now on we will identify $\mathrm{Str}\, A$ with its $\iota$-image. One can then write

(3.5) $$\mathrm{Str}\, A = G(V) \amalg (-\theta)G(V).$$

3.2. We will preserve the notations in §2. We want to study the structure of the automorphism group Aut $\underline{g}$ of the real simple Lie algebra $\underline{g}$. Let $\underline{g}^C$ be the complexification of $\underline{g}$ and let $\sigma$ be the conjugation of $\underline{g}^C$ with respect to $\underline{g}$. Let $\underline{h}$ be a Cartan subalgebra of $\underline{g}$ containing $\underline{a}$. Let $\tilde{\Pi}$ be a $\sigma$-fundamental system of the root system $\Delta(\underline{g}^C, \underline{h}^C)$ of $\underline{g}^C$ with respect to the complexified Cartan subalgebra $\underline{h}^C$. Then the automorphism group $\mathrm{Aut}(\tilde{\Pi}, \sigma)$ of the Satake diagram $(\tilde{\Pi}, \sigma)$ of $\underline{g}$ extends naturally to the subgroup $A_\sigma$ of Aut $\underline{g}$ (Matsumoto [13]). Let $\Pi = \{\alpha_1, \cdots, \alpha_r\}$ be the restricted fundamental system on

$\underline{a}$ associated with $(\tilde{\Pi},\sigma)$. Let $\{\varepsilon_1,\cdots,\varepsilon_r\}$ be the basis of $\underline{a}$ dual to $\Pi$ with respect to the Killing form of $\underline{g}$. Put $\omega_j = \mathrm{Ad}\exp\pi i\varepsilon_j$ $(1\leq j\leq r)$, and let $Q_1$ be the finite abelian subgroup of $\mathrm{Aut}\,\underline{g}$ generated by $\omega_1$, $\cdots,\omega_r$. Put $Q_0 = Q_1\cap \mathrm{Ad}\,\underline{g}$. Then we have [13]

(3.6) $$\mathrm{Aut}\,\underline{g}/\mathrm{Ad}\,\underline{g} \simeq A_\sigma Q_1/Q_0.$$

Lemma 3.2. Let $\mathrm{Aut}_{gr}\underline{g}$ denote the subgroup of $\mathrm{Aut}\,\underline{g}$ of all grade-preserving automorphisms of $\underline{g}$. Then

(3.7) $$\mathrm{Aut}\,\underline{g} = (\mathrm{Aut}_{gr}\,\underline{g})\,\mathrm{Ad}\,\underline{g}.$$

Proof. By (3.6), it suffices to show that $A_\sigma$ and $Q_1$ are contained in $\mathrm{Aut}_{gr}\,\underline{g}$. Let $\Delta(\underline{g},\underline{a})$ be the restricted root system corresponding to $\Pi$ (cf. (2.6)). Then we have

(3.8) $$\underline{g} = \underline{c}(\underline{a}) + \sum_{\gamma\in\Delta(\underline{g},\underline{a})} \underline{g}^\gamma,$$

where $\underline{c}(\underline{a})$ is the centralizer of $\underline{a}$ in $\underline{g}$. By the definition of $\omega_j$, it is easy to see that $\omega_j$ leaves $\underline{c}(\underline{a})$ and each root space stable. The graded subspace $g_0$ is spanned by $\underline{c}(\underline{a})$ and some of the root spaces $g^\gamma$, while $g_1$ and $g_{-1}$ are spanned by some of the remaining root spaces, respectively ([8]). Therefore $\omega_j$ leaves $g_0$ and $g_{\pm1}$ stable, which implies $Q_1\subset \mathrm{Aut}_{gr}\,\underline{g}$. To proceed further, we should first note that $\underline{g}$ is the Lie algebra of the full holomorphic automorphism group $G(D(V))$ of the irreducible symmetric Siegel domain $D(V)$ (Koecher [11]). By checking the Satake diagram of $\underline{g}$ for each domain $D(V)$, we see that $A_\sigma \neq (1)$ if and only if $\underline{g}$ is either $su(r,r)$, $r\geq 2$ or $so(2,2m-2)$, $m\geq 3$.

Consider first the case $g = su(r,r), r\geq 3$. Here we exclude the case $r=2$, because of the isomorphism $su(2,2) \simeq so(2,4)$. We realize $su(r,r)$ in the following form:

(3.9) $$su(r,r) = \{X\in sl(2r,C):\ {}^t\overline{X}J + JX = 0\},$$

where $J = (a_{ij})$ is the $2r\times 2r$ matrix with $a_{ij} = \delta_{i,2r+1-j}$. From the classification of simple graded Lie algebra [8], it follows that, for $su(r,r)$ realized in the form (3.9), the gradation (2.3) is given by the one in the Table I [8]. In that case, $\underline{a}$ and $\underline{h}^C$ are given by certain sets of diagonal matrices. We choose the parameters $x_1,\cdots,x_{2r}$ such that $\Delta(\underline{g}^C,\underline{h}^C) = \{\pm(x_i - x_j): 1\leq i<j\leq 2r\}$. Let $\widetilde{\Pi} = \{\alpha_1,\cdots,\alpha_{2r-1}\}$, $\alpha_i = x_i - x_{i+1}$. The Satake diagram $(\widetilde{\Pi},\sigma)$ is given by

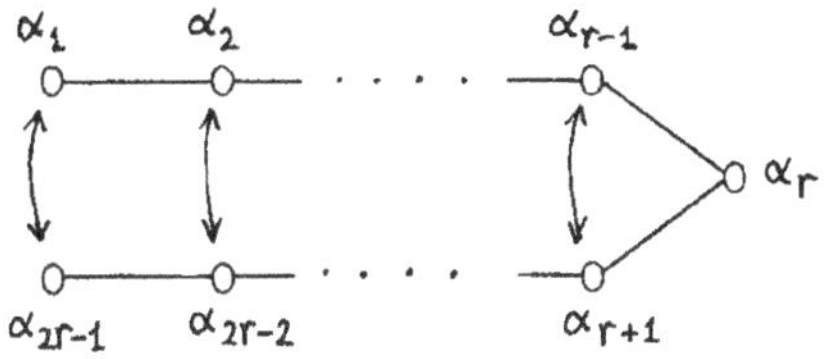


$\mathrm{Aut}(\widetilde{\Pi},\sigma)$ is generated by the permutation s defined by $s(\alpha_i) = \alpha_{2r-i}$ $(1\leq i\leq 2r-1)$ or equivalently $s(x_i) = -x_{2r+1-i}$. It can be seen that the element $\xi_s \in A_\sigma$ corresponding to s is given by

$$(3.10) \qquad \xi_s(X) = -J^tXJ, \qquad X \in su(r,r).$$

Therefore we conclude easily that $\xi_s$ is grade-preserving.

Next we consider the case $g = so(2,2m-2), m\geq 3$. We realize $so(2,2m-2)$ in the form:

$$(3.11) \qquad so(2,2m-2) = \{X \in gl(2m,R): {}^tXA + AX = 0\},$$

where $A = (a_{ij})$ with $a_{ij} = \delta_{i,2m+1-j}$ for $i = 1,2,2m-1,2m$ and $a_{ij} = \delta_{ij}$ for $3\leq i\leq 2m-2$. With respect to the realization (3.11) of $so(2,2m-2)$, the gradation (2.3) is found in the Table I [8]. In that case, $\underline{a}$ and $\underline{h}^C$ are given by certain sets of diagonal matrices. We choose the parameters $x_1,\cdots,x_m$ such that $\Delta(\underline{g}^C,\underline{h}^C)$ is given by the set $\{\pm(x_i - x_j), \pm(x_i + x_j): 1\leq i<j\leq m\}$. Let $\widetilde{\Pi} = \{\alpha_1,\cdots,\alpha_m\}$, $\alpha_i = x_i - x_{i+1}$ $(1\leq i\leq m-1)$, $\alpha_m = x_{m-1} + x_m$. The Satake diagram $(\widetilde{\Pi},\sigma)$ is given by

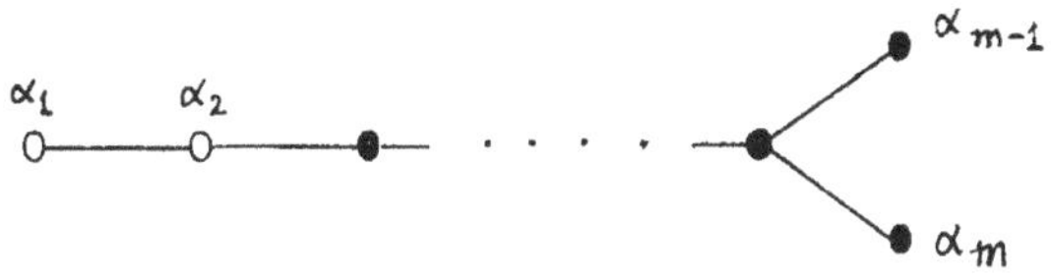


$\mathrm{Aut}(\widetilde{\Pi},\sigma)$ is generated by the transposition s which interchanges

$\alpha_{m-1}$ with $\alpha_m$. This action is rewritten as $s(x_i) = x_i$ $(1 \leq i \leq m-1)$, $s(x_m) = -x_m$. It can be seen that the action of the element $\xi_s \in A_\sigma$ corresponding to s is the operation which interchanges the m-th column with the (m+1)-th column and the m-th row with the (m+1)-th row in each matrix in so(2,2m-2). Therefore we can conclude that $\xi_s$ is grade-preserving. q.e.d.

3.3. We shall go back to the situation in 3.1. Let I(D(V)) denote the full isometry group of D(V) with respect to the Bergman metric. I(D(V)) contains the full holomorphic automorphism group G(D(V)) as an open subgroup. Thus, from what was noted in the proof of Lemma 3.2, it follows that Lie I(D(V)) = $\underline{g}$. Consider the adjoint representation $Ad_{\underline{g}}$ of I(D(V)) on $\underline{g}$. Then, by a theorem of E. Cartan (see [17]), one has

$$(3.12)\qquad I(D(V)) \simeq Ad_{\underline{g}} I(D(V)) = \operatorname{Aut} \underline{g}\,.$$

On the other hand, G(V) is a subgroup of G(D(V)). The conjugation $-\theta$ leaves D(V) stable, and it can be seen that $-\theta \in I(D(V))$. Therefore Str A is a subgroup of I(D(V)) (cf. (3.5)).

Lemma 3.3. $Ad_{\underline{g}} \operatorname{Str} A = \operatorname{Aut}_{gr} \underline{g}$.

Proof. Note that $L_{-e} \in \underline{p}_0$ is a symmetric operator on A with respect to $t_A$. Hence it follows from (1.4) that $L_{-e}$ is the characteristic element ([8]) of the graded Lie algebra $\underline{g}$. Let $h \in \operatorname{Str} A$. Then one has $hL_{-e}h^{-1} = L_{-e}$, or equivalently, $(Ad_{\underline{g}} h)L_{-e} = L_{-e}$, which implies that $Ad_{\underline{g}} h \in \operatorname{Aut}_{gr} \underline{g}$. To prove the converse inclusion, let $g \in \operatorname{Aut}_{gr} \underline{g}$, and set $g_\pm = g|_{\underline{g}_{\pm 1}}$ and $g_0 = g|_{\underline{g}_0}$. Then, by Lemma1.1 we have

$$(3.13)\qquad g_- \in \operatorname{Str} A\,, \quad g_0 = Ad_{\underline{g}_0} g_-, \quad g_+ = \tau((g_-)^*)^{-1}\tau\,.$$

Consider the Lie subgroup $(\exp \underline{g}_{-1}) \operatorname{Str} A$ of the affine transformation group $\operatorname{Aff}(\underline{g}_{-1}) = (\exp \underline{g}_{-1}) GL(\underline{g}_{-1})$. Then the Lie algebra of that subgroup coincides with the subalgebra $\underline{g}_{-1} + \underline{g}_0$. In view of this, it follows that

$$(3.14)\qquad g_-a = (\mathrm{Ad}_{\underline{g}_{-1}} g_-)a, \qquad a \in g_{-1}.$$

Let B be the Killing form of $\underline{g}$. Then we have ([17])

$$(3.15)\qquad B(a, \tau(b)) = -4t_A(a,b), \qquad a,b \in \underline{g}_{-1} = A.$$

For $\lambda \in \mathrm{End}\, g_{-1}$, we denote by $\lambda^{\vee}$ the adjoint operator of $\lambda$ with respect to B. Since B, restricted to $\underline{g}_{-1} \times \underline{g}_1$, is nondegenerate, it follows that

$$(3.16)\qquad \lambda^{\vee} = \tau\lambda^*\tau .$$

By using (3.13), (3.14), (3.16) and the invariance of B under Aut $\underline{g}$, we obtain

$$(3.17)\qquad g_+ = \tau((g_-)^*)^{-1}\tau = \tau((\mathrm{Ad}_{\underline{g}_{-1}} g_-)^*)^{-1}\tau$$
$$= ((\mathrm{Ad}_{g_{-1}} g_-)^{\vee})^{-1} = \mathrm{Ad}_{\underline{g}_1} g_- .$$

(3.13), (3.14) and (3.17) imply that $g \in \mathrm{Ad}_{\underline{g}}\mathrm{Str}\, A$. q.e.d.

Proposition 3.4. Let V be the cone of positive definite elements in a compact simple Jordan algebra A. Let D(V) be the Siegel domain of the first kind given in (3.4). Let G(D(V)) and I(D(V)) be, respectively, the full holomorphic automorphism group and the full isometry group of D(V) (with respect to the Bergman metric). Then we have

$$(3.18)\qquad I(D(V)) = (\mathrm{Str}\, A)\, G^0(D(V)) = G(D(V)) \amalg (-\theta)G(D(V))$$
$$= G(D(V))\{1,-\theta\} \qquad \text{(semi-direct).}$$

Here $G^0(D(V))$ is the identity component of G(D(V)), and $-\theta$ is the anti-holomorphic automorphism of D(V) defined by $(-\theta)(z) = -\bar{z}$. Also we have

$$(3.19)\qquad G(D(V)) = G(V)G^0(D(V)).$$

Proof. The first equality of (3.18) follows from (3.7),

(3.12) and Lemma 3.3. The set $G(V)$ (resp. $(-\theta)G(V)$) is the totality of C-linear (resp. anti-linear) automorphisms of $D(V)$. Hence, from (3.5) we have

$$\mathrm{Str}\,A \cap G(D(V)) = G(V). \tag{3.20}$$

And $I(D(V))/G(D(V)) = (\mathrm{Str}\,A)\,G(D(V))/G(D(V)) \simeq \mathrm{Str}\,A/G(V) \simeq \{1,-\theta\}$. (3.19) follows from (3.18) and (3.20). q.e.d.

## §4. The $\xi$-equivariant action of the isometry group

4.1. Let $A$ be a compact simple Jordan algebra of degree $r$, and let $\underline{g}$ be the simple graded Lie algebra given in (2.3). We denote the subalgebra $\underline{g}_0 + \underline{g}_1$ by $\underline{u}$. Consider the complexification $\underline{g}^C$ of $\underline{g}$:

$$\underline{g}^C = \underline{g}_{-1}^C + \underline{g}_0^C + \underline{g}_1^C, \tag{4.1}$$

which is a complex simple graded Lie algebra. Let $\bar{G} = \mathrm{Ad}\,\underline{g}^C$ and let $\bar{U}$ denote the normalizer $N_{\bar{G}}(\underline{u}^C)$ of $\underline{u}^C = \underline{g}_0^C + \underline{g}_1^C$ in $\bar{G}$, that is, $N_{\bar{G}}(\underline{u}^C) = \{g \in \bar{G}: g\underline{u}^C = \underline{u}^C\}$. $\bar{U}$ is connected and $\mathrm{Lie}\,\bar{U} = \underline{u}^C$. The complex coset space $M = \bar{G}/\bar{U}$ is the compact irreducible Hermitian symmetric space dual to the symmetric Siegel domain $D(V)$. Let $o_M$ be the origin of the coset space $M$. Let $\underline{g}^{CR}$ be the real simple Lie algebra obtained from $\underline{g}^C$ by the scalar restriction to R. Let $\tilde{G}(M)$ be the group of all $\pm$-holomorphic (i.e. holomorphic or anti-holomorphic) automorphisms of $M$, and $G(M)$ be the subgroup of $\tilde{G}(M)$ of all holomorphic automorphisms. Then $\tilde{G}(M)$ and $G(M)$ are realized in $\mathrm{Aut}\,\underline{g}^{CR}$ and $\mathrm{Aut}\,\underline{g}^C$, respectively, as the following subgroups ([22]):

$$\tilde{G}(M) = (\mathrm{Aut}_{gr}\,\underline{g}^{CR})\bar{G}, \qquad G(M) = (\mathrm{Aut}_{gr}\,\underline{g}^{C})\bar{G}, \tag{4.2}$$

where $\mathrm{Aut}_{gr}$ denotes the subgroup of all grade-preserving automorphisms. Let $\tilde{U}(M)$ and $U(M)$ be the isotropy subgroups of $\tilde{G}(M)$ and $G(M)$ at $o_M$, respectively. Then, considering $\bar{U} = N_{\bar{G}}(\underline{u}^C)$ and (4.2), we have

$$(4.3)\qquad \begin{aligned} \tilde{U}(M) &= N_{\mathrm{Aut}\underline{g}^{CR}}(\underline{u}^C) = (\mathrm{Aut}_{gr}\underline{g}^{CR})\exp \underline{g}_1^C, \\ U(M) &= N_{\mathrm{Aut}\underline{g}^C}(\underline{u}^C) = (\mathrm{Aut}_{gr}\underline{g}^C)\exp \underline{g}_1^C, \end{aligned}$$

where $N_{\mathrm{Aut}\underline{g}^{CR}}(\underline{u}^C)$ and $N_{\mathrm{Aut}\underline{g}^C}(\underline{u}^C)$ are the normalizers of $\underline{u}^C$ in $\mathrm{Aut}\,\underline{g}^{CR}$ and $\mathrm{Aut}\,\underline{g}^C$, respectively. M can be expressed as

$$(4.4)\qquad M = \tilde{G}(M)/\tilde{U}(M) = G(M)/U(M).$$

Since $-\theta$ lies in $\mathrm{Str}\,A \subset I(D(V))$ (cf. (3.5),(3.18)), one can consider

$$(4.5)\qquad \tilde{\theta} = \mathrm{Ad}_{\underline{g}^{CR}}(-\theta).$$

As was shown in 3.1, $\mathrm{Str}\,A$ is viewed as a subgroup of $\mathrm{Str}\,A^{CR}$, and $G(V)$ is viewed as a subgroup of $\mathrm{Str}\,A^C$. Note that $\underline{g}(A^C) = \underline{g}^C$ and $\underline{g}(A^{CR}) = \underline{g}^{CR}$. Then, from (1.7) and the proof of Lemma 3.3, it follows that

$$(4.6)\qquad \varphi_{A^{CR}}(g) = \mathrm{Ad}_{\underline{g}^{CR}}(g), \qquad g \in \mathrm{Str}\,A.$$

Lemma 4.1. The following equalities are valid:

$$(4.7)\qquad \mathrm{Aut}_{gr}\underline{g}^{CR} = \mathrm{Aut}_{gr}\underline{g}^C \amalg \tilde{\theta}(\mathrm{Aut}_{gr}\underline{g}^C),$$

$$(4.8)\qquad \mathrm{Aut}_{gr}\underline{g}^{CR} \cap G(M) = \mathrm{Aut}_{gr}\underline{g}^C,$$

$$(4.9)\qquad \tilde{G}(M) = G(M) \amalg \tilde{\theta}G(M) = G(M)\{1, \tilde{\theta}\} \qquad \text{(semi-direct)}.$$

Proof. Applying the isomorphism $\varphi_{A^{CR}}$ to the both sides of (3.3), we have (4.7) by Lemma 1.1. The inclusion $\supset$ in (4.8) is trivial. Therefore, by (4.7) we have

$$[\mathrm{Aut}_{gr}\underline{g}^{CR} : \mathrm{Aut}_{gr}\underline{g}^{CR} \cap G(M)] \leq [\mathrm{Aut}_{gr}\underline{g}^{CR} : \mathrm{Aut}_{gr}\underline{g}^{C}] = 2.$$

If the value of the term in the left side is equal to 1, then $\mathrm{Aut}_{gr}\underline{g}^{CR}$ is a subgroup of $G(M)$, and hence it lies in $\mathrm{Aut}\,\underline{g}^{C}$. This is a contradiction. Consequently the value of the term in the left side is equal to 2, and so we have (4.8). It follows then that

$$\tilde{G}(M)/G(M) = (\mathrm{Aut}_{gr}\underline{g}^{CR})G(M)/G(M) \simeq \mathrm{Aut}_{gr}\underline{g}^{CR}/\mathrm{Aut}_{gr}\underline{g}^{CR} \cap G(M)$$

$$= \mathrm{Aut}_{gr}\underline{g}^{CR}/\mathrm{Aut}_{gr}\underline{g}^{C} \simeq \{1, \tilde{\theta}\},$$

which proves (4.9). q.e.d.

Lemma 4.2. The representation $\mathrm{Ad}_{\underline{g}^{CR}}$ is an injective homomorphism of $I(D(V))$ into $\tilde{G}(M)$. Moreover we have

(4.10) $$\mathrm{Ad}_{\underline{g}^{CR}} G(D(V)) \subset G(M).$$

Proof. The injectivity of $\mathrm{Ad}_{\underline{g}^{CR}}$ was already noted in (3.12). By (3.18),(4.6) and Lemma 1.1, we have

$$\mathrm{Ad}_{\underline{g}^{CR}} I(D(V)) = (\mathrm{Ad}_{\underline{g}^{CR}} \mathrm{Str}\, A)\mathrm{Ad}_{\underline{g}^{CR}} G^{0}(D(V))$$

$$\subset \varphi_{A^{CR}}(\mathrm{Str}\, A^{CR})\mathrm{Ad}_{\underline{g}^{CR}}\underline{g} \subset (\mathrm{Aut}_{gr}\underline{g}^{CR})\mathrm{Ad}\,\underline{g}^{C} = \tilde{G}(M).$$

Similarly, by using (3.19) instead of (3.18), we get (4.10). q.e.d.

Later on, for simplicity, we will write $\rho$ for $\mathrm{Ad}_{\underline{g}^{CR}}$.

4.2. The unit element $e$ of $A$ lies in the cone $V$. Let $K(V)$ be the isotropy subgroup of $G(V)$ at $e$, and let $K(D(V))$ be the isotropy subgroup of $G(D(V))$ at the point $ie \in D(V)$.

Lemma 4.3. $K(D(V)) = K(V)K^{0}(D(V))$, where $K^{0}(D(V))$ denotes the identity component of $K(D(V))$.

Proof. Choose a maximal connected R-triangular subgroup $T_1$

of G(V) such that

$$(4.11)\qquad G(V) = K(V)\,T_1 , \qquad K(V)\cap T_1 = (1).$$

Let T be an Iwasawa subgroup ([6]) of $G^0(D(V))$ containing $T_1$. Then we have

$$(4.12)\qquad G(D(V)) = K(D(V))\,T = T\,K(D(V)) , \quad K(D(V))\cap T = (1).$$

Let $k \in K(D(V))$. One can write it in the form $k = g_1g_2$, where $g_1 \in G(V)$ and $g_2 \in G^0(D(V))$ (cf. (3.19)). Moreover, by using (4.11) and (4.12), one can write $g_1$ and $g_2$ as $g_1 = k_1t_1$, $g_2 = k_2t_2$, where $k_1 \in K(V)$, $t_1 \in T_1$, $k_2 \in K^0(D(V))$, $t_2 \in T$. Therefore $k = k_1(t_1k_2)t_2$. Furthermore, $t_1k_2 \in T_1K^0(D(V)) \subset TK^0(D(V)) = K^0(D(V))T$. Hence there exist the elements $k_2' \in K^0(D(V))$, $t_1' \in T$ such that $t_1k_2 = k_2't_1'$. Consequently, $k = k_1(k_2't_1')t_2 = (k_1k_2')(t_1't_2)$. So we have $(k_1k_2')^{-1}k = t_1't_2$; but the left-hand side lies in K(D(V)), while the right-hand side lies in T. Thus $(k_1k_2')^{-1}k$ lies in $K(D(V))\cap T = (1)$, which implies $k = k_1k_2' \in K(V)K^0(D(V))$. The proof was now completed. q.e.d.

Lemma 4.4. $\rho(K(D(V))) \subset \rho(\exp ie)U(M)\rho(\exp -ie)$.

Proof. We have already known(Tanaka [24])

$$(4.13)\qquad \rho(K^0(D(V))) \subset \rho(\exp ie)U(M)\rho(\exp -ie).$$

So, in view of Lemma 4.3, it is enough to prove

$$(4.14)\qquad \rho(K(V)) \subset \rho(\exp ie)U(M)\rho(\exp -ie).$$

By using (4.3), the right-hand side can be rewritten as $N_{\mathrm{Aut}\underline{g}^C}(\rho(\exp ie)\underline{u}^C)$. Let $a \in K(V)$. Then we have (cf. (3.14))

$$a(\exp ie)a^{-1} = \exp((\mathrm{Ad}\, a)(ie)) = \exp i(\mathrm{Ad}\, a)e = \exp iae = \exp ie,$$

which implies that a commutes with exp ie. On the other hand, by using (4.6) and Lemma 1.1, we have

$$\rho(a) \in \rho(K(V)) \subset \rho(G(V)) \subset \varphi_{A^{CR}}(\mathrm{Str}\, A^C) = \mathrm{Aut}_{gr}\, \underline{g}^C .$$

From this it follows that $\rho(a)\underline{u}^C = \underline{u}^C$. Thus

$$\rho(a)(\rho(\exp ie)\underline{u}^C) = \rho(\exp ie)\rho(a)\underline{u}^C = \rho(\exp ie)\underline{u}^C .$$

Therefore we have $\rho(a) \in N_{\mathrm{Aut}\,\underline{g}^C}(\rho(\exp ie)\underline{u}^C)$. This proves the lemma. q.e.d.

Let L(D(V)) be the isotropy subgroup of I(D(V)) at the point $ie \in D(V)$. By using (3.18), we see

$$(4.15) \qquad L(D(V)) = K(D(V)) \amalg (-\theta)K(D(V))$$

$$= K(D(V))\{1,-\theta\} \qquad \text{(semi-direct)}.$$

Lemma 4.5. $\rho(L(D(V)) \subset \rho(\exp ie)\tilde{U}(M)\rho(\exp -ie)$.

Proof. By (4.15) and Lemma 4.4, it suffices to show that

$$(4.16) \qquad \tilde{\theta} = \rho(-\theta) \in \rho(\exp ie)\tilde{U}(M)\rho(\exp -ie).$$

Since $-\theta$ leaves the point ie fixed, it follows that $-\theta$ commutes with exp ie (cf. the proof of Lemma 4.4). On the other hand, since $-\theta \in \mathrm{Str}\, A^{CR}$, we have that $\tilde{\theta} \in \mathrm{Aut}_{gr}\, \underline{g}^{CR}$ (cf. Lemma 1.1). Thus $\tilde{\theta}\,\underline{u}^C = \underline{u}^C$. From these two, we can go along the same line as in the proof of Lemma 4.4 to get $\tilde{\theta} \in N_{\mathrm{Aut}\underline{g}^{CR}}(\rho(\exp ie)\underline{u}^C)$. So, considering (4.3), we obtain the assertion of the lemma. q.e.d.

Let us define the map $\xi$ of $\underline{g}^C_{-1}$ (= $A^C$) to M by putting

$$(4.17) \qquad \xi(X) = \rho(\exp X)\, o_M , \qquad X \in \underline{g}^C_{-1}.$$

It is known [24] that $\xi$ is a holomorphic open dense imbedding of $\underline{g}^C_{-1}$ into M. $\xi$ is the so-called Tanaka imbedding. We will prove that $\xi$ is equivariant on D(V) under the group I(D(V)).

Proposition 4.6. Let D(V) be an irreducible symmetric Siegel domain of the first kind and M be its compact dual. Let $\rho$ be the injective homomorphism of the isometry group I(D(V)) into the group $\tilde{G}(M)$ of $\pm$-holomorphic automorphisms of M (cf. 4.1). Let $\xi$ be the Tanaka imbedding of the ambient vector space $\underline{g}^C_{-1}$ into M. Then we have

$$(4.18)\qquad \xi(gp) = \rho(g)\xi(p), \qquad g \in I(D(V)),\ p \in D(V).$$

Proof. The isotropy subgroup of $\tilde{G}(M)$ at the point $\xi(ie) = \rho(\exp ie)\, o_M$ is given by $\rho(\exp ie)\tilde{U}(M)\rho(\exp -ie)$. By Lemma 4.5, we have

$$(4.19)\qquad \rho(L(D(V)))\xi(ie) = \xi(ie)$$

On the other hand, similarly as in (4.12), we get

$$(4.20)\qquad I(D(V)) = L(D(V))\,T = T\,L(D(V)),$$

$$L(D(V)) \cap T = (1).$$

By using (4.19) and (4.20), we can make the similar arguments as in the proof of Lemma 3.4 in Tanaka [24] so as to obtain (4.18). q.e.d.

For later considerations we need the following

Lemma 4.7. $\xi(-\theta)(X) = \tilde{\theta}\xi(X)$ holds for each $X \in \underline{g}^C_{-1}$.

Proof. As was remarked in the proof of Lemma 4.5, we have $\tilde{\theta}\underline{u}^C = \underline{u}^C$. This implies that $\tilde{\theta} \in N_{\mathrm{Aut}\,\underline{g}^{CR}}(\underline{u}^C) = \tilde{U}(M)$. In view of this,

$$\tilde{\theta}\xi(X) = \rho(-\theta)\rho(\exp X)\, o_M = \rho(-\theta)\rho(\exp X)\rho(-\theta)^{-1} o_M$$

$$= \rho((-\theta)(\exp X)(-\theta)^{-1})\, o_M = \rho(\exp(\mathrm{Ad}\ -\theta)X)\, o_M$$

$$= \rho(\exp(-\theta)X)\, o_M = \xi((-\theta)X). \quad \text{q.e.d.}$$

## §5. Causal structures on Silov boundaries

5.1. Let $D_{can}$ be a canonically realized irreducible bounded symmetric domain of tube type, that is, the Harish-Chandra realization of an irreducible noncompact Hermitian symmetric space whose restricted root system is of type C. Let $S_{can}$ be its Silov boundary. $D_{can}$ is holomorphically equivalent to a Siegel domain D(V) of the first kind, where V is the cone of positive definite elements in a compact simple Jordan algebra A. Let us consider the domain

$$D = \xi(D(V)) \tag{5.1}$$

in the compact Hermitian symmetric space M (cf. §4) dual to $D_{can}$. By Proposition 4.6, the group $\rho(G(D(V)))$ acts on D as the full group of holomorphic automorphisms. The group $\rho(I(D(V)))$ acts on D as the full isometry group with respect to the metric induced by the Bergman metric on D(V). In the sequel, we will write G(D) and I(D) for $\rho(G(D(V)))$ and $\rho(I(D(V)))$, respectively. Let $G^0(D)$ denote the identity component of G(D). Consider the $G^0(D)$-orbit through the origin $o_M$ of M:

$$S = G^0(D)\, o_M \ . \tag{5.2}$$

Then S is a compact submanifold of M ([14]). The pair (D,S) is also obtained from the pair $(D_{can}, S_{can})$ by applying the inverse mapping of the Harish-Chandra realization succeeded by the Cayley transform (Korányi-Wolf [12]). Hence S is the Silov boundary of the domain D. Originally we wish to work on the pair $(D_{can}, S_{can})$. But, in view of the above situation, we may work on the pair (D,S) rather that $(D_{can}, S_{can})$.

Lemma 5.1. The action of I(D) on M leaves the Silov boundary S stable. Let B(D) and H(D) be the isotropy subgroups of I(D) and G(D) at $o_M \in S$, respectively. Then

$$B(D) = N_{I(D)}(\underline{u}) = (\mathrm{Aut}_{gr}\underline{g})(\mathrm{Ad}_{\underline{g}} \exp \underline{g}_1) \ , \tag{5.3}$$

$$H(D) = N_{G(D)}(\underline{u}) = (\mathrm{Ad}_{\underline{g}} G(V))(\mathrm{Ad}_{\underline{g}} \exp \underline{g}_1) \ , \tag{5.4}$$

where $N_{I(D)}(\underline{u})$ (resp. $N_{G(D)}(\underline{u})$) is the normalizer of $\underline{u}$ in I(D) (resp. G(D)).

Proof. Let $g \in I(D)$. Since the subgroup I(D) of $\tilde{G}(M)$ normalizes $G^0(D)$, we have $gS = gG^0(D)o_M = G^0(D)go_M$. The subset gS of M is a $G^0(D)$-orbit which is diffeomorphic to S. It follows from a result of Takeuchi [21] that S is a unique closed $G^0(D)$-orbit in M. Therefore we get gS = S. To prove (5.3), note that $I(D) = \rho(I(D(V)))$ $= \operatorname{Aut} \underline{g}$ (cf. (3.12)). Then we have

$$B(D) = I(D) \cap \tilde{U}(M) = \operatorname{Aut} \underline{g} \cap N_{\operatorname{Aut}\underline{g}^{CR}}(\underline{u}^C) = N_{\operatorname{Aut}\underline{g}}(\underline{u}^C) = N_{\operatorname{Aut}\underline{g}}(\underline{u}) .$$

It is easy to see that the last one is identical with the group $(\operatorname{Aut}_{gr}\underline{g})(\operatorname{Ad}_{\underline{g}}\exp \underline{g}_1)$. To prove (5.4), let us recall the inclusions $G(D) = \rho(G(D(V))) \subset G(M) \subset \operatorname{Aut} \underline{g}^C$ (cf. (4.10)). We have then

$$H(D) = G(D) \cap U(M) = G(D) \cap N_{\operatorname{Aut}\underline{g}^C}(\underline{u}^C) = N_{G(D)}(\underline{u}^C) = N_{G(D)}(\underline{u}).$$

Consequently we have to prove

$$(5.5) \qquad N_{G(D)}(\underline{u}) = (\operatorname{Ad}_{\underline{g}}G(V))(\operatorname{Ad}_{\underline{g}}\exp \underline{g}_1 ).$$

Let $g \in N_{G(D)}(\underline{u})$, and write it as $g = g_1 g_2$, where $g_1 \in \operatorname{Ad}_{\underline{g}}G(V)$, $g_2 \in \operatorname{Ad} \underline{g}$ (cf. (3.19)). Then, by using (5.3), one has

$$\begin{aligned} g_2 = g_1^{-1} g \in N_{\operatorname{Aut}\underline{g}}(\underline{u}) \cap \operatorname{Ad}\underline{g} &= (\operatorname{Aut}_{gr}\underline{g})(\operatorname{Ad}_{\underline{g}}\exp \underline{g}_1) \cap \operatorname{Ad}\underline{g} \\ &= (\operatorname{Aut}_{gr}\underline{g} \cap \operatorname{Ad}\underline{g}) \operatorname{Ad}_{\underline{g}}\exp \underline{g}_1 \subset (\operatorname{Aut}_{gr}\underline{g} \cap \operatorname{Ad}_{\underline{g}}G(D(V)))\operatorname{Ad}_{\underline{g}}\exp \underline{g}_1 . \end{aligned}$$

In view of (3.20) and Lemma 1.1, it follows that $\operatorname{Aut}_{gr}\underline{g} \cap \operatorname{Ad}_{\underline{g}}G(D(V))$ $= \operatorname{Ad}_{\underline{g}}G(V)$. This proves (5.5). q.e.d.

As a special case of a result of Tanaka [25], we obtain the following lemma from Lemma 5.1.

Lemma 5.2. The action of I(D) on S is effective. Thus I(D) and consequently G(D) are subgroups of the group Diffeo(S) of diffeomorphisms of S.

5.2. We give here basic facts on causal structures (cf. Segal [20]). Let W be a finite-dimensional vector space over R. A subset C of W is called a causal cone, if C is a nonzero closed convex cone and satisfies the condition $C \cap (-C) = (0)$. By the automorphism group of C we mean the group

$$G(C) = \{ g \in GL(W) : gC = C \}. \tag{5.6}$$

Let C be a causal cone in W. Hereafter we shall always assume that the interior Int C of C is not empty and that Int C is described by a finite number of inequalities, that is,

$$\operatorname{Int} C = \{ x \in W : f_1(x) > 0, \cdots, f_s(x) > 0 \}, \tag{5.7}$$

where $f_1, \cdots, f_s$ are smooth functions on W. It follows from (2.18) and Lemma 2.7 (3) that the closure $\overline{V}$ of the cone $V \subset A$ considered in §2 is a causal cone. In this case we have $\operatorname{Int} \overline{V} = V$ (cf. Theorem 2.8), and V satisfies the condition (5.7) in which $f_1, \cdots, f_s$ can be chosen to be certain polynomial functions on A (cf. Vinberg [30]).

Let M be a smooth manifold. We say that $\underline{C}$ is a causal structure on M, if $\underline{C}$ is a smooth assignment to each point $p \in M$ of a causal cone $C_p$ in the tangent space $T_p(M)$ at p. Here, by the terminology "smooth" we mean that the functions which describe Int $C_p$ in the sense of (5.7) depend smoothly on the points p. Sometimes we write $\underline{C} = \{C_p\}_{p \in M}$ for $\underline{C}$. The pair $(M, \underline{C})$ is called a causal manifold. Let $(M, \underline{C})$ and $(M', \underline{C}')$ be two causal manifolds, and let $\underline{C} = \{C_p\}_{p \in M}$ and $C' = \{C'_q\}_{q \in M'}$. Let $\varphi$ be a smooth map of M to N. $\varphi$ is said to be a causal map (resp. anti-causal map), if, for each point $p \in M$, $\varphi_{*p} C_p = C'_{\varphi(p)}$ (resp. $\varphi_{*p} C_p = -C_{\varphi(p)}$) holds.. A causal (resp. anti-causal) diffeomorphism $\varphi$ of $(M, \underline{C})$ onto itself is called a causal automorphism (resp. anti-causal automorphism). In this case, we also say that the causal structure $\underline{C}$ is $\varphi$-invariant. The totality of causal automorphisms of $(M, \underline{C})$ forms a group, which is called the causal automorphism group and is denoted by $G(M, \underline{C})$. The set of all anti-causal automorphisms of $(M, \underline{C})$ is denoted by $G_-(M, \underline{C})$.

5.3. Let M = G/H be a homogeneous space of a Lie group G by a closed subgroup H. Let o be the origin of M. Let $\sigma$ be the linear isotropy representation of H on the tangent space $T_o(M)$. Suppose that we are given a causal cone C in $T_o(M)$. Then we have

Lemma 5.3. Suppose that the linear isotropy group $\sigma(H)$ is contained in the automorphism group G(C). Then there exists a unique G-invariant causal structure $\underline{C} = \{C_p\}_{p\in M}$ satisfying $C_o = C$.

Proof. Let $p \in M$ and choose $g \in G$ such that $p = g\,o$. We define the causal cone $C_p$ to be $g_{*o}(C)$. We have to verify that $C_p$ is independent of the choice of $g \in G$. Choose another $g' \in G$ such that $p = g'o$. Then $h := g'^{-1}g \in H$, and we have

$$(g'_*)^{-1}g_*(C) = (g'^{-1})_*g_*(C) = (g'^{-1}g)_*(C) = h_*(C) = \sigma(h)C.$$

But, since $\sigma(h)$ lies in G(C) by the assumption, $\sigma(h)C$ is equal to C. This implies that $g_{*o}(C) = g'_{*o}(C)$, and $C_p$ is well-defined. The assignment $\underline{C}: p \mapsto C_p$ is clearly smooth, and also $\underline{C}$ is G-invariant. q.e.d.

The G-invariant causal structure $\underline{C}$ on M just constructed is said to be modelled after C. We will now introduce a causal structure on the Silov boundary of an irreducible symmetric bounded domain of tube type.

Proposition 5.4. The Silov boundary S has a unique G(D)-invariant causal structure $\underline{C}$ modelled after the causal cone $\overline{V}$.

Proof. By Lemma 5.1, we can express S as the coset space

$$(5.8) \qquad S = G(D)/\mathrm{Ad}_{\underline{g}}G(V)\mathrm{Ad}_{\underline{g}}\exp \underline{g}_1$$
$$= \mathrm{Ad}_{\underline{g}}G(D(V))/\mathrm{Ad}_{\underline{g}}G(V)\mathrm{Ad}_{\underline{g}}\exp \underline{g}_1 .$$

If we identify the Lie algebra Lie G(D) with $\underline{g}$, then the Lie algebra of the isotropy subgroup is identified with $\underline{g}_0 + \underline{g}_1$. So the tangent space $T_{o_M}(S)$ can be identified with $\underline{g}_{-1}$. Under this identification, the linear isotropy group $\sigma(\mathrm{Ad}_{\underline{g}}G(V)\mathrm{Ad}_{\underline{g}}\exp \underline{g}_1)$ acts on $\underline{g}_{-1}$ in the following manner

$$(5.9) \qquad \sigma((\mathrm{Ad}_{\underline{g}} a)(\mathrm{Ad}_{\underline{g}} \exp Y))X = aX, \qquad X \in \underline{g}_{-1},$$

where $a \in G(V)$ and $Y \in \underline{g}_1$ (cf. (3.14)). This implies that the linear isotropy group is no other than $G(V)$, which coincides with $G(\overline{V})$. The proposition is now a direct consequence of Lemma 5.3. q.e.d.

## §6. Causal automorphism groups of Silov boundaries

6.1. Let $S$ be the Silov boundary expressed as the coset space (5.8). By Lemma 5.2, the group $G(D)$ is a subgroup of Diffeo(S). Let us consider the frame bundle $F(S)$ of $S$. Let $x \in S$. Then the fiber of $F(S)$ over $x$ is viewed as the totality of linear isomorphisms of $\underline{g}_{-1}$ onto the tangent space $T_x(S)$ at $x$. Let $z_0 \in F(S)$ be the natural linear isomorphism in the proof of Proposition 5.4, through which we have identified $T_{o_M}(S)$ with $\underline{g}_{-1}$. The action of an element $g \in G(D)$ on $S$ is lifted to a bundle automorphism of $F(S)$, which will be denoted by $\bar{g}$. By Tanaka [23], the $G(D)$-orbit $Q$ through the point $z_0 \in F(S)$ is a $G(V)$-structure on $S$, $G(V)$ being the linear isotropy group of the coset space $S = G(D)/H(D)$ (cf. 5.3). We then have ([23])

$$(6.1) \qquad \bar{h}(z_0) = z_0\sigma(h), \qquad h \in H(D),$$

where $\sigma$ is the linear isotropy representation of $H(D)$ at $o_M$. One can consider the automorphism group $\mathrm{Aut}(S,Q)$ of the $G(V)$-structure $Q$ on $S$.

Lemma 6.1. Suppose $\dim D > 1$. Then

$$(6.2) \qquad \mathrm{Aut}(S,Q) = G(D).$$

Proof. This is a direct consequence of the results of Tanaka (cf. Theorem 2.7 [26], Proposition 9.3 [23] and Lemma 1.12 [27]). We note that originally his results are valid for the $I(D)$-orbit through $z_0$, but they also work for an orbit under an open subgroup of $I(D)$. We note in addition that the Tanaka's results [23], [26] are valid under the condition that $S$ is not diffeomorphic to a real or complex projective space. But this condition is equivalent to saying in our case that $\dim D > 1$. q.e.d.

Theorem 6.2. Let D be an irreducible symmetric bounded domain of tube type realized as the domain (5.1) in its compact dual, and G(D) be the full holomorphic automorphism group of D. Let S be the Silov boundary of D and $\underline{C}$ be the causal structure on S introduced in Proposition 5.4. Let $G(S,\underline{C})$ be the causal automorphism group. Suppose $\dim D > 1$. Then we have

$$G(S,\underline{C}) = G(D), \tag{6.3}$$

where the both sides are viewed as subgroups of Diffeo(S).

Proof.*) By the definition of the causal structure $\underline{C}$, we have $G(D) \subset G(S,\underline{C})$. Let $g \in G(S,\underline{C})$. Since G(D) acts transitively on S, one can find an element $a \in G(D)$ such that $ag(o_M) = o_M$. Put $g' = ag \in G(S,\underline{C})$. It follows that $g'_{*o_M} \overline{V} = \overline{V}$, from which we have $g'_{*o_M} \in G(V)$. Therefore there exists $h \in H(D)$ such that $h_{*o_M} = g'_{*o_M}$. This implies that $\overline{a}\overline{g}(z_0) = \overline{g}'(z_0) = \overline{h}(z_0)$. Consequently $\overline{g}(z_0) = \overline{a}^{-1}\overline{h}(z_0) \in Q$. Since G(D) acts transitively on S, this implies that $\overline{g}(Q) = Q$, or equivalently, $g \in \mathrm{Aut}(S,Q)$. (6.3) now follows from Lemma 6.1. q.e.d.

We remark that when S is a unitary group U(n), Segal [20] showed that the group SU(n,n)/center is contained in $G(U(n),\underline{C})$.

6.2. We want to consider the relation between anti-holomorphic automorphisms of D and anti-causal automorphisms of S. Consider the vector space $\underline{g}_{-1}$, which is the Silov boundary of the Siegel domain D(V). Let $\overline{V}_X$ be the parallel transport of the causal cone $\overline{V}$ along the vector $X \in \underline{g}_{-1}$. Let $\underline{C}_{-1}$ be the causal structure on $\underline{g}_{-1}$ obtained by assigning $\overline{V}_X$ to each point $X \in \underline{g}_{-1}$. Let Aff(D(V)) denote the group of all affine automorphisms of the Siegel domain D(V). Then we have

$$\mathrm{Aff}(D(V)) = (\exp \underline{g}_{-1})\, G(V). \qquad \text{(semi-direct)} \tag{6.4}$$

*) The proof given here was inspired by Takeuchi [22].

It follows easily from (6.4) that the causal structure $\underline{C}_{-1}$ is invariant under Aff(D(V)). We note ([14]) that

(6.5) $$\xi(\underline{g}_{-1}) = S \cap \xi(\underline{g}_{-1}^{C}).$$

Hence the image $\xi(\underline{g}_{-1})$ is open dense in S.

Lemma 6.3. The map $\xi|_{\underline{g}_{-1}}$ of $(\underline{g}_{-1}, \underline{C}_{-1})$ into $(S, \underline{C})$ is an injective causal diffeomorphism.

Proof. Let $\underline{C} = \{C_x\}_{x \in S}$, where $C_{o_M} = \overline{V}$. Let $\pi$ be the natural projection of G(D) onto S = G(D)/H(D). Note that $\xi(X) = \pi(\rho(\exp X))$, $X \in \underline{g}_{-1}$. If we identify $\underline{g}$ with the tangent space to G(D) at the identity and $\underline{g}_{-1}$ with the tangent space $T_{o_M}(S)$, then the differential $\xi_{*o}$ of $\xi$ at the origin $o \in \underline{g}_{-1}$ coincides with the identity mapping on $\underline{g}_{-1}$. Denoting by $\tau_X$ the parallel trnslation in $\underline{g}_{-1}$ along a vector $X \in \underline{g}_{-1}$, we have that $\xi\tau_X = \rho(\exp X)\xi$ on $\underline{g}_{-1}$. Therefore we get

(6.6) $$\xi_{*X}(\overline{V}_X) = \xi_{*X}(\tau_X(\overline{V})) = \rho(\exp X)_{*o_M}\xi_{*o}(\overline{V}) = \rho(\exp X)_{*o_M}(\overline{V})$$
$$= C_{\rho(\exp X)o_M} = C_{\xi(X)},$$

which shows that $\xi$ is a causal diffeomorphism of $\underline{g}_{-1}$ into S. q.e.d.

Lemma 6.4. Let $\varphi$ be a diffeomorphism of S onto itself which satisfies $\varphi(o_M) = o_M$ and normalizes $G^0(D)$. If $\varphi_{*o_M}\overline{V} = \overline{V}$ (resp. $= -\overline{V}$) holds, then $\varphi$ is a causal (resp. an anti-causal) automorphism of $(S, \underline{C})$.

Proof. Suppose that $\varphi_{*o_M}\overline{V} = -\overline{V}$. Let $\underline{C} = \{C_x\}_{x \in S}$, and let $p \in S$. Then there exist two elements $a, g \in G^0(D)$ such that $p = ao_M$ and $\varphi(p) = \varphi ao_M = gao_M$. Since $b := \varphi a \varphi^{-1}$ lies in $G^0(D)$, we have

(6.7) $$\varphi_{*p}C_p = \varphi_{*p}a_{*o_M}\overline{V} = (b\varphi)_{*o_M}\overline{V} = b_{*o_M}\varphi_{*o_M}\overline{V} = -b_{*o_M}\overline{V}.$$

On the other hand, since $\underline{C}$ is G(D)-invariant, we have

$$(6.8)\qquad C_{\varphi(p)} = C_{gao_M} = g_{*p}a_{*o_M}\overline{V}.$$

Also we have $gao_M = \varphi ao_M = b\varphi o_M = bo_M$, and consequently $g_{*p}a_{*o_M}\overline{V} = b_{*o_M}\overline{V}$. Hence, from (6.7) and (6.8) it follows that $\varphi_{*p}C_p = -C_{\varphi(p)}$, which implies that $\varphi$ is anti-causal. q.e.d.

Lemma 6.5. $\tilde{\theta}$ acts on S as an anti-causal automorphism.

Proof. Since $\tilde{\theta} \in I(D)$ (cf. (3.18),(4.5)), $\tilde{\theta}$ normalizes $G^0(D) = I^0(D)$ (= the identity component of I(D)). $-\theta$ leaves the origin 0 $\in \underline{g}_{-1}$ fixed, and $\xi(0) = o_M$ holds. Consequently, by Lemma 4.7, $\tilde{\theta}$ fixes the origin $o_M \in S$. Again, by Lemma 4.7, we have $\tilde{\theta}_{*o_M} = \xi_{*0}(-\theta)_{*0}(\xi^{-1})_{*o_M} = -id$, which implies that $\tilde{\theta}_{*o_M}\overline{V} = -\overline{V}$. Therefore the assertion follows from Lemma 6.4. q.e.d.

Theorem 6.6. Let D and $(S,\underline{C})$ be the same as in Theorem 6.2. Let $G_-(D)$ be the set of anti-holomorphic automorphisms of D, and $G_-(S,\underline{C})$ be the set of anti-causal automorphisms of $(S,\underline{C})$. Suppose $\dim D > 1$. Then we have

$$(6.9)\qquad G_-(S,\underline{C}) = G_-(D),$$

where the both sides are being regarded as subsets of Diffeo(S).

Proof. Since $\tilde{\theta}$ acts on D anti-holomorphically (cf.(4.9)), we have $G_-(D) = \tilde{\theta}G(D)$. From Lemma 6.5, (6.3) and the fact that $\tilde{\theta}^2 = 1$, it follows that $\tilde{\theta}G(D) = G_-(S,\underline{C})$. q.e.d.

6.3. We will give a characterization of the affine automorphism group of the domain D(V) in terms of the causal structure $\underline{C}_{-1}$ on its Silov boundary $\underline{g}_{-1}$.

Lemma 6.7. The map $\xi|_{\underline{g}_{-1}}$ of $\underline{g}_{-1}$ into S is equivariant under Aff(D(V)).

Proof. By Lemma 5.1, we have $\rho(G(V))\xi(0) = \xi(0)$. Also we have $\xi\tau_X = \rho(\exp X)\xi$ ( cf. the proof of Lemma 6.3). From these two the lemma follows. q.e.d.

In the sequel, we will identify $\underline{g}_{-1}$ with its $\xi$-image. Lemma 6.3 then implies that the causal structure $\underline{C}_{-1}$ is just the restriction of the causal structure $\underline{C}$ on S to $\underline{g}_{-1}$. In other words, $(S,\underline{C})$ is a causal compactification of $(\underline{g}_{-1},\underline{C}_{-1})$. By Lemma 6.7, the action of Aff(D(V)) on $\underline{g}_{-1}$ is the restriction of its action on S to $\underline{g}_{-1}$. Let Q' be the restriction of the bundle Q to $\underline{g}_{-1}$. Q' is a G(V)-structure on $\underline{g}_{-1}$. The group Aff(D(V)) acts on Q as a subgroup of G(D).

Lemma 6.8. Aff(D(V)) acts on Q' simply transitively.

Proof. Let $\varpi$ be the projection of Q' onto $\underline{g}_{-1}$. Let $z \in Q'$ and let $\varpi(z) = X$. If we denote by $\tau_{-X}$ the parallel translation in $\underline{g}_{-1}$ along -X, then $\overline{\tau_{-X}}(z)$ lies in the fiber through the reference point $z_0$. There exists an element $h \in G(V)$ such that $\bar{h}(z_0) = z_0\sigma(h) = \overline{\tau_{-X}}(z)$ (cf. (6.1)). Therefore the point z lies in the orbit $\mathrm{Aff}(D(V))z_0$ (cf. (6.4)). One has $\underline{g}_{-1} = \varpi(Q') \subset \varpi(\mathrm{Aff}(D(V))z_0) = \mathrm{Aff}(D(V))0 = \underline{g}_{-1}$, which shows that $Q' = \mathrm{Aff}(D(V))z_0$. The simple transitivity of Aff(D(V)) follows from (5.9),(6.4) and (6.1). q.e.d.

Lemma 6.9. The causal automorphism group $G(\underline{g}_{-1},\underline{C}_{-1})$ is a subgroup of the automorphism group $\mathrm{Aut}(\underline{g}_{-1},Q')$ of the G(V)-structure Q'.

Proof. The proof is quite similar to the proof of the inclusion $G(S,\underline{C}) \subset \mathrm{Aut}(S,Q)$ (cf. the proof of Theorem 6.2), if we replace G(D) by Aff(D(V)). q.e.d.

Theorem 6.10. Let $\underline{g}_{-1} = A$ be a compact simple Jordan algebra, and let V be the cone of positive definite elements of $\underline{g}_{-1}$. Suppose $\dim A > 1$. Then the causal automorphism group $G(\underline{g}_{-1},\underline{C}_{-1})$ coincides with the affine automorphism group Aff(D(V)) of the Siegel domain D(V) of the first kind.

Proof. Here we will not identify $\underline{g}_{-1}$ with $\xi(\underline{g}_{-1})$. By a theorem of Tanaka [23], every element of $\xi G(\underline{g}_{-1},\underline{C}_{-1})\xi^{-1}$ extends uniquely to an element of $G(S,\underline{C}) = G(D)$ (cf. Theorem 6.2). Therefore, taking Lemma 6.3 into account, we have

$$(6.10) \qquad \xi G(\underline{g}_{-1},\underline{C}_{-1})\xi^{-1} = \{a \in G(D)\colon a(\xi(\underline{g}_{-1})) = \xi(\underline{g}_{-1})\}.$$

In other words, the group $G(\underline{g}_{-1},\underline{C}_{-1})$ consists of all the elements $g \in G(D(V))$ which are holomorphic on the closure $\overline{D(V)}$ and satisfy $g(\underline{g}_{-1}) = \underline{g}_{-1}$. Such an element $g \in G(D(V))$ necessarily lies in $\mathrm{Aff}(D(V))$ by a theorem of Pyatetskii-Shapiro [16]. The inclusion $\mathrm{Aff}(D(V)) \subset G(\underline{g}_{-1},\underline{C}_{-1})$ is trivial. q.e.d.

Remark 6.12. Let M be a smooth manifold and let $\underline{K}$ be a smooth assignment to each point of M of a generalized light cone (§2) in the tangent space at that point. By a light ray on M we mean a smooth curve whose tangent vector at each point lies on the generalized light cone belonging to $\underline{K}$ at that point. $\underline{K}$ is called a light ray structure on M. An automorphism of the light ray structure $\underline{K}$ on M is defined in the analogous way as for the causal structures. We can consider the group of automorphisms of $\underline{K}$, denoted by $G(M,\underline{K})$. The Silov boundary S of an irreducible symmetric bounded domain D of tube type has a natural G(D)-invariant light ray structure. Also one can define the light ray structure $\underline{K}_{-1}$ on $\underline{g}_{-1}$ in the same way as the causal structure $\underline{C}_{-1}$. Then it can be proved that Theorems 6.2 and 6.10 are still valid, if we replace $G(S,\underline{C})$ by $G(S,\underline{K})$ and $G(\underline{g}_{-1},\underline{C}_{-1})$ by $G(\underline{g}_{-1},\underline{K}_{-1})$.

Added in proof. The following comment was kindly conveyed by the referee: Lemmas 3.2 and 4.1 follow also from Lemma 1.2 and §4 of Takeuchi [22], respectively.

Bibliography

[1] A.D. Alexandrov, Mappings of ordered spaces I, Proc. Steklov Inst. Math., 128(1972), 1-21.

[2] H. Asano and S. Kaneyuki, On compact generalized Jordan triple systems of the second kind, Tokyo J. Math., 11(1988), 105-118.

[3] A. Ash, D. Mumford, M. Rapoport and Y. Tai, Smooth Compactification of Locally Symmetric Varieties, Math. Sci. Press, Brookline, 1975.

[4] H. Braun and M. Koecher, Jordan-Algebren, Springer, Berlin-Heidelberg-New York, 1966.

[5] N. Jacobson, Some groups of linear transformations defined by Jordan algebras I, J. Reine Angew. Math., 201(1959), 178-195.

[6] S. Kaneyuki, Homogeneous Bounded Domains and Siegel Domains, Lecture Notes in Math., 241, Springer, Berlin-Heidelberg-New York, 1971.

[7] S. Kaneyuki, The Sylvester's law of inertia for Jordan algebras, Proc. Japan Acad., Ser.A 64(1988), 311-313.

[8] S. Kaneyuki and H. Asano, Graded Lie algebras and generalized Jordan triple systems, Nagoya Math. J., 112(1988), 81-115.

[9] I.L. Kantor, Transitive-differential groups and invariant connections on homogeneous spaces, Trudy Sem. Vekt. Tenz. Anal., 13(1966), 310-398.

[10] M. Koecher, Imbeddings of Jordan algebras into Lie algebras.I., Amer. J. Math., 89(1967), 787-816, II, 90(1968), 476-510.

[11] M. Koecher, An Elementary Approach to Bounded Symmetric Domains, Lect. Notes, Rice Univ., Houston, 1969.

[12] A. Korányi and J.A. Wolf, Realization of Hermitian symmetric spaces as generalized half-planes, Ann. of Math., 81(1965), 265-288.

[13] H. Matsumoto, Quelques remarques sur les groupes de Lie algébriques réels, J. Math. Soc. Japan, 16(1964), 419-446.

[14] K. Nakajima, On Tanaka's imbeddings of Siegel domains, J. Math. Kyoto Univ., 14(1974), 533-548.

[15] E. Neher, Klassifikation der einfachen reellen speziellen Jordan-Tripelsysteme, Manuscripta Math., 31(1980), 197-215.

[16] I.I. Pyatetskii-Shapiro, Automorphic Functions and the Geometry of Classical Domains, Gordon and Breach, New York, 1969.

[17] I. Satake, Algebraic Structure of Symmetric Domains, Iwanami Shoten, Tokyo and Princeton Univ. Press, Princeton, 1980.

[18] I. Satake, A formula in simple Jordan algebras, Tohoku Math. J., 36(1984), 611-622.

[19] I. Satake, On zeta functions associated with self-dual homogeneous cones; Reports on Symposium of Geometry and Automorphic Functions, Tohoku Univ., Sendai, 1988, 145-168.

[20] I.E. Segal, Mathematical Cosmology and Extragalactic Astronomy, Academic Press, New York, 1976.

[21] M. Takeuchi, On orbits in a compact Hermitian symmetric space, Amer. J. Math., 90(1968), 657-680.

[22] M. Takeuchi, Basic transformations of symmetric R-spaces, Osaka J. Math., 25(1988), 259-297.

[23] N. Tanaka, On the equivalence problems associated with a certain class of homogeneous spaces, J. Math. Soc. Japan 17(1965), 103-139.

[24] N. Tanaka, On infinitesimal automorphisms of Siegel domains, J. Math. Soc. Japan, 22(1970), 180-212.

[25] N. Tanaka, On non-degenerate real hypersurfaces, graded Lie algebras and Cartan connections, Japan. J. Math.,2(1976), 131-190.

[26] N. Tanaka, On the equivalence problems associated with simple graded Lie algebras, Hokkaido Math. J., 8(1979), 23-84.

[27] N. Tanaka, On affine symmetric spaces and the automorphism groups of product manifolds, Hokkaido Math. J., 14(1985), 277-351.

[28] A.E. Tumanov and G.M. Khenkin, Local characterization of holomorphic automorphisms of Siegel domains, Func. Anal. Appl., 17(1983), 285-294.

[29] E.B. Vinberg, Homogeneous cones, Soviet Math. Dokl., 1(1961), 787-790.

[30] E.B. Vinberg, The theory of homogeneous convex cones, Trans. Moscow Math. Soc., 1963, 340-403.

[31] E.C. Zeeman, Causality implies the Lorentz group, J. Math. Physics, 5(1964), 490-493.

# The behavior of the extremal length function on arbitrary Riemann surface

BY MASAHIKO TANIGUCHI

Department of Mathematics, Kyoto university, Kyoto 606, Japan

**Introduction.** In the case of compact Riemann surfaces, there are so many important conformal invariants, such as the geodesic length functions, the period matrices, the energy of the harmonic maps and so on. But some of such basic quantities are hard to handle in case of Riemann surfaces of infinite type. An exception is the extremal length of a suitable curve family. In this note, we will consider the extremal length of a 1-cycle on an arbitrary Riemann surface. And as applications of several second variational formulas obtained in [**10**], we will investigate the behavior of this fundamental conformal invariant under basic kinds of deformation of the surface.

**1. Known facts on the extremal length of a 1-cycle.** First we recall definition of the extremal length (cf. [**2**, Ch.4]). Let $R$ be an arbitrary Riemann surface, not necessarily of finite type. Fix a 1-cycle $C$ on $R$. Then a conformal linear density $\rho(z)|dz|$ on $R$ is called to be admissible for $C$ if

$$\int_c \rho(z)|dz| \geq 1$$

for every loop $c$ belonging to the curve family $C$, where $z = x + iy$ is a generic local parameter on $R$. Set

$$M(C) = \inf_\rho \iint_R \rho(z)^2 dxdy,$$

where $\rho(z)|dz|$ moves all admissible density for $C$. This $M(C)$ is called the modulus of $C$ on $R$, and $\lambda(C) = 1/M(C)$ is called the extremal length of $C$ on $R$.

Here Accola's theorem ([**1**]) is fundamental. To state it, we need some definitions. Let $\Gamma(R)$ be the Hilbert space consisting of all real square integrable 1-forms on $R$ with the inner product

$$(\alpha, \beta)_R = \iint_R \alpha \wedge^* \beta.$$

Let $\Gamma_h(R)$ be the subspace of $\Gamma(R)$ consisting of all harmonic 1-forms. For the given 1-cycle $C$ on $R$, let $\sigma = \sigma(C, R)$ be the period reproducer of $C$ in $\Gamma_h(R)$, i.e. the 1-form in $\Gamma_h(R)$ such that

$$(\alpha, \sigma)_R = \int_C \alpha$$

for every $\alpha$ in $\Gamma_h(R)$. Then Accola's theorem states that

$$\lambda(C) = \|\sigma\|_R^2 (= (\sigma, \sigma)_R),$$

i.e., the extremal length $\lambda(C)$ is equal to the Dirichlet energy of the reproducer $\sigma$ corresponding to $C$.

Next we will recall some variational formulas for $\lambda(C)$ under quasiconformal deformation of $R$. Also see [3] and [10].

Let $\{\mu(x)\}$ be a family of Beltrami coefficients, i.e. bounded (-1,1)-forms whose sup-norms are less than one, on $R$ depending smoothly on a real or complex parameter $x$ such that $\mu(0) = 0$. For every $x$, let $f^x = f^{\mu(x)}$ be a quasiconformal mapping of $R$ to another surface $R_x$ with the complex dilatation $\mu(x)$, $C_x$ be the 1-cycle on $R_x$ corresponding to $C$ under $f^x$ and $\lambda(x)$ be the extremal length of $C_x$ on $R_x$.

When $x$ is a real parameter, we have

$$\lambda'(0) = \mathrm{Re} \iint_R (\mu'(0)\phi) \wedge^* \phi, \tag{1.1}$$

where $\phi = \sigma(C,R) + i^*\sigma(C,R)$ and $\cdot' = d\cdot/dx$. Next let $\Phi(x)$ be the pull-back of $\sigma(C_x,R_x) + i^*\sigma(C_x,R_x)$ by $f^x$ for every $x$. Then $\Phi(x)$ is a (not necessarily smooth) closed 1-form on $R$ for every $x$ and differentiable at $x = 0$ in $\Gamma(R)$. And we have ([10, Theorem 1])

$$\lambda''(0) = \mathrm{Re} \iint_R (\mu''(0)\phi) \wedge^* \phi + \Phi'(0) \wedge^* \Phi'(0). \tag{1.2}$$

Next when $x$ is a complex parameter and $\mu(x)$ depends holomorphically on $x$, we have

$$(\Phi^{0,1})_x(0) = \mu'(0)\phi, \tag{1.3}$$

$$\lambda_x(0) = \frac{1}{2} \iint_R \phi \wedge^* \overline{\Phi_{\bar{x}}(0)} \tag{1.4}$$

and

$$\Delta\lambda(0) = 4 \iint_R \Phi_{\bar{x}}(0) \wedge^* \overline{\Phi_{\bar{x}}(0)} \geq 0, \tag{1.5}$$

where $\Phi^{0,1}$ is the (0,1)-part of $\Phi$, $\cdot_x = \partial\cdot/\partial x$ , $\cdot_{\bar{x}} = \partial\cdot/\partial\bar{x}$ and $\Delta = 4\partial^2/\partial x\partial\bar{x}$. In particular, we have the following convexity results. (Compare with [4].)

**Theorem 1** ([**3**, Formula 2], [**10**, Corollary]). *Let $\{\mu(t)\}$ be a family of Beltrami coefficients on $R$ with $\mu(0) = 0$, depending holomorphically on the complex vector-valued parameter $t$ in a neighborhood $D$ of the origin. For every $t$, let $R_t$ and $C_t$ be as above, and denote by $\lambda(t)$ the extremal length $\lambda(C_t)$ of $C_t$ on $R_t$. Then $\lambda(t)$ is continuous and plurisubharmonic on $D$.*

Here $\lambda$ is either positive or identically zero on $D$. So we further assume that $\lambda$ is positive on $D$. Then we have the following

Corollary 1. *The modulus $M(t) = 1/\lambda(t)$ is continuous and plurisuperharmonic on $D$.*

Proof: Fix a point $t_0$ in $D$ and a complex line $L$ passing through $t_0$. Let $x$ be a complex parameter of $L$ such that $x = 0$ corresponds to $t_0$. Then a direct computation shows that

$$\Delta M = -\lambda^{-2} \cdot \Delta\lambda + 2\lambda^{-3} \cdot ({\lambda_u}^2 + {\lambda_v}^2) \qquad (x = u + iv).$$

Since $\lambda$ is real and

$$8 \cdot |\lambda_x(0)|^2 \leq \lambda(0) \cdot \Delta\lambda(0)$$

by (1.4) and (1.5), we have

$$\lambda^3(0) \cdot \Delta M(0) = 8 \cdot |\lambda_x(0)|^2 - \lambda(0) \cdot \Delta\lambda(0) \leq 0.$$

Since $t_0$ and $L$ are arbitrary, we can show the assertion. ∎

**Remark.** As a corollary, the extremal length function of a fixed 1-cycle gives a continuous non-negative plurisubharmonic function on the Teichmüller space of any, possibly finitely punctured, compact Riemann surface.

**2. Deformation by pinching loops.** As another basic kind of deformation of a Riemann surface, we consider deformation by pinching a finite number of simple loops. This is a generalization of the classical degeneration of a compact surface, and has been used, for instance, by H. Masur and S. Wolpert in case of compact Riemann surfaces. (also see [**8**] and [**9**].)

Let $R_0$ be an arbitrary Riemann surface, again not necessarily of finite type, with a finite number of nodes. Denote by $N(R_0)$ the set $\{p_j\}_{j=1}^n$ of all nodes of $R_0$, and set $R_0' = R_0 - N(R_0)$. Namely, $R_0'$ is a union of ordinary Riemann surfaces whose universal coverings are conformally equivalent to the unit disc, and each $p_j$ has a neighborhood homeomorphic to the subset $\{|z| < 1, |w| < 1, zw = 0\}$ in $\mathbf{C}^2$.

For every $j$, we fix a neighborhood $U_j$ of $p_j$ on $R_0$ such that each component, say $U_{j,v} (v = 1, 2)$, of $U_j - \{p_j\}$ is mapped conformally onto $D_0 = \{0 < |z| < 1\}$ by mapping, say, $z = z_{j,v}(p)$. Also we suppose that $\{\overline{U_j}\}_{j=1}^n$ are mutually disjoint.

Next let $\{\mu(t)\}$ be a family of Beltrami coefficients, with $\mu(0) = 0$, supported on $R_0 - \overline{U}$ and depending holomorphically on $t$ in $(B_1)^m$, where $U$ is the union of all $U_j$ and $B_1$ is the unit disc $\{|\zeta| < 1\}$. Let $f^t$ be a quasiconformal mapping of $R_0'$ onto another union $R_t'$ of Riemann surfaces with the complex dilatation $\mu(t)$. Finally, for every $t$ in $(B_1)^m$ and $s = (s_1, \cdots, s_n)$ in $(B_{1/2})^n$, where $B_{1/2} = \{|\zeta| < 1/2\}$, let $R_{t,s}$ be the Riemann surface, possibly with nodes, obtained from $R_t'$ by deleting two punctured disks ${z_{j,v,t}}^{-1}(\{0 < |z| < |s_j|^{1/2}\}) \quad (v = 1, 2)$, and by identifying the resulting borderes into a loop $C_{j,t,s}$ under the mapping

$${z_{j,2,t}}^{-1}(s_j / z_{j,1,t}(p))$$

for every $j$, where $z_{j,v,t} = z_{j,v} \circ (f^t)^{-1}$. Here we regard that $R_{t,s}$ has a node corresponding to $p_j$ for every such $j$ that $s_j = 0$. Then we have a family $\{R_{t,s}\}$ of Riemann surfaces, possibly with nodes, which we call the complex pinching deformation family with the center $R_0$ and the deformation deta $(\{\mu(t)\}, U)$.

Now fix a simple closed curve $C$ on $R_0 - \overline{U}$. For every point $(t,s)$ in $B^* = (B_1)^m \times (B^*_{1/2})^n$, where $B^*_{1/2} = B_{1/2} - \{0\}$, let $C_{t,s}$ be the 1-cycle on $R_{t,s}$ corresponding to $C$ and $\lambda(t,s)$ be the extremal length of $C_{t,s}$ on $R_{t,s}$. Then recalling that, for any point $(T,S)$ in $B^*$, the family $\{R_{t,s}\}$ can be considered as a quasiconformal deformation family depending holomorphically on $(t,s)$ in some neighborhood of $(T,S)$ (cf. Lemma 1 in **4**), we can see by Theorem 1 and Corollary 1 the following

THEOREM 2. *The extremal length function $\lambda(t,s)$ is non-negative plurisubharmonic on $B^*$, and is continuous on $B = (B_1)^m \times (B_{1/2})^n$.*

*When $\lambda(t,s)$ is positive on $B^*$, the mudulus function $M(t,s) = 1/\lambda(t,s)$ is continuous and plurisuperharmonic on $B^*$.*

Here continuity of $\lambda$ on $B$ follows by [**5**, Proposition 4].

**Remark.** When $\lambda(t,s)$ is pluriharmonic on $B^*$, $\lambda(t,s)$ should be a constant on $B^*$. In fact, if so, (1.5) implies that $\Phi_{\overline{x}}(0) = 0$ for every complex line. Then by (1.4), $\lambda_x(0) = 0$ for every complex line. Thus we can see the assertion.

As an example, we consider the extremal length $\lambda_j(t,s)$ of the j-th pinching loop $C_{j,t,s}$ on $R_{t,s}$ for every $j$ and $(t,s)$ on $B$. Then it is clear that $\lambda_j$ vanishes on $\{s_j = 0\}$. Moreover we have the following

COROLLARY 2. *Suppose that every $\lambda_j$ is positive on $B^*$ and set $\lambda_N = \prod_{j=1}^n \lambda_j$. Then $\lambda_N$ is non-negative continuous plurisubharmonic function on $B$ which defines the singular locus $B - B^*$. (Namely, $\lambda_N(t,s) = 0$ if and only if $(t,s)$ belongs to $B - B^*$.)*

PROOF: Since $\lambda_N$ is non-negative and continuous on $B$ and vanishes exactly on $B - B^*$, it suffices to show that $\lambda_N$ is plurisubharmonic on $B^*$. For this purpose, fix a point $(T,S)$ in $B^*$ and a complex line $L$ passing through $(T,S)$. Let $x$ be a complex parameter of $L$ such that $x = 0$ corresponds to $(T,S)$. Then since

$$(\lambda_N)_{\xi\xi} = \lambda_N \left( \left( \sum_{j=1}^n (\lambda_j)_\xi/\lambda_j \right)^2 + \sum_{j=1}^n \left( (\lambda_j)_{\xi\xi}/\lambda_j - (\lambda_j)_\xi{}^2/\lambda_j{}^2 \right) \right),$$

where $\xi$ is either Re $x$ or Im $x$, we have

$$\Delta\lambda_N \geq \lambda_N \sum_{j=1}^n \left( (\Delta\lambda_j/\lambda_j) - 4|(\lambda_j)_x|^2/\lambda_j{}^2 \right).$$

Thus we can conclude the assertion by (1.4) and (1.5). ∎

**3. Behavior near the singular locus.** Next as another application of the variational formulas in **1**, we will give the order estimate of partial derivatives of the extremal length $\lambda(t,s)$ near the origin of $B$.

In the sequel, fix a relatively compact neighborhood $\Omega$ of $(0,0)$ in $B$ arbitrarily, and assume that constants $A$ are always independent of $(t,s)$ in $\Omega \cap B^*$.

THEOREM 3. *There is a constant A such that, on $\Omega \cap B^*$, we have*

$$\left|\frac{\partial \lambda}{\partial t_k}(t,s)\right| \le A$$

*for every $k = 1, \cdots, m$, and*

$$\left|\frac{\partial \lambda}{\partial s_j}(t,s)\right| \le \frac{1}{2\pi|s_j|}|a_{j,t,s}|^2 + A \tag{3.1}$$

*for every $j (= 1, \cdots, n)$, where we set*

$$a_{j,t,s} = \int_{C_{j,t,s}} \sigma(C_{t,s}, R_{t,s}).$$

The proof will be given in **4**. Note that $|\partial\lambda/\partial x| = |\partial\lambda/\partial\bar{x}|$ for every complex parameter $x$, since $\lambda$ is real. Also recall that $\partial\lambda/\partial t_k$ are continuous on the whole $\Omega$, as is seen from (1.1) and [**5**,Proposition 4].

**Remark 1.** If some pinching loop $C_J$ is essentially trivial, namely, $C_{J,t,s}$, with suitable orientation, bounds a parabolic part of $R_{t,s}$ for every $(t,s)$ in $B^*$ (cf. [**6**, 513p]), $a_{J,t,s}$ vanishes identically on $B^*$. (A typical example is a dividing loop on a compact Riemann surface.) Hence in this case, the bound in (3.1) of Theorem 3 can be replaced by $A$ for $j = J$.

THEOREM 4. *There is a constant A such that, on $\Omega \cap B^*$, we have*

$$\left|\frac{\partial^2 \lambda}{\partial (Re\ t_k)^2}(t,s)\right| \le A$$

*for every $k (= 1, \cdots, m)$, and*

$$\left|\frac{\partial^2 \lambda}{\partial (Re\ s_j)^2}(t,s)\right| \le \frac{A}{|s_j|^2} b_{j,t,s} \tag{3.2}$$

*for every $j (= 1, \cdots, n)$, where we set*

$$b_{j,t,s} = \|\sigma(C_{t,s}, R_{t,s})\|^2_{W_{j,t,s}}$$

*with $W_{j,t,s} = {z_{j,1,t}}^{-1}(\{1/2 < |z| < 1\})$ considered as a subset of $R_{t,s}$.*

The proof will be given in **4**. Note that $b_{j,t,s}$ is bounded on $\Omega \cap B^*$ (cf. Lemma 2 in **4**) and that similarly we can obtain the order estimate of the other second partial derivatives.

Now a rough estimate of $|a_{j,t,s}|$ is $O\left((\log(1/|s_j|))^{-1/2}\right)$ as is seen by Lemma 3 in **4**. But the best possible one may be $O\left((\log(1/|s_j|))^{-1}\right)$. This is true when we further assume, for instance, that $\{C_j\}_{j=1}^n$ is free, namely, no subset of $\{C_{j,t,s}\}_{j=1}^n$, with any orientations, bounds a parabolic part of $R_{t,s}$ for every $(t,s)$ in $B^*$ (cf. [**6**, 513p]).

Corollary 3. *If $\{C_j\}$ is free, the bound in (3.1) of Theorem 3 can be replaced by*

$$\frac{A}{|s_j|(\log(1/|s_j|))^2} \tag{3.3}$$

Proof: First note that

$$a_{j,t,s} = \lambda_j(t,s)\int_{C_{t,s}} \phi_j(t,s), \tag{3.4}$$

where $\lambda_j(t,s)$ is as in **2** and

$$\phi_j(t,s) = \lambda_j(t,s)^{-1}\left(\sigma(C_{j,t,s}, R_{t,s}) + i^*\sigma(C_{j,t,s}, R_{t,s})\right)$$

for every $j$.

Since [**6**,Theorems 5 and 6] still holds even in the complex case, we can show the following facts:

(I) For every $j$, there is a constant $A_j$ such that

$$\log(1/|s_j|) + A_j \geq \frac{\pi}{\lambda_j(t,s)} \geq \log(1/|s_j|)$$

on $\Omega \cap B^*$.

(II) Functions

$$\|\phi_j(t,s)\|^2_{E_{t,s}} = 2\lambda_j(t,s)^{-2}\|\sigma(C_{j,t,s}, R_{t,s})\|^2_{E_{t,s}}$$

and

$$\int_{C_{t,s}} \phi_j(t,s) = \lambda_j(t,s)^{-1} a_{j,t,s}$$

are continuous on $\Omega$, where $E_{t,s}$ is the union of all $W_{j,t,s}$ and the image $R''_{t,s}$ of $R_0 - U$ by $f^t$ (considered as a subset of $R_{t,s}$).

Thus we have the assertion by Theorem 3. ∎

**Remark 2.** The order of the bound (3.3) in Corollary 3 is the best possible one, as is seen from the proof.

Next when $\lambda(t,s)$ is positive on the whole $B$, the first partial derivatives of $M(t,s) = 1/\lambda(t,s)$ have the same order estimates as those for $\lambda(t,s)$. So we consider the case that $\lambda$ is positive on $B^*$ but $\lambda(0,0) = 0$. Then $\sigma(C_{t,s}, R_{t,s})$ is equivalent to

$$\sum_{j=1}^{n} \epsilon_j \sigma(C_{j,t,s}, R_{t,s}),$$

where every $\epsilon_j$ is an integer independent of $(t,s)$ (cf. [**6**, 513p]). Hence we will consider only the case that $C$ is equivalent to one of the pinching loop, say $C_J$, e.g. the case that $\lambda = \lambda_J$.

**Remark 3.** In the above case, the bound (3.2) in Theorem 4 can be replaced by

$$\frac{A}{|s_j|^2(\log(1/|s_J|))^2} \quad (j = 1, \cdots, n).$$

In fact, the assertion follows by Theorem 4 and the facts (I) and (II) in the proof of Corollary 3.

THEOREM 5. *Suppose that $\{C_j\}$ is free and $C$ is equivalent to $C_J$. Set $M(t,s) = 1/\lambda_J(t,s)$. Then there is a constant $A$ such that, on $\Omega \cap B^*$, we have*

$$\left|\frac{\partial M}{\partial t_k}(t,s)\right| \leq A \quad (k = 1, \cdots, m), \tag{3.6}$$

$$\left|\frac{\partial M}{\partial s_j}(t,s)\right| \leq \frac{A}{|s_j|} \quad (j = 1, \cdots, n), \tag{3.7}$$

$$\left|\frac{\partial^2 M}{\partial (\mathrm{Re}\ t_k)^2}(t,s)\right| \leq A \quad (k = 1, \cdots, m), \tag{3.8}$$

$$\left|\frac{\partial^2 M}{\partial (\mathrm{Re}\ s_j)^2}(t,s)\right| \leq \frac{A}{|s_j|^2} \quad (j = 1, \cdots, n). \tag{3.9}$$

PROOF: We can see by (1.1) and the fact (II) that

$$|\partial(\lambda_J)/\partial t_k| \leq A_0(\lambda_J)^2 \quad (k = 1, \cdots, m)$$

on $\Omega \cap B^*$ with a suitable constant $A_0$, hence that (3.6) follows. Next (3.7) follows by (3.1) and the fact (II) (cf. the proof of Theorem 3).

Finally let $\xi$ be any one of $\{\mathrm{Re}\ t_k, \mathrm{Re}\ s_j\}$. Then since

$$M_{\xi\xi} = -(\lambda_J)_{\xi\xi}/(\lambda_J)^2 + 2((\lambda_J)_\xi)^2/(\lambda_J)^3,$$

we can conclude (3.8) and (3.9) by (1.2), (4.3) in **4**, the fact (II), (3.6) and (3.7) (cf. the proof of Theorem 4). ∎

**4. Proofs.** As before, for every $(t,s)$, we consider $f^t$ also as a mapping of $R_0'' = R_0 - U$ into $R_{t,s}$, and hence consider the image $R_{t,s}''$ of $R_0''$ by $f^t$ as a subset of $R_{t,s}$. Next we denote by $U_{j,t,s}$ the component of $R_{t,s} - R_{t,s}''$ corresponding to $U_j$ for every $j$. Then $z_{j,1,t}$ can be considered as a conformal mapping of $U_{j,t,s}$ onto the domain $W_j(s) = \{|s_j| < |z| < 1\}$, and we have the following

LEMMA 1. *Fix a pont $(T,S)$ in $B^*$. Set*

$$g^{t,s} = \begin{cases} f^t \circ (f^T)^{-1} & \text{on} \quad R_{T,S}'' \\ (z_{j,1,t})^{-1} \circ F_{j,s} \circ z_{j,1,T} & \text{on} \quad U_{j,T,S} \quad (j = 1, \cdots, n), \end{cases}$$

where

$$F_{j,s}(z) = \begin{cases} z & \text{on} \quad \{3/4 \leq |z| < 1\} \\ z \cdot \exp\left(\left(1 - \frac{\log(2|z|)}{\log(3/2)}\right)\log(s_j/S_j)\right) & \text{on} \quad \{1/2 < |z| < 3/4\} \\ (s_j/S_j) \cdot z & \text{on} \quad \{|S_j| < |z| \leq 1/2\}. \end{cases}$$

*(We take the branch such that $\log 1 = 0$ .)*

*Then $g^{t,s}$ is a quasiconformal mapping of $R_{T,S}$ to $R_{t,s}$ for every $(t,s)$ in some neighborhood $D$ of $(T,S)$ in $B^*$, and the complex dilatation $\nu(t,s)$ of $g^{t,s}$ depends holomorphically on $(t,s)$ in $D$.*

Proof is given by direct computation. In particular, we have

$$\frac{\partial \nu}{\partial t_k}(T,S) = \left(\frac{\partial \mu}{\partial t_k}(T) \cdot (1-|\mu(T)|^2)^{-1} \cdot \frac{\overline{(f^T)_w}}{(f^T)_w}\right) \circ (f^T)^{-1} \tag{4.1}$$

for every $k$, where $w$ is a generic local parameter on $R_{T,S}$, and every $\frac{\partial \nu}{\partial s_j}(T,S)$ has its support in $W_{j,T,S}$ (defined in Theorem 4) and corresponding to the Beltrami coefficient

$$\frac{1}{2S_j \cdot \log(2/3)} \chi_{[1/2,3/4]}(|z|) \frac{z d\bar{z}}{\bar{z} dz}, \tag{4.2}$$

on $W_j(s)$, where $\chi_{[1/2,3/4]}$ is the characteristic function of the interval $[1/2, 3/4]$ on $\mathbf{R}$.

Now let $C$ and $\Omega$ be as in **2** and **3**, respectively. To show Theorem 3, we need the following well-known facts.

LEMMA 2. *There is a constant $A_1$ such that*

$$\|\sigma(C_{t,s}, R_{t,s})\|^2_{R_{t,s}} \leq A_1$$

*for every $(t,s)$ in $\Omega$.*

PROOF: Without loss of generality, we may assume that $C$ is simple. Take a doubly connected neighborhood $W$ of $C$ on $R_0 - \overline{U}$, and fix a smooth function $e_C(p)$ on $W$ such that $0 \leq e_C(p) \leq 1$, $de_C$ has a compact support on $W$, and

$$\int_C \omega = \iint_W \omega \wedge de_C$$

for every smooth closed square integrable differential $\omega$ on $W$.

Then for every $(t,s)$ in $B$, ${}^*\sigma(C_{t,s}, R_{t,s})$ is the projection of the pull-back of $de_C$ by $(f^t)^{-1}$ into $\Gamma_h(R_{t,s})$. Hence we can show the assertion. ∎

LEMMA 3. *For every $j$ and every $(t,s)$ in $\Omega \cap B^*$,*

$$|a_{j,t,s}|^2 \leq 2\pi A_1 (\log(1/|s_j|))^{-1}.$$

PROOF: Set

$$e_j(z) = \log(|z|/|s_j|) \cdot (\log(1/|s_j|))^{-1} \quad \text{on } W_j(s).$$

Then

$$\begin{aligned} |a_{j,t,s}|^2 &= \left| \iint_{U_{j,t,s}} \sigma(C_{t,s}, R_{t,s}) \wedge de_j \circ {z_{j,1,t}}^{-1} \right|^2 \\ &\leq A_1 \|de_j\|^2_{W_j(s)} = 2\pi A_1 \left(\log(1/|s_j|)\right)^{-1}. \end{aligned}$$

∎

LEMMA 4. *For every $j$ and every $(t,s)$ in $\Omega \cap B^*$, write*

$$\phi_{t,s} = \sigma(C_{t,s}, R_{t,s}) + i^*\sigma(C_{t,s}, R_{t,s})$$

*as*

$$\phi_{t,s}(z)dz = \sum_{n=-\infty}^{+\infty} a_n(j,t,s) z^n dz$$

*on $W_j(s)$. Then there is a constant $A_2$ independent of $j$ and $(t,s)$ such that*

$$|a_n(j,t,s)| \le A_2 \quad \text{for every } n \ge 0,$$

*and*

$$|a_{-n}(j,t,s)| \le A_2 |s_j|^{n/2} \quad \text{for every } n \ge 2.$$

PROOF: Set

$$\psi_{t,s}(z)dz = \phi_{t,s}(z)dz - a_{j,t,s}\frac{dz}{2\pi i z}.$$

Here recall that $a_{j,t,s} = 2\pi i \cdot a_{-1}(j,t,s)$. By Lemmas 2 and 3,

$$\|\psi_{t,s}(z)dz\|^2_{W_j(s)} \left( = \iint_{W_j(s)} |\psi_{t,s}(z)|^2 i dz \wedge d\bar{z} \right) \le 8A_1.$$

Hence by [6, Lemma 1], $|\psi_{t,s}(z)|$ is bounded by a constant, independent of $j$ and $(t,s)$ in $\Omega$, on $\{|z| = |s_j|^{1/2}\}$. Moreover, Lemma 1 implies that the same assertion holds also on $\{|z| = 1\}$. Hence by the maximal principle, $\sup_{\{|s_j|^{1/2} < |z| < 1\}} |\psi_{t,s}(z)|$ is uniformly bounded on $\Omega \cap B^*$. Thus the assertion follows by Cauchy's estimates. ∎

PROOF OF THEOREM 3: By (1.1), (4.1) and Lemma 2, the first assertion follows.

To show the second assertion, fix $j$ and $(t,s)$ in $\Omega \cap B^*$. Then by (1.1) and (4.2), $\left|\frac{\partial\lambda}{\partial s_j}(t,s)\right|$ can be written as

$$I_{t,s} = \frac{1}{2|s_j|\log(3/2)} \left| \iint_{W_0} \frac{z}{\bar{z}} \phi_{t,s}(z) d\bar{z} \wedge \phi_{t,s}(z)dz \right|,$$

where $W_0 = \{1/2 < |z| < 3/4\}$ and $\phi_{t,s}(z)$ is as in Lemma 4. Here simple computation shows that

$$\begin{aligned} I_{t,s} &= \frac{2\pi}{|s_j|} \left| \sum_{n=-\infty}^{+\infty} a_n(j,t,s) \cdot a_{-n-2}(j,t,s) \right| \\ &\le \frac{2\pi}{|s_j|} \left( |(1/2\pi) a_{j,t,s}|^2 + 2 \sum_{n=1}^{+\infty} |a_{n-1}(j,t,s)| \cdot |a_{-n-1}(j,t,s)| \right). \end{aligned}$$

Since

$$\sum_{n=1}^{+\infty} |a_{n-1}(j,t,s)| \cdot |a_{-n-1}(j,t,s)| \le (A_2)^2 |s_j| / (1 - |s_j|^{1/2})$$

by Lemma 4, we conclude the second assertion. ∎

PROOF OF THEOREM 4: Fix $(T,S)$ in $B^*$ and a real line $L$ passing through $(T,S)$. Let $x$ be a real parameter of $L$ such that $x=0$ corresponds to $(T,S)$. Let $\Phi(x)$ be as in 1, where we take as $\{\mu(x)\}$ the family $\{\nu(t,s)\}$ in Lemma 1, restricted on $L$.

Let $\Phi^{1,0}(x)$ and $\Phi^{0,1}(x)$ be the (1,0)- and (0,1)-part, respectively, of $\Phi(x)$. Then the argument in the proof of [**10**, Lemma 1] shows that

$$\|(\Phi^{1,0})'(0)\|_{R_{T,S}} = \|(\Phi^{0,1})'(0)\|_{R_{T,S}}.$$

Hence by (1.3) we have

$$\|\Phi'(0)\|_{R_{T,S}} \leq 2 \cdot \|\mu'(0)\|_\infty \|\phi_{T,S}\|_E, \tag{4.3}$$

where $E$ is any set containing the support of $\mu'(0)$.

Now the first assertion follows by (1.2), (4.3) and Lemma 2. Next we can see from (4.2) that

$$\frac{\partial \nu^2}{\partial(\mathrm{Re}\ s_j)^2}(T,S) = -(S_j)^{-1}\frac{\partial \nu}{\partial(\mathrm{Re}\ s_j)}(T,S)$$

for every $j$. Hence the second assertion follows by (1.2), (4.2), (4.3), Lemma 2 and the fact that the support of $\partial\nu/\partial(\mathrm{Re}\ s_j)$ is contained in $W_{j,t,s}$. ∎

## REFERENCES


1. R. D. M. Accola, *Differentials and extremal length on Riemann surfaces*, Proc. Nat. Acad. Sci. U. S. A. **46** (1960), 540–543.
2. L. Ahlfors, "Conformal Invariants," McGraw-Hill Inc., 1973.
3. F. Maitani, *Variations of meromorphic differentials under quasiconformal deformations*, J. Math. Kyoto Univ. **24** (1984), 49–66.
4. H. L. Royden, *The variation of harmonic differentials and their periods*, in "Complex Analysis," Birkhäuser Verlag, 1988, pp. 211–223.
5. M. Taniguchi, *Square integrable harmonic differentials on arbitrary Riemann surfaces with a finite number of nodes*, J. Math. Kyoto Univ. **25** (1985), 597–617.
6. M. Taniguchi, *Variational formulas on arbitrary Riemann surfaces under pinching deformation*, ibid. **27** (1987), 507–530.
7. M. Taniguchi, *Supplements to my previous papers; a refinement and applications*, ibid. **28** (1988), 81–86.
8. M. Taniguchi, *Abelian differentials with normal behavior and complex pinching deformation*, ibid. **29** (1989), 45–56.
9. M. Taniguchi, *Pinching deformation of arbitrary Riemann surfaces and variational formulas for abelian differentials*, in "Analytic function theory of one complex variable," Longman Sci. Techn., 1989, pp. 330–345.
10. M. Taniguchi, *A note on the second variational formulas of functionals on Riemann surfaces*, Kodai Math. J. **12** (1989), 283–295.
11. M. Taniguchi, *On the singularity of the periods of abelian differentials with normal behavior under pinching deformation*, J. Math. Kyoto Univ. (to appear).

# A Strong Harmonic Representation Theorem on Complex Spaces with Isolated Singularities

By

Takeo OHSAWA

1. Let $X$ be a reduced irreducible complex space, let $X'$ be the set of regular points of $X$, and let $d$ be the exterior derivative acting on the set of smooth differential forms say $C(X')$. Given a Hermitian metric $ds^2_{X'}$ on $X'$, the set of compactly supported elements of $C(X')$, say $C_0(X')$, is canonically provided with a structure of a pre-Hilbert space, and $d|_{C_0(X')}$ has closed extensions to the completion $\overline{C_0(X')}$. The minimal closed extension $d_{min}$ of $d|_{C_0(X')}$ is then given by $(d|_{C_0(X')})^{**}$, where $(\cdot)^*$ denotes the adjoint of $\cdot$, and the maximal closed extension $d_{max}$ is given by $((d|_{C_0(X')})^*|_{C_0(X')})^*$. The $L^2$ cohomology group $H_{(2)}(X',ds^2_{X'})$ is then defined as $\mathrm{Ker}\, d_{max}/\ \mathrm{Im}\, d_{max}$. In case $X$ is compact and $ds^2_{X'}$ is quasi-isometrically equivalent to a Hermitian metric of $X$, say $ds^2_X$, $H_{(2)}(X',ds^2_{X'})$ represents a diffeomorphism invariant, so that it will be denoted by $H_{(2)}(X)$, briefly. Cheeger-Goreski-MacPherson [1] has conjectured that the $L^2$ cohomology group $H_{(2)}(X)$ is canonically isomorphic to a topological invariant of $X$, the so called the intersection cohomology group of $X$ at the middle perversity. In the author's previous articles it has been shown that their conjecture is true if the singular points of $X$ are isolated (cf.[3],[4] and [5]). Here we would like to know whether $H_{(2)}(X)$ is isomorphic to the kernel

of the operator $\Delta_{max}$, the maximal closed extension of the Laplacian $\Delta$ with respect to $ds_X^2$. Unfortunately it is not ture in general, as $\log|z|$ is an $L^2$ harmonic function on the Riemann sphere minus two points $\{z=0\}$ and $\{z=\infty\}$. Nevertheless it is clear that under a suitable growth assumption, the solutions of $\Delta_{max}u = 0$ will be contained in $\mathrm{Ker}(d_{max}+d_{max}^*)$. Hence the right thing to be looked for will be a growth condition which characterizes the subspace $\mathrm{Ker}(d_{max}+d_{max}^*)$ in $\mathrm{Ker}\,\Delta_{max}$. Let $\delta$ be the distance function measured by $ds_X^2$ from the set of singular points of $X$, and let $I(X) := \{u \in \overline{C_0(X')};\ \|(\delta\log(\delta^{-1}+1))^{-1}u\| < \infty\}$, where $\|\ \|$ denotes the $L^2$ norm. The r-th component of $H_{(2)}(X)$ (resp. $I(X)$) will be denoted by $H^r_{(2)}(X)$ (resp. $I^r(X)$).

<u>Theorem 1</u> Let $X$ be a compact irreducible complex space of dimension $n$ whose singular points are isolated. Then $H^r_{(2)}(X)$ is canonically isomorphic to $\mathrm{Ker}\,\Delta_{max} \cap I^r(X)$ if $r \neq n$. Moreover there is a canonical injective homomorphism from $\mathrm{Ker}\,\Delta_{max} \cap I^n(X)$ into $H^n_{(2)}(X)$.

Thus we are left with the following

<u>Conjecture</u> Under the above situation, $\dim H^n_{(2)}(X) = \dim \mathrm{Ker}\,\Delta_{max} \cap I^n(X)$.

2. Let $V \subset \mathbb{C}^N$ be an irreducible complex analytic set of dimension $n$ containing the origin as an isolated singular point, let $z = (z_1,\dots,z_N)$ be the coordinate of $\mathbb{C}^N$, and let $\|z\|^2 = \sum_{i=1}^{N} |z_i|^2$. By Milnor's lemma, the function $\|z\|_{V\setminus\{0\}}$ has no critical points near $0$, say on $K_\varepsilon := \{z \in V;\ 0 < \|z\| \leq \varepsilon\}$ (cf. [2], §3). Let $U := \mathring{K}_\varepsilon$, and let $H_{(2)}(U)$ be the $L^2$ cohomology group of $U$ with

respect to the restriction of the euclidean metric. Moreover, let $\Phi := \bigcup_{\varepsilon'<\varepsilon}\{K;\ K \subset K_{\varepsilon'}\}$, $S := \{u \in \overline{C_0(U)};\ \text{supp}\ u \in \Phi\}$, and let $H_{(2),0}(U) := \text{Ker}\ d_{max} \cap S / \text{Im}\ d_{max} \cap S$. Let $H^r_{(2)}(U)$ (resp. $H^r_{(2),0}(U)$) be the r-th component of $H_{(2)}(U)$ (resp. $H_{(2),0}(U)$).

<u>Lemma 1</u> Under the above situation

(1) $$H^r_{(2)}(U) = \{0\} \quad \text{if} \quad r < n$$

and

(2) $$H^r_{(2),0}(U) = \{0\} \quad \text{if} \quad r < n.$$

For the proof of (1) and (2), see [3] and [5], respectively.

In what follows $X$ will denote a compact irreducible complex space of dimension $n$ with isolated singularities and $ds^2_X$ will be a Hermitian metric of $X$. Recall that a straightforward consequence of Lemma 1 is the following (cf.[3] or [4]).

<u>Proposition 2</u>

$$H^r_{(2)}(X) \cong \begin{cases} H^r(X') & \text{if} \quad r < n-1 \\ \\ H^r_0(X') & \text{if} \quad r > n+1 \end{cases}$$

canonically, and the canonical homomorphism $H^{n+1}_0(X') \longrightarrow H^{n+1}_{(2)}(X)$ (resp. $H^{n-1}_{(2)}(X) \longrightarrow H^{n-1}(X')$) is surjective (resp. injective). Here $H^r(X')$ (resp. $H^r_0(X')$) denotes the r-th de Rham cohomology group (resp. —— with compact support) of $X'$. Moreover $\text{Im}\ d_{max}$ is a closed subspace of $\overline{C_0(X')}$.

Let $\psi$ be any real valued $C^\infty$ function on $X'$ such that $ds^2_X + \partial\bar{\partial}\psi$ is a complete Hermitian metric on $X'$ satisfying

$|\partial\psi| < 1$, where $\partial\bar{\partial}\psi$ is canonically identified with a Hermitian form on X'. For each $\varepsilon \in [0,1]$, let $\|\ \|_\varepsilon$ (resp. $(\ ,\ )_\varepsilon$) denote the $L^2$ norm (resp. the inner product) with respect to $ds^2_\varepsilon := ds^2_X + \varepsilon\partial\bar{\partial}\psi$, and let $d^\varepsilon$ be the formal adjoint of d with respect to $ds^2_\varepsilon$.

Proposition 3 There exist a compact subset $K_0 \subset X'$, a positive number C and a family of $C^\infty$ functions $\{\varphi_\varepsilon\}_{\varepsilon \geq 0}$ on X' which converges to $\varphi_0 = \delta\log(\delta^{-1}+1)$ uniformly on compact subsets of X' as $\varepsilon \longrightarrow 0$, such that

$$(3) \qquad \|\varphi_\varepsilon^{-1}u\|_\varepsilon \leq C(\|u_{K_0}\|_0 + \|du\|_\varepsilon + \|d^\varepsilon u\|_\varepsilon)$$

for all $u \in C_0^r(X')$ with $r \neq n$ if $\varepsilon \in [0,1]$. Here $C_0^r(X')$ denotes the r-th component of $C_0(X')$, and $u_{K_0}$ the trivial extension of $u|_{K_0}$ to X'.

Proof is similar as that of Proposition 10 in [4].

In virtue of a theorem of Saper [6], the function $\psi$ can be chosen so that $H_{(2)}(X', ds^2_X + \partial\bar{\partial}\psi)$ represents the intersection cohomology group of X at the middle perversity. Combining this fact with an approximation argument based on the estimate as in the proof of Runge's approximation theorem, one obtains

$$(4) \qquad \mathrm{Ker}(d_{max}+d_{max}^*) = \mathrm{Ker}(d_{min}+d_{min}^*) \cong H_{(2)}(X).$$

For the logic see [4], Proposition 1 and Proof of Main Theorem.

3. Let $f \in \mathrm{Ker}(d_{max}+d_{max}^*) \cap \overline{C_0^r(X')}$ with $r \neq n$. By (4), we have $f \in \mathrm{Ker}(d_{max}+d_{max}^*) \cap \mathrm{Ker}(d_{min}+d_{min}^*) = \mathrm{Ker}(d_{min}+d^0{}_{min})$. Hence there exists a sequence $f_\mu \in C_0^r(X')$ which converges to f

with respect to the graph norm of $d + d^0$. Since $r \neq n$,

$$\|(\delta \log(\delta^{-1}+1))^{-1} f_\mu\|_0 \leq C( \|f_{\mu K_0}\|_0 + \|df_\mu\|_0 + \|d^0 f_\mu\|_0)$$

by (3). Letting $\mu \longrightarrow \infty$ we obtain

$$\|(\delta \log(\delta^{-1}+1))^{-1} f\|_0 \leq C( \|f_{K_0}\|_0 + \|d_{max} f\|_0 + \|d_{max}{}^* f\|_0)$$

which proves that $\mathrm{Ker}(d_{max}+d_{max}{}^*) \cap \overline{C_0^r(X')} \subset \mathrm{Ker}\Delta_{max} \cap I^r(X)$ for $r \neq n$.

On the other hand, suppose that $f \in \mathrm{Ker}\Delta_{max} \cap I(X)$. Then $f \in C(X')$ by the strong ellipticity of $\Delta$. Since the function $(t\log(t^{-1}+1))^{-1}$ is a continuous non-integrable function of t on $(0,1]$, there exists a family of compactly supported $C^\infty$ functions $\chi_\mu(t)$ ($\mu = 1,2,\ldots$) on $(0,\infty)$ such that

$$\begin{cases} \chi_\mu(t) \leq \chi_{\mu+1}(t) \leq 1 \\ |\chi_\mu'(t)| \leq (t\log(t^{-1}+1))^{-1} \end{cases} \qquad \text{for all } \mu,$$

and

$$\bigcap_{\mu=1}^{\infty} \mathrm{supp}(\chi_\mu - 1) = \emptyset .$$

We put $\rho_\mu = \chi_\mu \circ \delta$. Then

$$\begin{aligned} 0 &= (\Delta_{max} f, \rho_\mu^2 f)_0 \\ &= \|\rho_\mu df\|_0^2 + \|\rho_\mu d^0 f\|_0^2 \\ &\qquad + (df, 2\rho_\mu d\rho_\mu \wedge f)_0 - (d^0 f, 2\rho_\mu (d\rho_\mu)^0 f)_0, \end{aligned}$$

where $(d\rho_\mu)^0$ denotes the adjoint of exterior multiplication by $d\rho_\mu$ from the left hand side. By Cauchy-Schwartz inequality,

$$|(df, 2\rho_\mu d\rho_\mu \wedge f)_0| \leq \tfrac{1}{2}\|\rho_\mu df\|_0^2 + 2\| d\rho_\mu \wedge f\|_0^2$$

and

$$\left|(df^0, 2\varrho_\mu(d\varrho_\mu)^0 f)_0\right| \leq \tfrac{1}{2}\|\varrho_\mu d^0 f\|_0^2 + 2\|(d\varrho_\mu)^0 f\|_0^2 .$$

Therefore

$$(5) \qquad 0 \geq \|\varrho_\mu df\|_0^2 + \|\varrho_\mu d^0 f\|_0^2 - \tfrac{1}{2}\|\varrho_\mu df\|_0^2 - 2\|d\varrho_\mu \wedge f\|_0^2$$
$$- \tfrac{1}{2}\|\varrho_\mu d^0 f\|_0^2 - 2\|(d\varrho_\mu)^0 f\|_0^2$$
$$\tfrac{1}{2}(\|\varrho_\mu d^0 f\|_0^2 + \|\varrho_\mu d^0 f\|_0^2)$$
$$- 4\|(\delta \log(\delta^{-1}+1))^{-1} f_{\mathrm{supp}(\varrho_\mu - 1)}\|_0^2 .$$

Since $f \in I(X)$, $\lim_{\mu\to\infty}\|(\delta\log(\delta^{-1}+1))^{-1} f_{\mathrm{supp}(\varrho_\mu - 1)}\|_0 = 0$. Thus we infer from (5) that $df = d^0 f = 0$. By the same argument we see that $\|d(\varrho_\mu f)\|_0 + \|d^0(\varrho_\mu f)\|_0 \longrightarrow 0$ as $\mu \longrightarrow \infty$. Hence we have $f \in \mathrm{Ker}(d_{min} + d^0{}_{min})$, which proves that $\mathrm{Ker}\,\Delta_{max} \cap I(X) \subset \mathrm{Ker}(d_{max} + d_{max}{}^*)$.

Combining (4) with the above mentioned proof, we obtain the desired conclusion as in Theorem 1.

4. As for the complex Laplacian $\square$, it is not hard to show that

$$\mathrm{Ker}\,\square_{max} \cap I^r(X) = \mathrm{Ker}(\bar\partial_{min} + \bar\partial^0{}_{min}) \cap \overline{C^r(X')} \quad \text{for} \quad r \neq n$$

and that $\mathrm{Ker}\,\square_{max} \cap I^n(X) \subset \mathrm{Ker}(\bar\partial_{min} + \bar\partial^0{}_{min}) \cap \overline{C^n(X')}$. Moreover it is true that $\mathrm{Ker}(\bar\partial_{min} + \bar\partial^0{}_{min}) \cap \overline{C^r(X')} = \mathrm{Ker}(\bar\partial_{max} + \bar\partial_{max}{}^*) \cap \overline{C^r(X')}$ if $r \neq n, n\pm 1$, which can be shown similarly as in [4] by using an $\bar\partial$-analogue of the estimate (3). However it is not true that $\mathrm{Ker}(\bar\partial_{max} + \bar\partial_{max}{}^*) = \mathrm{Ker}(\bar\partial_{min} + \bar\partial^0{}_{min})$ in general. For instance let $X = \{(\zeta_0:\zeta_1:\zeta_2) \in \mathbb{CP}^2;\ \zeta_0\zeta_1{}^2 + \zeta_2{}^3 = 0\}$ and let $t = \zeta_1/\zeta_2$. Then $\zeta_1/\zeta_0 = t^3$ and $\zeta_2/\zeta_0 = t^2$ on $X$, so that $t^{-1}$ is square integrable near $t=0$ with respect to the restriction of the Fubini-Study metric. Since $t^{-1}$ is holomorphic on $X\setminus\{t=0\}$, we obtain $t^{-1} \in \mathrm{Ker}\bar{\partial}_{max} \cap \overline{C_0^0(X')} \subset \mathrm{Ker}(\bar{\partial}_{max} + \bar{\partial}_{max}{}^*)$. But obviously $\|(\delta\log\delta^{-1}+1))^{-1}t^{-1}\| = \infty$. Hence $t^{-1} \notin \mathrm{Ker}(\bar{\partial}_{min}+\bar{\partial}^0{}_{min})$. Therefore the condition $r \neq n, n\pm 1$ cannot be dropped if one wants to talk about the $\bar{\partial}$-analogue of Theorem 1.

References

[1] Cheeger,J., Goreski,M. and MacPherson,R.,
$L^2$-cohomology and intersection homology for singular varieties, Ann.Math.Stud. 102 Seminar on Differential Geometry, 1982, 303-340.

[2] Milnor,J.,
Singular points of complex hypersurfaces, Ann.Math.Stud. 61, 1968.

[3] Ohsawa,T.,
Hodge spectral sequence on compact Kähler spaces, Publ. RIMS, Kyoto Univ. 23 (1987),613-625.

[4] ———,
Cheeger-Goreski-MacPherson's conjecture for the varieties with isolated singularities, to appear in Math.Zeit.

[5] ———,
Supplement to "Hodge spectral sequence on compact Kähler spaces", to appear in Publ. RIMS, Kyoto Univ.

[6] Saper,L.,
$L^2$-cohomology and intersection homology of certain algebraic varieties with isolted singularities, Invent.Math. 82 (1985), 207-255, $L^2$-cohomology of isolated singularities, preprint.

# Mordell-Weil lattices of type $E_8$ and deformation of singularities

Dedicated to Professor F. Hirzebruch

Tetsuji Shioda
Department of Mathematics, Rikkyo University, Tokyo

## 0. Introduction

In our talk at the Kyoto Symposium, August 1989, we introduced the notion of the Mordell-Weil lattice of an elliptic surface and explained the basic results on the Mordell-Weil lattices of rational elliptic surfaces (see [S1] for the summary).

In this note, we discuss an application of this theory to the study of deformation of singularities. From this new viewpoint, we can reprove the well-known results on the deformation of rational double points of type $E_8$, $E_7$ or $E_6$, due to Briekorn, Tjurina and others (cf.[B1,2],[DPT],[L],[O],[S1]), in a more global, refined form. In particular, we have a purely algebraic proof of the surjectivity of the monodromy, which says that the fundamental group of the complement of the discriminant locus in the parameter space maps onto the Weyl group $W(E_r)$. Further we obtain a very precise description of the stratification of the parameter space according to the type of singularities.

As a typical case, we treat here the case of $E_8$-singularity, but the same method can be applied to the other cases as well.

Actually our theory works over any ground field k. For instance, for $k = \mathbb{Q}$, the results will have application to number theory and arithmetic geometry (cf.[S2]), and for a field k with positive characteristic, similar arguments can be applied to the singularities in characteristic p. In this note, however, we mainly work over the complex number field $\mathbb{C}$ for the sake of simplicity.

## 1. Mordell-Weil lattices of type $E_8$.

We consider the universal deformation of the rational double point of type $E_8$ : $y^2 = x^3 + t^5$, defined by the weighted homogeneous equation:

(1.1) $$y^2 = x^3 + t^5 + x( \Sigma_{i=0}^{3} p_i t^i) + ( \Sigma_{i=0}^{3} q_i t^i),$$

with the parameter $\lambda = (p_0,\dots,p_3,q_0,\dots,q_1) \in \mathbb{C}^8$. The weights are defined as follows:

(1.2)

| y | x | t | $p_0$ | $p_1$ | $p_2$ | $p_3$ | $q_0$ | $q_1$ | $q_2$ | $q_3$ |
|---|---|---|---|---|---|---|---|---|---|---|
| 15 | 10 | 6 | 20 | 14 | 8 | 2 | 30 | 24 | 18 | 12 |

For a fixed $\lambda$, we can regard (1.1) as the defining equation of:

(1.3) $X_\lambda$ = an affine surface in $\mathbb{C}^3$,

(1.4) $E_\lambda$ = an elliptic curve defined over $\mathbb{C}(t)$,

or

(1.5) $S_\lambda$ = a smooth projective elliptic surface with relatively minimal fibration $f: S_\lambda \to \mathbb{P}^1$.

The relation of these three objects are described as follows. First $E_\lambda$ is the generic fibre of $f: S_\lambda \to \mathbb{P}^1$, and $S_\lambda$ is the Kodaira-Néron model of the elliptic curve $E_\lambda$ over $K = \mathbb{C}(t)$. We denote by $E_\lambda(K)$ the Mordell-Weil group of K-rational points of $E_\lambda$:

(1.6) $$E_\lambda(K) = \{O\} \cup \{P = (x, y) \mid x, y \in K \text{ satisfy } (1.1)\}$$

where the origin O is the point at infinity (x:y:1) = (0:1:0). This group can be naturally identified with the group of holomorphic sections of $f: S_\lambda \to \mathbb{P}^1$, and we use the same notation P,.. for a K-rational point of $E_\lambda$ and the corresponding section of f. We denote by (P) the curve in $S_\lambda$ defined as the image of the section P: $\mathbb{P}^1 \to S_\lambda$.

For any $\lambda$, the fibre $f^{-1}(\infty)$ of $f: S_\lambda \to \mathbb{P}^1$ over $t = \infty$ is a singular fibre of type II (a rational curve with a cusp);see (1.13) below. Let

(1.7) $$S'_\lambda = S_\lambda - (O) - f^{-1}(\infty).$$

Then $S'_\lambda$ is a smooth surface giving the minimal resolution of the affine surface $X_\lambda$. See [K],[N] or [T] for the above. In particular, we have

**Lemma 1.1** The affine surface $X_\lambda$ is smooth if and only if the elliptic surface $f: S_\lambda \to \mathbb{P}^1$ has no reducible fibres.

Indeed, if $X_\lambda$ is smooth, then $S'_\lambda = X_\lambda$ is affine and hence any irreducible component of a fibre $f^{-1}(t)$ over $t \neq \infty$ must meet the section (O). Thus f has no reducible fibres. The converse is easy.

Let $D \subset \mathbb{C}^8$ denote the discriminant locus:

(1.8) $$D = \{ \lambda \in \mathbb{C}^8 \mid X_\lambda \text{ is not smooth } \}.$$

**Corollary 1.2** For $\lambda \in \mathbb{C}^8 - D$, the elliptic surface $f: S_\lambda \to \mathbb{P}^1$ has no reducible fibres, and conversely.

Note that the discriminant $\Delta$ of the elliptic curve $E_\lambda$ is given by

(1.9) $$\Delta = 4 ( \Sigma_{i=0}^{3} p_i t^i )^3 + 27 ( \Sigma_{i=0}^{3} q_i t^i + t^5 )^2,$$

which is a polynomial of degree 10 in t. The fibre $f^{-1}(v)$ at $t = v$ $(v \neq \infty)$ is irreducible if and only if (i) $\Delta(v) \neq 0$ (a smooth fibre), or (ii) $\Delta$ has a simple zero at v (type $I_1$, a rational curve with a node), or $\Delta$ has a zero of order 2 at v and $\Sigma_{i=0}^{3} p_i v^i = 0$ (type II, a rational curve with a cusp); cf.[K], [T]. For general choice of $p_i$ and $q_j$, $\Delta$ has 10 simple zeros, and hence $S_\lambda$ has 10 singular fibres of type $I_1$ and one of type II (at $\infty$); thus $\lambda \in \mathbb{C}^8 - D$.

Now we apply the theory of Mordell-Weil lattices to this situation to obtain:

**Proposition 1.3** Assume $\lambda \in \mathbb{C}^8 - D$. Then the Mordell-Weil group $E_\lambda(K)$ is a torsion-free abelian group of rank 8, and the structure of the Mordell-Weil lattice on it is the root lattice $E_8$, i.e., the unique positive-definite even unimodular lattice of rank 8.

**Proposition 1.4** Under the same assumption, there are exactly 240 minimal sections $P_i$ in $E_\lambda(K)$ with $\langle P_i, P_i \rangle = 2$, which are of the form $P_i = (x, y)$ with

(1.10) $$\begin{cases} x = g t^2 + a t + b \\ y = h t^3 + c t^2 + d t + e \end{cases} \qquad (a,b,\dots,g,h \in \mathbb{C}).$$

They correspond to the 240 roots of $E_8$.

See Theorems 2.1, 2.2 in [S1].

Next we consider the specialization homomorphism

(1.11) $\quad sp_\infty : E_\lambda(\mathbb{C}(t)) \to \mathbb{C} \subset f^{-1}(\infty)$,

defined as follows. For each P, $sp_\infty(P)$ is the unique intersection point of the section (P) and $f^{-1}(\infty)$. With respect to the coordinates

(1.12) $\quad X = x/t^2,\ Y = y/t^3,\ T = 1/t$,

the singular fibre $f^{-1}(\infty)$ of type II is defined by

(1.13) $\quad Y^2 = X^3 \qquad (T = 0)$,

which has the cusp at $(X, Y) = (0,0)$. The smooth part of $f^{-1}(\infty)$ is naturally identified with $\mathbb{C}$, via the coordinate $u = X/Y$, in such a way that the group law on the elliptic curve $f^{-1}(t)$ specializes to the usual additive group law on $\mathbb{C}$. Since a section meets any fibre at a smooth point of the fibre, we have $sp_\infty(P) \in \mathbb{C}$. For instance, for the minimal section $P_1$ given by (1.10), $(P_1)$ meets $Y^2 = X^3$ at $(X, Y) = (g, h)$. Hence we have $g \neq 0$, $h \neq 0$, and

(1.14) $\quad sp_\infty(P_1) = g/h \neq 0$.

Note that the coefficients a,b,.. in (1.10) have weights compatible with (1.2):

(1.15)

| a | b | c | d | e | g | h | u = g/h |
|---|---|---|---|---|---|---|---|
| 4 | 10 | 3 | 9 | 15 | -2 | -3 | 1 |

This suggests that u be the most important parameter. Indeed, this is the case. Namely we have

(1.16) $\quad g = u^{-2}, \qquad h = u^{-3}$,

and, for generic $\lambda$, a,b,...,e are determined rationally from u. Moreover u itself is an interesting algebraic function of the parameter $\lambda = (p_0,\dots,q_3)$, as we see in the next section.

## 2. The universal polynomial of type $E_8$

In this section, we assume that $\lambda = (p_0,\ldots,q_3) \in \mathbb{C}^8$ is generic, that is, $p_0,\ldots,q_3$ are algebraically independent over $\mathbb{Q}$, unless otherwise mentioned. As we saw in §1, the Mordell-Weil lattice $E_\lambda(\mathbb{C}(t))$ contains 240 minimal sections $P_i$ ($1 \le i \le 240$) corresponding to the roots of $E_8$.

Define the polynomial of degree 240 in the indeterminate u:

$$\Phi(u, \lambda) = \prod_{i=1}^{240} (u - u_i), \quad u_i = sp_\infty(P_i). \tag{2.1}$$

This will be called the *universal polynomial of type* $E_8$. Let $P_1,\ldots,P_8$ form a basis in the sense of root system, and arrange $\{P_i\}$ so that $P_1,\ldots,P_{120}$ correspond to the positive roots, i.e., they can be written as $\mathbb{Z}$-linear combination of $P_1,\ldots,P_8$ with non-negative coefficients (cf.[B]).

**Theorem 2.1** The polynomial $\Phi(u, \lambda)$ has coefficients in the polynomial ring $\mathbb{Q}[\lambda] = \mathbb{Q}[p_0,\ldots,q_3]$, and it is irreducible over the rational function field $\mathbb{Q}(\lambda) = \mathbb{Q}(p_0,\ldots,q_3)$. The splitting field $\mathcal{K} = \mathbb{Q}(\lambda)(u_1,\ldots,u_{240})$ is a Galois extension of $\mathbb{Q}(\lambda)$ such that

$$\mathrm{Gal}(\mathcal{K}/\mathbb{Q}(\lambda)) = W(E_8) \quad \text{(the Weyl group of type } E_8) \tag{2.2}$$

and

$$\mathcal{K} = \mathbb{Q}(u_1,\ldots,u_8) \quad \text{(a purely transcendental extension of } \mathbb{Q}). \tag{2.3}$$

Furthermore we have

$$\mathbb{Q}[u_1,\ldots,u_8]^{W(E_8)} = \mathbb{Q}[p_0,\ldots,q_3]. \tag{2.4}$$

(N.B. If the reader prefers it, $\mathbb{Q}$ can be replaced by $\mathbb{C}$ everywhere in the above statements, by taking $p_0,\ldots,q_3$ to be algebraically independent over $\mathbb{C}$. This makes the theorem a little weaker, though.)

Proof(outline). The key idea is to compute $\Phi(u, \lambda)$ in two different ways. On the one hand, we have, by definition,

$$(2.5) \qquad \Phi(u, \lambda) = u^{240} + \sum_{n=1}^{240} (-1)^n I_n \cdot u^{240-n}$$

where $I_n$ is the n-th elementary symmetric polynomial of $u_i$ $(1 \le i \le 240)$. Obviouly each $I_n$ is invariant under the Weyl group $W(E_8)$ and $I_n = 0$ for n odd. Thus we have

$$(2.6) \qquad \mathbb{Q}[I_2, \ldots, I_{240}] \subset \mathbb{Q}[u_1, \ldots, u_8]^{W(E_8)},$$

in which $\subset$ will turn out to be equality below.

On the other hand, we obtain another expression of $\Phi$ by means of elimination. For that, substitute (1.10) into the equation (1.1) and look at the coefficients of $t^m$ for $m = 6,5,\ldots,0$. Then we get 7 polynomial relations among $a,b,\ldots,g,h$ over $\mathbb{Q}[p_0,\ldots,q_3]$:

$$(2.7) \qquad h^2 = g^3, \quad 2\, c \cdot h = 3\, a^2 g + 1, \ \ldots,$$
$$e^2 = b^3 + b \cdot p_0 + q_0.$$

Now, we set $u = g/h$ in view of (1.14) and determine g,h by (1.16). Then, eliminating c,d,e,b and a in this order from the relations in (2.7), we finally obtain a monic polynomial of degree 240 in u with coefficients in $\mathbb{Q}[p_0,\ldots,q_3]$. Since $u = g/h = sp_\infty(P_i)$, this polynomial must coincide with $\Phi(u, \lambda)$ defined by (2.1). This proves the first assertion in Theorem 2.1.

Next we compare the coefficients of $u^d$ in the two expression of $\Phi$ for d = 2,8,12,14,18,20,24,30, which are the weights of the fundamental invariants of the Weyl group $W(E_8)$ (cf.[B] and (1.2)). Then we find the following explicit formulas:

$$(2.8) \qquad I_2 = 60\, p_3, \qquad\qquad I_8 = 720\, p_2 + 478170\, p_3^4,$$

$I_{12} = 15120\, q_3 + \dots, \quad I_{14} = 79200\, p_1 + \dots,$

$I_{18} = 2620800\, q_2 + \dots, \quad I_{20} = 11040480\, p_0 + \dots,$

$I_{24} = 419237280\, q_1 + \dots, I_{30} = 65945880000\, q_0 + \dots,$

where ... stands for a sum of terms in $p_i$ or $q_j$ of lower weights. Obviously it follows that

(2.9) $\quad \mathbb{Q}[I_2, \dots, I_{30}] = \mathbb{Q}[p_3, p_2, q_3, p_1, q_2, p_0, q_1, q_0].$

This shows that both $\{I_2, \dots, I_{30}\}$ and $\{p_0, \dots, q_3\}$ form the fundamental invariants of $W(E_8)$ because they are algebraically independent elements with the right weights. This proves (2.4), i.e.,

$$\mathbb{Q}[u_1, \dots, u_8]^{W(E_8)} = \mathbb{Q}[p_0, \dots, q_3].$$

Note, in particular, that $u_1, \dots, u_8$ are algebraically independent. By Galois theory, it is immediate that $\mathbb{Q}(u_1, \dots, u_8)$ is a Galois extension of $\mathbb{Q}(p_0, \dots, q_3)$ with Galois group $W(E_8)$. Moreover this Galois group acts transitively on the 240 roots $u_i$ of the polynomial $\Phi(u, \lambda)$, since the Weyl group $W(E_8)$ acts transitively on the "roots" of $E_8$. This proves the irreducibility of $\Phi$ over $\mathbb{Q}(p_0, \dots, q_3)$. q.e.d.

**Corollary 2.2** For $\lambda$ generic, the specialization map

$$sp_\infty : E_\lambda(\mathbb{C}(t)) \to \mathbb{C}$$

is an injective homomorphism, with the image $\Sigma_{i=0}^{8}\, \mathbb{Z}\, u_i$ being a submodule of rank 8 in $\mathcal{X} = \mathbb{Q}(u_1, \dots, u_8)$. In particular, each minimal section $P_i$ is uniquely determined by $u_i = sp_\infty(P_i)$; in other words, the coefficients a, b, .. of $P_i$ in (1.10') are uniquely determined by $u_i$.

This is obvious, since $u_1, \dots, u_8$ are linearly independent over $\mathbb{Q}$ (recall that they are even algebraically independent).

N.B. We take this opportunity to cancel Lemma 3.4 of [S1, II] which wrongly claimed that the specialization map $sp_\infty$ is injective as far as the Mordell-Weil lattice $E_\lambda(K)$ is of type $E_8$.

## 3. Specialization of $\Phi(u, \lambda)$

Let us extend the definition of $\Phi(u, \lambda_1)$ to arbitrary $\lambda_1 \in \mathbb{C}^8$. For that purpose, consider the Néron-Severi lattice NS(S) of $S = S_{\lambda_1}$ (with respect to the intersection pairing) and the sublattice $\langle (O), F \rangle$ spanned by the zero section and the fibre class F. Its orthogonal complement, say L, is a negative-definite even unimodular lattice of rank 8 (i.e., $L \simeq E_8^-$); indeed L is isomorphic to the opposite lattice of the Mordell-Weil lattice of type $E_8$ for a generic $\lambda$. We propose to call L the $E_8$-*frame* of the rational elliptic surface $f: S \to \mathbf{P}^1$ with the zero section O.

Now L contains the 240 "roots", i.e. the divisor classes $\{D_i\}$ on the surface S such that

(3.1) $\quad (D_i^2) = -2, \quad D_i \perp (O), F \quad (1 \le i \le 240).$

Letting $T \subset NS(S)$ be the "trivial lattice" spanned by (O) and all the fibre components of f, we consider the composite map

(3.2) $\quad \sigma : L \subset NS(S) \to NS(S)/T \simeq E(K) \xrightarrow{sp_\infty} \mathbb{C},$

where the middle isomorphism is the canonical one ([S1,Theorem 1.1]).

Define the extended polynomial by

(3.3) $\quad \Phi(u, \lambda_1) = \prod_{i=1}^{240} ( u - \sigma(D_i)).$

**Proposition 3.1** (i) If $\lambda = \lambda_1$ is generic, then both definitions of $\Phi(u, \lambda)$ coincide. (ii) For any specialization $\lambda \to \lambda_1$ over some field $\kappa \subset \mathbb{C}$ (cf.[W]), $\Phi(u, \lambda)$ is uniquely specialized to $\Phi(u, \lambda_1)$.

Proof. For $\lambda$ generic (or more generally, for $\lambda_1$ with the Mordell-Weil lattice $E_8$), the 240 roots in the $E_8$-frame are given by

(3.4) $D_i = (P_i)-(0)- F \quad (1\le i\le 240)$

where $P_i$ are the 240 minimal sections, and we have

(3.5) $\sigma(D_i) = sp_\infty(P_i)$.

This proves (i).

Next, for (ii), we consider any specialization of $NS(S_\lambda)$ to $NS(S_{\lambda_1})$ over $\lambda \to \lambda_1$. Obviously the $E_8$-frame of $S_\lambda$ specializes to that of $S_{\lambda_1}$ and the 240 roots in the former specialize bijectively to those in the latter by the invariance of the intersection number under specialization. Hence the assertion.

**Remark 3.2** In a more geometric term, the above (ii) can be rephrased as follows. If we connect $\lambda$ and $\lambda_1$ by a path in $\mathbb{C}^8$, then we can deform the 240 roots $\{D_i\}$ in the $E_8$-frame of $S_\lambda$ in a unique, continuous way to the $\{D_i'\}$ in the $E_8$-frame of $S_{\lambda_1}$. A subtle point here is that the family $\{S_\lambda | \lambda \in \mathbb{C}^8\}$ is *not* a smooth family, though its restriction over $\mathbb{C}^8 - D$ is. See the next section.

**Lemma 3.3** For $\lambda_1 \in \mathbb{C}^8$, $\Phi(u, \lambda_1)$ has zero at $u = 0$ if and only if $S_{\lambda_1}$ has a reducible fibre. In other words, the discriminant locus $D \subset \mathbb{C}^8$ is defined by the equation $I_{240} = 0$.

Proof. Suppose that $S_{\lambda_1}$ has a reducible fibre $f^{-1}(v)$. Take an irreducible component $\theta$ not meeting the zero section (O). Then it satisfies $(\theta^2) = -2$, $\theta \perp (O)$, F and $\sigma(\theta) = 0$. Hence $\Phi(u, \lambda_1)$ has zero at $u = 0$. Conversely, if $S_{\lambda_1}$ has no reducible fibre, then there are 240 $D_i$ with $\sigma(D_i) = sp_\infty(P_i) \neq 0$ by (3.4),(3.5) and (1.14), hence the constant term $I_{240}$ of $\Phi$ does not vanish. The last statement follows from Corollary 1.2. q.e.d.

## 4. Monodromy of the Mordell-Weil lattice

Let us describe the geometric consequence of Theorem 2.1.

First, by (2.4), tensored by $\mathbb{C}$ over $\mathbb{Q}$, we have

(4.1) $$\mathbb{C}[u_1,\ldots,u_8] \supset \mathbb{C}[u_1,\ldots,u_8]^{W(E_8)} = \mathbb{C}[p_0,\ldots,q_3].$$

Geometrically this means that the quotient variety (or orbit space) of the affine space $\mathbb{C}^8$ by the action of the Weyl group $W(E_8)$ is isomorphic to the affine space $\mathbb{C}^8$ with coordinates $(p_0,\ldots,q_3)$. Let $\pi$ be the quotient morphism:

(4.2) $$\pi : \mathbb{C}^8 \to \mathbb{C}^8/W(E_8) \simeq \mathbb{C}^8.$$

Then it is wellknown (cf.[B,Ch.5]) that its restriction

(4.3) $$\pi' : \mathbb{C}^8 - \pi^{-1}(D) \to \mathbb{C}^8 - D$$

is an unramified covering. Recall that D is defined by the vanishing of $I_{240} = \prod_{i=1}^{240} u_i = \left(\prod_{i=1}^{120} u_i\right)^2$.

Now let $\mathfrak{X}$ denote the hypersurface in $\mathbb{C}^3 \times \mathbb{C}^8$ defined by the equation (1.1), and $\varphi : \mathfrak{X} \to \mathbb{C}^8$ the projection.

(4.4) $$\begin{array}{ccccc} \mathfrak{X} & \subset & \mathbb{C}^3 \times \mathbb{C}^8 & \qquad & (x,y,t;p_0,\ldots,q_3) \\ \varphi\downarrow & & \downarrow & & \downarrow \\ \mathbb{C}^8 & = & \mathbb{C}^8 & & \lambda = (p_0,\ldots,q_3). \end{array}$$

Then $\varphi$ defines a flat family of affine surfaces $X_\lambda = \varphi^{-1}(\lambda)$ in $\mathbb{C}^3$, parametrized by $\mathbb{C}^8$, which is smooth over $\mathbb{C}^8 - D$. For $\lambda \in \mathbb{C}^8 - D$, the elliptic surface $S_\lambda$ is obtained from $X_\lambda$ by adding the zero section (O) and the singular fibre of type II at $t = \infty$. It is easy to see that $\{S_\lambda | \lambda \in \mathbb{C}^8 - D\}$ forms also a smooth family. Therefore the Mordell-Weil lattice (and the $E_8$-frame) of $S_\lambda$ ($\lambda \in \mathbb{C}^8 - D$) defines a local system of rank 8 lattice over $\mathbb{C}^8 - D$ in the classical (as well as in the etale) topology:

$$(4.5) \qquad \mathcal{L} = \bigcup_{\lambda \in \mathbb{C}^8 - D} E_\lambda(K).$$

Take a base point $\lambda_o \in \mathbb{C}^8 - D$, and consider the monodromy representation

$$(4.6) \qquad \rho : \pi_1(\mathbb{C}^8 - D, \lambda_o) \to \mathrm{Aut}(E_8) = W(E_8)$$

where we identify the Mordell-Weil lattice (or the $E_8$-frame) of $S_{\lambda_o}$ with the root lattice $E_8$. Then Theorem 2.1 implies:

**Proposition 4.1** The map $\rho$ is surjective.

Indeed we have an explicit covering (4.3) of $\mathbb{C}^8 - D$ with the Galois group $W(E_8)$.

**Proposition 4.2** The 240 minimal sections of the Mordell-Weil lattice $E_\lambda(K)$ of type $E_8$ ($\lambda \in \mathbb{C}^8 - D$) form a connected unramified covering of degree 240 of $\mathbb{C}^8 - D$.

Indeed this is nothing but the locus of the 240 roots in the local system $\mathcal{L}$. It is connected because of the irreducibility of $\Phi$ stated in Theorem 2.1.

Now we consider the pull-back of the family $\mathfrak{X} \to \mathbb{C}^8$ via $\pi$ :

$$(4.7)\qquad \begin{array}{ccc} \mathfrak{Y} & \to & \mathfrak{X} \\ \psi\downarrow & & \varphi\downarrow \\ \mathbb{C}^8 & \xrightarrow{\pi} & \mathbb{C}^8 \end{array}$$

More explicitly, $\mathfrak{Y}$ is the hypersurface in $\mathbb{C}^3 \times \mathbb{C}^8$ with coordinates $(x,y,t;u_1,\dots,u_8)$ defined by

$$(4.8)\qquad y^2 = x^3 + t^5 + x(\textstyle\sum_{i=0}^{3} p_i t^i) + (\textstyle\sum_{i=0}^{3} q_i t^i),$$

in which $p_i$, $q_j$ are polynomials in $u_1,\dots,u_8$ determined by (2.4), or rather by (2.8), (2.9). For each $z = (u_1,\dots,u_8) \in \mathbb{C}^8$, $Y_z = \psi^{-1}(z)$ is equal to $X_\lambda = \varphi^{-1}(\lambda)$ with $\lambda = \pi(z)$, and thus it is smooth if and only if $z \in \mathbb{C}^8 - \pi^{-1}(D)$. We denote by $S_z$ the elliptic surface defined by (4.8), which is isomorphic to $S_\lambda = S_{\pi(z)}$. By (1.7), $S_\lambda$ is a minimal resolution of $X_\lambda$, compactified by adjoining (0) and $f^{-1}(\infty)$. Thus the same is true for $S_z$ and $Y_z$, but there is a significant difference. Namely, according to the theory of simultaneous resolution (see [B1], [B2],[P],[S1], cf. appendix by Saito and Naruki in [Sa]), we have:

**Proposition 4.3** $\{S_z \mid z \in \mathbb{C}^8\}$ forms a smooth family. In other words, if we let $S_z' = S_z-(0)-f^{-1}(\infty)$, then $S_z'$ gives a simultaneous resolution of $Y_z$ for all $z \in \mathbb{C}^8$.

We only note here that the obstruction for the simultaneous resolution vanishes for the smooth family $\{S_z \mid z \in \mathbb{C}^8 - \pi^{-1}(D)\}$, since $\pi_1(\mathbb{C}^8 - \pi^{-1}(D)) = \mathrm{Ker}(\rho)$ has the trivial monodromy representation.

**Remark 4.4** It will be of some interest to prove Proposition 4.3 directly, that is, to carry out the resolution of (4.8) over the ring

$\mathbb{Q}[u_1,\ldots,u_8]$. This should be possible and it will afford, we hope, the best concrete solution to Brieskorn's program for the simultaneous resolution for $E_8$-singularity (cf.[B2]).

At any rate, the field $\mathcal{K} = \mathbb{Q}(u_1,\ldots,u_8)$ appearing in Theorem 2.1 will be the smallest extension of the coefficient field $\mathbb{Q}(p_0,\ldots,q_3)$ over which the simultaneous resolution of (1.1) can be realized.

## 5. Stratification of the parameter space

We can describe the stratification of the parameter space $\mathbb{C}^8$ according to the type of the singularities of $X_\lambda$, using the invariants of the Weyl group $W(E_8)$ which are the coefficients of the universal polynomial $\Phi(u, \lambda)$. Recall that

(5.1) $\Phi(u, \lambda) = u^{240} + I_2 \cdot u^{238} + \cdots + I_{236} \cdot u^4 + I_{238} \cdot u^2 + I_{240}$

with

(5.2) $I_2, \ldots, I_{240} \in \mathbb{Q}[u_1,\ldots,u_8]^{W(E_8)} = \mathbb{Q}[p_0,\ldots,q_3]$.

For simplicity, we give the explicit description of stratification only for the first few strata, but it will be clear that our method allows one to continue further.

**Theorem 5.1** For $\lambda = (p_0,\ldots,q_3) \in \mathbb{C}^8$, let $X_\lambda$ be the affine surface defined by the equation (1.1). Then we have

(i) $X_\lambda$ is smooth $\iff$ $I_{240}(\lambda) \neq 0$

(ii) $X_\lambda$ has $A_1$-singularity $\iff$ $I_{240}(\lambda) = 0$, $I_{238}(\lambda) \neq 0$.

(iii) $X_\lambda$ has 2 $A_1$-sing. $\iff$ $I_{240}(\lambda) = I_{238}(\lambda) = 0$, $I_{236}(\lambda) \neq 0$.

(iv) $X_\lambda$ has either (a) $A_2$-sing. or (b) 3 $A_1$-sing.

$\iff$ $I_{240}(\lambda) = I_{238}(\lambda) = I_{236}(\lambda) = 0$, $I_{234}(\lambda) \neq 0$.

. . . . .

(f) $X_\lambda$ has $E_8$-sing. <=> $I_n(\lambda) = 0$, all n <=> $\lambda = 0$.

(Here 2 $A_1$-sing. means 2 singular points of type $A_1$, for example.)

Furthermore, the polynomial $\Phi(u, \lambda)$ has the following factorization in each case ( $F_d(u)$ stands for a polynomial of degree d in u):

(ii) $\Phi(u, \lambda) = u^2 F_{56}(u)^2 F_{126}(u)$

(iii) $u^4 F_{12}(u)^4 F_{64}(u)^2 F_{60}(u)$

(iv-a) $u^6 F_{54}(u)^3 F_{72}(u)$

(iv-b) $u^6 F_2(u)^8 F_{24}(u)^4 F_{48}(u)^2 F_{26}(u)$

The above decomposition type will distinguish the subcases (a),(b) in (iv). In particular, for general $\lambda$ in the locus of $X_\lambda$ with the $A_2$-singularity, $\Phi(u, \lambda)$ has 72 simple roots, while it has only 26 simple roots for general $\lambda$ in the locus of $X_\lambda$ having 3 $A_1$-singularities.

Proof. The key idea is to keep track of the 240 roots in the $E_8$-frame of the corresponding elliptic surface $S_\lambda$ under specialization, which is conveniently expressed by the polynomial $\Phi(u, \lambda)$ studied in §3. First we have:

**Lemma 5.2** Let $M_\lambda = E_\lambda(K)$ be the Mordell-Weil lattice of $S_\lambda$ and $M_\lambda^o = E_\lambda(K)^o$ the narrow Mordell-Weil lattice ( $K = \mathbb{C}(t)$). Then

(i) $X_\lambda$ : smooth <=> $M_\lambda \simeq E_8$

(ii) $X_\lambda$ : $A_1$-sing. <=> $M_\lambda \simeq E_7^*$ <=> $M_\lambda^o \simeq E_7$

(iii) $X_\lambda$ : 2 $A_1$-sing. <=> $M_\lambda \simeq D_6^*$ <=> $M_\lambda^o \simeq D_6$

(iv) $X_\lambda$ : $A_2$-sing. <=> $M_\lambda \simeq E_6^*$ <=> $M_\lambda^o \simeq E_6$

(v) $X_\lambda$ : 3 $A_1$-sing. <=> $M_\lambda \simeq D_4^* \oplus A_1^*$ <=> $M_\lambda^o \simeq D_4 \oplus A_1$.

Proof. The singularities of $X_\lambda$ correspond to the reducible fibres of the elliptic surface $f: S_\lambda \to \mathbb{P}^1$, which determines and is determined by the Mordell-Weil lattice $M_\lambda$ in case the Mordell-Weil rank $\geq 5$ (see [S1],[OS]). For instance, $X_\lambda$ has $A_1$- singularity if and only if $S_\lambda$ has only one reducible fibre with 2 irreducible components. In this case, the trivial lattice is $\langle(0),F\rangle \oplus A_1^-$ ($A_1^-$ = the opposite of the root lattice $A_1$), hence its orthogonal complement in $NS(S_\lambda)$ is $E_7^-$. It follows that the narrow Mordell-Weil lattice $M_\lambda^0$ is $E_7$ and that the Mordell-Weil lattice $M_\lambda$ is the dual lattice $E_7^*$. The other cases are similar. q.e.d.

We go back to the proof of Theorem 5.1. The case (i) is treated in Lemma 3.3. For the case (ii), let $f^{-1}(v) = \Theta_0 + \Theta_1$ be the reducible fibre where $\Theta_0$ is the identity component. To abbreviate the notation, set

(5.3) $\quad D(P) = (P) - (0) - F \quad (P \in M_\lambda).$

Then the following divisors represent the 240 roots in the $E_8$-frame of $S_\lambda$ in case (ii):

$$\pm\, \Theta_1$$

(5.4) $\quad D(P),\ D(P) + \Theta_1 \quad$ for 56 $P \in M_\lambda \simeq E_7^*$ with $\langle P,P\rangle=3/2$.

$\quad D(Q) \quad$ for 126 $Q \in M_\lambda^0 \simeq E_7$ with $\langle Q,Q\rangle=2$.

Therefore we have by (3.3)

$$\Phi(u, \lambda) = u^2\, F_{56}(u)^2\, F_{126}(u)$$

where we set

$$F_{56}(u) = \prod_P (u - sp_\infty(P)), \quad F_{126}(u) = \prod_Q (u - sp_\infty(Q)).$$

Since the constant terms of $F_{56}(u)$ and $F_{126}(u)$ are not zero (by the same argument as (1.14)), $\Phi(u, \lambda)$ has zero of exact order 2 at $u = 0$. Conversely, if $\Phi(u, \lambda)$ has order m at $u = 0$, then the number of the roots in the trivial lattice must be m in view of (3.2). Hence, if $m = 2$, $S_\lambda$ has only one reducible fibre with 2 irreducible components, so that its Mordell-Weil lattice $M_\lambda$ is $E_7^*$. This proves all the assertion for the case (ii).

The other cases can be proven in the same way. So we only write down the roots in the $E_8$-frame in the remaining cases. Let the reducible fibres of $S_\lambda$ be given as follows:

case (iii) $f^{-1}(v_j) = \theta_0^{(j)} + \theta_1^{(j)}$ $\quad ( j = 1,2)$

case (iv-a) $f^{-1}(v) = \theta_0 + \theta_1 + \theta_2$

case (iv-b) $f^{-1}(v_j) = \theta_0^{(j)} + \theta_1^{(j)}$ $\quad ( j = 1,2,3)$

with the identity component $\theta_0$ or $\theta_0^{(j)}$.

We list the 240 roots in the $E_8$-frame of $S_\lambda$:

(5.5) case (iii):

$\pm \theta_1^{(j)}$ $\quad (j = 1,2)$,

$D(P)$, $D(P) + \theta_1^{(1)}$, $D(P) + \theta_1^{(2)}$, $D(P) + \theta_1^{(1)} + \theta_1^{(2)}$

for 12 $P \in M_\lambda \simeq D_6^*$ with $\langle P,P\rangle=1$,

$D(P)$, $D(P) + \theta_1^{(1)}$, $D(P) + \theta_1^{(2)}$

for 64 $P \in M_\lambda \simeq D_6^*$ with $\langle P,P\rangle=3/2$,

$D(Q)$ for 60 $Q \in M_\lambda^0 \simeq D_6$ with $\langle Q,Q\rangle=2$.

(5.6) case (iv-a):

$\pm \theta_1$, $\pm \theta_2$, $\pm(\theta_1 + \theta_2)$

$D(P)$, $D(P) + \theta_1$, $D(P) + \theta_2$

for 54 $P \in M_\lambda \simeq E_6^*$ with $\langle P,P\rangle=4/3$,

$D(Q)$ for 72 $Q \in M_\lambda^0 \simeq E_6$ with $\langle Q,Q\rangle=2$.

(5.7) case (iv-b):

$\pm\ \theta_1^{(j)}$ $(j = 1,2,3)$,

$D(P)$, $D(P) + \theta_1^{(j)}$, $D(P) + \theta_1^{(j)} + \theta_1^{(k)}$, $D(P) + \theta_1^{(1)} + \theta_1^{(2)} + \theta_1^{(3)}$

for 2 $P \in A_1^* \subset M_\lambda$ with $\langle P,P\rangle = 1/2$ $(1 \le j < k \le 3)$,

$D(P')$, $D(P') + \theta_1^{(j)}$, $D(P') + \theta_1^{(k)}$, $D(P') + \theta_1^{(j)} + \theta_1^{(k)}$

for 24 $P' \in D_4^* \subset M_\lambda$ with $\langle P',P'\rangle = 1$ (some $j,k$),

$D(P")$, $D(P") + \theta_1^{(j)}$

for 48 $P" \in M_\lambda$ with $\langle P",P"\rangle = 3/2$ (some $j$)

$D(Q)$ for 26 $Q \in M_\lambda^0 \simeq D_4 \oplus A_1$ with $\langle Q,Q\rangle = 2$.

Perhaps it is easier to understand the above if we draw the picture. For instance, in case (iv-b), the typical sections P, P' or P" interesect the fibre components $\theta$ of the elliptic surface $S_\lambda$ as in the figure 1, because we have, with the notation of [S1,§1],

$$\mathrm{Contr}_{v_j}(P) = \begin{cases} 1/2 & \text{if } (P\theta_1^{(j)}) = 1, \\ 0 & \text{otherwise} \end{cases}$$

so that (noting $\chi = 1$ for a rational surface)

$$\langle P, P\rangle = 2 - 1/2 - 1/2 - 1/2 = 1/2,$$

$$\langle P',P'\rangle = 2 - 1/2 - 1/2 = 1,$$

$$\langle P",P"\rangle = 2 - 1/2 = 3/2.$$

Passing to the dual graph, we then obtain the diadrams as in the figure 2. We have only to find all the "roots" in these Dynkin diagrams which contain the "root" D(P) with multiplicity one.

This completes the proof of Theorem 5.1.

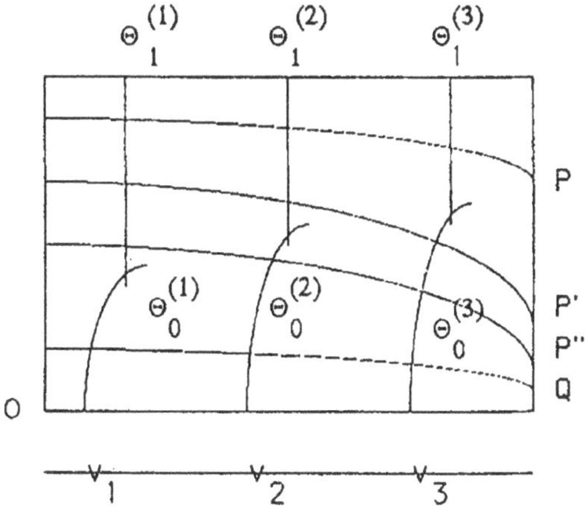


Figure 1

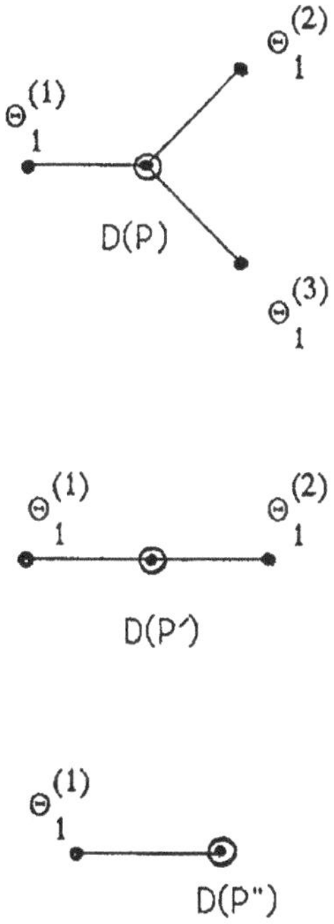


Figure 2

## 6. Explicit behavior of vanishing cycles

By now it must be clear to the reader that the Mordell-Weil lattice in our approach is playing the role of what is called the Milnor lattice in the topology of singularities. It should be noted, however, that the Mordell-Weil lattice has a finer structure than the Milnor lattice and contains more information. Let us illustrate this in the case of $E_8$-singularity by describing the behavior of the "vanishing cycles" under specific specialization. In fact, this is just a translation of the results of the previous sections.

The dictionary is:

Milnor lattice $\simeq$ $E_8$-frame $\simeq$ Mordell-Weil lattice

vanishing cycles $\leftrightarrow$ roots $\leftrightarrow$ minimal sections

Let $\lambda \in \mathbb{C}^8$ be generic and $\lambda_1 \in D \subset \mathbb{C}^8$. Consider a specialization $\gamma: \lambda \to \lambda_1$ (in algebraic context) or a path $\gamma$ in $\mathbb{C}^8$ connecting $\lambda$ and $\lambda_1$ (in topological context). Thus the nonsingular surface $X_\lambda$ degenerates to $X_{\lambda_1}$ acquiring some singularities, and, correspondingly, the Mordell-Weil lattice $M_\lambda \simeq E_8$ "degenerates" to the Mordell-Weil lattice $M_{\lambda_1}$ of smaller rank. As in (3.2), we have a homomorphism induced by $\gamma$:

(6.1) $\qquad \gamma_* : M_\lambda \to M_{\lambda_1}$

whose kernel is the root sublattice corresponding to the singularity of $X_{\lambda_1}$ (equivalently of the reducible fibres of $S_{\lambda_1}$).

First, let us consider the case where $X_{\lambda_1}$ has $A_1$-singularity, i.e., $\lambda_1$ is a point in D belonging to the first stratum. Then, by keeping track of the 240 roots $\{D(P_i)\}$ $(P_i \in M_\lambda)$ in the $E_8$-frame along $\gamma$ and using the result (5.4) ($\lambda$ there is to be replaced by $\lambda_1$ here), we see that the minimal sections $\{P_i\}$ can be rearranged and classified into 3 types as follows:

(6.2) i) 56 pairs $\{P_i, P_i'\}$ ($1 \le i \le 56$) such that $\gamma_*(P_i) = \gamma_*(P_i')$ are the minimal sections of $M_{\lambda_1} \simeq E_7^*$,

ii) 126 $Q_j$ such that $\gamma_*(Q_j)$ ($1 \le j \le 126$) are the minimal sections of $M_{\lambda_1}^0 \simeq E_7$,

iii) $R_1$, $R_2 = - R_1$ such that $\gamma_*(R_k) = 0$.

Therefore the 57 independent "vanishing cycles" in $NS(S_\lambda) = H_2(S_\lambda, \mathbf{Z})$

(6.3) $\quad (P_i) - (P_i') \quad (1 \le i \le 56)$ and $D(R_1)$

*literally vanish* under the specialization (or along the path) $\gamma$ from $X_\lambda$ to $X_{\lambda_1}$, since they are mapped by $\gamma_*$ to $\pm \theta_1$ in the $E_8$-frame of $S_{\lambda_1}$ and $\theta_1$ is nothing but the exceptional curve arising from the $A_1$-singularity of $X_{\lambda_1}$. This is visualized in the figure 3 below.

The above case for $A_1$-singularity is the situation occuring in the theory of Lefschetz pencils where only the simplest type of singularity is allowed.

In our approach, more complicated singularities can be treated in the same way. For example, consider the case of $A_2$-singularity. Then, by (5.6), the 240 $\{P_i\}$ in $M_\lambda$ are rearranged so that there are 54 tpiples $\{P_i, P_i', P_i''\}$ mapped to the 54 minimal sections of $M_{\lambda_1} \simeq E_6^*$ and 72 $Q_j$ mapped bijectively to the minimal sections of $M_{\lambda_1}^0 \simeq E_6$, and the 6 $R_k$ belonging to the kernel of $\gamma_*$. See the figure 4.

The other cases are similar, and so we leave it to the reader as an exercise.

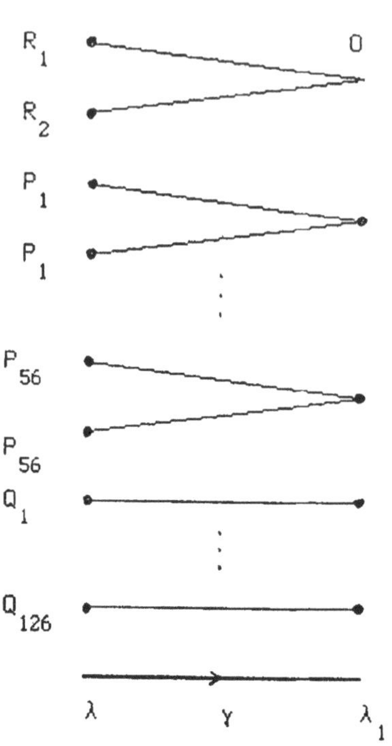


Figure 3

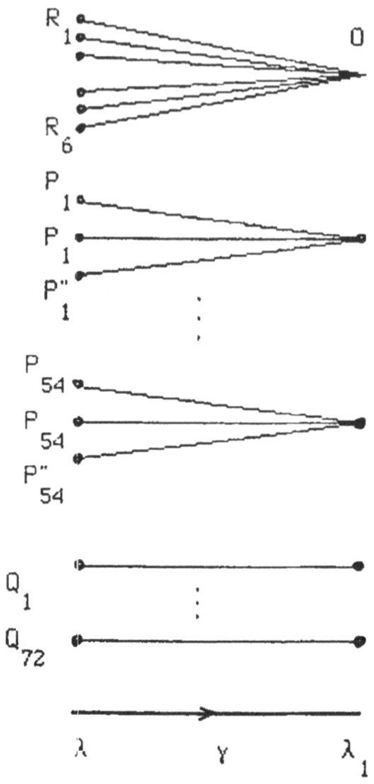


Figure 4

## 7. Examples

By restricting our attention to some subvariety V of the parameter space $\mathbb{C}^8$, we obtain a family of $X_\lambda$ or $S_\lambda$ parametrized by V. For instance, let us take as V the 1-dimensional affine space $\mathbb{C}^1$ which is the coordinate axis of the variable $s = p_i$ or $q_j$, with all the other coordinates being set to be zero. For brevity, we call such a family the "$p_i$-model" or "$q_j$-model" (of type $E_8$). Explicitly, the $p_i$-model is defined by

(7.1) $\quad y^2 = x^3 + p_i t^i x + t^5,$

and the $q_j$-model by

(7.2) $y^2 = x^3 + q_j t^j + t^5$

(cf. [S1, §4]).

Fix one of such models and write $X_s$ $(s \in \mathbb{C})$ for the affine surface corresponding to the parameter value $p_i$ (or $q_j$) = s. Of course the central member $X_0$ has the $E_8$-singularity. The singularity of $X_s$ has one and the same type for all $s \neq 0$, which can be easily determined. The results are summarized in the following table, together with the corresponding Mordell-Weil lattice $M_s$.

(7.3)

| model | $p_0$, $q_0$, $q_1$ | $p_1$ | $q_2$ | $p_2$, $q_3$ | $p_3$ |
|---|---|---|---|---|---|
| sing. of $X_s$ | $\phi$ | $A_1$ | $A_2$ | $D_4$ | $E_7$ |
| $M_s$ | $E_8$ | $E_7^*$ | $E_6^*$ | $D_4^*$ | $A_1^*$ |

For each model, the degeneration of $X_s$ to $X_0$ along a path $s \to 0$ (or a specialization) can be desribed as before. For the $p_0$-model (also for $q_0$ or $q_1$-model), $X_s$ is nonsingular for $s \neq 0$ and the 240 minimal sections of $M_s$ *vanish altogether* under any specialization $s \to 0$. For the other models, the general member $X_s$ is singular. For instance, the $p_3$-model gives a degeneration of $E_7$-singularity to $E_8$-singularity inside the discriminant locus $D \subset \mathbb{C}^8$.

The universal polynomial $\Phi(u) = \Phi(u, \lambda)$ introduced in §2 and §3 can be written down explicitly for each model. We mention here only the case of $q_0$-model and $p_3$-model.

**Proposition 7.1** For the $p_3$-model, the universal polynomial $\Phi(u)$ takes the form:

(7.4) $\Phi(u) = u^{126} (u^2 + p_3)^{56} (u^2 + 4 p_3)$,

where the 2nd or 3rd factor corresponds to the minimal sections of $M_s$

$\simeq A_1^*$ or of $M_s^0 \simeq A_1$.

**Proposition 7.2** For the $q_0$-model, the polynomial $\Phi(u)$ is a monic polynomial in $u^{30}$ of degree 8 with coefficients in $\mathbf{Z}[q_0]$, which decomposes into a product of two quartic polynomials. Explicitly,

$$(7.5) \qquad \Phi(u) = f_1(X/q_0)\ f_2(X/q_0)\ q_0^8 \qquad (X = u^{30})$$

with

$$f_1(X) = X^4 - 135432000\ X^3 + 56473225380000\ X^2 + 2176717249713600000\ X + 583200000,$$

$$f_2(X) = X^4 + 66081312000\ X^3 - 4811512860000\ X^2 + 1167566400000\ X + 583200000.$$

Both $f_1$ and $f_2$ are irreducible over $\mathbb{Q}$ and split in the cyclotomic field $\mathbb{Q}(\zeta_{30})$ ($\zeta_m = e^{2\pi i/m}$). The ratio of any two roots of $f_1 \cdot f_2$ is a 30-th power in $\mathbb{Q}(\zeta_{30})$.

We omit the proof. Note that 30 is the Coxeter number for $E_8$. Brieskorn [B2] studied the pull-back of the $q_0$-model via $q_0 = v^{30}$:

$$(7.6) \qquad y^2 = x^3 + t^5 + v^{30}$$

to obtain the simultaneous resolution of $E_8$-singularity over the 1-dimensional base. The smallest field over which such a simultaneous resolution of (7.6) can be done is the extension

$$(7.7) \qquad \mathbb{Q}(\zeta_{30})(\xi^{1/30}) \qquad (\xi\text{: a root of } f_1 \text{ or } f_2).$$

Thus this field would have appeared in [B2] if every step of the resolution is written out explicitly. Notice that the cyclotomic field $\mathbb{Q}(\zeta_{12})$ or $\mathbb{Q}(\zeta_{18})$ appeared, in the same vein, for the resolution of $E_6$- or $E_7$-singularity in [B1], the number 12 or 18 being the Coxeter number.

In closing this paper, we note that our method will allow one to study the differential equations arising from deformation of rational double points, from a global point of view (cf.[O, Appendix]). We hope to come back to this subject in some other occasion.

Acknowledgement During the preparation of this work, I have had helpful discussions with many people. I would like to thank all of them, especially, E. Horikawa, I. Nakamura, I. Naruki, T. Oda, K. Oguiso, K. Saito and I. Shimada.

## References

[B] Bourbaki, N.: Groupes et Algèbres de Lie, Chap. 4, 5 et 6, Hermann, Paris (1968).

[B1] Brieskorn, E.: Über die Auflösung gewisser Singularitäten von holomorphen Abbildungen, Math. Ann. 166, 76-102 (1966).

[B2] Brieskorn, E.: Die Auflösung der rationalen Singularitäten holomorpher Abbildungen, Math. Ann. 178, 255-270 (1968).

[CS] Conway, J., Sloane, N.: Sphere Packings, Lattices and Groups, Grund. Math. Wiss. 290, Springer-Verlag (1988).

[DPT] Demazure, M., Pinkham, H., Teissier, B.: Séminaire sur les Singularités des Surfaces, Lect. Notes in Math. 777 (1980).

[K] Kodaira, K.: On compact analytic surfaces II-III, Ann. of Math. 77, 563-626(1963); 78, 1-40(1963); Collected Works, vol.III, Iwanami and Princeton Univ. Press, 1269-1372 (1975).

[L] Looijenga,E.: On the semi-universal deformation of a simple-elliptic hypersurface singularity II:the discriminant, Topology 17, 23-40 (1978).

[N] Néron, A.: Modèles minimaux des variétés abéliennes sur les corps locaux et globaux, Publ. Math. I.H.E.S. 21(1964).

[O] Oda, T.: Introduction to algebraic singularities, (to appear).

[OS] Oguiso, K., Shioda, T.: The Mordell-Weil lattice of a rational elliptic surface, (in preparation).

[P] Pinkham, H.: Résolution simultanée de points doubles rationells, in:[DPT], 179-203 (1980).

[Sa] Saito, K.: Algebraic surfaces for regular systems of weights, in: Algebraic Geometry and Commutative Algebra, vol.II, Kinòkuniya, Tokyo, 517-612 (1988).

[S1] Shioda, T.: Mordell-Weil lattices and Galois representation, I, II,III, Proc. Japan Acad. 65A, 267-271, 296-299, 300-303 (1989).

[S2] Shioda, T.: Construction of elliptic curves with high rank via the invariants of the Weyl groups, (in preparation).

[Sl] Slodowy, P.: Simple singularities and simple algebraic groups, Lect. Notes in Math. 815 (1980).

[T] Tate, J.: Algorithm for determining the type of a singular fiber in an elliptic pencil, Lect. Notes in Math. 476, 33-52(1975).

[W] Weil, A.: Foundation of Algebraic Geometry, AMS (1962).

# THE SPECTRUM OF A RIEMANN SURFACE WITH A CUSP

Scott A. Wolpert*
Mathematics Department, University of Maryland
College Park, Maryland 20742

An intriguing question concerns the nature of the spectrum for a finite area hyperbolic Riemann surface $R$ with cusps. It is known that the spectrum of the Laplacian acting on $L^2(R)$ is the union of the band $[\frac{1}{4},\infty)$ with multiplicity the number of cusps and a discrete set corresponding to the square integrable eigenfunctions. Surprisingly little is known about the relative asymptotic contribution of the two components.

Selberg showed in 1954 that for congruence subgroups of $PSL(2,\mathbb{Z})$ the discrete spectrum alone already satisfies Weyl's asymptotic law. In contrast Deshouillers, Iwaniec, Phillips and Sarnak jointly conjecture that the generic Riemann surface with cusps will only have a finite discrete spectrum. The conjecture is supported by their joint work and also by the earlier work of Colin de Verdiere and of Phillips-Sarnak, showing that eigenvalues can *disappear* after deformation of the surface. The conjecture is also supported by numerical evidence [Hj4,Wn] as well as the corresponding result in the function-field case, [Ef].

Our plan is to 1) give a sketch of Colin de Verdiere's and Phillips-Sarnak's arguments that eigenvalues can *disappear* after deformation, and 2) an analysis of how the disappearance of an eigenvalue corresponds to the creation of a pole-zero pair on the critical line of the Eisenstein series with the pole moving off to the left and the zero moving off to the right. The exposition provides an opportunity for extending the Phillips-Sarnak result.

* Partially supported by a National Science Foundation grant.

To wit, if $\psi$ is an eigenfunction with eigenvalue $\lambda = \frac{1}{4} + r^2 \geq \frac{1}{4}$ and $\varphi$ is a holomorphic quadratic differential, we show that $\psi$ disappears for a first order deformation in the $\varphi$-direction if and only if $L_{\varphi * \psi}(2+ir) \neq 0$ for the L-function, the Rankin-Selberg convolution of $\varphi$ and $\psi$. In fact, the combination of a Phillips-Sarnak formula and our formula (2.5) provides that the special value of the L-function is (up to a factor) the first derivative of the coefficient which determines if the variation of the eigenfunction $\psi$ lies in $L^2$. That the variation of $\psi$ lies in $L^2$ is precisely the condition that the eigenvalue *remain* after perturbation. The second matter is the variation of a pole (of a pole-zero pair from the critical line) of the scattering matrix. The first variation necessarily vanishes. And in general we find that the variation of the pole away from the critical line (to lowest deformation order) is simply the square of the critical coefficient of $\psi$. In particular the second variation is given (up to a factor) by the square of the special value of the L-function. This is simply Fermi's Golden Rule for the dissolution under perturbation of an eigenvalue embedded in the continuum, [Si, sec.4]. The non-vanishing of the special value of the L-function is indeed a fundamental problem.

Many questions come to mind. Will higher variations help? Are there geometric conditions for the surface, the quadratic differential and the eigenfunction which guarantee nonvanishing of the special value? Can degeneration arguments provide any information? Would statistical methods help? At this point we do not know the answers. A great deal remains to be done.

I would like to thank the Taniguchi Foundation for their invitation to visit Japan and also for their most generous hospitality. I would also like to thank Dennis Hejhal for many interesting discussions as well as for his encouragement.

1. The spectrum of a surface with a cusp, Colin de Verdiere's pseudo Laplacian and the Eisenstein series.

1.1. Our purpose is to review the basic spectral theory for a Riemann surface with a cusp. We only give a sketch. The interested reader should consult [Hj1, Hj3, Kb, Ln, LP].

1.2. We start with a surface R with hyperbolic metric and a single cusp. R has a uniformisation group $\Gamma \subset PSL(2;\mathbb{R})$ with a unique conjugacy class of a maximal parabolic subgroup. Assume that the cusp is represented at $\infty$ and $z \longrightarrow z+1$ generates the $\Gamma$-stabilizer $\Gamma_\infty$ of $\infty$.

We will be considering Laplace-like operators $\Delta$, each acting on an $L^2$-type space $\mathcal{L}$. Recall that the spectrum is simply $\mathrm{Spec}(\Delta) = \{\lambda \in \mathbb{C} \mid (\Delta-\lambda)$ does not have a closed, bounded inverse$\}$ and the eigenvalues are $\mathrm{Ev}(\Delta) = \{\lambda \in \mathbb{C} \mid \psi \in \mathcal{L},\ \Delta\psi = \lambda\psi\}$. For us the basic case is $D = -y^2\left(\frac{\partial^2}{\partial x^2} + \frac{\partial^2}{\partial y^2}\right)$, $z = x+iy$, $D$ acting on $L^2(H/\Gamma)$; $H$ the upper half plane.

For R with a cusp and D acting on $L^2(H/\Gamma)$, Spec(D) contains the continuous band $[\frac{1}{4},\infty)$. The spectral projection associated to the band is given in terms of the Eisenstein series $E(z;s)$. To motivate the construction of E, first note that $y^s$ is invariant for translations $z \longrightarrow z+a$ and satisfies $Dy^s = s(1-s)y^s$. Now for $\mathrm{Re}\ s > 1$ the series

$$E(z;s) = \sum_{\gamma\in\Gamma_\infty\backslash\Gamma} (\mathrm{Im}\ \gamma(z))^s$$

i) converges absolutely; is analytic in s,

ii) is $\Gamma$-invariant, and

iii) satisfies $DE(z;s) = s(1-s)E(z;s)$.

Fourier series expansions are a basic part of our considerations. Start with a function $f \in L^2_{loc}(H/\Gamma)$. The function is $\Gamma$, hence $\Gamma_\infty$ invariant, thus

$$f(z) = \sum_n f_n(y)e^{2\pi inx}, \quad z = x+iy \in H.$$

We are interested in the situation

$$\begin{cases} Df = s(1-s)f & \text{for } y > a \geq 0 \\ f-f_0 \in L^2(\{\mathrm{Im}\, z \geq a\}/\Gamma_\infty) \end{cases}$$

then by the method of separation of variables

$$f-f_0 = \sum_n c_n y^{1/2} K_{ir}(2\pi ny)e^{2\pi inx}$$

for the K-bessel function $K_{ir}$ and $s(1-s) = \frac{1}{4}+r^2$. It also follows that the $0^{th}$ Fourier coefficient is given as $f_0 = \alpha y^s + \beta y^{1-s}$ for $\alpha, \beta \in \mathbb{C}$ and that $f-f_0$ is $O(e^{-cy})$ for $c > 0$, $y \longrightarrow \infty$. In particular we have

$$E(z;s) = y^s + \varphi(s)y^{1-s} + O(e^{-cy}).$$

Note: the identity coset of the Eisenstein sum precisely contributes the $y^s$.

The Eisenstein series $E(z;s)$ and coefficient $\varphi(s)$ are central to our considerations.

Basic Facts, [CV1, CV2, Hj1, Hj3, Kb, Ln, LP].

i) $E(z;s)$ has a meromorphic continuation to $\mathbb{C}$; poles independent of $z$,

ii) $E(z;\bar{s}) = \overline{E(z;s)}$; $\varphi(\bar{s}) = \overline{\varphi(s)}$,

iii) $E(z;s) = \varphi(s)E(z;1-s)$; $\varphi(s)\varphi(1-s) = 1$.

A sketch of Colin de Verdiere's proof of the analytic continuation will be given in section 1.6.

Example. $\Gamma = PSL(2;\mathbb{Z})$

$$E(z;s) = \frac{1}{2} \sum_{\substack{c,d \\ \text{rel.prime}}} \frac{y^s}{|cz+d|^s}, \quad \text{Re } s > 1$$

and of course $\varphi(s) = \pi^{2s-1} \frac{\Gamma(1-s)}{\Gamma(s)} \frac{\zeta(2-2s)}{\zeta(2s)}$ for the Euler $\Gamma$-function and the Riemann $\zeta$-function.

The spectrum of $D$ acting on $L^2(H/\Gamma)$ may also contain a discrete part (in addition to the 0-eigenspace). In fact, any eigenfunctions $\psi$ with eigenvalue $\lambda \geq \frac{1}{4}$ must be of a very special form.

Definition 1.1 (provisional). A function $f \in L^2(H/\Gamma)$ is cuspidal provided $f_0 \equiv 0$.

An elementary observation is that an eigenfunction $\psi$ with eigenvalue $\lambda \geq \frac{1}{4}$ is necessarily cuspidal. This essentially follows from two observations: $\psi_0 = \alpha y^{1/2+ir} + \beta y^{1/2-ir}$, $r \in \mathbb{R}$ and that $y^{1/2-ir} \notin L^2([1,\infty), \frac{dy}{y^2})$.

The spectral decomposition of $D$ acting on $L^2(H/\Gamma)$ consists of a discrete part and a continuous part. In particular, there exists an orthonormal system of eigenfunctions $\psi_n$, $D\psi_n = \lambda_n\psi_n$ such that for $f \in C_0^2(H/\Gamma)$, $\langle\ ,\ \rangle$ the Hermitian pairing and $\hat{f}(s) = \langle f, E(z;s)\rangle$ the Eisenstein transform, then

$$(1.1) \qquad f(z) = \sum_n \langle f,\psi_n\rangle\psi_n(z) + \frac{1}{4\pi}\int_{-\infty}^{\infty} \hat{f}(\tfrac{1}{2}+ir)E(z;\tfrac{1}{2}+ir)dr$$

and

$$Df(z) = \sum_n \langle f,\psi_n\rangle\lambda_n\psi_n + \frac{1}{4\pi}\int_{-\infty}^{\infty} \hat{f}(\tfrac{1}{2}+ir)(\tfrac{1}{4}+r^2)E(z;\tfrac{1}{2}+ir)dr,$$

[Hj1, Hj3, Kb, Ln, LP]. Note: $E(z;\frac{1}{2}+ir)$ is not in $L^2(H/\Gamma)$; the domain of the pairing $\langle\ ,\ \rangle$ is extended to include absolutely integrable integrands.

The action of the Laplacian on $L^2(H/\Gamma)$ has been diagonalized.

**1.3.** The fundamental question on the spectrum concerns the relative size of the discrete and continuous components. The analog of Weyl's law provides an overall measure of the size of the spectrum. Specifically for $N(R) = \#\{n|\lambda_n \leq R^2\}$ and $M(R) = -\frac{1}{4\pi}\int_{-R}^{R}\frac{\varphi'}{\varphi}(\frac{1}{2}+ir)dr$, then

$$M(R) + N(R) \sim \frac{\mathrm{Vol}(H/\Gamma)}{4\pi}R^2, \qquad \text{[HJ1, LP].} \tag{1.2}$$

The first term, the winding number of the scattering matrix [LP], measures the asymptotic density of the continuous spectrum, and of course the second term measures the asymptotic density of the discrete spectrum. The relative sizes of $M$ and $N$ are the open problem. What is the size of each term?

The question has two forms. The first is for a particular surface; this has a number theoretic air. Selberg has shown for $\Gamma \subset PSL(2;\mathbb{Z})$, a congruence subgroup, that

$$N(R) = \frac{\mathrm{Vol}(H/\Gamma)}{4\pi}R^2 + O(R \log R)$$

and had conjectured that for the general modular subgroup $\Gamma$ there would be a constant $\eta_\Gamma > 0$ such that $M(R) = O(R^{2-\eta})$, [S1]. In particular for a congruence subgroup the discrete spectrum alone already satisfies Weyl's asymptotic law. The second form of the size of $M$ versus $N$ question is for the generic surface. One can introduce variational considerations and consider how the size of $M$ versus $N$ changes under deformation of the surface. Colin de Verdiere first considered this matter for deformations through non-constant curvature metrics, [CV1, CV2]; Phillips-Sarnak followed with a result that cuspidal eigenfunctions can *disappear* after deformation of the surface through curvature $-1$ metrics, [PS1]. To date the optimal result is due to Deshouillers, Iwaniec, Phillips and Sarnak. They show (assuming the extended

Lindelöf hypothesis) for $\Gamma_0(p)$ a Hecke level-p congruence subgroup, p a prime, that of the $\sim R^2$ orthonormal eigenfunctions $\psi_n$ with $\lambda_n \le R^2$ that $\sim R^{2-\varepsilon}$ of them disappear for the generic deformation, [DIPS]. They then conjecture that for the general group the discrete spectrum will be finite dimensional; the spectrum could be almost pure continuous.

How can an eigenfunction disappear? It might seem that if there is a nice perturbation of the Laplacian, then there should be nice variations of the eigenfunctions. Actually in the sense of solutions of partial differential equations this is correct; the distinction though is that a function solving a PDE is a local condition, while a function lying in $L^2$ is a global condition. For instance the Eisenstein series $E = E(z;\frac{1}{2}+ir)$ satisfies $DE = (\frac{1}{4}+r^2)E$ and $E \notin L^2(H/\Gamma)$; the variation of an $L^2$-eigenfunction could be like the Eisenstein series.

**1.4.** We present an integration by parts formula in the style of Maass-Selberg. Choose a finite-sided fundamental domain $\mathcal{F}$ and a value a such that $\mathcal{F} \cap \{\mathrm{Im}\, z \ge a\}$ is a half infinite rectangle.

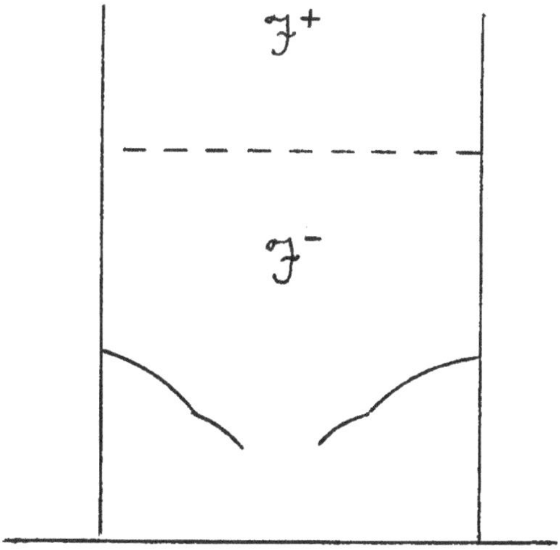

Let $f$ and $g$ be $\Gamma$-invariant functions satisfying:

i) $f$ and $g$ are $C^2$ on the closures of $\mathcal{F}^+$ and $\mathcal{F}^-$,

ii) $f - f_0$ and $g - g_0$ are $C^2$ in a neighborhood of $\operatorname{Im} z = a$,

iii) $f$, $f_x$, $f_y$ are $O(e^{-cy})$, $c > 0$ as $y \to \infty$ and $g$, $g_x$, $g_y$ are $O(y^{c'})$ as $y \to \infty$,

i.e., $f$ and $g$ are almost $C^1$ except that their $0^{th}$ Fourier coefficients (or derivatives thereof) may jump at $y = a$.

<u>Lemma 1.2</u>. $f$ and $g$ as above. For $D = -y^2\left(\frac{\partial^2}{\partial x^2} + \frac{\partial^2}{\partial y^2}\right)$ and $dA$ the hyperbolic area element

$$\int_{\mathcal{F}} fDg - gDf \; dA = \left[f_0 \frac{\partial g_0}{\partial y} - g_0 \frac{\partial f_0}{\partial y}\right]\Bigg|_{y=a^-}^{y=a^+} .$$

<u>Proof</u>. Start with the Euclidean form $\Delta = -\left(\frac{\partial^2}{\partial x^2} + \frac{\partial^2}{\partial y^2}\right)$, $dE$ Euclidean area, $\Omega$ a region and

$$\int_{\Omega} g\Delta f - f\Delta g \; dE = \int_{\partial\Omega} f\frac{\partial g}{\partial n} - g\frac{\partial f}{\partial n} ds.$$

First apply this with $\Omega = \mathcal{F}^+$; the vertical edges cancel and the integrand is $O(e^{-\varepsilon y})$. The right hand side reduces to

$$\int_{\substack{0 \le x \le 1 \\ y = a^+}} g\frac{\partial f}{\partial y} - f\frac{\partial g}{\partial y} dx.$$

Next consider $\Omega = \mathcal{F}^-$; again edges cancel and we have for the right hand side

$$\int_{\substack{0 \le x \le 1 \\ y = a^-}} f\frac{\partial g}{\partial y} - g\frac{\partial f}{\partial y} dx.$$

Now we insert the Fourier expansions for $f$ and $g$; by the orthogonality of exponentials the integrands reduce to $\sum_k f_{-k}\frac{\partial g_k}{\partial n} - g_{-k}\frac{\partial f_k}{\partial n}$. Now except possibly for the $k = 0$ term, the remaining terms have the same value on the upper and lower edge. Only the edge difference of the $0^{th}$ term remains; inserting factors of $y^2$ on the left hand side now completes the proof.

1.5. We review the basics for Colin de Verdiere's pseudo Laplacian: its spectral decomposition and his proof of the analytic continuation of the Eisenstein series, [CV1, CV2].

A most basic matter is the domain of an operator. Consider $\mathcal{H}$ a closed subspace of $H^1(R)$, the Sobolev space of square-integrable functions with square-integrable gradient. Associated to $\mathcal{H}$ is a positive self-adjoint operator $D_{\mathcal{H}}$ on the closure of $\mathcal{H}$ in $L^2(R)$; the operator is associated by the Friedrichs procedure to the restriction to $\mathcal{H}$ of the quadratic form $\int_R \|\text{grad } f\|^2$, $f \in H^1(R)$. We are interested in particular subspaces of $H^1(R)$: $\mathcal{H}_a = \{f \mid f \in H^1(R),\ f_0(y) = 0,\ a \le y < \infty\}$. Colin de Verdiere shows that the domain of $D_a$ consists of those $f \in \mathcal{H}_a$ such that $D_{dist} f - \alpha\delta_a(y) \in L^2(R)$, where $D_{dist}$ is the distributional Laplacian, $\alpha \in \mathbb{C}$ and $\delta_a$ is the Dirac measure at a. Note: for $\mathcal{H} = H^1(R)$, we get the standard Laplacian, denoted simply as D.

Quite often all that will be required is the subspace $C_a^2 \subset \text{Domain}(D_a)$: $f \in C_a^2$ provided $f$ is $C^2$ on the closures of $\mathcal{F}^+$ and $\mathcal{F}^-$, $f - f_0$ is $C^2$ in a neighborhood of $\text{Im } z = a$, and $f_0$ is continuous with $f_0(y) = 0$ for $a \le y < \infty$.

<u>Theorem 1.3</u> [CV2]. $D_a$ is self-adjoint with compact resolvent.

In particular $\text{Spec}(D_a)$ is discrete and the classical theory can be used for perturbations of $D_a$. The spectral decomposition of $D_a$ involves the Eisenstein series $E(z;s)$ and its coefficient $\varphi(s)$. Technical matter:

assume $a \geq 1$, thus guaranteeing that the projections of horocycles to $R$ with height at least $a$ are topological circles. The following truncation operator will be useful.

Definition 1.4. For $f \in L^2(R)$ set

$$\hat{f}(z) = f(z) - \chi_{[a,\infty)}(\operatorname{Im} z) f_0(z).$$

Theorem 1.5 [CV2]. The spectrum of $D_a$ is the union with appropriate multiplicities of two sequences:

1) values $\lambda_m$, where $D$ has a cuspidal eigenfunction for this eigenvalue, and

2) values $\mu_j(a)$ of multiplicity 1, where $\mu_j(a) = s_j(1-s_j)$ with either

i) $\operatorname{Im} s_j \geq 0$, $\operatorname{Re} s_j \geq \frac{1}{2}$, $s_j \neq \frac{1}{2}$ and $a^{s_j} + \varphi(s_j)a^{1-s_j} = 0$, in which case the eigenfunction is $\hat{E}(z;s_j)$, or

ii) $s_j = \frac{1}{2}$, provided $\varphi(\frac{1}{2}) = -1$, $\varphi'(\frac{1}{2}) = -2 \log a$ and the associated eigenfunction is $\frac{d}{ds} \hat{E}(z;s_j)$.

Furthermore, $0 < \mu_j(a) < \mu_{j+1}(a)$; as $a \to \infty$ the initial $\mu(a)$'s converge down to the initial noncuspidal eigenvalues of $D$, while the remaining $\mu(a)$'s converge to $\frac{1}{4}$.

*In brief* the spectral decomposition of $D_a$ consists of the cuspidal eigenfunctions for $D$ and a discrete approximation to the band: $\lambda \in [\frac{1}{4},\infty)$, $E(z;\sqrt{\lambda-\frac{1}{4}})$.

1.6. The next matter is Colin de Verdiere's argument for the analytic continuation of the Eisenstein series. First we fix a number of conventions for the discussion. Choose constants $a$ and $b$: $1 < b < a-1$. Fix an approximate step function $h(Y)$ such that $h(Y) = 0$ for $Y < b$ and $h(Y) = 1$ for $Y > b+1$. Define a function $Y$ on $R$ such that on $\mathcal{F}$: $Y(z) = \operatorname{Im} z$ for $\operatorname{Im} z$

$\geq 1$ and $Y(z) = 1$ otherwise.

Lemma 1.6. For $s$ satisfying $\mathrm{Re}\, s > \frac{1}{2}$, $s(1-s) \notin \mathrm{Spec}(D)$ there exists a unique function $\mathcal{E}(z;s)$, holomorphic in $s$, such that

i) $D-s(1-s))\mathcal{E}(z;s) = 0$

ii) $\mathcal{E}(z;s) - h(Y)Y^s \in L^2(R)$.

Proof. Write $\mathcal{E}(z;s) = h(Y)Y^s + g(z;s)$; by condition i) $(D-s(1-s))g(z;s) = -(D-s(1-s))h(Y)Y^s$. The right hand side has support contained in $\mathrm{supp}(h)$, thus has compact support. Now by the hypothesis for $s$ the operator $(D-s(1-s))$ is uniquely invertible in $L^2(R)$; $g$ is uniquely determined. The proof is complete.

Now the Eisenstein series satisfies the conditions of the lemma for $\mathrm{Re}\, s > 1$; the analytic continuation to $\mathrm{Re}\, s > \frac{1}{2}$, $s(1-s) \notin \mathrm{Spec}(D)$ is established.

Theorem 1.7. $E(z;s)$ has a meromorphic continuation to $\mathbb{C}$ and is pole-free on $\mathrm{Re}\, s = \frac{1}{2}$. For $E_0(z;s) = y^s + \varphi(s)y^{1-s}$

i) $E(z;s) = \varphi(s)E(z;1-s)$

ii) $\varphi(s)\varphi(1-s) = 1$.

Proof [CV1]. The continuation will be given in terms of

$$(1.3) \qquad F(z;s) = h(Y)Y^s - (D_a-s(1-s))^{-1}(D-s(1-s))h(Y)Y^s.$$

In particular, $(D-s(1-s))h(Y)Y^s$ has support contained in $\{Y < a\}$, thus lies in $\mathrm{Domain}(D_a)$. By the basic theory of the resolvent the right hand side of (1.3) is defined and analytic for $s$, $s(1-s) \notin \mathrm{Spec}(D_a)$, a discrete set. The next matter concerns the $0^{th}$ Fourier coefficient $F_0(z;s) = A(s)y^s + B(s)y^{1-s}$ for $b+1 < y < a$. In particular, $F(z;s) - h(Y)Y^s \in \mathrm{Domain}(D_a)$ thus for $\mathrm{Im}\, z = a$, $F_0(a;s) - h(a)a^s = A(s)a^s + B(s)a^{1-s} = 0$ for

$s(1-s) \notin \mathrm{Spec}(D_a)$. Thus $A(s)a^s + B(s)a^{1-s} = a^s$ and for each $s$, $s(1-s) \notin \mathrm{Spec}(D_a)$ at least one of $A(s)$, $B(s)$ is nonzero. The next matter is to naturally extend in $Y$ the $0^{th}$ Fourier coefficient of $F$: set

$$\tilde{F}(z;s) = F + \chi_{[a,\infty)}(Y)(A(s)Y^s + B(s)Y^{1-s} - Y^s).$$

The $0^{th}$ Fourier coefficient $\tilde{F}_0$ is now $A(s)Y^s + B(s)Y^{1-s}$ for $Y > b+1$, and more to the point $\tilde{F}$ satisfies $(D-s(1-s))\tilde{F} = 0$ and $\tilde{F} = A(s)h(Y)Y^s + g$, $g \in L^2(R)$. By the preceding lemma $\tilde{F}(z;s) = A(s)E(z;s)$ and $B(s) = A(s)\varphi(s)$; similarly $\tilde{F}(z;s) = B(s)E(z;1-s)$ and $A(s) = B(s)\varphi(1-s)$. Thus for $s(1-s) \notin \mathrm{Spec}(D_a)$ $\varphi(s) = B(s)/A(s)$ and $\varphi(s)\varphi(1-s) = 1$; since $F(z;\bar{s}) = \overline{F(z;s)}$ then $\varphi(\bar{s}) = \overline{\varphi(s)}$ and $|\varphi(s)|^2 = 1$ for $\mathrm{Re}\, s = \frac{1}{2}$. All that remains is the behavior of $E$ on the $\mathrm{Re}\, s = \frac{1}{2}$ line. Lemma 1.2 can be applied to $\hat{E}(z;\frac{1}{2}+ir)$ to obtain the Maass-Selberg relation $\|\hat{E}(z;\frac{1}{2}+ir)\|^2 =$ $2\log a - \frac{\varphi'}{\varphi}(\frac{1}{2}+ir) - \frac{1}{r}\,\mathrm{Im}(\varphi(\frac{1}{2}+ir)a^{-2ir})$; the right hand side is clearly finite, hence $E$ is analytic on the $\mathrm{Re}\, s = \frac{1}{2}$ line. The proof is complete.

2. <u>A sketch of the Colin de Verdiere and Phillips-Sarnak arguments</u>.

2.1. The general scheme is as follows: showing that a cuspidal eigenfunction with eigenvalue $\lambda$ disappears under deformation comes down to establishing that special integrals of the resolvent kernel for $(D-\lambda)^{-1}$ are nonzero. In order to emphasize ideas we present the outline of the argument first. Let $\Delta^{\varepsilon}$, $\psi^{\varepsilon}$, $\lambda^{\varepsilon}$ be respectively a Laplace-like operator, an eigenfunction and eigenvalue, all depending nicely on a parameter $\varepsilon$. Assume that $\mathrm{Domain}(\Delta^{\varepsilon})$ is an inner product space and that the variation is normalized by $\langle\psi^{\varepsilon},\psi^{\varepsilon}\rangle = 1$. Then $(\Delta^{\varepsilon}-\lambda^{\varepsilon})\psi^{\varepsilon} = 0$ and the first variational derivative is simply $(\dot{\Delta}-\dot{\lambda})\psi + (\Delta-\lambda)\dot{\psi} = 0$. Now we write $(\Delta-\lambda)^{-1}_{\mathrm{rel}}$ for the operator

$$\begin{cases} 0 & \lambda\text{-eigenspace} \\ (\Delta-\lambda)^{-1} & \text{orthogonal complement} \end{cases}$$

and $\dot{\psi}$ is simply

$$-(\Delta-\lambda)^{-1}_{\mathrm{rel}}\ \dot{\Delta}\psi. \tag{2.1}$$

That the variation $\psi^{\varepsilon}$ remains cuspidal to first order in $\varepsilon$ will come down (in the appropriate setting) to showing that $\dot{\psi}_0(y) = 0$ for $y$ in a neighborhood of $a$. We are primarily interested in the situation of $\psi^0$ cuspidal; the essential quantity is then the $0^{\mathrm{th}}$ Fourier coefficient

$$((\Delta-\lambda)^{-1}_{\mathrm{rel}}\ \dot{\Delta}\psi)_0(y). \tag{2.2}$$

This is the key expression. Several basic difficulties are immediately apparent: the operator $(\Delta-\lambda)$ does not have a maximum principle, and since $\lambda$ is in the midst of the spectrum, any expression for $(\Delta-\lambda)^{-1}_{\mathrm{rel}}$ necessarily involves analytic continuation in the spectral parameter.

Colin de Verdiere addresses the issue by considering variations of a special form, thus simplifying $\dot{\Delta}$ and the expression (2.1). On the other hand, Phillips and Sarnak observe that (2.2) is in fact given by a special value of an L-function. This observation transforms the question.

**2.2.** We sketch the essential steps for Colin de Verdiere's result. He considers compactly supported conformal deformations: $ds^2_f = e^f\, ds^2_{\mathrm{hyp}}$, $f \in C^{\infty}_0(R)$. Denote by $D^f$ the $ds^2_f$-Laplacian. The appropriate notion of cuspidal is as follows.

<u>Definition 2.1</u>. $\psi$ a $D^f$-eigenfunction is cuspidal provided $\psi_0(y) = 0$ for $y$ in a neighborhood of a.

Colin de Verdiere's result is the following.

Theorem 2.2 [CV2]. $K \subset R$ with nonempty interior, compact. For a set $f \in C_0^\infty(K)$ of second Baire category the spectral resolution of $D^f$ has no cuspidal eigenfunctions; $D^f$ has no discrete spectrum in $[\frac{1}{4},\infty)$.

We assume that the variation function $f$ satisfies $\mathrm{supp}(f) \subset K$, $Y(K) \subset [1,b]$, $b < a$. To make the variation we shall need the $ds_f^2$-pseudo Laplacian $D_a^f$. Note: the variation is trivial for $Y > b$; the pseudo Laplacian is defined as before. Reminder: $D_a$ is the hyperbolic pseudo Laplacian.

Preliminary Lemma 2.3. Given $U \subset R$ open, and $\lambda \in \mathbb{R}$, there exists $h \in \mathrm{Domain}(D_a)$ such that $h$ is noncuspidal and $(D_a-\lambda)h \in C_0^\infty(U)$.

Proof of Lemma. The idea is to give $h$ directly in terms of the resolvent kernel $R(w,z)$ for $(D_a-\lambda)_{rel}^{-1}$; consider $R(w,z)$ as a function of $z$ with parameter $w$. As such $R(w,z)_0 = A(w)y^s + B(w)y^{1-s}$ for $\lambda = s(1-s)$ and $\mathrm{Im}\, z = y$. Claim: $R(w,z)$ is a nontrivial function of $w$ for each $z$; by contradiction, since $(D_a-\lambda)_{rel}^{-1} k(z) = \int R(w,z)k(w)dA(w)$ it follows taking $0^{th}$ Fourier coefficients in $z$, that the $0^{th}$ Fourier coefficient of the left hand side would be zero at $z$ with no hypothesis for $k$. This is not possible: take $k$ as $(D_a-\lambda)F$ for $F$ with $F_0(z) \neq 0$. The claim is established.

Now by symmetry $R(w,z)_0$ satisfies $(D_a-\lambda)R(w,z)_0 = 0$ as a function of $w$. Thus, by general PDE considerations, $R(w,z)_0$ is nonzero at some point $w_0$ of the open set $U$. Now let $\chi(z)$ be an approximate characteristic function of a small neighborhood of $w_0$, contained in $U$, and set $h(z) = (1-\chi(z))R(w_0,z)$. The reader will check that $h$ is smooth; for $z$ away from $w_0$, $h(z) = R(w_0,z)$, thus $\mathrm{supp}((D_a-\lambda)h) \subset U$ and generically for $\mathrm{Im}\, z \sim a$, $\mathrm{Im}\, z < a$ (in the cusp) $h_0(z) = R(w_0,z)_0 = A(w_0)y^s + B(w_0)y^{1-s} \neq 0$. The argument is complete.

We will now use the lemma to show that given a compact set $K$ with interior and a cuspidal eigenfunction $\psi$ with eigenvalue $\lambda$, there is a variation of the metric $e^{\varepsilon f}ds^2$ such that $\dot{\psi}$ is noncuspidal. Technical matter: assume that the eigenvalue $\lambda$ has multiplicity one.

<u>Proof of Theorem</u>. Pick $U \subset K$ such that $\psi$ is nowhere zero on $U$ and $k$ nontrivial with $\mathrm{supp}(k) \subset U$. Choose $h$ as in the lemma, satisfying $(D_a-\lambda)h = k$; define $f$ by the relation $f\lambda\psi = k$. Now the metric deformation is $e^{\varepsilon f}ds^2$ with pseudo Laplacian $D_a^{\varepsilon} = e^{-\varepsilon f}D_a$. $D_a^{\varepsilon}$ has compact resolvent; the eigenpair has a variation $(\psi^{\varepsilon},\lambda^{\varepsilon})$. Now $\dot{D}_a = -fD_a$; $(D_a-\lambda)\dot{\psi} + (\dot{D}_a-\dot{\lambda})\psi = 0$ and thus $(D_a-\lambda)\dot{\psi} = f\lambda\psi + \dot{\lambda}\psi$. Substituting in $h$, we have $(D_a-\lambda)(\dot{\psi}-h) = \dot{\lambda}\psi$; taking the product $\langle\ ,\psi\rangle$ and integrating by parts gives $\dot{\lambda} = 0$. Thus $(D_a-\lambda)(\dot{\psi}-h) = 0$, hence $(\dot{\psi}-h)$ is a $\lambda$-eigenfunction. Thus $\dot{\psi} = h + C\psi$ and $\dot{\psi}_0 = h_0 + C\psi_0$. This provides the desired conclusion: $\psi_0$ is trivial, $h_0$ is nontrivial and thus $\dot{\psi}_0$ is nontrivial. The argument is complete.

It is interesting to compare the above to (2.1): $\dot{D}\psi = -fD\psi = k$ and thus $h$ plays the role of $-(\Delta-\lambda)^{-1}_{\mathrm{rel}}\dot{\Delta}\psi$.

2.3. Unfortunately Colin de Verdiere's approach is not directly suitable for the case of deformations of $R$ through constant curvature $-1$ metrics. For this case the variation of the Laplacian consists of two terms: one for the change in the conformal structure and a second for the change in the area element.

We start with a brief sketch of the relevant deformation theory. A variation of the conformal structure is specified by a Beltrami differential, a tensor of type $\frac{\partial}{\partial z} \otimes d\bar{z}$ for $z$ a conformal coordinate. In particular, a metric for the (new) $\mu$-conformal structure is given as $ds^2 = \rho(z)^2|dz+\mu d\bar{z}|^2$ (we assume $\|\mu\|_{\infty} < 1$). We are interested in deformations through constant curvature $-1$ metrics. If $ds^2 = \rho(z)^2|dz|^2$ indeed has curvature $-1$, then by a standard calculation the $\varepsilon\mu$-conformal structure hyperbolic metric satisfies

$$(2.3) \qquad ds_\varepsilon^2 = \rho(z)^2|dz + \varepsilon\mu d\bar{z}|^2(1+\varepsilon 8(D-2)^{-1}\mathrm{Re}\ K_{-1}K_{-2}\mu + O(\varepsilon^2)),$$

where $K_r = \rho^{r-1}\frac{\partial}{\partial z}\rho^{-r}$ is the Maass derivative (hypothesis: $\mu$ is $C^2$, and $\mu$ and its derivatives tend to zero in the cusp; note: $(D-2)$ is invertible in $L^2$, has a maximum principle and $(D-2)^{-1}$ is given by a convergent Poincaré series), [W].

The simplest Beltrami differentials are those which are harmonic-tensors with respect to the hyperbolic metric. It is elementary that $\mu$ is harmonic $\Leftrightarrow K_{-2}\mu = 0 \Leftrightarrow$ there exists a holomorphic quadratic differential $\varphi$, such that $\mu = (ds^2_{hyp})^{-1} \otimes \bar{\varphi}$ for $ds^2_{hyp}$ the hyperbolic metric. Note that (2.3) is particularly simple for the harmonic Beltrami differentials: $ds_\varepsilon^2 = \rho(z)^2|dz + \varepsilon\mu d\bar{z}|^2(1+O(\varepsilon^2))$.

Phillips-Sarnak use a slight modification of the harmonic differentials. Just as with Colin de Verdiere's argument the basic setup requires that the identity map of $R$ induce a map from Domain$(D_a^\varepsilon)$ to Domain $(D_a)$. This essentially forces one to require that $ds_\varepsilon^2(z) = ds^2(z)$ for $Y(z) > b$. For the remaining discussion we bypass this matter; we presume that a deformation through hyperbolic metrics exists with $ds_\varepsilon^2 = \rho(z)^2|dz + \varepsilon\mu d\bar{z}|^2(1+O(\varepsilon^2))$ and $ds_\varepsilon^2 = ds^2$ for $Y(z) > b$ (see [PS1] pp. 348-355 for a proper treatment).

Now assume that $\psi$ is a cuspidal eigenfunction with eigenvalue $\lambda$ and that the $ds_\varepsilon^2$-variation is $(D_a^\varepsilon - \lambda^\varepsilon)\psi^\varepsilon = 0$, giving $(D_a - \lambda)\dot{\psi} + (\dot{D}_a - \dot{\lambda})\psi = 0$. Phillips' and Sarnak's next step is to take a product with $<\ ,E(z;\frac{1}{2}+ir)>$, $\lambda = \frac{1}{4}+r^2$ (this is completely natural since the Eisenstein series is part of the resolvent kernel). We now have $<(D_a-\lambda)\dot{\psi},E> + <(\dot{D}_a-\dot{\lambda})\psi,E> = 0$, and it is standard that $\langle\psi, E\rangle = 0$. At this point Phillips and Sarnak assume that $\dot{\psi}$ is cuspidal, thus $D_a\dot{\psi} = D\dot{\psi}$, and they then integrate by parts. An alternate approach is the following: first note that $\dot{\psi}$ and $E$ satisfy the hypothesis of Lemma 1.2. Now on $\mathcal{F}^+$ and $\mathcal{F}^-$ the local expression for $D_a$ coincides with that of $D$; we apply Lemma 1.2 to conclude

$$\langle(D_a-\lambda)\dot{\psi}, E\rangle = -E_0(a;\tfrac{1}{2}-ir)\frac{\partial\psi_0(a^-)}{\partial y}, \tag{2.4}$$

and, since $\psi$ is cuspidal, $\dot{D}_a\psi = \dot{D}\psi$, thus, on combining our considerations

$$\langle\dot{D}\psi, E\rangle = E_0(a;\tfrac{1}{2}-ir)\frac{\partial\dot{\psi}_0(a^-)}{\partial y}. \tag{2.5}$$

A few remarks and notes are in order.

The parameter $a$ was only restricted by the inequality $a > 2$; we reformulate: given a cuspidal eigenvalue $\lambda$, $\lambda = \frac{1}{4}+r^2$, choose $a$ such that $E_0(a;\frac{1}{2}-ir) \neq 0$. Now the inner product vanishes simultaneously with $\frac{\partial}{\partial y}\dot{\psi}_0(a^-)$. What is the behavior of $\dot{\psi}_0(y)$ for $y$ near $a$? First recall that $ds_\varepsilon^2 = ds^2$ for $Y(z) > b$ thus in this range $\dot{D}$ is the trivial operator; for $Y(z) > b$, $D_a$ is $z$-translation invariant. By separation of variables for $Y(z) > b$, we have from the variational equation $(D_a-\lambda)\dot{\psi}_0 - \dot{\lambda}\psi_0 = 0$ and of course $\psi_0 = 0$. Thus $\dot{\psi}_0$ satisfies the ODE $(-y^2\frac{d^2}{dy^2} - \lambda)\dot{\psi}_0 = 0$ for $b < y \leq a$. Now $\psi_0^\varepsilon(a) = 0$ for all $\varepsilon$, thus $\dot{\psi}_0(a) = 0$ and since we have a second order ODE: $\dot{\psi}_0$ is trivial on $b < y \leq a$ (hence $\dot{\psi}$ is cuspidal) if and only if $\frac{\partial}{\partial y}\dot{\psi}_0(a^-) = 0$. This completes the proof of the following.

<u>Theorem 2.4</u>. $\psi$ a cuspidal eigenfunction for the simple eigenvalue $\lambda = \frac{1}{4}+r^2$. Assume the parameter $a$ is chosen such that $E_0(a;\frac{1}{2}-ir) \neq 0$, then the variation $\psi^\varepsilon$ is cuspidal to first order in $\varepsilon$ if and only if $\langle\dot{D}\psi, E(\frac{1}{2}+ir)\rangle = 0$.

The Phillips-Sarnak result is the essential case: a nonvanishing inner product implies a noncuspidal variation [PS1, CV3]. They then evaluate the inner product.

Theorem 2.5 [PS1]. Consider the Beltrami differential $(ds^2_{hyp})^{-1} \otimes \bar{\varphi}$, $\varphi$ a holomorphic quadratic differential. The associated first variation of the cuspidal eigenfunction $\psi$ is noncuspidal provided $L_{\varphi*\psi}(2+ir) \neq 0$, where $L_{\varphi*\psi}$ is the Rankin-Selberg convolution of $\varphi$ and $\psi$; $\psi$ has eigenvalue $\frac{1}{4}+r^2$.

Idea of proof. Start with $\langle\dot{D}\psi, E(z;\sigma)\rangle$, Re $\sigma > 1$. Unfold the series, integrate by hand and analytically continue. The result is a product of Gamma-factors and the desired L-function.

This transforms the problem to one of showing that special values of L-functions are nonzero. Phillips-Sarnak then find nine nonzero special values for the congruence subgroup $\Gamma_1(24) \subset SL(2;\mathbb{Z})$. And in a joint effort, Deshouillers, Iwaniec, Phillips and Sarnak show that there is a nontrivial asymptotic distribution of such special values [DIPS].

Remark 2.6. The techniques for (2.6) can be applied to show

$$2\langle(\dot{D}_a-\dot{\lambda})\dot{\psi}, E(\tfrac{1}{2}+ir)\rangle + \langle(\ddot{D}_a-\ddot{\lambda})\psi, E(\tfrac{1}{2}+ir)\rangle = E_0(a;\tfrac{1}{2}-ir)\,\frac{\partial\ddot{\psi}_0(a^-)}{\partial y}.$$

Thus for $E_0(a;\frac{1}{2}-ir) \neq 0$ and $\psi_0(a) = \frac{\partial\dot{\psi}_0}{\partial y}(a^-) = 0$, vanishing of the left hand side is equivalent to $\psi^\varepsilon$ being cuspidal to second order in $\varepsilon$.

## 3. What happens when a cusp form disappears?

3.1. The asymptotic formula (1.2) shows that the winding number of the scattering matrix is closely related to the eigenvalue count. What happens to the winding number and the Eisenstein series when a cuspidal eigenvalue disappears? We will see that a zero-pole pair is created on the critical line of the scattering matrix. The pole will move to the left and the zero will move to the right. We assume for the entire chapter that the cuspidal eigenvalue is not $\frac{1}{4}$. The parametrization $\lambda = s(1-s)$ is branched at $s = \frac{1}{2}$; the $s = \frac{1}{2}$ case must be treated separately.

3.2. Colin de Verdiere's description of the Eisenstein series is ideal for this question. Recall that

$$\begin{cases} F(z;s) = h(Y)Y^s-(D_a-s(1-s))^{-1}(D-s(1-s))h(Y)Y^s \\ F(z;s) \text{ is holomorphic in } s \text{ for } s(1-s) \notin \mathrm{Spec}(D_a) \\ F_0(z;s) = A(s)Y^s+B(s)Y^{1-s}, \quad Y < a \\ \tilde{F}(z;s) = A(s)E(z;s) \text{ and } B(s) = \varphi(s)A(s). \end{cases}$$

Let $\psi$ be a $D_a$-eigenfunction with eigenvalue $\lambda_0 = s_0(1-s_0)$, which for the sake of exposition we assume to be simple.

<u>Lemma 3.1</u>. With the above setup and writing $\mathbb{P}$ for the principal part at $s_0$ of a meromorphic function

$$\mathbb{P}F_0(z;s) = \mathbb{P}A(s)Y^s + \mathbb{P}B(s)Y^{1-s} = \frac{a^{s_0}}{(s-s_0)(1-2s_0)}\psi_0(Y)\ \frac{\partial\psi_0}{\partial Y}(a^-).$$

<u>Proof</u>. We start with the formula for $F$, expand and drop terms holomorphic near $s_0$ (we write $f(s) \equiv g(s)$ if $\mathbb{P}f = \mathbb{P}g$). Consider the resolvent kernel $R(w,z)$ for $(D_a-s(1-s))^{-1}$. We have

$$R(w,z) \equiv \frac{\psi(w)\psi(z)}{s_0(1-s_0)-s(1-s)}.$$

Thus

$$\begin{aligned} F &\equiv -(D_a-s(1-s))^{-1}(D-s(1-s))h(Y)Y^s \\ &\equiv \frac{\langle (D-s(1-s))h(Y)Y^s,\psi\rangle\psi(z)}{(s-s_0)(1-2s_0)} \end{aligned}$$

and we apply Lemma 1.2, integrate by parts and obtain (note: $h(a) = 1$)

$$\frac{a^{s_0}}{(s-s_0)(1-2s_0)}\psi(z)\,\frac{\partial\psi_0(a^-)}{\partial Y}.$$

The desired formula follows on taking $0^{th}$ Fourier coefficients.

The interpretation of the formula will reveal a great deal. First assume that $\psi$ is in fact cuspidal then $\psi_0$ is trivial and $F$ is regular at $s_0$. Consider next that $\psi$ is noncuspidal with $\psi_0(y) = \alpha y^s + \beta y^{1-s}$ (for $y < a$), and at least one of $\alpha, \beta$ is nonzero (also $\frac{\partial\psi_0}{\partial Y}(a^-)$ is nonzero as noted prior to Theorem 2.4). But $\psi_0(a) = 0$ hence both $\alpha$ and $\beta$ must be nonzero. It follows that *both* $A$ and $B$ have a simple pole at $s_0$ (this provides an alternate proof that $\varphi$ is regular at $s_0$).

3.3. The next matter is the variation of $F$ and $E$. Recall that $D_a$ has compact resolvent and as noted in [CV2] Kato's perturbation theorem can be applied to conclude that the spectral resolution for $D_a^\varepsilon$ varies real analytically in the parameter $\varepsilon$, [Kt]. In particular, we are now in the context that $F^\varepsilon, E^\varepsilon, \psi^\varepsilon$ and $\lambda^\varepsilon = s_\varepsilon(1-s_\varepsilon)$ vary real analytically in $\varepsilon$. We modify our notation and write $\mathbb{P}$ now for the principal part at $s_\varepsilon$. Thus

$$\mathbb{P}F_0^\varepsilon(z;s) = \mathbb{P}A^\varepsilon(s)Y^s + \mathbb{P}B^\varepsilon(s)Y^{1-s} = \frac{a^{s_\varepsilon}}{(s-s_\varepsilon)(1-2s_\varepsilon)}\psi_0^\varepsilon(Y)\,\frac{\partial\psi_0^\varepsilon(a^-)}{\partial Y}. \tag{3.1}$$

To study the variation we first pick $\varepsilon_0$, $\delta > 0$, such that for $|\varepsilon| < \varepsilon_0$, $s_\varepsilon(1-s_\varepsilon)$ is the sole eigenvalue of $D_a^\varepsilon$ satisfying $|s_\varepsilon - s_0| < \delta$. We are interested in the situation: $\psi^0$ is cuspidal and $\psi^\varepsilon$ is noncuspidal for $\varepsilon \neq 0$, $\varepsilon$ small. Claim: $A^0$ and $B^0$ are in fact both regular nonzero at $s_0$. By the above formula both are regular. Recall that $A(s)a^s + B(s)a^{1-s} = a^s$, thus at least one is nonvanishing at $s_0$. Finally $B(s) = A(s)\varphi(s)$ with $\varphi$ regular nonzero at $s_0$; both $A^0$ and $B^0$ are nonzero at $s_0$. Further restrict $\delta$ to insure that $A^0$ and $B^0$ are regular nonzero on $|s-s_0| \le \delta$. Their winding number on $|s-s_0| = \delta$ is zero. By now further restricting $\varepsilon_0$ we can insure that the winding numbers of $A^\varepsilon$ and $B^\varepsilon$ on $|s-s_0| = \delta$ are also each zero.

Now for $\varepsilon \neq 0$, $A^\varepsilon$ and $B^\varepsilon$ each have a simple pole at $s_\varepsilon$. Thus each has exactly one zero in $|s-s_0| < \delta$. We consider the question of where the zeros can lie. Suppressing superscripts, if $A(s_A)$ vanishes with $\mathrm{Re}\, s_A > \frac{1}{2}$ then $F_0(z;s_A) = B(s_A)Y^{1-s_A}$ and $F(z;s_A) \in L^2(R)$, thus $F(a;s_A)$ is an $L^2$-eigenfunction and $s_A(1-s_A) \in \mathrm{Spec}(D)$. But $s_\varepsilon(1-s_\varepsilon)$ is the sole $D_a^\varepsilon$-eigenvalue satisfying $|s-s_0| < \delta$. It follows that the zero $s_A$ of $A^\varepsilon$ in $|s-s_0| < \delta$, must satisfy $\mathrm{Re}\, s_A \le \frac{1}{2}$. Similarly the zero $s_B$ of $B^\varepsilon$ in $|s-s_0| < \delta$ must satisfy $\mathrm{Re}\, s_B \ge \frac{1}{2}$. As noted above though, $A^\varepsilon$ and $B^\varepsilon$ must be zero-free on $\mathrm{Re}\, s = \frac{1}{2}$; the new zeros are *forced* to move *off* the critical line: $\mathrm{Re}\, s_A < \frac{1}{2}$ and $\mathrm{Re}\, s_B > \frac{1}{2}$.

To now summarize: for $\varepsilon = 0$, $A^0, B^0$ are regular nonzero in $|s-s_0| < \delta$. For $\varepsilon \neq 0$, $A^\varepsilon$ has a pole at $s_\varepsilon$ and a zero at $s_A^\varepsilon$, $\mathrm{Re}\, s_A^\varepsilon < \frac{1}{2}$, and $B^\varepsilon$ has a pole at $s_\varepsilon$ and a zero at $s_B^\varepsilon$, $\mathrm{Re}\, s_B^\varepsilon > \frac{1}{2}$. Finally, since $B^\varepsilon(s) = \varphi^\varepsilon(s)A^\varepsilon(s)$, then for $\varepsilon \neq 0$, $\varphi^\varepsilon$ has a pole at $s_A^\varepsilon$ and a zero at $s_B^\varepsilon$, with $\mathrm{Re}\, s_A^\varepsilon, \mathrm{Re}\, s_B^\varepsilon \longrightarrow \frac{1}{2}$ as $\varepsilon \longrightarrow 0$. This provides the desired conclusion: when a cusp form disappears a zero-pole pair is created for the scattering matrix. The pole moves properly to the left and the zero moves properly to the right.

And so the disappearance of a cusp form corresponds to the appearance of a zero-pole pair for the scattering matrix. There is an immediate question: does the variation of the zero or pole provide any further information? We will see that this is not the case. The variation of the zero and pole are similar. We consider the variation of the pole as evidenced by the variation of the A-coefficient of $F_0(z;s)$.

Lemma 3.2. Let $\psi$ be a cuspidal eigenfunction for the simple eigenvalue $\lambda$. Suppose that the variation $\psi^\varepsilon$ with eigenvalue $s_\varepsilon(1-s_\varepsilon)$ is noncuspidal precisely at order $n$: $\psi_0^\varepsilon = \varepsilon^n(\alpha Y^{s_\varepsilon} + \beta Y^{1-s_\varepsilon}) + O(\varepsilon^{n+1})$, at least one of $\alpha,\beta$ nonzero. Then

$$\mathbb{P}A^\varepsilon(s) = \frac{-a}{(s-s_\varepsilon)(2s_\varepsilon-1)^2}\left[\frac{\partial\psi_0^\varepsilon(a^-)}{\partial Y}\right]^2 + O(\varepsilon^{2n+1})$$

for all small $\varepsilon$.

Proof. The first point is the expansion

$$\frac{\partial\psi_0^\varepsilon(a^-)}{\partial Y} = \varepsilon^n(2s_\varepsilon-1)\alpha a^{s_\varepsilon-1} + O(\varepsilon^{n+1}). \tag{3.2}$$

This follows on substituting the condition $\psi_0^\varepsilon(a) = \alpha a^{s_\varepsilon} + \beta a^{1-s_\varepsilon} = 0$ into the formula $\frac{\partial\psi_0^\varepsilon(a^-)}{\partial Y} = \varepsilon^n(\alpha s_\varepsilon a^{s_\varepsilon-1} + \beta(1-s_\varepsilon)a^{-s_\varepsilon}) + O(\varepsilon^{n+1})$. Now from (3.1) we have

$$\mathbb{P}A^\varepsilon(s)Y^s + \mathbb{P}B^\varepsilon(s)Y^{1-s} = \frac{a^{s_\varepsilon}}{(s-s_\varepsilon)(1-2s_\varepsilon)}\psi_0^\varepsilon(Y)\,\frac{\partial\psi_0^\varepsilon(a^-)}{\partial Y}$$

and on substituting the expansion for $\psi_0^\varepsilon(Y)$ we find

$$\mathbb{P}A^\varepsilon(s) = \frac{a^{s_\varepsilon}}{(s-s_\varepsilon)(1-2s_\varepsilon)}\varepsilon^n\alpha\,\frac{\partial\psi_0^\varepsilon(a^-)}{\partial Y} + O(\varepsilon^{2n+1}).$$

The desired formula now follows on substituting (3.2). The proof is complete.

Thus considering the variation of $A^\varepsilon$ simply leads us back to the quantity $\frac{\partial\psi_0^\varepsilon(a^-)}{\partial Y}$. Recall (2.5), Theorem 2.4 and Remark 2.6. The suggestion is clear: $\frac{\partial\psi_0^\varepsilon(a^-)}{\partial Y}$ and the special value of the L-function are indeed the fundamental quantities.

Remark 3.3. Equation (2.5) and Lemma 3.2 provide a formula for the variation of the pole in terms of the square of the inner product $\langle \dot{D}\psi, E\rangle$. This is Fermi's Golden Rule: the variation of an embedded eigenvalue to a resonance is given by the square of the inner product of: the product of the perturbation of the operator and the eigenfunction with the *continuum eigenfuntion*, [Si, pp.260-262].

## References


[Ah] L.V. Ahlfors, Some remarks on Teichmüller's space of Riemann surfaces, Ann. of Math. 74, 171-191 (1961).

[CV1] Y. Colin de Verdiere, Pseudo Laplacians I, Ann. Inst. Fourier 32, 275-286 (1982).

[CV2] Y. Colin de Verdiere, Pseudo Laplacians II, Ann. Inst. Fourier 33, 87-113 (1983).

[CV3] Y. Colin de Verdiere, Resonances, in Seminaire de Theorie Spectrale et Geometrie, Universite de Grenoble, 1984-1985.

[DIPS] J.M. Deshouillers, H. Iwaniec, R.S. Phillips and P. Sarnak, Maass cusp forms, Proc. Nat. Acad. Sci. 82, 3533-3534 (1985).

[Ef] I. Efrat, On the existence of cusp forms over function fields, J. Reine Angew. Math. 399, 173-187 (1989).

[Fy] J. Fay, Perturbation of the spectrum of a compact Riemann surface, preprint.

[Hj1] D.A. Hejhal, The Selberg trace formula for $PSL(2;\mathbb{R})$, vol. 2, Springer Lecture Notes, Vol. 1001, 1983.

[Hj2] D.A. Hejhal, Continuity method, Proc. 1983, Durham symposium on Modular Forms and Functions.

[Hj3] D.A. Hejhal, The Selberg trace formula and the Riemann zeta function, Duke Math. J. 43, 441-482 (1976).

[Hj4] D.A. Hejhal, Eigenvalues of the Laplacian for Hecke triangle groups, preprint.

[Kt] T. Kato, Perturbation theory for linear operators, Springer, Berlin, 1966.

[Kb] T. Kubota, Elementary theory of Eisenstein series, Kodansha, Tokyo, 1973.

[Ln] S. Lang, SL(2;ℝ), Springer, New York, 1985.

[LP] P. Lax and R.S. Phillips, Scattering theory for automorphic functions, Ann. Math. Stud. 87.

[PS1] R.S. Phillips and P. Sarnak, On cusp forms for cofinite subgroups of PSL(2;ℝ), Invent. Math. 80, 339-364 (1985).

[PS2] R.S. Phillips and P. Sarnak, The Weyl theorem and the deformation of discrete groups, preprint.

[Sr] P. Sarnak, On cusp forms, preprint.

[Sl] A. Selberg, Göttingen lectures, 1954.

[Si] B. Simon, Resonances in N-Body Quantum Systems, Annals of Math. 97, 247-274 (1973).

[Vn] A.B. Venkov, Spectral theory of automorphic functions, Proc. Steklov Inst. of Math. 4, 163 (1982).

[Wn] A. Winkler, Cusp forms and Hecke groups, J. Reine Angew. Math. 386, 187-204 (1988).

[W] S.A. Wolpert, The hyperbolic metric and the geometry of the universal curve, J. Diff. Geom., to appear.

# Moduli Spaces of Harmonic and Holomorphic Mappings and Diophantine Geometry

Toshiki Miyano and Junjiro Noguchi*)

**Introduction.** In this paper we first show some new results on the structure of the moduli space of harmonic (resp. holomorphic) mappings into a Riemannian (resp. Kähler) manifold with non–positive sectional curvature, and then, applying these results, give a survey on recent developments in the theory of Diophantine geometry. Let $N$ be a compact analytic Riemannian manifold with non–positive sectional curvatures and $M$ a compact Riemannian manifold. We denote by $\text{Harm}(M, N)$ the space of all harmonic mappings of $M$ into $N$ endowed with compact–open topology, which is called the *moduli space of harmonic mappings* of $M$ into $N$. Let $f: M \longrightarrow N$ be a smooth mapping and put

$$\text{Harm}(M, N; f) = \{g \in \text{Harm}(M, N), g \sim f(\text{homotopic})\} .$$

Then Schoen–Yau [SY] proved that $\text{Harm}(M, N; f)$ carries a structure of a compact Riemannian manifold such that the evaluation mapping

$$\Phi_{1p} : g \in \text{Harm}(M, N; f) \longrightarrow g(p) \in N$$

with an arbitrarily fixed point $p \in M$ is an isometric immersion onto a totally geodesic submanifold of $N$ (see also [$S_2$] for the case of locally symmetric $N$). We put $X = \text{Harm}(M, N; f)$ and consider $X$ as a domain and $M$ as a parameter space. That is, we put

$$Y = \text{Harm}(X, N; \Phi_{1,p}) ,$$

$$\Phi_2 : (x, y) \in X \times Y \longrightarrow y(x) \in N .$$

These $Y$ and $\Phi_2(x, \cdot)$ have properties similar to $X$ and $\Phi_{1p}$. Naturally, we have a smooth mapping $p \in M \longrightarrow \Phi_{1p} \in Y$ and the following commutative diagram:

---

*) The second author stayed at Max–Planck–Institut für Mathematik, Bonn during the preparation of this paper. He expresses his sincere gratitude for the hospitality of the institute.

$$\begin{array}{ccc}
p \in M & \xrightarrow{\quad x \quad} & x(p) \\
\uparrow & & \| \\
(x,\ p) \in X \times M & \longrightarrow & \Phi_{1p}(x) \in N \\
\downarrow & & \| \\
(x,\ \Phi_{1p}) \in X \times Y & \xrightarrow[\Phi_2]{} & \Phi_2(x,\ \Phi_{1p})
\end{array}$$

Our first theorem is

**Theorem (1.14).** *Assume that the groups of isometries of* $X$ *and* $Y$ *are finite (in particular, the Ricci curvature of* $N$ *is negative,* $\dim X \geq 2$, *and* $\dim Y \geq 2$*) . Then the second evaluation mapping*

$$\Phi_2 : (x,\ y) \in X \times Y \longrightarrow y\,(\mathrm{x}) \in N$$

*is an immersion onto a totally geodesic submanifold of* $N$ *and the pull–backed metric by* $\Phi_2$ *is isometric to the product metric on* $X \times Y$.

In the complex category, we deal with the case where $N$ is not necessarily compact, but a complete Kähler manifold such that the (Riemannian) sectional curvatures are non–positive and the holomorphic sectional curvatures are bounded from above by a negative constant. Moreover, we assume that $N$ is a Zariski open subset of a complex projective variety $\overline{N}$ such that $N$ is hyperbolically imbedded into $\overline{N}$. Let $M$ be a Zariski open subset of a compact Kähler manifold $\overline{M}$. We denote by $\mathrm{Hol}(M,\ N)$ the moduli space of all holomorphic mappings of $M$ into $N$ endowed with compact–open topology. By $[\mathrm{N}_3]$ $\mathrm{Hol}(M,\ N)$ has a structure of a Zariski open subset of a compact complex space, and non–singular. Let $X \subset \mathrm{Hol}(M,\ N)$ be a connected component and put

$$\Phi_1 : (x,\ p) \in X \times M \longrightarrow x\,(p) \in N\,.$$

Let $Y \subset \mathrm{Hol}(X,\ N)$ be the connected component containing $\Phi_1(\ \cdot\ ,\ p)(p \in M)$. Then we have

**Theorem (2.15).** *The second evaluation mapping*

$$\Phi_2 : (x,\ y) \in X \times Y \longrightarrow y(x) \in N$$

*is a proper holomorphic immersion onto a complex totally geodesic submanifold of* $N$ , *and the pull–backed metric by* $\Phi_2$ *is isometric to the product metric on* $X \times Y$.

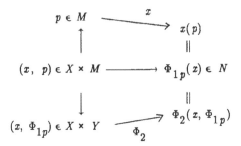

It is the most important and interesting case when $N$ is a quotient $\Gamma \backslash \mathbb{D}$ of a bounded symmetric domain $\mathbb{D}$ by an arithmetic discrete subgroup $\Gamma$ of the holomorphic automorphism group $\mathrm{Aut}(\mathbb{D})$.

In the course of the proof of the above theorem we show

**Theorem (2.4).** *Let $S$ be a complete hyperbolic manifold such that $S$ is a Zariski open subset of a compact complex space $\bar{S}$ and $S$ is hyperbolically imbedded into $\bar{S}$. Then* $\mathrm{Aut}(S)$ *is a finite group.*

The first and the second sections are devoted to the proof of the above results, and to the preparations for the latter sections.

In § 3 we discuss the higher dimensional Mordell's conjecture over function fields and related topics. In § 4 we deal with the Parshin–Arakelov–type theorems for curves, Abelian varieties and K3–surfaces. We will give a new proof to the Parshin–Arakelov theorem for curves, based on our results.

Table of Contents



## § 1. Moduli space of harmonic mappings

Let $M$ and $N$ be Riemannian manifolds and $TM$ (resp. $T^*M$) denote the tangent (resp. cotangent) bundle over $M$. Let $f: M \longrightarrow N$ be a $C^\infty$–mapping. The energy functional $E(f)$ is defined by

$$E(f) = \int_M < df, df > dV_M .$$

We call $f$ a *harmonic* mapping if for any smooth family $f_t: M \longrightarrow N$, $-1 \leq t \leq 1$, of $C^\infty$-mappings such that $f_0 = f$ on $M$ and $f = f_t$, $-1 \leq t \leq 1$, outside a compact subset $K \subset M$,

$$\frac{d}{dt}\Big|_{t=0} \int_K < df_t, df_t > dV_M = 0 .$$

It is known that $f$ is harmonic if and only if

$$\text{(1.1)} \qquad \operatorname{trace} {}^{T^*M \otimes f^{-1}TN}\nabla\, df = 0 ,$$

where ${}^{T^*M \otimes f^{-1}TN}\nabla$ is the Riemannian connection of $T^*M \otimes f^{-1}TN$ induced from those on $TM$ and $TN$.

If $f: M \longrightarrow N$ is an immersion and if $f \circ c: [a,b] \longrightarrow N$ are geodesic for all geodesic curves $c: [a, b] \longrightarrow M$ with $[a, b] \subset \mathbb{R}$, then $f$ is called a *totally geodesic* immersion; an immersion $f$ is totally geodesic if and only if

$$\text{(1.2)} \qquad {}^{T^*M \otimes f^{-1}TN}\nabla\, df = 0 .$$

Hence a totally geodesic immersion is harmonic. We recall the following theorem due to Schoen–Yau [SY, p. 371], which will play a key role.

**(1.3) Theorem.** *Assume the following conditions:*

a) *$M$ is complete and has a finite volume;*

b) *$N$ is complete and has non-positive sectional curvature.*

*Let $f, g: M \longrightarrow N$ be harmonic mappings with finite energies which are mutually homotopic by $H: [0, 1] \times M \longrightarrow N$ with $H(0, \cdot) = f$ and $H(1, \cdot) = g$. Then there exists a 1-parameter family $F(t, p) = f_t(p)$, $t \in [0, 1]$ of harmonic mappings $f_t: M \longrightarrow N$ with $f_0 = f$ and $f_1 = g$ such that*

i) *the homotopy $F(t, \cdot)$ is equivalent to $H(t, \cdot)$,*

ii) *for any $p \in M$, the curve $\{f_t(p); t \in [0, 1]\}$ is a geodesic with constant speed independent of $p \in M$,*

iii) *the section $p \in M \longrightarrow F_*((\partial/\partial t)_{(t,p)}) \in f_t^{-1}TN$ is parallel.*

Moreover, we obtain the following Lemma from the proof of the above theorem ([SY, p. 370]).

**(1.4) Lemma.** *Let the notation be as in Theorem (1.3). Then $f_{t*}(v)$ is parallel along the geodesic $\{f_t(p); t \in [0, 1]\}$ for all $p \in M$ and $v \in T_pM$.*

**(1.5) Corollary.** *Let $f, g : M \longrightarrow N$ be as in Theorem (1.3). Then $f$ is an immersion (resp. isometric immersion) if so is $g$.*

The proof is clear.

**(1.6) Proposition.** *Let $M$ be a compact Riemannian manifold, $N$ a compact analytic Riemannian manifold with non-positive sectional curvatures, and $f: M \longrightarrow N$ a totally geodesic immersion. Then every harmonic mapping from $M$ to $N$ which is homotopic to $f$ is a totally geodesic immersion.*

Proof. Put $X = \mathrm{Harm}(M, N; f)$ and let $\Phi_1 : (g, p) \in X \times M \longrightarrow g(p) \in N$ be the evaluation mapping. We endow with $X$ the pull–backed Riemannian metric by $\Phi_1(\cdot\ , p)$. We show that the image of any geodesic in $M$ by any $g \in X$ is a geodesic in $N$. Let $c : t \in [0, \varepsilon] \longrightarrow c(t) \in M (\varepsilon > 0)$ be a geodesic in $M$. Then $f \circ c : [0, \varepsilon] \longrightarrow N$ is a geodesic in $N$. We may assume that $\|(f \circ c)_*(\partial / \partial t)\| = 1$ and $\varepsilon < \mathrm{Inj}(N)/2$, where $\mathrm{Inj}(N)$ is the injective radius of $N$. Note that $\Phi_1(\cdot\ , c(0))$ and $\Phi_1(\cdot\ , c(\varepsilon))$ are isometric immersions from $X$ onto totally geodesic submanifolds of $N$. Therefore they are harmonic and $\Phi_1(\cdot\ , c(t))$ gives a homotopy between them. Deforming $\Phi_1(\cdot\ , c(t))$, we have by Theorem (1.3) a 1–parameter family

$$\Omega_s : X \longrightarrow N,\ 0 \leq s \leq 1$$

of harmonic mappings with $\Omega_0 = \Phi_1(\cdot\ , c(0))$ and $\Omega_1 = \Phi_1(\cdot\ , c(\varepsilon))$. Moreover, for any $g \in X$, the curve $\{\Omega_s(g);\ s \in [0, 1]\}$ is a geodesic with constant speed, independent of $g$. Since $\{\Omega_s(f);\ s \in [0, 1]\}$ is a geodesic which coincides with $\{\Phi_1(f, c(t));\ t \in [0, \varepsilon]\}$, we change the parameter $s$ of $\Omega_s$ so that

$$(1.8)\qquad \begin{cases} \Omega_t(f) = \Phi_1(f, c(t)),\ 0 \leq t \leq \varepsilon, \\ \Omega_0 = \Phi_1(\cdot\ , c(0)),\quad \Omega_\varepsilon = \Phi_1(\cdot\ , c(\varepsilon)). \end{cases}$$

By Lemma (1.4) we see that $\{\Phi_1(g, c(t));\ 0 \leq t \leq \varepsilon\}$ is a curve with arclength parametrization. Therefore we have

$$(1.9)\qquad d_N(\Phi_1(g, c(t)), \Omega_t(g)) \leq 2\varepsilon < \mathrm{Inj}(N).$$

Using Theorem (1.3) again, we see that $d_N(\Phi_1(h, c(t)), \Omega_t(h))$ is a constant $(\leq 2\varepsilon)$ in $h \in X$.

Since $\Phi_1(f, c(t)) = \Omega_t(f)$ by (1.8),

$$\Phi_1(g, c(t)) = \Omega_t(g)$$

for all $t$. Hence $\{\Phi_1(g, c(t));\ 0 \leq t \leq \varepsilon\}$ is a geodesic. Q.E.D.

Let $f: M \longrightarrow N$ be as in Proposition (1.6), $X = \mathrm{Harm}(M, N; f)$ and

$$\Phi_1 : (g, p) \in X \times M \longrightarrow g(p) \in N .$$

We endow $X$ with the pull–backed Riemannian metric by $\Phi_1(\cdot\, , p)$. Then we have the natural mapping $p \in M \longrightarrow \Phi_1(\cdot\, , p) \in \mathrm{Harm}(X, N)$. Put

$$Y = \mathrm{Harm}(X, N; \Phi_1(\cdot\, , p)) ,$$

$$\Phi_2 : (g, y) \in X \times Y \longrightarrow y(g) \in N .$$

We endow $Y$ with the pull–backed Riemannian metric by $\Phi_2(g, \cdot)$.

**(1.10) Corollary.** *For any* $y \in Y$

$$\Phi_2(\cdot\, , y) : g \in X \longrightarrow y(g) \in N$$

*is a totally geodesic isometric immersion.*

This immediately follows from Corollary (1.5) and Proposition (1.6). By the construction we see also that

(1.11) for any $v_g \in T_gX$ (resp. $v_y \in T_yY$) the section $z \in Y \longrightarrow d\Phi_2(g, z)(v_g) \in \Phi_2(g\, , \cdot)^{-1}\, TN$ (resp. $x \in X \longrightarrow d\Phi_2(x, y)(v_y) \in \Phi_2(\cdot\, , y)^{-1}\, TN$) is parallel.

We denote by $\mathrm{Is}(X)$ the group of all isometries of $X$.

**(1.12) Lemma.** *Let* $\Phi_2 : X \times Y \longrightarrow N$ *be as above.*

(i) *If* $\mathrm{Is}(X)$ *or* $\mathrm{Is}(Y)$ *is finite, then* $\Phi_2$ *is an immersion.*
(ii) *If* $\mathrm{Is}(X)$ *and* $\mathrm{Is}(Y)$ *are finite, then the pull–backed metric on* $X \times Y$ *by* $\Phi_2$ *is the product metric of those on* $X$ *and* $Y$.

**Proof.** (i) Note that $\Phi_2(\cdot\, , y)$ and $\Phi_2(x, \cdot)$ are totally geodesic isometric immersions for $y \in Y$ and $x \in X$. Suppose that $\Phi_2$ is not an immersion. Then there are non–zero vectors $v_g \in T_gX$ and $v_y \in T_yY$ such that $d\Phi_2(g, y)(v_g + v_y) = 0$, so that

(1.13) $$d\Phi_2(g, y)(v_g) = -\, d\Phi_2(g, y)(v_y) \neq 0 .$$

Now, consider the vector field $\{d\Phi_2(x, y)(v_y);\, x \in X\}$ in $\Phi_2(\cdot\, , y)^{-1}\, TN$. It follows from (1.11) and (1.13) that $d\Phi_2(x, y)(v_y)$ is paralled and tangent to the image $\Phi_2(X, y)$. Therefore we have a non–zero parallel vector field on $X$, which generates a 1–parameter subgroup of $\mathrm{Is}(X)$. In the same way, we have a 1–parameter subgroup of $\mathrm{Is}(Y)$. Hence, if one of $\mathrm{Is}(X)$ and $\mathrm{Is}(Y)$ is finite, $\Phi_2$ is an immersion.

(ii) Take an arbitrary $v_x \in T_x X$. Then by (1.11), $y \in Y \longrightarrow d\Phi_2(x, y)(v_x)$ is parallel. Suppose that $d\Phi_2(x, y)(v_x)$ is not perpendicular to $\Phi_2(x, Y)$ at some $y_0 \in Y$. Then $d\Phi_2(x, y)(v_x)$ are not perpendicular to $\Phi_2(x, Y)$ at all $y \in Y$. Let $u_y$ denote the orthogonal projection of $d\Phi_2(x, y)(v_x)$ to $T\Phi_2(x, Y)$ in $TN$. We get a parallel vector field on $Y$, which generates a 1–parameter subgroup of $\mathrm{Is}(Y)$. Therefore we obtain our assertion. Q.E.D.

**(1.14) Theorem.** *Let* $\Phi_2 : X \times Y \longrightarrow N$ *be as above. Assume that* $\mathrm{Is}(X)$ *and* $\mathrm{Is}(Y)$ *are finite (especially, the Ricci curvature of* $N$ *is negative,* $\dim X \geq 2$, *and* $\dim Y \geq 2$). *Then* $\Phi_2$ *is a totally geodesic isometric immersion into* $N$.

**Proof.** It remains to show that

$$^{E}\nabla d\Phi_2 \equiv 0 ,$$

where $E = T^*(X \times Y) \otimes \Phi_2^{-1} TN$ with the naturally induced metric and connection $^{E}\nabla$. Put

$$r = \dim X , \quad s = \dim Y , \quad n = \dim N .$$

Take an arbitrary point $(x, y) \in X \times Y$. Let $(x^1, \dots, x^r)$ (resp. $(y^1, \dots, y^s)$) be a local coordinate system around $x$ (resp. $y$). Since $\Phi_2$ is an immersion, we can take a local coordinate system $(w^1, \dots, w^n)$ around $\Phi_2(x, y)$ such that

$$\Phi_2(x^1, \dots, x^n, y^1, \dots, y^s) = (x^1, \dots, x^n, y^1, \dots, y^s, 0, \dots, 0) .$$

We use indices as follows.

$$i, j, k = 1, \dots, r ,$$
$$\mu, \nu, \tau = 1, \dots, s ,$$
$$\alpha, \beta, \gamma = 1, \dots, n .$$

Put $\varphi = \Phi_2(x, \cdot) : Y \longrightarrow N.$ Then $\varphi$ is a totally geodesic immersion, so that $^{T^*Y \otimes \varphi^{-1}TN}\nabla d\varphi \equiv 0$ ; in terms of local coordinates,

$$\frac{\partial^2 \varphi^\alpha}{\partial y^\mu \partial y^\nu} - {}^{Y}\Gamma^\tau_{\mu\nu}(y) \frac{\partial \varphi^\alpha}{\partial y^\tau} + {}^{N}\Gamma^\alpha_{\beta\gamma}(y(x)) \frac{\partial \varphi^\beta}{\partial y^\mu} \frac{\partial \varphi^\gamma}{\partial y^\nu} \equiv 0 ,$$

where ${}^{Y}\Gamma^{\tau}_{\mu\nu}$ (resp. ${}^{N}\Gamma^{\alpha}_{\beta\gamma}$) are Christoffel symbols of $Y$ (resp. $N$) and Einstein's summation convention is used. Note that

$$\frac{\partial\varphi^{\alpha}}{\partial y^{\mu}} = \delta^{\alpha}_{r+\mu} .$$

Thus we have

$${}^{N}\Gamma^{\alpha}_{r+\mu\ r+\nu}(y(x)) = 0 , \quad 1 \le \alpha \le r ,$$

$$-\,{}^{Y}\Gamma^{\alpha-r}_{\mu\nu}(x) + {}^{N}\Gamma^{\alpha}_{r+\mu\ r+\nu}(y(x)) = 0 , \quad r < \alpha \le s .$$

Since $\Phi_2$ is isometric, we have

$$(1.15) \qquad \begin{cases} {}^{X\times Y}\Gamma^{k}_{\mu\nu}(x,y) = {}^{N}\Gamma^{k}_{r+\mu\ r+\nu}(y(x)) = 0 , \\ {}^{X\times Y}\Gamma^{\tau}_{\mu\nu}(x,y) = {}^{N}\Gamma^{r+\tau}_{r+\mu\ r+\nu}(y(x)) = {}^{Y}\Gamma^{\tau}_{\mu\nu}(y) . \end{cases}$$

In the same way, we obtain

$$(1.16) \qquad \begin{cases} {}^{X\times Y}\Gamma^{k}_{ij}(x,\, y) = {}^{N}\Gamma^{k}_{i\,j}(y(x)) = {}^{X}\Gamma^{k}_{ij}(x) \\ {}^{X\times Y}\Gamma^{\tau}_{ij}(x,\, y) = {}^{N}\Gamma^{r+\tau}_{i\,j}(y(x)) = 0 . \end{cases}$$

The section $x' \in X \longrightarrow d\Phi_2(x',\, y)(\partial/\partial y^{\mu}) \in \Phi_2(\cdot\,,\, y)^{-1}TN$ is parallel by (1.11), and hence

$${}^{\Phi_2(\cdot\,,\,y)^{-1}TN}\nabla_{\frac{\partial}{\partial x^{i}}}\left[d\Phi_2(x,\, y)\left[\frac{\partial}{\partial y^{\mu}}\right]\right] = 0 .$$

It follows that

$${}^{TN}\nabla_{d\Phi_2(\partial/\partial x^{i})}\left[d\Phi_2(x,\, y)\left[\frac{\partial}{\partial y^{\mu}}\right]\right] = 0 .$$

On the other hand

$$\frac{\partial}{\partial w^{i}} = d\Phi_2\left[\frac{\partial}{\partial x^{i}}\right] ,$$

$$\frac{\partial}{\partial w^{r+\mu}} = d\Phi_2\left[\frac{\partial}{\partial y^\mu}\right].$$

Therefore

$$^{TN}\nabla_{\frac{\partial}{\partial w^i}}\left[\frac{\partial}{\partial w^{r+\mu}}\right]_{y(x)} = 0\,.$$

We get

(1.17) $$^{N}\Gamma^{\alpha}_{i\,r+\mu}(y(x)) = 0\,,$$

and so

(1.18) $$\begin{cases} {}^{X\times Y}\Gamma^{k}_{i\mu}(x) = {}^{N}\Gamma^{k}_{i\,r+\mu}(y(x)) = 0\,, \\ {}^{X\times Y}\Gamma^{\tau}_{i\mu}(x) = {}^{N}\Gamma^{r+\tau}_{i\,r+\mu}(y(x)) = 0\,. \end{cases}$$

By the choices of local coordinate systems, we have

(1.19) $$\frac{\partial\Phi_2^\alpha}{\partial x^i} = \delta^\alpha_i\,,\ \frac{\partial\Phi_2^\alpha}{\partial y^\mu} = \delta^\alpha_{r+\mu}\,.$$

By making use of (1.15) – (1.19), we compute ${}^{E}\nabla d\Phi_2$ at $(x, y)$. The $dx^i\otimes dx^j\otimes\Phi_2^{-1}(\partial/\partial w^\alpha)$ component of ${}^{E}\nabla d\Phi_2$ at $(x, y)$ is as follows:

$$\frac{\partial^2\Phi_2^\alpha}{\partial x^i\partial y^j} - {}^{X\times Y}\Gamma^{k}_{ij}(x,y)\frac{\partial\Phi_2^\alpha}{\partial x^k} - {}^{X\times Y}\Gamma^{\tau}_{ij}(x,y)\frac{\partial\Phi_2^\alpha}{\partial y^\tau} + {}^{N}\Gamma^{\alpha}_{\beta\gamma}(y(x))\frac{\partial\Phi_2^\beta}{\partial x^i}\frac{\partial\Phi_2^\gamma}{\partial x^j}$$

$$= \frac{\partial^2\Phi_2^\alpha}{\partial x^i\partial y^j} - {}^{X}\Gamma^{k}_{ij}(x)\frac{\partial\Phi_2^\alpha}{\partial x^k} + {}^{N}\Gamma^{\alpha}_{\beta\gamma}(y(x))\frac{\partial\Phi_2^\beta}{\partial x^i}\frac{\partial\Phi_2^\gamma}{\partial x^j} = 0\,.$$

Here one reminds that $\Phi_2(\cdot\,, y)$ is totally geodesic. In the same way as above, we see that $dy^\mu\otimes dy^\nu\otimes\Phi_2^{-1}(\partial/\partial w^\alpha)$ component of ${}^{E}\nabla d\Phi_2$ at $(x, y)$ vanishes. The cross–term, $dx^i\otimes dy^\mu\otimes\Phi_2^{-1}(\partial/\partial w^\alpha)$ component of ${}^{E}\nabla d\Phi_2$ at $(x, y)$ is computed as follows:

$$\frac{\partial^2\Phi_2^\alpha}{\partial x^i\partial y^\mu} - {}^{X\times Y}\Gamma^{k}_{i\mu}(x, y)\frac{\partial\Phi_2^\alpha}{\partial x^k} - {}^{X\times Y}\Gamma^{\tau}_{i\mu}(x,y)\frac{\partial\Phi_2^\alpha}{\partial y^\tau} + {}^{N}\Gamma^{\alpha}_{\beta\gamma}(y(x))\frac{\partial\Phi_2^\beta}{\partial x^i}\frac{\partial\Phi_2^\gamma}{\partial x^\mu}$$

$$= {}^N\Gamma^{\alpha}_{i\,r+\mu}(y(x)) = 0 \,.$$

Q.E.D.

§ 2. Moduli space of holomorphic mappings

In this section we deal with moduli spaces of holomorphic mappings between complex manifolds, especially, those into complete hyperbolic manifolds. For fundamental facts on hyperbolic manifolds (or hyperbolic complex spaces) we refer to $[K_1, K_3]$ and [NO].

Roughly speaking, we have the following correspondences between the cases of Riemannian and complex manifolds:

| | | |
|---|---|---|
| Riemannian manifolds with non–positive or negative curvatures | $\longleftrightarrow$ | hyperbolic manifolds, |
| harmonic mappings | $\longleftrightarrow$ | holomorphic mappings, |
| isometries | $\longleftrightarrow$ | holomorphic automorphisms, |
| parallel vector fields | $\longleftrightarrow$ | holomorphic vector fields. |

There are two theorems which are essential in our arguments. One is Theorem (1.3) due to Schoen–Yau [SY], and another is the following:

**(2.1) Theorem** **($[N_3]$).** *Let $N$ be a complete hyperbolic complex space such that $N$ is hyperbolically imbedded into a compact complex space as a Zariski open subset. Let $M$ be a complex manifold which is a Zariski open subset of a compact complex space. Then the moduli space* $\mathrm{Hol}(M, N)$ *of all holomorphic mappings from $M$ into $N$ endowed with compact–open topology carries a structure of a Zariski open subset of a compact complex space such that*

i) *the evaluation mapping*

$$\Phi_1 : (f, p) \in \mathrm{Hol}(M, N) \times M \longrightarrow f(p) \in N$$

*is holomorphic,*

ii) *if $F: T \times M \longrightarrow N$ is a holomorphic mapping with a complex space $T$, then the mapping, $t \in T \longrightarrow F(t, \cdot) \in \mathrm{Hol}(M, N)$ is holomorphic.*

**(2.2) Remark.** Let $\bar{M}$ be a compactification of $M$ such that $\partial M = \bar{M} - M$ is a hypersurface with only normal crossings. Then $\mathrm{Hol}(M, N)$ is imbedded into $\mathrm{Hol}(\bar{M}, \bar{N})$ by holomorphic extension so that the closure $\overline{\mathrm{Hol}(M, N)}$ in $\mathrm{Hol}(\bar{M}, \bar{N})$ is a compact complex space and $\mathrm{Hol}(M, N)$ is a Zariski open subset of $\overline{\mathrm{Hol}(M, N)}$. The above evaluation mapping $\Phi_1$ extends to a holomorphic mapping from $\overline{\mathrm{Hol}(M, N)} \times \bar{M}$ into $\bar{N}$.

For $f \in \mathrm{Hol}(M, N)$ we set

$$\operatorname{rank} f = \sup\{\dim M - \dim_p f^{-1}f(p);\ p \in M\}\ ,$$

$$\mathrm{Hol}(k; M, N) = \{f \in \mathrm{Hol}(M, N);\ \operatorname{rank} f = k\}\ .$$

The notion, rank $f$, is similarly defined for holomorphic mappings between complex spaces. We give some general fact:

**(2.3) Proposition.** *Let* $Y_1$ *and* $Y_2$ *be compact complex spaces with reduced structure. Then* $\mathrm{Hol}(k, Y_1, Y_2)$ *are open and closed in* $\mathrm{Hol}(Y_1, Y_2)$ *for all* $k$.

The proof is quite elementary (see [$N_3$, Lemma (2.17)]).

**(2.4) Theorem.** *Let* $N$ *be as in Theorem* (2.1). *Assume that* $N$ *is non-singular. Then the group* $\mathrm{Aut}(N)$ *of holomorphic automorphisms of* $N$ *is finite.*

**Remark.** In the case where $N$ is compact, this was obtained by Kobayashi ([$K_1$, p. 70]).

**Proof.** Let $\bar{N}$ be a compactification of $N$ such that $\bar{N} - N$ is a hypersurface with only normal crossings. Put $n = \dim N$. By Remark (2.2) we see that $\mathrm{Hol}(n; N, N)$ is imbedded into $\mathrm{Hol}(n; \bar{N}, \bar{N})$ and the closure $\overline{\mathrm{Hol}(n; N, N)}$ in $\mathrm{Hol}(\bar{N}, \bar{N})$ is a compact complex space. We first claim that $\mathrm{Hol}(n; N, N)$ is compact in $\mathrm{Hol}(\bar{N}, \bar{N})$. Let $f_\nu \in \mathrm{Hol}(n; N, N)$, $\nu = 1, 2, \dots$, be any sequence. Since $\mathrm{Hol}(N, N)$ is relatively compact in $\mathrm{Hol}(\bar{N}, \bar{N})$, we may assume that the extended mapping $\bar{f}_\nu \in \mathrm{Hol}(\bar{N}, \bar{N})$ converges to $\bar{f} \in \mathrm{Hol}(\bar{N}, \bar{N})$. Since $\operatorname{rank} \bar{f}_\nu = n$, $\operatorname{rank} \bar{f} = n$ by Proposition (2.3). Thus the image $\bar{f}(\bar{N})$ contains a non-empty open subset. Take a point $p_0 \in N$ such that $\bar{f}(p_0) = q_0 \in N$. For an arbitrary point $p \in N$ we have

$$f_\nu(p) = \bar{f}_\nu(p) \longrightarrow \bar{f}(p)\ (\nu \longrightarrow \infty)\ ,$$

$$d_N(f_\nu(p), f_\nu(p_0)) \le d_N(p, p_0)\ .$$

Since $f_\nu(p_0) \longrightarrow q_0$ and $N$ is complete hyperbolic, $f_\nu(p)$ stay in a compact subset of $N$. Therefore we have that $\bar{f}(p) \in N$. The restriction $f = \bar{f} \mid N$ belongs to $\mathrm{Hol}(n;\, N,\, N)$ and is the limit of $f_\nu$, $\nu = 1, 2, \ldots$ . Put

$$X = \{f \in \mathrm{Hol}(n;\, N,\, N)\, ;\, \bar{f}(\bar{N} - N) \subset \bar{N} - N\}\,.$$

Then $X$ is an analytic subset of $\mathrm{Hol}(n;\, N,\, N)$. Since $N$ is measure–hyperbolic (cf. [$K_1$, Chapter IX]) and its total measure is positive and finite, we easily see that $\deg \bar{f} = 1$ for all $f \in \mathrm{Hol}(n;\, N,\, N)$. Let $f \in X$. Then the inverse $\bar{f}^{-1} : \bar{N} \longrightarrow \bar{N}$ of $\bar{f}: \bar{N} \longrightarrow \bar{N}$ is meromorphic and satisfies

$$g = \bar{f}^{-1} \mid N : N \longrightarrow N\,.$$

Since $N$ is complete hyperbolic, $g$ must be holomorphic (see [$K_1$, p. 90]), so that $g = f^{-1}$. Therefore $X = \mathrm{Aut}(N)$ and it follows that $\mathrm{Aut}(N)$ is a compact complex Lie group. Since $N$ is hyperbolic, there is no 1–parameter subgroup of $\mathrm{Aut}(N)$. Thus $\mathrm{Aut}(N)$ is finite.

Q.E.D.

In the rest of this section we assume the following conditions for $N$:

(2.5) (i) $N$ is a complete Kähler manifold with non–positive (Riemannian) sectional curvatures and with negative holomorphic sectional curvatures bounded away from 0,

(ii) $N$ is quasi–projective algebraic and carries a projective compactification $\bar{N}$ such that $N$ is hyperbolically imbedded into $\bar{N}$.

**(2.6) Remark.** It follows from (2.5), (i) that $N$ is complete hyperbolic; moreover, there is a constant $C > 0$ such that

$$\sqrt{h(v,\, v)} \le C\, F_N(v)\,, \quad v \in TN\,,$$

where $h$ denotes Kähler metric on $N$ and $F_N$ the infinitesimal Kobayashi metric. Therefore we have

$$\mathrm{Vol}_h(N) = \int_N dV_h < \infty\,.$$

Now, let $\overline{M}$ be a compact Kähler manifold and $M$ a Zariski open subset of $\overline{M}$.

**(2.7) Lemma.** *Let $M$ be as above. Then there is a complete Kähler metric $g$ on $M$ with finite volume such that*

$$F_M \leq C_1 \sqrt{g},$$

*where $C_1 > 0$ is a constant.*

**Proof.** Take a modification $\tilde{M} \longrightarrow \overline{M}$ with center contained in $\overline{M} - M$ so that $\tilde{M}$ is Kähler and $\tilde{M} - M$ is a hypersurface with only simple normal crossings. Then as in the proof of Proposition (6.2) of [GK], one can construct such a metric $g$. Q.E.D.

By Theorem (2.1) we see that $\mathrm{Hol}(M, N)$ carries a structure of a Zariski open subset of a compact complex space, so that

(2.8) $\mathrm{Hol}(M, N)$ is locally arcwise connected.

We endow $M$ with a Kähler metric in Lemma (2.7). The choices of the metrics on $M$ and $N$, and the decreasing principle $f^* F_N \leq F_M$ for $f \in \mathrm{Hol}(M, N)$ imply that

(2.9) $E(f) < \infty$ for all $f \in \mathrm{Hol}(M, N)$.

Since $M$ and $N$ are Kähler,

(2.10) all $f \in \mathrm{Hol}(M, N)$ are harmonic.

By making use of Theorem (1.3), we have the following lemma (cf. [$N_3$, p. 29]):

**(2.11) Lemma.** *Let $f : M \longrightarrow N$ be a harmonic mapping, homotopic to a holomorphic mapping from $M$ into $N$. Then $f$ is holomorphic, too.*

**Remark.** In the case where $M$ is compact, this lemma holds without the assumption of non-positive curvatures on $N$, and is called the Lichnerowicz theorem. In the present case, we have to use that curvature assumption to get some estimate on the boundary behavior of the homotopy between $f$ and the holomorphic mapping.

Even though the global injectivity radius of $N$ is zero, by making use of (2.8), (2.9) and Lemma (2.11), we can apply the arguments of Schoen–Yau [SY, p. 372] to infer that the evaluation mapping $\Phi_1 : \mathrm{Hol}(M, N) \times M \longrightarrow N$ has the following properties:

**(2.12) Theorem.** i) *For an arbitrary point* $p \in M$ , *the holomorphic mapping* $\Phi_1(\cdot, p):$ $f \in \mathrm{Hol}(M, N) \longrightarrow f(p) \in N$ *is a proper immersion onto a complex totally geodesic submanifold of* $N$.

ii) *The pull backed metric* $\Phi_1(\cdot, p)^* h$ *is independent of* $p$ .

Thus, $\mathrm{Hol}(M, N)$ is non-singular. We take a connected component $X$ of $\mathrm{Hol}(M, N)$ . We may assume that $\partial M = \overline{M} - M$ is a hypersurface with only normal corssings. Then we have the natural imbedding

$$f \in \mathrm{Hol}(M, N) \longrightarrow \bar{f} \in \mathrm{Hol}(\overline{M}, \overline{N})$$

which is an into–homeomorphism (cf. [$N_3$, Theorem (1.19)] ). Let $\overline{X}$ denote the closure of $X$ in $\mathrm{Hol}(\overline{M}, \overline{N})$ . By Remark (2.2) $\Phi(\cdot, p)(p \in M)$ naturally extends to a holomorphic mapping

$$\overline{\Phi}(\cdot, p) : \overline{X} \longrightarrow \overline{N} .$$

**(2.13) Lemma.** i) *$X$ is complete hyperbolic and hyperbolically imbedded into* $\overline{X}$ .

ii) *$X$ is quasi–projective.*

**Proof.** i) It is easy to show that $X$ is complete hyperbolic. Take distinct $F_1$ , $F_2 \in \overline{X} - X$ . Take any sequences $\{f_{1\nu}\}_{\nu=1}^{\infty}$ , $\{f_{2\nu}\}_{\nu=1}^{\infty}$ in $X$ such that $\bar{f}_{1\nu} \longrightarrow F_1$ and $\bar{f}_{2\nu} \longrightarrow F_2$ in $\mathrm{Hol}(\overline{M}, \overline{N})$ . Since $F_1 \neq F_2$ , there is a point $p \in M$ such that $F_1(p) \neq F_2(p)$ and $F_i(p) \in N$ , $i = 1, 2$ . Then

$$f_{1\nu}(p) \longrightarrow F_1(p) \in N ,$$

$$f_{2\nu}(p) \longrightarrow F_2(p) \in N .$$

Therefore we have

$$d_X(f_{1\nu}, f_{2\nu}) \geq d_N(f_{1\nu}(p), f_{2\nu}(p)) > \tfrac{1}{2}\, d_N(F_1(p), F_2(p)) > 0$$

for all large $\nu$ . This shows that $X \hookrightarrow \overline{X}$ is a hyperbolic imbedding.

ii) This follows from the fact that $\Phi_1(\cdot, p) : X \longrightarrow \Phi_1(X, p)(\subset N)$ with $p \in M$ is finite and $\overline{\Phi_1(X, p)}(\subset N)$ is projective. Q.E.D.

**(2.14) Corollary.** Aut$(X)$ *is finite.*

This follows from Lemma (2.13) and Theorem (2.4).

As in § 1, we now consider $X$ as a domain space and $M$ as a parameter space. Note that by Lemma (2.13) $X$ satisfies the conditions put on $M$. Let $Y \subset \mathrm{Hol}(X, N)$ be the connected component containing $\Phi_1(\cdot\,, p)(p \in M)$. Then we have the evalutation mapping

$$\Phi_2 : X \times Y \longrightarrow N.$$

By Corollary (2.14), Aut$(Y)$ is finite, too. To obtain a complex analytic version of Theorem (1.14), we first note that the topology of $X$ (resp. $Y$) is defined by the compact–open topology on the mapping space from $M$ (resp. $X$) into $N$, and that $X$ and $Y$ are locally arcwise connected. These makes us possible to carry out the same arguments as in the proof of Theorem (1.14) in compact subsets of $N$, where, of course, their injectivity radii remain positive. We endow $X$ (resp. $Y$) with the pull–backed Kähler metric $\Phi_1(\cdot\,, p)^* h$ (resp. $\Phi_2(x, \cdot)^* h$) with $p \in M$ (resp. $x \in X$), and $X \times Y$ with the product metric.

**(2.15) Theorem.** *Let* $\Phi_2 : X \times Y \longrightarrow N$ *be as above. Then* $\Phi_2$ *is a proper, holomorphic, totally geodesic and isometric immersion.*

**Remark.** $X$ is a connected component of Hol$(Y, N)$ which contains $\Phi_2(x, \cdot)(x \in X)$. For the composition of the holomorphic mapping $p \in M \longrightarrow \Phi_1(\cdot\,, p) \in Y$ and any $g \in \mathrm{Hol}(Y, N)$ is a holomorphic mapping from $M$ into $N$ which is homotopic to $\Phi_1(x, \cdot) : M \longrightarrow N$.

## § 3. Higher dimensional Mordell's conjecture over function fields and related topics

By making use of the results obtained in § 2, we discuss the Diophantus geometry in the present and the next sections. We start with the following theorem due to Manin [M] and Grauert [G].

**(3.1) Theorem (Mordell's conjecture over function fields).** *Let* $K$ *be a function field over an algebraically closed field* $k$ *with* char $k = 0$. *Let* $C$ *be a smooth curve defined over* $K$, *of which genus is not less than* 2. *Then either the set* $C(K)$ *of* $K$*–rational points of* $C$ *is finite, or* $C$ *is* $K$*–isomorphic to a curve* $C_0$ *defined over* $k$ *and* $C_0(K) - C_0(k)$ *is finite.*

**Remark.** The original conjecture over number fields was solved by G. Faltings $[F_2]$.

In geometric terms, the above Theorem (3.1) is equivalent to the following.

**(3.2) Theorem.** *Let* $X \xrightarrow{\pi} R$ *be a smooth fiber space over* $k$ *such that the genus of the fibers* $X_t\,(t \in R)$ *is not less than* 2. *Then the set* $\Sigma$ *of rational sections* $\sigma : R \longrightarrow X\,(\pi \circ \sigma = id)$ *is finite, or* $X$ *is isomorphic to the product* $R \times X_{t_0}$ *and there are only finitely many non–constant rational sections (non–constant mappings from* $R$ *to* $X_{t_0}$*).*

Related to this theorem, S. Lang $[L_1]$ gave two conjectures.

**(3.3) Conjecture.** *Let* $X \longrightarrow R$ *be an algebraic fiber space over* $\mathbb{C}$ *with hyperbolic fibers* $X_t$. *If the set* $\Sigma$ *of rational sections of* $X \longrightarrow R$ *is infinite, then* $X \longrightarrow R$ *contains a splitting fiber subspace.*

**(3.4) Conjecture.** *Let* $N$ *be a complex projective manifold such that* $N$ *is hyperbolic. Let* $M$ *be another complex projective manifold. Then there are only finitely many surjective holomorphic mappings from* $M$ *onto* $N$.

For a moment we discuss the latter conjecture. Instead of the hyperbolicity assumption on $N$, Kobayashi–Ochiai $[KO_2]$ assumed that $N$ is of general type, and showed a finiteness theorem:

**(3.5) Theorem** **($[KO_2]$).** *Let* $N$ *be a complex manifold of general type. Then there are only finitely many surjective meromorphic mappings from* $M$ *onto* $N$.

It is natural to expect (cf. $[K_3]$)

**(3.6) Conjecture.** *A complex projective algebraic hyperbolic manifold is of general type.*

This is true for curves, and for surfaces by Mori–Mukai [MM]; in fact, they proved that a complex projective algebraic surface is of general type if and only if it is measure–hyperbolic.

Let $N$ be a complex projective manifold. According to the theory of Mori [Mo], if there is a curve $C \subset N$ such that the intersection number $C \cdot K_N$ of $C$ and the canonical bundle $K_N$ of $N$ is negative, then $C$ is deformed to a sum of curves, which contains a rational curve. Thus, if $N$ is hyperbolic, then $C \cdot K_N \geq 0$ for all curves $C \subset N$. Making use of this fact, Horst $[H_1, H_2]$ solved Conjecture (3.4):

**(3.7) Theorem** **($[H_1, H_2]$).** *Let* $N$ *be a hyperbolic Kähler manifold and* $M$ *a complex manifold. Then there are only finitely many surjective holomorphic mappings from* $M$ *onto* $N$.

**Remark**. Assuming additionally that $K_N$ carries a metric with non-positive curvature form, Noguchi [$N_2$] proved the above finiteness theorem.

There are many finiteness theorems of various types. Cf. [$S_1$], [$S_2$], [I], [NS], [BN] and [KSW].

As for Conjecture (3.3), Noguchi first proved the following by employing the idea of Grauert [G]:

**(3.8) Theorem** ([$N_1$]). *Let* $X \longrightarrow R$ *be an algebraic smooth fiber space such that the holomorphic tangent bundle* $T(X_{t_0})$ *is negative* ($T^*(X_{t_0})$ *is ample) for some point* $t_0 \in R$. *Assume that* $\Sigma(t_0) = \{\sigma(t_0); \sigma \in \Sigma\}$ *is Zariski dense in* $X_{t_0}$. *Then there is a Zariski open neighborhood* $R' \subset R$ *of* $t_0$ *such that* $X|R' \cong R' \times X_{t_0}$, *and there are only finitely many non-constant rational sections.*

If $T(X_{t_0})$ is negative, then $X_{t_0}$ is hyperbolic ([$K_2$]); but the converse is not true. For instance, if $C_i$, $i = 1, 2,$ are smooth algebraic curves with genus $\geq 2$, then $C_1 \times C_2$ is hyperbolic, but $T(C_1 \times C_2)$ is not negative.

In general, let $\pi : X \longrightarrow R$ be a proper fiber space, of which general fibers are irreducible. We call $(X, \pi, R)$ a *hyperbolic fiber space* if $X_t$ are hyperbolic complex spaces for all $t \in R$. Assume that $(X, \pi, R)$ has a compactification $(\overline{X}, \overline{\pi}, \overline{R})$; that is, $\overline{\pi} : \overline{X} \longrightarrow \overline{R}$ is a compact fiber space such that $R$ is a Zariski open subset of $\overline{R}$, $X = \overline{X}|R$ and $\overline{\pi}|X = \pi$. Now we consider the relative version of the notion of hyperbolic imbedding.

**(3.9) Definition**. Let $(X, \pi, R)$ be a hyperbolic fiber space and $(\overline{X}, \overline{\pi}, \overline{R})$ its compactification. We say that $(X, \pi, R)$ is hyperbolically imbedded into $(\overline{X}, \overline{\pi}, \overline{R})$ along $\partial R = \overline{R} - R$ if for any point $t \in \partial R$ there are small neigborhoods $U$ and $V$ of $t$ in $\overline{R}$ such that $V \subset U$, $V$ is relatively compact in $U$ and $X|(V - \partial R)$ is hyperbolically imbedded into $\overline{X}|U$.

**(3.10) Theorem**. *Let* $(X, \pi, R)$ *be a hyperbolic fiber space with a compactification* $(\overline{X}, \overline{\pi}, \overline{R})$ *such that* $(X, \pi, R)$ *is hyperbolically imbedded into* $(\overline{X}, \overline{\pi}, \overline{R})$ *along* $\partial R$. *Assume that* $R$ *is smooth and that there is a point* $t_0 \in R$ *such that* $\Sigma(t_0)$ *is Zariski dense in* $X_{t_0}$. *Then*

i) *there is a finite Galois covering* $\tilde{R} \longrightarrow R$ *such that* $\tilde{R} \times_R X \cong \tilde{R} \times X_{t_0}$;

ii) *if* $X_{t_0}$ *is a Kähler manifold, then* $X \cong R \times X_{t_0}$ (*i.e.*, $\tilde{R} = R$) .

The assertion i) was proved by $[N_2]$. Combining the argument of the proof of i) with Theorem (3.7), we have ii) (see $[N_2, \S 3]$).

**(3.11) Theorem.** *Let* $(X, \pi, R)$ *be a smooth fiber space of curves with genus* $g \geq 2$ *and* $\dim R = 1$ . *Then there is a compactification* $(\overline{X}, \overline{\pi}, \overline{R})$ *of* $(X, \pi, R)$ *such that* $(X, \pi, R)$ *is hyperbolically imbedded into* $(\overline{X}, \overline{\pi}, \overline{R})$ .

Combining Theorem (3.11) with Theorem (3.10), we have Theorem (3.2) in the case of $\dim R = 1$ ; this is an essential case. For the proof of Theorem (3.11), see $[N_2, \S 5]$. It has been informed to the authors that there is some incomplete part in the proof of Theorem (3.11). We here make the point clear. Since the construction of $(\overline{X}, \overline{\pi}, \overline{R})$ is local around points of $\partial R$ , we assume that $\overline{R} = \Delta$ (the unit disc in $\mathbb{C}$) and $R = \Delta^*(= \Delta - \{0\})$ . It was first showed that there is a finite Galois covering $S \longrightarrow \Delta^*$ with cyclic covering group $G$ such that $(X_S, \eta, S)$ is hyperbolically imbedded into some compactification $(\overline{X}_S, \overline{\eta}, \overline{S})$ along $\overline{S} - S$ , where $X_S = S \times_{\Delta^*} X$ and $\eta : X_S \longrightarrow S$ is the projection. The group $G$ acts holomorphically on $X_S$ and the action extends meromorphically on $\overline{X}_S$ . The point is that this extended action is not necessarily holomorphic. But we proceed as follows. Consider the following imbedding*)

$$\alpha : x \in X_S \longrightarrow (\ldots, g(x), \ldots)_{g \in G} \in \prod_{\# G} X_S \subset \prod_{\# G} \overline{X}_S .$$

Note that $\prod_{\# G} X_S$ is hyperbolically imbedded into $\prod_{\# G} \overline{X}_S$ along $\overline{S} - S$ . The action of $G$ on $X_S$ is transformed to the exchanges of variables of $\prod_{\# G} X_S$ , which extend holomorphically on $\prod_{\# G} \overline{X}_S$ . Put $Y = \alpha(X_S)$ . Then the closure $\overline{Y}$ in $\prod_{\# G} \overline{X}_S$ is an analytic subspace which is invariant by the action of $G$ . Identify $X_S$ with $Y$ through $\alpha$ . Then the quotient $G \backslash \overline{Y}$ provides the desired compactification.

Now we assume that $X \xrightarrow{\pi} R$ is a proper smooth fiber space. Brody [B] proved that if a fiber $X_{t_0}$ is hyperbolic, then there is a neighborhood $U$ of $t_0$ (with respect to the differential

---

*) This trick was suggested by C.T.C. Wall.

topology) such that $X|U$ is hyperbolic, and so $X_t$ are hyperbolic for all $t \in U$. In connection with Theorem (3.10), it is interesting to ask

**(3.12) Question (Lang).** *Let* $X \xrightarrow{\pi} R$ *be an algebraic fiber space. Then does the set* $\{t \in R;\ X_t$ *is hyperbolic*$\}$ *form a Zariski open subset of* $R$?

Lang has asked also

**(3.13) Question.** *If* $(X, \pi, R)$ *is a hyperbolic algebraic fiber space, then does there exist a compactification* $(\overline{X}, \overline{\pi}, \overline{R})$ *of* $(X, \pi, R)$ *such that* $(X, \pi, R)$ *is hyperbolically imbedded into* $(\overline{X}, \overline{\pi}, \overline{R})$ *along* $\partial R$?

It is also interesting to point out that Parshin $[P_2]$ gave a proof of the following theorem due to Raynaud [R] which is based on the Kobayashi distance.

**(3.14) Theorem ([R, $P_2$]).** *Let* $X \subset A$ *be a subvariety of an Abelian variety* $A$, *defined over a function field* $K = \mathbb{C}(R)$ *of a curve* $R$. *If* $X$ *does not contain any translation of a non-trivial Abelian subvariety, then the set* $X(K)$ *of* $K$*-rational points on* $X$ *is finite modulo the* $(K/\mathbb{C}-)$ *trace* $A_0$.

Here it is known that there exists a unique maximal Abelian subvariety $A_0$ of $A$ defined over $\mathbb{C}$, and $A_0$ is called the $(K/\mathbb{C}-)$ trace (see $[L_2]$).

On the hyperbolicity of a subvariety of an Abelian variety or of a complex torus we know ([Gr])

**(3.15) Proposition.** *Let* $T$ *be a complex torus and* $X$ *an analytic subspace of* $T$. *Then* $X$ *is hyperbolic if and only if* $X$ *does not contain any translation of a positive dimensional subtorus of* $T$.

Lately, Faltings $[F_3]$ proved a very surprising result:

**(3.16) Theorem.** *Let* $A$ *be an Abelian variety over a number field and* $X \subset A$ *a subvariety. If* $X$ *does not contain any translation of a positive dimensional Abelian subvariety, then* $X$ *contains only finitely many* $k$*-rational points.*

Now it is interesting to recall the following conjecture of Lang:

**(3.17) Conjecture.** *Let* $X$ *be a projective algebraic variety defined over a number field* $k$. *Assume that for some imbedding* $k \subset \mathbb{C}$, $X$ *is hyperbolic as a complex manifold. Then the number of*

*k–rational points of $X$ is finite.*

It is also interesting to ask (cf. [$L_3$])

**(3.18) Question.** *Let $X$ be a projective algebraic variety defined over a number field $k$. Assume that for some imbedding $k \subset \mathbb{C}$, $X$ is hyperbolic. Then is $X$ hyperbolic for any other imbedding $k \subset \mathbb{C}$?*

## § 4. Parshin–Arakelov–type theorems

Let $\overline{R}$ be a smooth compact Riemann surface, $S \subset \overline{R}$ a finite set of points of $\overline{R}$ and put $R = \overline{R} - S$. Let $g \geq 0$ be an integer and $\mathbf{M}(\overline{R}, S, g)$ denote the set of all fiber spaces $\overline{X} \xrightarrow{\overline{\pi}} \overline{R}$ of compact Riemann surfaces such that

i) $\pi : X \longrightarrow R$ are smooth, where $X = \overline{X} | R$ and $\pi = \overline{\pi} | X$,
ii) the genus of $X_t \, (t \in R)$ are $g$,
iii) $\pi : X \longrightarrow R$ are not locally trivial.

In the case of $S = \phi$ Parshin [$P_1$], and in general Arakelov [A] proved the following theorem which had been conjectured by Shafarevich.

**(4.1) Theorem.** *If* $g \geq 1$, *then* $\mathbf{M}(\overline{R}, S, g)$ *is finite.*

The case of $g = 1$ is not difficult and was somehow already known. The case of $g \geq 2$ is of our interest. Parshin [$P_1$] proved that the finiteness of $\mathbf{M}(\overline{R}, S, g)$ $(g \geq 2)$ implies Mordell's conjecture over function fields (Theorem (3.1)), and observed that if the same holds over the ring of integers of an algebraic number field, Mordell's conjecture follows. Falting's solution of Mordell's conjecture was carried out along this line.

Imayoshi and Shiga [IS] lately proved Theorem (4.1) by a purely function theoretic method. The proof of such a finiteness theorem is, in general, divided into two parts, boundedness and rigidity. They first proved the compactness of $\mathbf{M}(\overline{R}, S, g)$, of which proof is rather hard. Combining our elementary result, Theorem (3.1) with their easier part of rigidity, we here give a proof of Theorem (4.1).

Let $\mathbf{T}_g$ be the Teichmüller space of compact Riemann surfaces with genus $g \geq 1$, and $\Pi_g$ the Teichmüller modular group. Royden [Ro] proved that the Teichmüller distance on $\mathbf{T}_g$ coincides with Kobayashi distance, so that

(4.2) $$\mathbf{T}_g \text{ is complete hyperbolic.}$$

It is known that $\Pi_g$ contains a normal subgroup $\Pi'_g$ of finite index which freely acts on $\mathbf{T}_g$. By making use of the Torelli mappings, we have (see [$N_3$])

**(4.3) Lemma.** *The quotient* $\Pi'_g\backslash\mathbf{T}_g$ *has a projective compactification* $\overline{\Pi'_g\backslash\mathbf{T}_g}$ *such that* $\Pi'_g\backslash\mathbf{T}_g$ *is hyperbolically imbedded into* $\overline{\Pi'_g\backslash\mathbf{T}_g}$.

Every element $\alpha = (\overline{X}, \overline{\pi}, \overline{R}) \in \mathbf{M}(\overline{R}, S, g)$ naturally defines a monodromy representation

$$\chi_\alpha : \pi_1(R) \longrightarrow \Pi_g,$$

which induces

$$[\chi_\alpha] : \pi_1(R) \longrightarrow \Pi_g/\Pi'_g.$$

Put $\pi_1(R)' = \mathrm{Ker}[\chi_\alpha]$ and let $R'_{[\chi_\alpha]} \longrightarrow R$ be a finite Galois covering with group $\pi_1(R)/\pi_1(R)'$. Then $\alpha$ naturally defines a non–constant holomorphic mapping

$$f_\alpha : R'_{[\chi_\alpha]} \longrightarrow \Pi'_g\backslash\mathbf{T}_g.$$

Since $\pi_1(R)$ is finitley generated, $\mathrm{Hom}(\pi_1(R), \Pi_g\backslash\Pi'_g)$ is finite. Note that for distinct $\alpha$, $\beta \in \mathbf{M}(\overline{R}, S, g)$ with $[\chi_\alpha] = [\chi_\beta]$, $f_\alpha \not\equiv f_\beta$. Therefore the proof of Theorem (4.1) is reduced to

**(4.4) Theorem.** *The space* $\mathrm{Hol}'(R, \Pi'_g\backslash\mathbf{T}_g)$ *of all non–constant holomorphic mappings from* $R$ *into* $\Pi'_g\backslash\mathbf{T}_g$ *is finite for* $g \geq 1$.

**Proof.** It follows from Theorem (2.1), (4.2) and Lemma (4.3) that $\mathrm{Hol}'(R, \Pi'_g\backslash\mathbf{T}_g)$ is a Zariski open subset of a compact complex space. Therefore there are only finitely many homotopy types of $f \in \mathrm{Hol}'(R, \Pi'_g\backslash\mathbf{T}_g)$. Let $\Delta$ be the unit disc in $\mathbf{C}$, $\Delta \longrightarrow R$ the universal covering and $\Gamma = \pi_1(R)$. Let $f$, $g \in \mathrm{Hol}'(R, \Pi'_g\backslash\mathbf{T}_g)$ belong to the same connected component. Then $f$ and $g$ are mutually homotopic. We claim

(4.5) $$f \equiv g.$$

There is a homomorphism $\chi \in \mathrm{Hom}(\Gamma, \Pi'_g)$ such that

$$\tilde{f} \circ \gamma = \chi(\gamma) \circ \tilde{f}$$

$$\tilde{g} \circ \gamma = \chi(\gamma) \circ \tilde{g}$$

for all $\gamma \in \Gamma$, where $\tilde{f} : \Delta \longrightarrow \mathbf{T}_g$ (resp. $\tilde{g} : \Delta \longrightarrow \mathbf{T}_g$) is a suitable lifting of $f$ (resp. $g$). By Ber's imbedding, $\mathbf{T}_g$ is realized as a bounded domain of $\mathbb{C}^{3g-3}$. Then $\tilde{f}$ and $\tilde{g}$ are represented by $(3g-3)$ bounded holomorphic functions on $\Delta$. By Fatou's theorem, $\tilde{f}$ and $\tilde{g}$ have non-tangential boundary values at almost all points of $\partial\Delta$. Assume that there is a subset $E \subset \partial\Delta$ with positive measure such that $\tilde{f}(z) \neq \tilde{g}(z)$ for $z \in E$. By the idea of the rigidity part of [IS] we see that for all most all $z \in E$, there are $\gamma_n \in \Gamma$, $n = 1, 2, \dots$, and $z_0 \in \Delta$ such that

$$\gamma_n(z_0) \longrightarrow z \text{, non-tangentially}$$

$$d_{\mathbf{T}_g}(\tilde{f}(\gamma_n(z_0)), \tilde{g}(\gamma_n(z_0)) \longrightarrow +\infty .$$

Since the Kobayashi hyperbolic distance $d_{\mathbf{T}_g}$ is invariant by holomorphic automorphisms,

$$\begin{aligned} d_{\mathbf{T}_g}(\tilde{f}(\gamma_n(z_0)), \tilde{g}(\gamma_n(z_0)) &= d_{\mathbf{T}_g}(\chi(\gamma_n) \circ \tilde{f}(z_0), \chi(\gamma_n) \circ \tilde{g}(z_0)) \\ &= d_{\mathbf{T}_g}(\tilde{f}(z_0), \tilde{g}(z_0)) < \infty . \end{aligned}$$

This is a contradiction. Q.E.D.

In $[F_1]$ Faltings dealt with Parshin–Arakelov–type theorem for principally polarized Abelian varieties. Let $\overline{R}$, $S$ and $g$ be as above and $A(\overline{R}, S, g)$ denote the set of all fiber spaces $\overline{A} \longrightarrow \overline{R}$ such that $A(= \overline{A} \mid R) \longrightarrow R$ are smooth, locally–nontrivial fiber spaces of $g$–dimensional Abelian varieties $A_t (t \in R)$ with principal polarizations. Faltings proved that $A(\overline{R}, S, g)$ forms a scheme of finite type over $\mathbb{C}$. Applying Theorem (2.12), we have the following theorem.

**(4.6) Theorem ([N, p. 32]).** *$A(\overline{R}, S, g)$ is quasi–projective and every connected component $Z$ of $A(\overline{R}, S, g)$ is a quotient of a symmetric bounded domain such that there is a proper, holomorphic, totally geodesic, isometric immersion $\varphi : Z \longrightarrow Sp(2g, \mathbb{Z})\backslash\mathbf{H}_g$, where $\mathbf{H}_g$ is the Siegel upper–half* *space of rank g* .

From now on we call such $\varphi$ a *Kuga–Satake immersion.* Faltings $[F_1]$ gave also a criterion of the rigidity of an element of $A(\overline{R}, S, g)$ . The both, rigid and non–rigid cases can happen. This contrasts to the arithmetic case where the rigidity always holds ($[F_1]$). His criterion was described in terms of Hodge structure, and was generalized by Peters [Pe] to more general Hodge structures (see also Saitoh–Zucker [SZ] for K3–surfaces) .

In general, let $\mathbb{D}$ be a symmetric bounded domain and $\Gamma \subset \mathrm{Aut}_0(\mathbb{D})$ an arithmetic discrete subgroup of the identity component $\mathrm{Aut}_0(\mathbb{D})$ of $\mathrm{Aut}(\mathbb{D})$ . Let $\overline{M}$ be a compact Kähler manifold and $M$ a Zariski open subset of $\overline{M}$ .

Let $\mathrm{Hol}_{\mathrm{lift}}(M, \Gamma\backslash\mathbb{D})$ be the space of all "liftable" holomorphic mapping $f: M \longrightarrow \Gamma\backslash\mathbb{D}$ ; that is, there are holomorphic mapping $\tilde{f}$ from the universal covering space $\tilde{M}$ of $M$ into $\mathbb{D}$ and a homomorphism $\chi : \pi_1(M) \longrightarrow \Gamma$ such that

$$\tilde{f} \circ \alpha = \chi(\alpha) \circ \tilde{f} \ , \ \ \alpha \in \pi_1(M)$$

and $\tilde{f}$ induces $f$. If $\Gamma$ is torsion free, then

$$\mathrm{Hol}_{\mathrm{lift}}(M, \Gamma\backslash\mathbb{D}) = \mathrm{Hol}(M, \Gamma\backslash\mathbb{D}) \ .$$

For simplicity, we assume in the sequel that $\Gamma$ is torsion free, but remark that the same results hold with a slight modification even in the case where $\Gamma$ contains a torsion element. We recall the following facts:

(4.7) i) The Bergman metric on $\Gamma\backslash\mathbb{D}$ is complete, of finite volume and has non–positive sectional curvature.

ii) $\Gamma\backslash\mathbb{D}$ is complete hyperbolic and hyperbolically imbedded into the Satake compactification $\overline{\Gamma\backslash\mathbb{D}}$ which is projective.

For ii), cf. Kobayashi–Ochiai $[KO_1]$. Hence we can apply the results in § 2 to $\mathrm{Hol}(M, \Gamma\backslash\mathbb{D})$ . Let $X$ be a connected component of $\mathrm{Hol}(M, \Gamma\backslash\mathbb{D})$ . Then $X$ is a non–singular quasi–projective manifold and represented by $\Gamma_1\backslash\mathbb{D}_1$ , where $\mathbb{D}_1$ is a symmetric bounded domain. Let

$$\Phi_1 : (\Gamma_1\backslash\mathbb{D}_1) \times M \longrightarrow \Gamma\backslash\mathbb{D}$$

be the evalution mapping. Then for any $p \in M$

$$\Phi_1(\cdot\ , p) : \Gamma_1\backslash\mathbb{D}_1 \longrightarrow \Gamma\backslash\mathbb{D}$$

is a Kuga–Satake immersion. Let $\ell(\mathbb{D})$ (resp. $\ell(\Gamma)$) denote the maximum dimension of proper (resp. $\Gamma$–rational) boundary components of $\mathbb{D}$. Note that $\operatorname{rank} f$ is constant in $f \in X$ (cf. Proposition (2.3)).

**(4.8) Theorem** ([$N_3$, §§ 3 and 4]). *Let the notation be as above.*

i) *If* $\operatorname{rank} f > \ell(\Gamma)$ *for* $f \in X$, $X$ *is compact.*

ii) *If* $\operatorname{rank} f > \ell(\mathbb{D})$ *for* $f \in X$, *then* $\dim X = 0$.

iii) *If* $\overline{f(M)} \subset \Gamma\backslash\mathbb{D}$ *for one (and all)* $f \in X$, *then* $\dim X \leq \ell(\Gamma)$, *where* $\overline{f(M)}$ *denote the closure of* $f(M)$ *in* $\overline{\Gamma\backslash\mathbb{D}}$.

iv) $\dim \operatorname{Hol}(M, \Gamma\backslash\mathbb{D}) \leq \ell(\mathbb{D})$.

Now, suppose that $\dim X > 0$. Let $Y$ be the connected component of $\operatorname{Hol}(X, \Gamma\backslash\mathbb{D})$ containing $\Phi_1(\cdot, p)$, $p \in M$. Then $Y = \Gamma_2\backslash\mathbb{D}_2$ as in the case of $X$. We have the evaluation mapping

$$\Phi_2 : (\Gamma_1\backslash\mathbb{D}_1) \times (\Gamma_2\backslash\mathbb{D}_2) \longrightarrow \Gamma\backslash\mathbb{D}.$$

**(4.9) Theorem.** *Let the notation be as above. Then* $\Phi_2$ *is a Kuga–Satake immersion.*

While this is just a special case of Theorem (2.15), it would be worth to state it separately, as it is one of our goals in this directions. It is interesting to note that starting from any $M$, we come to a Kuga–Satake immersion, provided that the moduli has a positive dimension.

Now we consider the Parshin–Arakelov–type theorem for polarized algebraic K3–surfaces. It is known that the moduli space of polarized algebraic K3–surfaces is represented by the quotient $\Gamma\backslash\mathbb{D}_{IV}$ of a symmetric bounded domain $\mathbb{D}_{IV}$ of type IV (cf., e.g. [SZ]):

$$(4.10)\qquad \begin{cases} D_{IV} = SO(2,19)/SO(2) \times SO(19), \\ \dim_{\mathbb{C}} D_{IV} = 19, \\ \ell(\mathbb{D}_{IV}) = \ell(\Gamma) = 1. \end{cases}$$

Let $X \xrightarrow{\pi} M$ be a smooth fiber space of locally non–trivial polarized algebraic K3–surfaces, and $f \in \operatorname{Hol}_{\mathrm{lift}}(M, \Gamma\backslash\mathbb{D})$ the corresponding holomorphic mapping. Since $\operatorname{Hol}_{\mathrm{lift}}(M, \Gamma\backslash\mathbb{D})$ is a finite sum of quasi–projective varieties,

(4.11) there are only finitely many rigid $(X, \pi, M)$.

Moreover,

(4.12) if $\operatorname{rank} f \geq 2$, then $(X, \pi, M)$ is rigid.

Assume that rank $f = 1$ and $(X, \pi, M)$ is not rigid (there is such an example). Then by Theorems (4.8), (4.9) and (4.10) the problem is reduced to investigate a Kuga–Satake immersion

$$\Phi : (\Gamma_1 \backslash \mathbb{H}) \times (\Gamma_2 \backslash \mathbb{H}) \longrightarrow \Gamma \backslash \mathbb{D}_{IV},$$

where $\mathbb{H}$ is the upper–half plane of $\mathbb{C}$. Saitoh–Zucker [SZ] classified all possible such $\Phi$.

## References


[A] S.Ju. Arakelov, Families of algebraic curves with fixed degeneracies, Izv. Akad. Nauk SSSR Ser. Mat. **35** (1971), 1277–1302.

[B] R. Brody, Compact manifolds and hyperbolicity, Trans. Amer. Math.. Soc. **235** (1978), 213–219.

[BN] A. Borel and R. Narasimhan, Uniqueness conditions for certain holomorphic mappings, Invent. Math. **2** (1967), 247–255.

[$F_1$] G. Faltings, Arakelov's theorem for Abelian varieties, Invent. Math. **73** (1983), 337–347.

[$F_2$] G. Faltings, Endlichkeitssätze für abelsche Varietäten über Zahlkörpern, Invent. Math. **73** (1983), 349–366.

[$F_3$] G. Faltings, Diophantine approximation on Abelian varieties, preprint.

[G] H. Grauert, Mordells Vermutung über rationale Punkte auf Algebraischen Kurven und Funktionenkörper, Publ. Math. IHES **25** (1965), 131–149.

[GK] P. Griffiths and J. King, Nevanlinna theory and holomorphic mappings between algebraic varieties, Acta Math. **130** (1973), 145–220.

[Gr] M. Green, Holomorphic maps to complex tori, Amer. J. Math. **100** (1978), 615–620.

[$H_1$] C. Horst, Compact varieties of surjective holomorphic mappings, Math. Z. **196** (1987), 259–269.

[$H_2$] C. Horst, A finiteness criterion for compact varieties of surjective holomorphic mappings, preprint.

[I] Y. Imayoshi, Generalizations of de Franchis theorem, Duke Math. J. **50** (1983), 393–408.

[IS] Y. Imayoshi and H. Shiga, A finiteness theorem for holomorphic families of Riemann surfaces. Holomorphic Functions and Moduli II, pp. 207–219, Springer–Verlag, New York–Berlin, 1988.

[$K_1$] S. Kobayashi, Hyperbolic Manifolds and Holomorphic Mappings, Marcel Dekker, New York 1970.

[$K_2$] S. Kobayashi, Negative vector bundles and complex Finsler structures, Nagoya Math. J. **57** (1975), 153–166.

[$K_3$] S. Kobayashi, Intrinsic distances, measures, and geometric function theory, Bull. Amer. Math. Soc. **82** (1976), 357–416.

[$KO_1$] S. Kobayashi and T. Ochiai, Satake compactification and the great Picard theorem, J. Math. Soc. Japan **23** (1971), 340–350.

[$KO_2$] S. Kobayashi and T. Ochiai, Meromorphic mappings onto compact complex spaces of general type, Invent. Math. **31** (1975), 7–16.

[KSW] M. Kalka, B. Schiffman and B. Wang, Finiteness and rigidity theorems for holomorphic mappings, Michigan Math. J. **28** (1981), 289–295.

[$L_1$] S. Lang, Higher dimensional Diophantine problems, Bull. Amer. Math. Soc. **80** (1974), 779–787.

[$L_2$] S. Lang, Fundamentals of Diophantine Geometry, Springer–Verlag, Berlin, 1983.

[$L_3$] S. Lang, Hyperbolic and Diophantine analysis, Bull. Amer. Math. Soc., 14 (1986), 159–205.

[M] Ju. Manin, Rational points of algebraic curves over function fields, Izv. Akad. Nauk. SSSR. Ser. Mat. **27** (1963), 1395–1440.

[Mo] S. Mori, Threefolds whose canonical bundles are not numerically effective, Ann. Math. **116** (1982), 133–176.

[MM] S. Mori and S. Mukai, The uniruledness of the moduli space of curves of genus 11, Algebraic Geometry, Proc. Japan–France Conf. Tokyo and Kyoto 1982, Lecture Notes in Math. **1016**, Springer–Verlag, Berlin–Heidelberg–New York, 1983.

[$N_1$] J. Noguchi, A higher dimensional analogue of Mordell's conjecture over function fields, Math. Ann. **258** (1981), 207–212.

[$N_2$] J. Noguchi, Hyperbolic fibre spaces and Mordell's conjecture over function fields, Publ. RIMS, Kyto University **21** (1985), 27–46.

[$N_3$] J. Noguchi, Moduli spaces of holomorphic mappings into hyperbolically imbedded complex spaces and locally symmetric spaces, Invent. Math. **93** (1988), 15–34.

[NO] J. Noguchi and T. Ochiai, Geometric Function Theory in Several Complex Variables, to appear from A.M.S. Monograph Series (the Japanese edition, Iwanami Shoten, Tokyo, 1984).

[NS] J. Noguchi and T. Sunada, Finiteness of the family of rational and meromorphic mappings into algebraic varieties, Amer. J. Math. **104** (1982), 887–900.

[$P_1$] A.N. Parshin, Algebraic curves over function fields, I, Izv. Akad. Nauk SSSR Ser. Mat., **32** (1968), 1145–1170.

[$P_2$] A.N. Parshin, Finiteness theorems and hyperbolic manifolds, preprint.

[Pe] C. Peters, Rigidity for variations of Hodge structure and Arakelov–type finiteness theorems, Compositio Math.

[R] M. Raynoud, Around the Mordell conjecture for function fields and a conjecture of Serge Lang, Lecture Notes in Math., vol. **1016**, pp. 1–19, Springer–Verlag, Berlin–New York, 1983.

[Ro] H.L. Royden, Automorphisms and isometries of Teichmüller spaces, Advances in the Theory of Riemann Surfaces, pp. 369–383, Ann. of Math. Studies **66**, Princeton Univ. Press, Princeton, New Jersey, 1971.

[SZ] M.–H. Saito and S. Zucker, Classification of non–rigid families of K3 surfaces and a finiteness theorem of Arakelov type, preprint.

[SY] R. Schoen and S.–T. Yau, Compact group actions and the topology of manifolds with non–positive curvature, Topology **18** (1979), 361–380.

[$S_1$] T. Sunada, Holomorphic mappings into a compact quotient of symmetric bounded domain, Nagoya Math. J. **64** (1976), 159–175.

[$S_2$] T. Sunada, Rigidity of certain harmonic mappings, Invent. Math. **51** (1979), 297–307.

Department of Mathematics
Tokyo Institute of Technology
O'okayama, Meguro
Tokyo 152
JAPAN

Max–Planck–Institut
für Mathematik
Gottfried–Claren–Str. 26
5300 Bonn 3
Federal Republic of Germany

# GLOBAL NONDEFORMABILITY OF THE COMPLEX PROJECTIVE SPACE

Yum–Tong Siu [1]

Every irreducible compact Hermitian symmetric manifold is rigid in the sense that if it is a fiber in a holomorphic family of compact complex manifolds then any neighboring fiber that is sufficiently close to it must be biholomorphic to it. This simply is the consequence of the result of Frölicher–Nijenhuis [F–N] and Calabi–Vesentini [C–V]. It is in general not known whether an irreducible compact Hermitian symmeteric manifold admits nontrivial global deformation. Kodaira and Spencer [K–S, p.464, Problem 8] posed the problem whether an arbitrary deformation of the complex projective n–space is again the complex projective n–space.

For the case of $n = 3$ Peternell [P1, P2] and Nakamura [N1, N2] have independently obtained the more general result that a Moishezon compact complex threefold homeomorphic to $\mathbb{P}_3$ must be biholomorphic to it. They also proved the same result for the case of the hyperquadric of complex dimension three. Nakamura's method needs the additional assumption that no powers of the canonical line bundle admit nonzero holomorphic sections.

In the essential case (which readily implies the general case) when the parameter space of deformation is the open unit disk and all fibers except possibly the one at the origin are all biholomorphic to the complex projective n–space, a proof for the above problem of Kodaira and Spencer was given in [S] using the method of $\mathbb{C}^*$ action introduced by Tsuji [T] and the method of counting the zeroes of holomorphic vector fields. Later Tsuji told me that he and Mabuchi discovered the following gap in the last step of the proof in [S]. On p. 218 of [S] the holomorphic family $\{G(w,t_\nu) \mid w \in W_0\}$ of complex lines of degree one may approach a family of complex curves $\{G(w,0) \mid w \in W_*\}$ whose parameter space $W_*$ is reducible. Let $W'$ be the branch of $W_*$ so that the irreducible rational curve $\gamma_0$ equals $G(w_0)$ for some $w_0$ in $W'$. The gap is that the union of $G(w,0)$ with $w$ in $W'$ may not contain a nonempty open subset of $M_0$. We present below a new version of the proof of the global nondeformability of the complex projective space that avoids this last step of the argument. This new version is more involved and uses the same method of counting the number of zeroes of vector fields but applies the

[1]Research partially supported by a grant from the National Science Foundation

counting not only to curves but also to higher dimenisonal subvarieties. I would like to thank T. Mabuchi for reading an earlier sketch of the proof presented in this paper and making useful comments on it.

**Main Theorem.** Let $\pi: M \to \Delta$ be a holomorphic family of compact complex manifolds (where $\Delta = \{t \in \mathbb{C} \mid |t| < 1\}$) such that $M_t := \pi^{-1}(t)$ is biholomorphic to $\mathbb{P}_n$ for $t \in \Delta - 0$. Then $M_0$ is biholomorphic to $\mathbb{P}_n$.

The proof of the Main Theorem in this paper is divided into two parts. For the sake of completeness, in the first part we reproduce here with very slight modificaion the reduction of the problem to the extendiblity of the $\mathbb{C}^*$ action which is given in [S]. This method of reduction of the problem to the extendibility of the $\mathbb{C}^*$ action is due to Tsuji [T]. The second part which is the heart of the paper uses the method of counting the zeroes of vector fields on higher-dimensional subvarieties.

Suppose $X_0$ is a compact complex manifold and $X_\nu$ $(1 \le \nu < \infty)$ is a sequence of compact complex manifolds approaching $X_0$ as $\nu \to \infty$ (in the sense that all $X_\nu$ for $0 \le \nu < \infty$ can be regarded as complex structures on the same smooth manifold $Y$ and the complex structure $X_\nu$ of $Y$ approaches the complex structure $X_0$ of $Y$ as $\nu \to \infty$). It is well-known that the Main Theorem readily implies that if each $X_\nu$ is biholomorphic to $\mathbb{P}_n$ for $\nu \ge 1$, then $X_0$ is also biholomorphic to $\mathbb{P}_n$. The simple reason is as follows. By the result of Kuranishi [Ku] we have a holomorphic family $p: X \to S$ of compact complex manifolds over a complex space $S$ so that, for some points $P_\nu$ $(0 \le \nu < \infty)$ of $S$ with $P_\nu$ approaching $P_0$ as $\nu \to \infty$, the fiber $p^{-1}(P_0)$ is biholomorphic to $X_0$ and each fiber $p^{-1}(P_\nu)$ is biholomorphic to $\mathbb{P}_n$ under some biholomorphism $f_\nu: \mathbb{P}_n \to p^{-1}(P_\nu)$ for $\nu \ge 1$. By considering the holomorphic deformation of the graph of $f_\nu$ as a subvariety in $\mathbb{P}_n \times X$ and using the local rigidity of $\mathbb{P}_n$, we conclude that there exists a subvariety $Z$ of complex codimension at least one in $S$ such that $p^{-1}(P)$ is biholomorphic to $\mathbb{P}_n$ for every point $P$ of $S - Z$. By using an irreducible local complex curve $C$ in $S$ passing through $P_0$ and not entirely contained in $Z$ and the normalization of $C$, we can assume without loss of generality that $S$ equals the open unit 1-disk $\Delta$ and $P_0$ is the origin of $\Delta$ and $Z$ consists only of the single point $P_0$. Then by the Main Theorem $p^{-1}(P_0)$ is biholomorphic to $\mathbb{P}_n$.

## Table of Contents



## Part I

### §1. *Hyperplane Section Line Bundle*

(1.1) We use the notations in the statement of the Main Theorem. By considering the line bundle $K_M$ over $M$ whose restriction to $M_t$ is the canonical line bundle of $M_t$ for $t \neq 0$ and applying the semicontinuity of $\Gamma(M_t, \overline{K_{M_t}^{-k}})$ in $t$, we conclude that $M_0$ is Moishezon. For a complex space $X$ we use $\mathcal{O}_X$ to denote the structure sheaf of $X$ and use $\mathcal{O}_X^*$ to denote the sheaf of germs of nowhere zero holomorphic functions on $X$. Since on $M_0$ we have the Hodge decomposition [U, p.99, Cor.9.1], it follows from the vanishing of the first Betti number of $M_0$ that $H^1(M_0, \mathcal{O}_{M_0})$ vanishes. Clearly $H^1(M_t, \mathcal{O}_{M_t})$ vanishes for $t \in \Delta - 0$. It follows from [R, p.49, (3.3)] and [G, p.20, Satz 5] that $H^1(M, \mathcal{O}_M)$ vanishes. From the exact sequence

$$0 \to \mathbb{Z} \to \mathcal{O}_M \to \mathcal{O}_M^* \to 0$$

we have the exact sequence

$$0 \to H^1(M,\mathcal{O}_M^*) \to H^2(M,\mathbb{Z}) \overset{\varphi}{\to} H^2(M,\mathcal{O}_M).$$

Since $H^2(M,\mathbb{Z}) \approx \mathbb{Z}$ and the first Chern classs of $K_M$ is a nonzero element of $H^2(M,\mathbb{Z})$ which is mapped to zero by $\varphi$, it follows that $\varphi$ is the zero map and $H^1(M,\mathcal{O}_M^*) \approx H^2(M,\mathbb{Z})$. Thus we have a holomorphic line bundle $L$ over $M$ with the property that $L_t := L|M_t$ is the positive hyperplane section line bundle of $M_t$ for $t \in \Delta - 0$. We note that the same argument applied to $M_0$ instead of $M$ yields the isomorphism $H^1(M_0,\mathcal{O}_{M_0}^*) \approx H^2(M_0,\mathbb{Z})$.

(1.2) We want to show that there exist $n+1$ holomorphic sections $s_0,\cdots,s_n$ of $L$ over $M$ such that for every $t \in \Delta$ (including $t = 0$), the holomorphic sections $s_0|M_t,\cdots,s_n|M_t$ are linearly independent over $\mathbb{C}$.

Let $\mathcal{O}_M(L)$ be the sheaf of germs of local holomorphic sections of $L$ over $M$. Consider the direct image sheaf $R^0\pi_*\mathcal{O}_M(L)$ over $\Delta$. Since clearly $R^0\pi_*\mathcal{O}_M(L)$ is torsion-free, it follows that $R^0\pi_*\mathcal{O}_M(L)$ is locally free. Its rank must be $n+1$ because outside $0$ its rank is $n+1$. Thus we have $R^0\pi_*\mathcal{O}_M(L) \approx \mathcal{O}_\Delta^{n+1}$ and we have $n+1$ holomorphic sections $s_0,\cdots,s_n$ of $L$ over $M$ which generate $R^0\pi_*\mathcal{O}_M(L)$. Clearly $s_0|M_t,\cdots,s_n|M_t$ are linearly independent over $\mathbb{C}$ when $t \neq 0$. For the case $t = 0$ we argue as follows.

Suppose the contrary. After relabelling $s_0,\cdots,s_n$, we can assume without loss of generality that $s_0 = \Sigma_{j=1}^n c_j s_j$ on $M_0$ for some complex numbers $c_1,\cdots,c_n$. Let $s = \frac{1}{t}(s_0 - \Sigma_{j=1}^n c_j s_j)$. Then $s \in \Gamma(M,\mathcal{O}_M(L))$. Let $u_j$ (respectively $u$) be the element of $\Gamma(\Delta,R^0\pi_*\mathcal{O}_M(L))$ corresponding to $s_j$ (respectively $s$) for $0 \leq j \leq n$. Then $u_0 = \Sigma_{j=1}^n c_j u_j + tu$. It follows that

$$R^0\pi_*\mathcal{O}_M(L) \subset \Sigma_{j=1}^n \mathcal{O}_\Delta u_j + t\, R^0\pi_*\mathcal{O}_M(L).$$

By Nakayama's lemma the stalk of $R^0\pi_*\mathcal{O}_M(L) / \Sigma_{j=1}^n \mathcal{O}_\Delta u_j$ at $0$ is $0$, contradicting that $R^0\pi_*\mathcal{O}_M(L)$ is locally free of rank $n+1$.

Actually for this paper we use only the statement that $s_i|M_0$ and $s_j|M_0$ are linearly independent for $0 \leq i < j \leq n$.

(1.3) The zero–set of $s_j|M_0$ $(0 \leq j \leq n)$ is irreducible and has multiplicity 1. The reason is as follows. The line bundle $[V]$ associated to any irreducible positive divisor $V$ of $M_0$ must be an integral multiple of $L$, because $H^1(M_0, \mathcal{O}^*_{M_0}) \approx H^2(M_0, \mathbb{Z})$. It must a positive multiple, otherwise the product of the canonical section of $[V]$ times a positive power of a nonzero holomorphic section of of $L$ is a holomorphic function on $M_0$ which vanishes somewhere but not identically zero. If the zero–set of $s_0$ is reducible or has multiplicity bigger than 1, the line bundle $L$ would be the tensor product of two line bundles associated to positive divisors and would be a positive multiple of itself with factor bigger than 1.

The common zero–set of $s_i$ and $s_j$ $(0 \leq i < j \leq n)$ in $M_0$ is of complex codimension at least two in $M_0$, otherwise from the fact that the zero–set of $s_i$ is irreducible and has multiplicity it fillows that $s_j$ would have to vanish identically on the zero–set of $s_i$ and $\frac{s_j}{s_i}$ would be a holomorphic function on $M_0$ and must be constant, contradicting the linear independence of $s_i$ and $s_j$.

## §2. *Euler Vector Fields*

(2.1) Let $H$ be the zero–set of $s_0$ and $H_t = H \cap \pi^{-1}(t)$. For $t \in \Delta - 0$, $H_t$ can serve as the hyperplane at infinity. By replacing $\Delta$ by a smaller disk centered at 0, we can assume without loss of generality that there is a holomorphic section $\tau : \Delta \to M$ of $\pi : M \to \Delta$ such that the image of $\tau$ is disjoint from $H$. By replacing $s_\nu$ by $s_\nu - \left[\frac{s_\nu}{s_0}(\tau(t))\right] s_0$ for $1 \leq \nu \leq n$, we can assume without loss of generality that $s_\nu$ vanishes at $\tau(\Delta)$ for $1 \leq \nu \leq n$. Consider the Euler vector field associated to the origin $\tau(t)$ and the hyperplane $H_t$ for $t \in \Delta - 0$. Let $z_\nu = \frac{s_\nu}{s_0}$ $(\nu > 0)$. The Euler vector field is $X_t = \Sigma_{\nu=1}^n z_\nu \frac{\partial}{\partial z_\nu}$ on $M_t$ for $t \in \Delta - 0$. This vector field is clearly holomorphic on $M_t - H_t$ for $t \in \Delta - 0$. It is also holomorphic at points of $H_t$ and, as a matter of fact, vanishes at every point of $H_t$. Let $X$ be the holomorphic vector field on $M \cap \{t \neq 0\}$ whose restriction to $M_t$ is $X_t$ for $t \in \Delta - 0$.

(2.2) We now extend the holomorphic Euler vector field $X$ on $\pi^{-1}(\Delta - 0)$ to a meromorphic vector field on all of $M$ as follows. Let $\zeta_1, \cdots, \zeta_n$ be a local holomorphic coordinate system of $M_0$ centered at $\tau(0)$ so that $\zeta_1, \cdots, \zeta_n, t$ form a local holomorphic coordinate system of $M$ centered at $\tau(0)$ with $\tau(\Delta) = \{\zeta_1 = \cdots = \zeta_n = 0\}$. Let $z_\nu = f_\nu(\zeta_1, \cdots, \zeta_n, t)$. Then for any fixed $t \in \Delta - 0$ and fixed $\mu$

$$\delta_{\nu\mu} = \frac{\partial z_\nu}{\partial z_\mu} = \Sigma_{\lambda=1}^n \frac{\partial f_\nu}{\partial \zeta_\lambda} \frac{\partial \zeta_\lambda}{\partial z_\mu}.$$

For fixed $\mu$ and for $1 \leq \nu \leq n$ consider the above equations as a system of $n$ linear equations in the unknowns $\frac{\partial \zeta_1}{\partial z_\mu}, \cdots, \frac{\partial \zeta_n}{\partial z_\mu}$ and solve for the unknowns by Cramer's rule. We conclude that $\frac{\partial \zeta_1}{\partial z_\mu}, \cdots, \frac{\partial \zeta_n}{\partial z_\mu}$ are meromorphic functions of $\zeta_1, \cdots, \zeta_n, t$. Hence

$$X = \Sigma_{\mu=1}^n z_\mu \frac{\partial}{\partial z_\mu} = \Sigma_{\lambda,\mu=1}^n f_\mu(\zeta_1, \cdots, \zeta_n, t) \frac{\partial \zeta_\lambda}{\partial z_\mu} \frac{\partial}{\partial \zeta_\lambda}$$

is a meromorphic vector field with respect to the coordinates $\zeta_1, \cdots, \zeta_n, t$. Since $X$ is holomorphic on $\pi^{-1}(\Delta - 0)$, there exists a unique integer $k$ such that $t^k X$ is a holomorphic vector field on some open neighborhood of $\tau(0)$ in $M$ and $t^k X$ is not identically zero on $M_0$. By Hartogs' theorem, $t^k X$ is a holomorphic vector field on all of $M$. Let $X^* = t^k X$ and $X_t^* = X^* | M_t$. Then $X_0^*$ is not identically zero on $M_0$.

(2.3) Let $Y = (s_0)^{-1} X^*$. Then $Y$ is an $L^{-1}$–valued holomorphic tangent vector field on $M$. We let $Y_t = Y | M_t$. The introduction of $Y$ is important for the method of counting the zeroes of vector fields in Part II. Just as we can construct integral curves in the case of a holomorphic vector field, we can also get integral curves from a line–bundle–valued holomorphic vector field as curves to which the line–bundle–valued holomorphic vector field is tangential. At the intersection point $P$ of two integral curves of $Y$ we must have the vanishing of $Y$ at $P$. The use in Part II of the integral curves for the line–bundle–valued holomorphic vector field $Y$ avoids the complication of having to investigate when $(s_0)^{-1} X^*$ still vanishes at the intersection point of two branches of an integral curve when one branch lies in the zero–set of $s_0$.

## §3. *The Special Case of the Extendibility of the* $\mathbb{C}^*$ *Action*

(3.1) First we consider the special case where the integer $k$ defined in (2.2) is nonpositive and finish the proof of the Main Theorem for this special case. This special case means that the holomorphic Euler vector field $X$ on $\pi^{-1}(\Delta - 0)$ can be extended holomorphically to all of $M$. We denote this extension also by $X$.

We have a $\mathbb{C}^*$ action on $M$ which acts on each $M_t$, because the integral complex curves of the Euler vector field in $\mathbb{C}^n$ (with coordinates $z_1, \cdots, z_n$) are $z_\nu = a_\nu e^t$ $(1 \leq \nu \leq n)$ so that $\frac{dz_\nu}{dt} = a_\nu e^t = z_\nu$ and the action obtained by integrating the Euler vector field has period $2\pi i$. For fixed $\zeta \in \mathbb{C}^*$, the action $z_\nu \to \zeta z_\nu$ (i.e. with $\zeta = e^t$), $1 \leq \nu \leq n$, in $\mathbb{C}^n$ is the $\mathbb{C}^*$ action. Thus the action on $\pi^{-1}(\Delta - 0)$ defined by integrating the holomorphic vector field $X$ has also period $2\pi i$ and, by continuity, the action on all of $M$ defined by integrating the holomorphic vector field $X$ has also period $2\pi i$. We have therefore a holomorphic $\mathbb{C}^*$ action $\Psi: \mathbb{C}^* \times M \to M$ on $M$.

Since the action $\Psi(\zeta, \cdot)$ on $M$ is holomorphic in $\zeta \in \mathbb{C}^*$, a point $P$ of $M$ is fixed under all $\Psi(\zeta, \cdot)$ for $\zeta \in \mathbb{C}^*$ if and only if it is fixed under $\Psi(\zeta, \cdot)$ for $\zeta$ belonging to the unit circle in $\mathbb{C}^*$, because of the identity theorem for holomorphic functions and the fact that $P$ being fixed means the vanishing of the local holomorphic functions $\Psi(\zeta, P) - P$ in $\zeta$. Thus the fixed point set of $M$ under the $\mathbb{C}^*$ action is the same as the fixed point set of $M$ under the circle group action.

(3.2) We now introduce a Riemannian metric in $M$. By averaging over the circle group, we can assume that the Riemannian metric is invariant under the circle group action. This is possible, because the circle group is a compact group. Now we consider the fixed point set of this circle group of isometries. The fixed point set is always totally geodesic, because when an isometry fixes a point and a tangent vector at that point, it fixes also the geodesic through that point and in that tangent direction. Totally geodesic sets are submanifolds. Now $X$ is zero on $H$. Since $H$ is already of pure codimension 1, it must coincide with a component of the fixed point set. The same argument concerning fixed point sets applies to $X | M_0$. Thus $(\pi | H): H \to \Delta$ is a regular family of compact complex manifolds whose fibers are $\mathbb{P}_{n-1}$ outside the zero fiber.

(3.3) We now finish the proof of the Main Theorem by induction on the dimension $n$ of the fiber of $\pi: M \to \Delta$. Under the induction hypothesis the zero-set $H_0$ of $s_0 | M_0$ is also $\mathbb{P}_{n-1}$ and $(\pi | H): H \to \Delta$ is trivial. Thus on $H$ the holomorphic line bundle $L | H$ admits $n$ holomorphic sections $s_1', \cdots, s_n'$ with no common zeroes. Consider the short exact sequence

$$0 \longrightarrow \mathcal{O}_M \xrightarrow{\psi} \mathcal{O}_M(L) \longrightarrow \mathcal{O}_H(L | H) \longrightarrow 0,$$

where $\psi$ is defined by multiplication by $s_0$ and $\mathcal{O}_H(L|H)$ is considered as a sheaf over $M$ by trivial extension. We have the long exact cohomology sequence

$$\Gamma(M, \mathcal{O}_M(L)) \longrightarrow \Gamma(H, \mathcal{O}_H(L | H)) \longrightarrow H^1(M, \mathcal{O}_M),$$

but we know that $H^1(M, \mathcal{O}_M) = 0$. Hence we can lift up the $n$ holomorphic sections $s_1', \cdots, s_n'$ of $L | H$ over $H$ to $n$ holomorphic sections of $L$ over $M$ which without loss of generality we can assume to be $s_1, \cdots, s_n$. Then on $H$ the sections $s_1, \cdots, s_n$, being equal to $s_1', \cdots, s_n'$, cannot have any common zeroes. On $M - H$ of course $s_0$ has no zero. Thus $s_0, s_1, \cdots, s_n$ have no common zeroes on $M$.

(3.4) Consider the holomorphic map $\Psi: M \to \Delta \times \mathbb{P}_n$ given by $\Psi = (t, [s_0, \cdots, s_n])$. This map is biholomorphic from $\pi^{-1}(\Delta - 0)$ to $(\Delta - 0) \times \mathbb{P}_n$. Let $Z$ be the set of points where the Jacobian determinant of $\Psi$ is zero. Then $Z \subset M_0$. If $Z$ is empty, then $\Psi$ is a local biholomorphism and is proper. Since $\Delta \times \mathbb{P}_n$ is simply connected, we must have $M$ biholomorphic to $\Delta \times \mathbb{P}_n$. Suppose $Z$ is nonempty. Then since $Z$ must be of pure codimension 1, we must have $Z = M_0$. By the properness of $\Psi$, the map $\Psi$ must be surjective and $\Psi(M_0) = 0 \times \mathbb{P}_n$. Thus the Jacobian determinant of $\Psi | M_0$ is not identically zero. This can only mean that $t$ vanishes to order at least 2 on $M$, which is a contradiction. This concludes the proof of the Main Theorem under the assumption that $k$ is nonpositive. From now on we assume that $k > 0$ and will prove the Main Theorem by deriving a contradiction from $k > 0$.

Part II

## §4. *Canonical Resolution of Singularities and the Lifting of Vector Fields.*

(4.1) By a holomorphic vector field on a local singularity we mean a holomorphic vector field on its regular part with the property that the local one-parameter subgroup obtained by its integration extends to local biholomorphisms of the local singularity. For example, if we have a holomorphic vector field on an ambient manifold of the local singularity which is tangential to the regular part of the singularity, then such a holomorphic vector field on the ambient manifold restricts to a holomorphic vector field on the local singularity.

By a canonical resolution of singularities we mean a method for resolving the local singularities so that the resolutions of two biholomorphic singularities will be canonically isomorphic. For example, the method of using Nash blow-ups to resolve the singularites of two-dimensional local singularities is a canonical resolutoin of singularites for dimension two (see [Sp]). Clearly for any method of canonical resolution of local singularieties, any holomorphic vector field on the local singularities can be lifted holomorphically to the resolution. The reason is the following. We consider two cases. In the first case the vector field does not vanish at that point. In that case the singularity is a product singularity according to a result of Rossi [R] and the vector field is along the regular factor. So clearly we have the lifting. In the second case the vector field vanishes at the point under consideration. Then the vector field generates a local holomorphic family of biholomorphisms of the singularity fixing the point under consideration. Since the resolution is canonical, this holomorphic family of biholomorphisms can be lifted to a holomorphic family of biholomorphisms on the resolution and the derivation of the family is the lifted vector field.

(4.2) When the normalization of a singularity is regular, we consider the normalization also as a canonical resolution of the singularity. Note that this argument on the lifting of vector fields provides an alternative proof of the following statement concerning the normalization of curves (Lemma 5.2 in [S]). Suppose $C$ is a complex curve defined on some open neighborhood $U$ of $0$ in $\mathbb{C}^n$ which is singular at $0$ and $\rho\colon \tilde{C} \to C$ is the normalization of $C$ with $\rho(P_0) = 0$ for some $P_0$ in $\tilde{C}$. Suppose that $T$ is a holomorphic vector field on $U$ which is tangential to $C$ (at every regular point of $C$). Then there exists a unique holomorphic vector field $\tilde{T}$ on $\tilde{C}$ with $(d\rho)(\tilde{T}(P)) = T(\rho(P))$ for $P \in \tilde{C}$ and, moreover, $\tilde{T}$ vanishes at $P_0$. Clearly $\tilde{T}$ vanishes at $P \in \tilde{C}$ whenever $T$

anishes at $\rho(P)$.

(4.3) For the other direction we have the following trivial observation about pushing forward a holomorphic vector field under a modification. Suppose we have a manifold $\tilde{W}$ and a vector field $\tilde{T}$ on $\tilde{W}$ and assume that $\tilde{W}$ can be blown down to $W$ *which is regular* and we have a proper modification $\rho: \tilde{W} \to W$ which is biholomorphic from $\tilde{W} - \tilde{V}$ to $W - V$ so that $\tilde{V}$ is a complex analytic subvariety of complex codimension one in $\tilde{W}$ and $V$ is a complex analytic subvariety of complex codimension at least two in $W$. Then the vector field $\tilde{T}|(\tilde{W} - \tilde{V})$ corresponds to a vector field $T$ on $W - V$ under $\rho$. Since $V$ is of complex codimension at least two in $W$, we know that $T$ can be extended to a holomorphic vector field on $W$.

## 5. *Strictly Plurisubharmonic Functions on Zariski Open Subsets*

(5.1) Let $u_\nu$ $(1 \leq \nu \leq N_\ell)$ be a basis of $\Gamma(M_0, L^\ell | M_0)$. Then, for $\ell$ sufficiently large, there exists a Zariski open subset $\Omega_\ell$ of $M_0$ such that at every point $P$ of $\Omega_\ell$ some $u_\mu$ is nonzero at $P$ and $d(\frac{u_\nu}{u_\mu})$ at $P$ for $1 \leq \nu \leq N_\ell$ span the cotangent space of $M_0$ at $P$. For $0 \leq j \leq n$ the function $\psi_j$ defined by $\psi_j = \Sigma_{1 \leq \nu \leq N_\ell} \left| \frac{u_\nu}{s_j^\ell} \right|^2$ is plurisubharmonic on $M_0 - \{s_j = 0\}$ and is strictly plurisubharmonic on $\Omega_\ell \cap \{s_j \neq 0\}$.

(5.2) Every irreducible compact complex curve $C$ in $M_0$ which intersects $\Omega_\ell$ for $\ell$ sufficiently large must intersect $s_j$ for every $0 \leq j \leq n$. Otherwise, the function $\psi_j$ which is plurisubharmonic in $M_0 \cap \{s_j \neq 0\}$ and strictly plurisubharmonic in $\Omega_\ell \cap \{s_j \neq 0\}$ would be subharmonic on the compact curve $C$ and strictly subharmonic at some point of $C$ which yields a contradiction by considering the point of $C$ where the maximum of $\psi_j$ is achieved.

## 6. *Regularity of Irreducible Orbital Curves.*

(6.1) Let $C$ be an irreducible curve in $M_0$ not entirely contained in the zero set of $X_0^*$ such that $C$ intersects $\Omega_\ell \cap (\cap_{j=0}^n \{s_j \neq 0\})$ for some sufficiently large $\ell$ and $X_0^*$ is tangential to $C$ at every point of $C$. We claim that $C$ is nonsingular and intersects $\{s_0 = 0\}$ normally at exactly one point.

First of all by (5.2) the curve $C$ must intersect $\{s_0 = 0\}$. We observe that $C$ cannot contain $\tau(0)$. The reason is as follows. The meromorphic function $(\frac{s_j}{s_0})|M_t$ is an eigenfunction for the Euler vector field $X_t$ for $t \neq 0$ with eigenvalue 1. Hence $\lim_{t\to 0} (t^k X)(\frac{s_j}{s_0}) = \lim_{t\to 0} t^k(\frac{s_j}{s_0}) = 0$ on $M_0$ and $X_0^*(\frac{s_j}{s_0}) = 0$ on $M_0$. Since $X_0^*$ is tangential to $C$, this means that $\frac{s_j}{s_0}$ is constant on $C \cap \{s_0 \neq 0\}$. Since $\frac{s_j}{s_0}$ is zero at $\tau(0)$ for $1 \leq j \leq n$, if $C$ contains $\tau(0)$, then the function $\frac{s_j}{s_0}$ $(1 \leq j \leq n)$ is identically zero on $C$, contradicting the fact that $C$ intersects $\cap_{j=0}^{n} \{s_j \neq 0\}$.

Consider the orbits of the $L^{-1}$-valued holomorphic tangent vector field $Y = (s_0)^{-1} X^*$ on $M$. For $t \neq 0$ the topological closures of the orbits of $Y|M_t$ are precisely all the projective lines in $M_t$ passing through the point $\tau(t)$. Consider the complex space $\mathscr{D}$ of all compact complex curves in $M$ and consider the subvariety $\mathscr{D}_0$ of $\mathscr{D}$ containing all points defined by projective lines in $M_t$ for $t \neq 0$ which contain $\tau(t)$. (For a discussion concerning the geneal construction of such a complex space $\mathscr{D}$ see Fujiki [F].) By using the complex space $\mathscr{D}_0$, we conclude that $C$ is a branch of the limit $V_0$ of a sequence of projective lines $V_t$ in $M_t$ containing $\tau(t)$ for $t \neq 0$ as $t \to 0$. Clearly $V_0$ contains $\tau(0)$ and the $L^{-1}$-valued holomorphic vector field $Y_0 = Y|M_0$ is tangential to every branch of $V_0$. Since $C$ does not contain $\tau(0)$ which is contained in $V_0$, some other branch $C'$ of $V_0$ must intersect $C$ at some point $Q_0'$ of $C$.

Let $\rho: \tilde{C} \to C$ be the normalization of $C$. We can pull back $Y_0 = (s_0)^{-1} X_0^*$ to a $\rho^* L^{-1}$-valued holomorphic vector field $\tilde{Y}_0$ on $\tilde{C}$. Let $Q_0$ be a point of $\tilde{C}$ with $\rho(Q_0) = Q_0'$. By (4.2) $\tilde{Y}_0$ vanishes at every point of $\tilde{C}$ which is mapped to a zero of $Y_0$ by $\rho$ and, in particular, vanishes at $Q_0$. Let $Q_1 + \cdots + Q_p$ be the divisor of $\rho^* s_0$ on $\tilde{C}$. We know that $p$ is at least one, because $C$ intersects $\{s_0 = 0\}$. Then $(\rho^* s_0)\, \tilde{Y}_0$ is a holomorphic vector field on the Riemann surface $\tilde{C}$ whose divisor is at least $Q_0 + \Sigma_{\nu=1}^{p} Q_\nu$. Since on any compact Riemann surface the number of zeroes of a holomorphic vector field is at most two, we conclude that $p = 1$ and the divisor of

$\rho^* s_0)\bar{Y}_0$ must be $Q_0 + Q_1$. This means that C intersects $\{s_0 = 0\}$ precisely at one oint where the intersection is normal. Moreover, it means also that $Y_0$ can have only ne zero on C which is $Q_0'$, because $\bar{Y}_0$ vanishes at every point of $\bar{C}$ which is mapped o a zero of $Y_0$ by $\rho$.

6.2) We know that the $L^{-1}$–valued holomorphic vector field $Y_0$ must vanish at the oint of intersection Q of C with $\{s_0 = 0\}$, otherwise we can find an irreducible ntegral curve C' of $Y_0$ so that C' also intersects $\Omega_\ell \cap (\cap_{j=0}^n \{s_j \neq 0\})$ and is not ontained entirely in the zero–set of $X_0^*$ but C' is disjoint from the base point set $s_0 = \cdots = s_n = 0\}$. One gets such a C' by choosing a point Q' on $\{s_0 = 0\}$ ufficiently close to Q such that Q' is not in the subvariety $\{s_0 = \cdots = s_n = 0\}$ of omplex codimension one in $\{s_0 = 0\}$ and then integrating $Y_0$ to get the topological losure C' of the integral curve of $Y_0$ that contains Q'. This yields a contradiction in he following way. Some $s_j$ with $1 \leq j \leq n$ does not vanish at Q'. Consider $\frac{s_j}{s_0}$ on $C' - \{Q'\}$. Since the linear function $\frac{s_j}{s_0}$ is annihilated by the vector field $X_0^*$ and since $X_0^*$ is tangential to C' and non identically zero there, it follows that $\frac{s_j}{s_0}$ is constant on $C' - \{Q'\}$. This constant must be nonzero, otherwise $s_j$ is identically zero on $C' - \{Q'\}$ nd by continuity $s_j$ would vanish at Q', which is a contradiction. So we conclude that $s_j$ is nowhere zero on C'. This contradicts the observation in (5.2) which says that there xists no irreducible curve in $M_0$ which intersects $\Omega_\ell$ but is disjoint from $\{s_j = 0\}$.

Since there is at most one zero for $Y_0$ on C, we know that $Y_0$ is nowhere zero on $C - \{Q\}$ from which we conclude that C is regular at every point of $C - \{Q\}$. On the ther hand C is shown earlier to be regular at Q. Thus we conclude that C is verywhere regular.

7. *Nonsingular Compact Surfaces with Two Commuting Vector Fields.*

7.1) For the time being we suspend our use of the notation X and Y and use X and Y o denote something else. Suppose we have a nonsingular compact complex surface W nd we have three proper subvarieties D', D, and E in W with $D' \subset D$ and we have wo holomorphic vector fields X and Y on W. Assume that the toplogical closure of the rbit in $W - D$ of every nonzero $\mathbb{C}$–linear combination of X and Y is a complex curve in W. Further assume the following.

(i) $W - D$ is biholomorphic to $\mathbb{C}^2$ under $f\colon W - D \to \mathbb{C}^2$,

(ii) X and Y correspond to $\frac{\partial}{\partial x}$ and $\frac{\partial}{\partial y}$ on $\mathbb{C}^2$ under f, where x and y are the coordinates of $\mathbb{C}^2$,

(iii) both X and Y vanish at every point of D',

(iv) every complex curve in W which does not lie entirely in E intersects D'.

Let $W_0$ be obtained from W by blowing down successively exceptional curves of the first kind in W. Then we claim that $W_0$ is biholomorphic to $\mathbb{P}_2$. Moreover, the blow-down from W to $W_0$ maps $W - D$ biholomorphically onto its image and the biholomorphism between $W_0$ and $\mathbb{P}_2$ is an extension of f when $W - D$ is identified with its image in $W_0$. Furthermore, with this biholomorphism between $W_0$ and $\mathbb{P}_2$ the vector fields X and Y correspond to the partial differentiation with respect to the two coordinates in some affine coordinate system of the affine part $\mathbb{C}^2$ of $\mathbb{P}_2$ corresponding to $W - D$.

(7.2) Let $\rho\colon W \to W_0$ be the holomorphic map obtained from blowing down successively exceptional curves of the first kind. Every exceptional curve C of the first kind in W must be contained in D, otherwise at a point P of $C - D$ one of the two vector fields X and Y must be nontangential to C and that vector field can give a nontrivial holomorphic deformation of C in W, contradicting the fact that C is exceptional. At each blowdown $\rho'\colon W \to W'$ of an exceptional curve C of first kind in W the conditions (i), (ii), (iii), and (iv) are satisfied when W, X, Y, D', D, E are replaced respectively by W', $(d\rho')(X)$, $(d\rho')(Y)$, $\rho'(D')$, $\rho'(D)$, and $\rho'(E)$. Thus for the proof of the claim we can assume without loss of generality that W is minimal in the sense that there is no exceptional curve of the first kind in W.

(7.3) Since W is a compactification of $\mathbb{C}^2$, we can use at this point the known results on the minimal compactification of $\mathbb{C}^2$ to conclude our claim. We can also argue directly as follows.

Let C be the closure of an orbit of X in $W - D$ not lying entirely in E. Since C intersects D' and therefore contains a zero of X, we know that C is a rational curve.

Consider the one–parameter subgroup $\varphi_{Y,s}$ of automorphisms of W generated by the vector field Y. Since Y is not tangent to C at points of C – D, the curves $C_s := \varphi_{Y,s}(C)$ are distinct from C for any nonzero sufficiently small s. On the other hand, since C intersects D' and therefore contains a zero P of Y, we know that $C_s$ intersects C at P. Since the curve $C_s$ defines the same cohomology class as C, we know that the self–intersection number of C is positive. By [K, p.757, Th.8] we know that W is algebraic.

The surface W does not admit any non identically zero holomorphic 1–form $\omega$, otherwise both holomorphic functions $\omega(X)$, $\omega(Y)$ on W are constant and in fact zero in view of the vanishing of X and Y somewhere on W, contradicting that X and Y are linearly independent at every point of W – D. Since W is Kähler, $H^1(W,\mathcal{O}_W)$ is zero. From the cohomology sequence of the exact sequence $0 \to \mathbb{Z} \to \mathcal{O}_W \to \mathcal{O}_W^* \to 0$ it follows that the map $H^1(W,\mathcal{O}_W^*) \to H^2(W,\mathbb{Z})$ is injective.

Let $[C_s]$ denote the line bundle associated to the divisor $C_s$. Clearly all the line bundles $[C_s]$ have the same Chern class for all s. By the injectivity of $H^1(W,\mathcal{O}_W^*) \to H^2(W,\mathbb{Z})$ it follows that all the line bundles $[C_s]$ are isomorphic for all s. As a consequence, the dimension $\dim|C|$ of the complete linear system containing the divisor C is at least one. By Noether's lemma [G–H, p.513] we know that W is rational. (We can also get the rationality of W directly from the result of Morrow [M, p.108, Th.11] that any compactification of $\mathbb{C}^2$ is rational.) We now finish our proof by looking at minimal rational surfaces admitting two such vector fields X and Y. Every minimal rational surface is either $S_n$ or $\mathbb{P}_2$ (see for example [G–H, p. 520]). Here $S_n$ is a Hirzebruch surface $p: W \to \mathbb{P}_1$ with $\mathbb{P}_1$ fiber (see Hirzebruch [H]). We have to show that the case of a Hirzebruch surface can be ruled out.

(7.4) Suppose we have a Hirzebruch surface $p: W \to \mathbb{P}_1$ whose fiber is $\mathbb{P}_1$. Consider the holomorphic tangent vector field X on W. The holomorphic vector field X generates a 1–parameter family of holomorphic transformations $\varphi_t$ of W. Take two distinct points P and Q of $\mathbb{P}_1$. Since $\varphi_t(p^{-1}(P))$ is a continuous (actually holomorphic) family of rational curves, the intersection number $\varphi_t(p^{-1}(P))\cdot p^{-1}(Q)$ should be independent of t and must be zero, because it is zero at $t = 0$. So the image of $\varphi_t(p^{-1}(P))$ under p is zero–dimensional and must be equal to a single point. We thus conclude that $\varphi_t$ is fiber–preserving. There are two cases. One case is that $\varphi_t$ induces the identity map in the base $\mathbb{P}_1$ of $p: W \to \mathbb{P}_1$ for every t. The second case is that $\varphi_t$ induces a nontrivial biholomorphism of the base for some $t$. The first case means that $X$ is tangential to every fiber. Since $X$ and $Y$ are linearly independent at every point of $W - D$, we know that not both $X$ and $Y$ can be tangential to every fiber. Without loss of generality we can assume that $X$ is not tangential to every fiber. So we have the second case. The vector field $X$ cannot have a zero at every fiber, otherwise the biholomorphism $\varphi_t$ would fix that zero point of every fiber, contradicting the fact that $\varphi_t$ induces a nontrivial biholomorphism of the base for some $t$. Thus the zero-set of $X$ is contained in a finite number of fibers. There exists at least one fiber $F$ not contained entirely in $D \cup E$ such that $X$ is nowhere zero on $F$. Now consider $Y$. Since $\mathbb{P}_1$ is of complex dimension one and the two non identically zero holomorphic vector fields $(dp)(X)$ and $(dp)(Y)$ on it have zero Lie bracket, it follows that $(dp)(Y)$ is a constant multiple of $(dp)(X)$ as one can easily see by writing $(dp)(X) = \alpha(z)\frac{d}{dz}$ and $(dp)(Y) = \beta(z)\frac{d}{dz}$ in terms of the local coordinate $z$ of $\mathbb{P}_1$ and computing their Lie bracket. By replacing $Y$ by $Y - cX$ where $c$ is the constant such that $(dp)(Y) = c\,(dp)(X)$, we can assume without loss of generality that $(dp)(Y)$ is identically zero and $Y$ is tangential to every fiber of $p$. This yields a contradiction, because the fiber $F$ which is the closure of an orbit of $Y$ does not contain a zero of $X$ and hence cannot intersect $D'$. (Since we know explicitly the identitiy component of the automorphism group of $S_n$, for example from [B], we can also use directly such an explicit description of the identity component of the automorphism group of $S_n$ in our argument instead.)

(7.5) Finally we have to verify that in the statement in (7.1) the vector fields $X$ and $Y$ correspond to the partial differentiation with respect to two coordinates of some affine coordinate system of the affine part $W - D = \mathbb{C}^2$ of $W$. Since every complex curve of $W$ which does not lie entirely in $E$ must intersect $D'$, it follows that the complex dimension of $D'$ is one. Since both $X$ and $Y$ vanish at every point of $D'$, it follows that the zero-set of the holomorphic section $X \wedge Y$ of the anticanonical line bundle of $W$ is of multiplicity at least two at every point of $D'$. Since the degree of the divisor of every non identically zero holomorphic section of the antiholomorphic line bundle of $\mathbb{P}_2$ is three, we know that $D'$ must be of pure complex dimension one and its degree must be one. Let $(u,v)$ be an affine coordinate system of $W - D' = \mathbb{C}^2$.

Every holomorphic vector field in $W = \mathbb{P}_2$ which vanishes at every point of the infinity line $D'$ is a $\mathbb{C}$-linear combination of the vector fields $\frac{\partial}{\partial u}$, $\frac{\partial}{\partial v}$, and $u\frac{\partial}{\partial u} + v\frac{\partial}{\partial v}$. One simple way to verify this elementary fact is to introduce another set of inhomogeneous coordinates $(\xi,\eta)$ for $\mathbb{P}_2$ with $\xi = \frac{1}{u}$ and $\eta = \frac{v}{u}$. A holomorphic vector field $A$ on $\mathbb{P}_2$ vanishing at every point of the infinity line $\{\xi = 0\}$ of $\mathbb{P}_2$ is of the form

$$P(\xi,\eta)\,\xi\frac{\partial}{\partial\xi}+Q(\xi,\eta)\,\xi\frac{\partial}{\partial\eta}=-u\,P(\tfrac{1}{u},\tfrac{v}{u})\frac{\partial}{\partial u}+(-v\,P(\tfrac{1}{u},\tfrac{v}{u})+Q(\tfrac{1}{u},\tfrac{v}{u}))\frac{\partial}{\partial v},$$

where $P(\xi,\eta)$ and $Q(\xi,\eta)$ are polynomials in $\xi$ and $\eta$. Since both $-u\,P(\frac{1}{u},\frac{v}{u})$ and $-v\,P(\frac{1}{u},\frac{v}{u})+Q(\frac{1}{u},\frac{v}{u})$ must be polynomials in u and v, it follows that both $P(\xi,\eta)$ and $Q(\xi,\eta)$ can only be of degree one in $\xi$ and $\eta$ and, moreover, $P(\xi,\eta)=a+b\,\xi$ and $Q(\xi,\eta)=c+b\,\eta$ for some complex numbers a, b, and c. So the vector field A equals $-a\,(u\frac{\partial}{\partial u}+v\frac{\partial}{\partial v})-b\frac{\partial}{\partial u}+c\frac{\partial}{\partial v}$.

Since X and Y both vanish at every point of the infinity line D' of W, by replacing X and Y by suitable $\mathbb{C}$–linear combinations, we can assume without loss of generality that $X=\frac{\partial}{\partial u}$ and $Y=\alpha\frac{\partial}{\partial v}+\beta\,(u\frac{\partial}{\partial u}+v\frac{\partial}{\partial v})$ for some complex numbers $\alpha$ and $\beta$. Since the Lie bracket of X and Y is zero, we can conclude that $\beta=0$. The vector fields X and Y therefore are equal to the differentiation with respect to the two coordinates in some affine coordinate system of $W-D'=\mathbb{C}^2$.

## §8. *Behavior of Vector Fields at Blowup*

Let $X_0$ and $Y_0$ be the holomorphic vector fields on $\mathbb{P}_2$ which equal $\frac{\partial}{\partial x}$ and $\frac{\partial}{\partial y}$ on the affine part $\mathbb{C}^2$ of $\mathbb{P}_2$ with affine coordinates x,y. Let H be the infinity line $\mathbb{P}_2-\mathbb{C}^2$ of $\mathbb{P}_2$. We observe that the liftings of $X_0$ and $Y_0$ to the manifold obtained by blowing up one point $P_0$ of H cannot be both zero at every point of the exceptional divisor. The reason is as follows. Let $[z_0,z_1,z_2]$ be the homogeneous coordinates of $\mathbb{P}_2$ so that $x=\frac{z_1}{z_0}$ and $y=\frac{z_2}{z_0}$. We can assume that $P_0$ is not $[z_0,z_1,z_2]=[0,0,1]$, otherwise we can switch $z_1$ and $z_2$. Use the affine coordinates $\xi=\frac{z_0}{z_1}$ and $\eta=\frac{z_2}{z_1}$ so that the new origin $(\xi,\eta)=(0,0)$ is the point at infinity contained in the closure of each orbit of $X=\frac{\partial}{\partial x}$ in $\mathbb{P}_2-H$. The vector field $Y=\frac{\partial}{\partial y}$ becomes $\xi\frac{\partial}{\partial\eta}$. Let $\xi'=\xi$ and $\eta'=\eta-\eta(P_0)$ so that $P_0$ is the origin for the coordinates $(\xi',\eta')$ and Y is $\xi'\frac{\partial}{\partial\eta'}$. Use the quadratic transformation $\xi'=u$, $\eta'=uv$ to blow up $P_0$. Then $\frac{\partial}{\partial\eta'}$ becomes $\frac{1}{u}\frac{\partial}{\partial v}$ and $\xi'\frac{\partial}{\partial\eta'}$ becomes $\frac{\partial}{\partial v}$ which is not zero at every point of the exceptional divisor $u=0$.

§9. *Singular Compact Surfaces With Two Commuting Vector Fields*

(9.1) We now consider the case when the nonsingular surface W in §7 is replaced by an irreducible reduced complex space $\tilde{W}$ of complex dimension two which may contain singularities. Let L be a holomorphic line bundle over $\tilde{W}$ with a non identically zero holomorphic section $s_0$. Let $\tilde{D}$ and $\tilde{E}$ be two proper subvarieties of $\tilde{W}$. Assume that there are two holomorphic vector fields $\tilde{X}$ and $\tilde{Y}$ on $\tilde{W}$ so that

(i) $\tilde{W} - \tilde{D}$ is biholomorphic to $\mathbb{C}^2$ under $\tilde{f}\colon \tilde{W} - \tilde{D} \to \mathbb{C}^2$,

(ii) $\tilde{X}$ and $\tilde{Y}$ correspond to $\frac{\partial}{\partial x}$ and $\frac{\partial}{\partial y}$ on $\mathbb{C}^2$ under $\tilde{f}$, where $\tilde{x}$ and $\tilde{y}$ are the coordinates of $\mathbb{C}^2$,

(iii) $\frac{1}{s_0}\tilde{X}$ and $\frac{1}{s_0}\tilde{Y}$ are $L^{-1}$–valued holomorphic vector fields on $\tilde{W}$.

Again we assume that the topological closure of the orbit in $\tilde{W} - \tilde{D}$ of any nonzero $\mathbb{C}$–linear combination of $\tilde{X}$ and $\tilde{Y}$ in $\tilde{W}$ is a closed complex curve in $\tilde{W}$. Further assume that any compact irreducible curve in $\tilde{W}$ which intersects $\tilde{W} - \tilde{E}$ must intersect $\{s_0 = 0\}$. We claim that $\tilde{W}$ is regular and is biholomorphic to $\mathbb{P}_2$ with $L|\tilde{W}$ biholomorphic to the hyperplane section line bundle of $\mathbb{P}_2$. Moreover, the biholomorphism between $\tilde{W}$ and $\mathbb{P}_2$ is an extension of $\tilde{f}$ and $\tilde{D}$ is equal to the zero set of $s_0$.

(9.2) Let us first consider the case where $\tilde{W}$ is normal. We take a canonical desingularization W of $\tilde{W}$ with the desingularization map $\xi\colon W \to \tilde{W}$. Let E' be the subvariety in W where $\xi$ is not locally biholomorphic. We lift the holomorphic vector fields $\tilde{X}$ and $\tilde{Y}$ on $\tilde{W}$ to holomorphic vector fields X and Y on W. The topological closure of the orbit of any $\mathbb{C}$–linear combination of X and Y in $\xi^{-1}(\tilde{W} - \tilde{D} - \tilde{E})$ is mapped to the topological closure, in $\tilde{W}$, of an orbit of a $\mathbb{C}$–linear combination of $\tilde{X}$ and $\tilde{Y}$ in $\tilde{W} - \tilde{D} - \tilde{E}$, which must intersect $\{s_0 = 0\}$. However, when we lift the vector fields X and Y through the desingularization map $\xi$, we can divide them first by $s_0$ and then lift the $L^{-1}$–valued holomorphic vector fields and then multiply such liftings by the pullback of $s_0$. Both $X$ and $Y$ vanish at every point of $\{s_0 = 0\}$. So by §7 we know that $W$ can be blown down to $\mathbb{P}_2$ by blowing down successively exceptional curves of the first kind so that $X$ and $Y$ correspond to the partial differentiation with respect to the two coordinates of some affine coordinate system of the affine part $\mathbb{C}^2$ of $\mathbb{P}_2$. Let $\eta: W \to \mathbb{P}_2$ be the blow–down map. We know that $\eta(\xi^{-1}(\{s_0 = 0\}))$ is the infinity line $H := \mathbb{P}_2 - \mathbb{C}^2$ of $\mathbb{P}_2$, because $\eta(\xi^{-1}(\{s_0 = 0\}))$ is of complex dimension one and is contained in the common zero–set of $(d\eta)(X)$ and $(d\eta)(Y)$ in $\mathbb{P}_2$. (The fact that $\eta(\xi^{-1}(\{s_0 = 0\}))$ equals $H$ is also given by the earlier arugment in (7.5).)

(9.3) To recover $W$ from $\mathbb{P}_2$ we look at the first time a point $P_0$ in $\mathbb{P}_2$ is blown up to a curve $\Gamma$. Let $\zeta: W' \to \mathbb{P}_2$ be the blow–down map so that $\Gamma = \zeta^{-1}(P_0)$. We have another blow–down map $\theta: W \to W'$ such that $\eta = \zeta\,\theta$. The point $P_0$ must be in the common zero set $Z$ of the two vector fields $(d\eta)(X)$ and $(d\eta)(Y)$ on $\mathbb{P}_2$, otherwise their liftings $(d\theta)(X)$ and $(d\theta)(Y)$ to $W'$ cannot be holomorphic vector fields. This common zero set $Z$ is equal to the infinity line $H$ of $\mathbb{P}_2$. By §8 we know that $\Gamma$ is not contained in the common zero set of the vector fields $(d\theta)(X)$ and $(d\theta)(Y)$ on $W'$, because they are the liftings of the vector fields $(d\eta)(X)$ and $(d\eta)(Y)$ on $\mathbb{P}_2$. Let $Q_1,\cdots,Q_\ell$ be all the points on $\Gamma$ which are to be blown up later when we continue our recovery of $W$ from $\mathbb{P}_2$. We know that $\Gamma - \{Q_1,\cdots,Q_\ell\}$ as a subset of $W$ cannot be mapped entirely to $\{s_0 = 0\}$ by $\xi$, otherwise it would be in the common zero set of $X$ and $Y$ and $\Gamma$ would be in the common zero–set of the vector fields $(d\theta)(X)$ and $(d\theta)(Y)$ on $W'$. Let $Q_{\ell+1},\cdots,Q_{\ell+m}$ be the set of all points of $\Gamma - \{Q_1,\cdots,Q_\ell\}$ (when regarded as a subset of $W$) which are mapped to $\{s_0 = 0\}$ by $\xi$.

Choose a projective line $T$ in $\mathbb{P}_2$ containing $P_0$ which is not contained entirely in $\eta(\xi^{-1}(\tilde{E}) \cup E')$ in $\mathbb{P}_2$ and is not equal to the infinity line $H$ of $\mathbb{P}_2$ so that the proper transform $T'$ of $T$ in $W'$ (with respect to $\zeta$) intersects $\Gamma$ at a point $Q$ not in $\{Q_1,\cdots,Q_{\ell+m}\}$. In other words, the direction of $T$ at $P_0$ is along the direction defined by $Q$. Since $Q$ is not in $\{Q_1,\cdots,Q_{\ell+m}\}$, it follows that $Q$ is not in $\eta(\xi^{-1}(\{s_0 = 0\}))$. We claim that the proper transform $\tilde{T}$ of $T$ in $W$ (with respect to $\eta$) is disjoint from $\xi^{-1}(\{s_0 = 0\})$. The reason is as follows.

If a point $P$ is in both $\tilde{T}$ and $\xi^{-1}(\{s_0 = 0\})$, then $\eta(P)$ is in both $T$ and the infinity line $H$ of $\mathbb{P}_2$, because $\eta(\xi^{-1}(\{s_0 = 0\})) = H$. It means that $\eta(P) = P_0$, because $P_0$ is the only point of intersection of $T$ and $H$. Thus $\theta(P)$ is in $T'$ and in $\zeta^{-1}(P_0) = \Gamma$, contradicting the fact that the intersection point $Q$ of $T'$ and $\Gamma$ is not in $\theta(\xi^{-1}(\{s_0 = 0\}))$.

Since $T$ is not entirely contained in $\eta(\xi^{-1}(\tilde{E}) \cup E')$, we know that $\tilde{T}$ is not contained entirely in $\xi^{-1}(\tilde{E}) \cup E'$. Since $\tilde{T}$ does not intersect $\xi^{-1}(\{s_0 = 0\})$, it follows that $\xi(\tilde{T})$ is a compact irreducible curve in $\tilde{W}$ which is not entirely in $\tilde{E}$ and which is disjoint from $\{s_0 = 0\}$. By assumption we know that such a compact irreducible curve $\xi(\tilde{T})$ cannot exist. So we know that there is no blow down of $W$. Hence $\tilde{W}$ is biholomorphic to $\mathbb{P}_2$ and $\xi^{-1}(\{s_0 = 0\})$ is the infinity line of $\mathbb{P}_2$. Since $\mathbb{P}_2$ does not contain any exceptional curve, it follows that $\xi\colon W \to \tilde{W}$ must be biholomorphic and $\tilde{W}$ is biholomorphic to $\mathbb{P}_2$. Clearly this biholomorphism between $\tilde{W}$ and $\mathbb{P}_2$ is an extension of $\tilde{f}$ and $\tilde{D}$ is equal to the zero set of $s_0$.

(9.4) Now we look at the case where $\tilde{W}$ may not be normal. Let $p\colon W \to \tilde{W}$ be the normalization of $\tilde{W}$. By applying the normal case to $W$ and the pullbacks of $\tilde{X}, \tilde{Y}, \tilde{D}, \tilde{E}$, $L$, $s_0$ through $p$, we conclude that $W$ is biholomorphic to $\mathbb{P}_2$ and $D := p^{-1}(\tilde{D})$ is the curve $\mathbb{P}_1$ of $\mathbb{P}_2$ at infinity. Let $X$ and $Y$ be respectively the liftings of $\tilde{X}$ and $\tilde{Y}$ to $W$. For notational simplicity we denote the lifting of $L$ and $s_0$ to $W$ also by $L$ and $s_0$. On the affine part $\mathbb{C}^2$ of $W = \mathbb{P}_2$ the vector fields $X$ and $Y$ are simply the partial differentiation with respect to the coordinates of $\mathbb{C}^2$. The line bundle $L$ on $W = \mathbb{P}_2$ is simply the line bundle associated to the infinity line $D = \mathbb{P}_2 - \mathbb{C}^2$. Hence the $L^{-1}$–valued vector field $\frac{1}{s_0}X$ is tangential to $D$ and has only a single zero which we denote by $P_1$. After we use a local trivialization of $L$, we can regard $\frac{1}{s_0}X$ and $\frac{1}{s_0}\tilde{X}$ as *local* holomorphic vector fields. Since the map $p\colon W \to \tilde{W}$ has only finite fibers, by considering the one–parameter subgroups generated by the *local* holomorphic vector fields $\frac{1}{s_0}X$ and $\frac{1}{s_0}\tilde{X}$, we conclude that $\frac{1}{s_0}X$ must vanish at the inverse image of the zero–set of $\frac{1}{s_0}\tilde{X}$

under p. Hence the $L^{-1}$–valued vector field $\frac{1}{s_0}\tilde{X}$ has only a single zero $p(P_1)$ in $\tilde{W}$. Likewise the $L^{-1}$–valued vector field $\frac{1}{s_0}\tilde{Y}$ has only a single zero $p(P_2)$ in $\tilde{W}$. We can assume without loss of generality that $p(P_1)$ is different from $p(P_2)$. The reason is the following. Since $(p|D)\colon D \to \tilde{D}$ has finite fibers, we can find a point $P_3$ in D such that $p(P_3)$ is distinct from $p(P_1)$. By replacing Y by a suitable $\mathbb{C}$–linear combination of X and Y we can assume that the only zero of $\frac{1}{s_0}Y$ is at $P_3$.

At any point Q of D either $\frac{1}{s_0}X$ or $\frac{1}{s_0}Y$ is nonzero at Q and both are tangential to D. At p(Q) either $\frac{1}{s_0}\tilde{X}$ or $\frac{1}{s_0}\tilde{Y}$ is nonzero at p(Q) and both are tangential to $\tilde{D}$. So we know that $p|D$ maps $\tilde{D}$ locally biholomorphically onto D. Thus at every point of $\tilde{D}$ we can find some local holomorphic function $\varphi$ on G such that $d\varphi$ pullbacked to D under $D \to \tilde{D}$ is a nonzero form on D. On the other hand we have $ds_0$ nonzero as a form on W at every point of D when $s_0$ is locally regarded as a holomorphic function. Since $ds_0$ is zero when pullbacked to D, it follows that $ds_0$ and $d\varphi$ are linearly independent at points of D. So we know that the map $W \to \tilde{W}$ has rank two over $\mathbb{C}$ at every point of D. Since X and Y are linearly independent at every point of $W - D$ and their corresponding vector fields $\tilde{X}$ and $\tilde{Y}$ are also linearly independent at every point of $\tilde{W} - \tilde{D}$, we conclude that $p\colon W \to \tilde{W}$ maps $W - D$ locally biholomorphically to $\tilde{W} - \tilde{D}$. So $p\colon W \to \tilde{W}$ is locally biholomorphic. Since every biholomorphism of $W = \mathbb{P}_2$ has at least one fixed point, $W = \mathbb{P}_2$ cannot be the universal cover of any nonsingular compact complex surface and $p\colon W \to \tilde{W}$ must be a biholomorphism. This concludes the proof of the claim for the general case.

## §10. *Maximal Projective Spaces in the Limit Fiber.*

(10.1) We now come back to the proof that $M_0$ is biholomorphic to $\mathbb{P}_n$. The notations X and Y now resume their meanings used before the interruption in §7. Recall that for each holomorphic path of origins $\tau(t)$ and for each holomorphic family of infinity hyperplane sections we have the Euler vector field $X_t$ on $M_t$ and a non identically zero holomorphic vector field $X^* = \lim_{t\to 0} t^k X_t$ on $M_0$ for some positive integer k. Consider the set $\Sigma$ of all paths $\tau(t)$ of origins. We use the same family of infinity hyperplane sections $\{s_0 = 0\}$ for each path of origins. So for each member $\tau$ of $\Sigma$ we have a positive integer $k(\tau)$ and have a limit of Euler vector fields after modification by $t^{k(\tau)}$ and we call the limit $X(\tau)$. Let $Y(\tau) = (s_0)^{-1} X(\tau)$. Then $Y(\tau)$ is an $L^{-1}$–valued holomorphic vector field on $M_0$.

We claim that the Lie bracket $[X(\tau_1),X(\tau_2)]$ of $X(\tau_1)$ and $X(\tau_2)$ is identically zero on $M_0$ for any two members $\tau_1$ and $\tau_2$ of $\Sigma$. The reason is as follows. For $t \neq 0$ let $z_1,\cdots,z_n$ be the affine coordinates of $M_t - \{s_0 = 0\}$ and let $\tilde{X}(\tau_\mu) = \Sigma_{\nu=1}^n (z_\nu - z_\nu(\tau_\mu(t))) \frac{\partial}{\partial z_\nu}$ for $\mu = 1,2$. Since $[\tilde{X}(\tau_1),\tilde{X}(\tau_2)] = \tilde{X}(\tau_1) - \tilde{X}(\tau_2)$, it follows that

$$[t^{k(\tau_1)}\tilde{X}(\tau_1),t^{k(\tau_2)}\tilde{X}(\tau_2)] = t^{k(\tau_2)}(t^{k(\tau_1)}\tilde{X}(\tau_1)) - t^{k(\tau_1)}(t^{k(\tau_2)}\tilde{X}(\tau_2)).$$

Since for $\mu = 1,2$ the limit of $t^{k(\tau_\mu)}\tilde{X}(\tau_\mu)$ is $X(\tau_\mu)$ as $t \to 0$, it follows from the positivity of $k(\tau_\mu)$ that $[X(\tau_1),X(\tau_2)]$ is identically zero on $M_0$.

(10.2) Let $q$ be the positive integer such that there exist members $\tau_1,\cdots,\tau_q$ of $\Sigma$ with the following two properties:

(i) For some point $P_0$ in $M_0$ the vector fields $X(\tau_1),\cdots,X(\tau_q)$ are linearly independent at $P_0$.

(ii) For any $q+1$ members $\tau_1',\cdots,\tau_{q+1}'$ of $\Sigma$ the vector fields $X(\tau_1'),\cdots,X(\tau_{q+1}')$ are linearly dependent at every point of $M_0$.

Without loss of generality we can pick the point $P_0$ so that $s_j$ is not zero at $P_0$ $(0 \leq j \leq n)$ and $P_0$ is in $\Omega_\ell$ for some sufficiently large $\ell$.

Each $Y(\tau_\nu) = \frac{1}{s_0} X(\tau_\nu)$, $1 \leq \nu \leq q$, is an $L^{-1}$–valued holomorphic vector field on $M_0$. By §6 any one–dimensional orbit of the vector field $X(\tau_\nu)$ in $M_0$, which intersects $\Omega_\ell \cap (\cap_{j=0}^n \{s_j \neq 0\})$ and on which $X(\tau_\nu)$ is not identically zero, is biholomorphic to $\mathbb{C}$ and under this biholomorphism the vector field $X(\tau_\nu)$ is equal to the partial differentiation with respect to the coordinate of $\mathbb{C}$. Let $G$ be the subvariety of $M_0$ where $X(\tau_1),\cdots,X(\tau_q)$ are linearly dependent. Since the Lie bracket of any two of the ector fields $X(\tau_1),\cdots,X(\tau_q)$ is identically zero, it follows that we can integrate them at he same time and get a submanifold V of G containing $P_0$ which is biholomoprhic to $^q$ under some biholomorphic map $\Psi: \mathbb{C}^q \to V$.

For $a = (a_1,\cdots,a_q) \in \mathbb{C}^q - 0$ we let $X(a) = \Sigma_{\nu=1}^q a_\nu X(\tau_\nu)$ and denote by C(a) he topological closure of its orbit containing $P_0$. Note that every X(a) is the limit at $= 0$ of a holomorphic vector field on M after modification by a suitable powers of t nd as a result the topological closure of every one-dimensional orbit of X(a) is a branch f the limit of complex curves of $M_t$ $(t \neq 0)$ and is itself a complex curve.

Let us first look at the case $q > 1$. When we have two a and a' which are not nearly dependent, we let $W_0(a,a') = \Psi(\mathbb{C}a + \mathbb{C}a')$. In other words, $W_0$ is the integral ubmanifold for the commuting vector fields X(a) and X(a') which contains $P_0$. Let V(a,a') be the topological closure of $W_0(a,a')$. Then W(a,a') is a subvariety of $M_0$. )ne way to see this is to consider the union W'(a,a') of the topological closure of the rbits of X(a') which intersect C(a). By considering the deformation of all curves which itersects C(a) and tangential to X(a'), we conclude that W'(a,a') is a subvariety of $I_0$ of complex dimension two. Since the Lie bracket of X(a) and X(a') is zero, it ollows that both X(a) and X(a') are tangential to W'(a,a') and $W_0(a,a')$ is contained W'(a,a'). Hence W(a,a') is a subvariety of $M_0$.

We now apply the result of §9 to $\bar{W} = W(a,a')$, $\bar{X} = X(a)$, $\bar{Y} = X(a')$, $s_0$, L, $= \bar{W} - \Omega_\ell$, $\bar{D} = W(a,a') - V$. By (5.2) every irreducible complex curve in $\bar{W}$ which oes not lie entirely in $\bar{E}$ must intersect $\{s_0 = 0\}$. By (9.1) W(a,a') is biholomorphic to $_2$. Moreover, in this biholomorphism the affine part $\mathbb{C}^2$ of $\mathbb{P}_2$ corresponds precisely to $V(a,a') \cap \{s_0 \neq 0\}$ and the two vector fields X(a) and X(a') correspond to partial ifferentiation with respect to the coordinates of $\mathbb{C}^2$. The two sets $W(a,a') - V$ and $V(a,a') \cap \{s_0 = 0\}$ agree. The submanifold W(a,a') can also described as the union of (a'') for $a'' \in \mathbb{C}a + \mathbb{C}a'$. Now we let a and let $S(\tau_1,\cdots,\tau_q,P_0)$ be the union of all the ırves C(a) for $a \in \mathbb{C}^q - 0$. We claim that $S(\tau_1,\cdots,\tau_q,P_0)$ is biholomorphic to $\mathbb{P}_q$.

Recall that $\Psi$ is the biholomorphic map from $\mathbb{C}^q$ to V with $\Psi(0) = P_0$. We want extend $\Psi$ to a holomorphic map from $\mathbb{P}_q$ to $M_0$. For every $a \in \mathbb{C}^q - 0$ we have a line a) in $\mathbb{P}_q$ given by the direction a. We map the infinity point $\ell(a) \cap (\mathbb{P}_q - \mathbb{C}^q)$ of $\ell(a)$

to the point $C(a) \cap \{s_0 = 0\}$ and get a map $\tilde{\Psi}$ from $\mathbb{P}_q$ to $M_0$. We have just seen that the restriction of the map $\tilde{\Psi}$ to any $\mathbb{P}_2$ in $\mathbb{P}_q$ containing $0$ maps $\mathbb{P}_2$ biholomorphically onto its image $\tilde{\Psi}(\mathbb{P}_2)$. In particular, $\tilde{\Psi}$ is injective. We want to conclude that $\tilde{\Psi}$ is holomorphic.

At first sight the most natural approach seems to be arguments involving Hartogs' type results on separate holomorphicity. However, it turns out to be easier to argue by using the method of deformation. Consider the moduli space $\mathscr{D}$ of all complex curves in $S(\tau_1,\cdots,\tau_q,P_0)$ passing through $P_0$. Let $\mathscr{D}'$ be the branch of $\mathscr{D}$ containing the point corresponding to some $C(a)$. Let $Z'$ be the Euler vector field $\Sigma_{\nu=1}^q z_\nu \frac{\partial}{\partial z_\nu}$ of $\mathbb{C}^q$ with coordinates $z_1,\cdots,z_q$ and let $Z = (d\Psi)(Z')$. Let $\mathscr{D}_0$ be the set of $\mathscr{D}'$ consisting of points represented by curves $C$ to which $Z$ is tangential at $P_0$ to *infinite* order. Clearly $\mathscr{D}_0$ is a subvariety of $\mathscr{D}'$, because the infinite-order tangency condition can be expressed in the following way. A point of $\mathscr{D}'$ is in $\mathscr{D}_0$ if and only if the curve $C$ corresponding to that point satisfies the condition that for any local holomorphic defining function $f_C$ for $C$ and for any positive integer $m$ the local holomorphic function $Z(f_C)$ vanishes at $P_0$ to order at least $m$. In general, this kind of infinite-order tangency condition is so stringent that $C(a)$ may give us only an isolated point in the moduli space $\mathscr{D}_0$. However, since it is already known that the restriction of the map $\tilde{\Psi}$ to any $\mathbb{P}_2$ in $\mathbb{P}_q$ containing $0$ is biholomorphic onto its image, we conclude that, given $a$ and $a'$ linearly independent in $\mathbb{C}^q$, the curve $C(a)$ can be deformed inside the moduli space $\mathscr{D}_0$ to $C(a')$ along the topological closure of $\Psi(\mathbb{C}a + \mathbb{C}a')$ in $M_0$. With the infinite-order tangency condition this implies that the moduli space $\mathscr{D}_0$ can only correspond to the set of all lines $\mathbb{P}_1$ in $\mathbb{P}_q$ containing $0$.

Now consider the total space $\mathscr{X}$ of all the curves for this moduli space $\mathscr{D}_0$ and consider the holomorphic map from the total space $\mathscr{X}$ to $M_0$. This holomorphic map from $\mathscr{X}$ to $M_0$ induces a holomorphic map from $\mathbb{P}_q$ to $M_0$ which clearly agrees with $\tilde{\Psi}$. So we can conclude that $\tilde{\Psi}$ is holomorphic. As we observed earlier the map $\tilde{\Psi}$ is injective. In particular, $\tilde{\Psi}\colon \mathbb{P}_q \to S(\tau_1,\cdots,\tau_q,P_0)$ is the normalization of $S(\tau_1,\cdots,\tau_q,P_0)$. Moreover, the injectivity of $\tilde{\Psi}$ and dimension considerations imply that $\tilde{\Psi}$ maps $\mathbb{P}_q - \mathbb{C}^q$ biholomorphically onto its image at least locally at some point. So $S(\tau_1,\cdots,\tau_q,P_0) \cap \{s_0 = 0\}$ is regular at some point. The only points of $\mathbb{P}_q$ where $\tilde{\Psi}\colon \mathbb{P}_q \to S(\tau_1,\cdots,\tau_q,P_0)$ can possibly fail to be biholomorphic are points of $\mathbb{P}_q - \mathbb{C}^q$, because $X(\tau_1),\cdots,X(\tau_q)$, which correspond to the partial differentiation with respect to the coordinates of $\mathbb{C}^q$, are linearly independent at every point of $S(\tau_1,\cdots,\tau_q,P_0) \cap \{s_0 \neq 0\}$ (by the equality of $W(a,a') - V$ and $W(a,a') \cap \{s_0 = 0\}$).

For any two distinct points $\tilde{Q}_1$ and $\tilde{Q}_2$ in $S(\tau_1,\cdots,\tau_q,P_0) \cap \{s_0 = 0\}$, we let $W$ denote the $\mathbb{P}_2$ in $\mathbb{P}_q$ containing $\tilde{\Psi}^{-1}(\tilde{Q}_1)$ and $\tilde{\Psi}^{-1}(\tilde{Q}_1)$ and the origin of $\mathbb{C}^q$. Let $\tilde{X}_1$ and $\tilde{X}_2$ be two linearly independent $\mathbb{C}$–linear combinations of $X(\tau_1),\cdots,X(\tau_q)$ which are tangential to $\tilde{\Psi}(W)$. For any holomorphic vector field $X$ on $M_0$ with $\frac{1}{s_0}X$ holomorphic, after we use a local trivialization of $L$, we can regard $\frac{1}{s_0}X$ as a *local* holomorphic vector field. Since $\tilde{\Psi}$ maps $W$ biholomorphically onto $\tilde{\Psi}(W)$, we conclude that a suitable composition of the local biholomorphisms generated by the *local* vector fields $\frac{1}{s_0}\tilde{X}_1$ and $\frac{1}{s_0}\tilde{X}_2$ maps $\tilde{Q}_1$ to $\tilde{Q}_2$. It follows that a suitable composition of the local biholomorphisms generated by the $\mathbb{C}$–linear combinations of the *local* holomorphic vector fields $\frac{1}{s_0}X(\tau_1),\cdots,\frac{1}{s_0}X(\tau_q)$ can map any point in $S(\tau_1,\cdots,\tau_q,P_0) \cap \{s_0 = 0\}$ to any other point in $S(\tau_1,\cdots,\tau_q,P_0) \cap \{s_0 = 0\}$. It follows that $S(\tau_1,\cdots,\tau_q,P_0) \cap \{s_0 = 0\}$ is equisingular at every point and hence is regular everywhere. Moreover, this argument of compositions of local biholomorphisms implies that the restriction of $\tilde{\Psi}$ to $\mathbb{P}_q - \mathbb{C}^q$ maps $\mathbb{P}_q - \mathbb{C}^q$ locally biholomorphically onto $S(\tau_1,\cdots,\tau_q,P_0) \cap \{s_0 = 0\}$. Locally at every point of $\tilde{\Psi}(\mathbb{P}_q - \mathbb{C}^q)$ we can find local holomorphic functions $\varphi_1,\cdots,\varphi_{q-1}$ on $M_0$ such that the pullbacks of $d\varphi_1,\cdots,d\varphi_{q-1}$ to $\mathbb{P}_q - \mathbb{C}^q$ under $\tilde{\Psi}$ are linearly independent. As in (9.4), we again combine this with $ds_0$ and conclude that the pullbacks of $ds_0,d\varphi_1,\cdots,d\varphi_{q-1}$ to $\mathbb{P}_q$ under $\tilde{\Psi}$ are linearly independent on $\mathbb{P}_q$ at every point of $\mathbb{P}_q - \mathbb{C}^q$. So we have the biholomorphism of $\tilde{\Psi}$ from $\mathbb{P}_q$ to $S(\tau_1,\cdots,\tau_q,P_0)$.

We now include the case $q = 1$. When $q = 1$, the subvariety $S(\tau_1,\cdots,\tau_q,P_0)$ is simply the closure of the orbit of $X(\tau_1)$ containing $P_0$ which we have shown in §6 to be a nonsingular rational curve. When $q = n$ we have $M_0$ biholomorphic to $\mathbb{P}_n$ and the proof ends. So we assume that $1 \leq q < n$.

(10.3) Recall that for a path $\tau$ of origins, $Y(\tau)$ means $\frac{1}{s_0}X(\tau)$. We now distinguish between the following two cases.

*Case 1.* The zero-set $Z(\tau_1,\cdots,\tau_q)$ of $Y(\tau_1)\wedge\cdots\wedge Y(\tau_q)$ is of codimension at least two in $M_0$. Take a member $\tau$ of $\Sigma$ so that the origin $\tau(0)$ for $t=0$ lies in $V$. Since $Y(\tau),Y(\tau_1),\cdots,Y(\tau_q)$ are linearly dependent at every point of $M_0$ and $Y(\tau_1),\cdots,Y(\tau_q)$ are linearly independent at every point of $M_0-Z(\tau_1,\cdots,\tau_q)$, by Cramer's rule we can find holomorphic functions $c_1,\cdots,c_q$ on $M_0-Z(\tau_1,\cdots,\tau_q)$ such that $Y(\tau)=\Sigma_{\nu=1}^q c_\nu Y(\tau_\nu)$ on $M_0-Z(\tau_1,\cdots,\tau_q)$. Since $Z(\tau_1,\cdots,\tau_q)$ is of complex codimension at least two in $M_0$, the functions $c_1,\cdots,c_q$ can be extended to holomorphic functions on all of $M_0$ and are therefore constant. From the vanishing of $Y(\tau)$ at $\sigma(0)$ and the linear independence of $Y(\tau_1),\cdots,Y(\tau_q)$ at $\sigma(0)$ we conclude that all the constants $c_1,\cdots,c_q$ must be zero, contradicting the fact that $Y(\tau)$ is not identically zero. The maximality property of $q$ is used only here to rule out Case 1 and is not used anywhere else.

*Case 2.* $Y(\tau_1)\wedge\cdots\wedge Y(\tau_q)$ vanishes on some hypersurface defined by $\rho=0$ in $M_0$, where $\rho$ is a holomorphic section of some power $L^r$ of $L$ with $r>0$. So $\frac{1}{\rho}Y(\tau_1)\wedge\cdots\wedge Y(\tau_q)$ is holomorphic on $M_0$. We know that $\frac{1}{\rho}Y(\tau_1)\wedge\cdots\wedge Y(\tau_q)$ cannot be nowhere zero on $S(\tau_1,\cdots,\tau_q,P_0)$, otherwise the integration of the distribution $\frac{1}{\rho}Y(\tau_1)\wedge\cdots\wedge Y(\tau_q)$ would make $S(\tau_1,\cdots,\tau_q,P_0)$ disjoint from $S(\tau_1,\cdots,\tau_q,P)$ for nearby $S(\tau_1,\cdots,\tau_q,P)$ and the normal bundle of $S(\tau_1,\cdots,\tau_q,P_0)$ in $M_0$ is trivial. Then for any $a\in\mathbb{C}^q-0$ the Chern class of the normal bundle of $C(a)$ in $M_0$ equals its Chern class of the normal bundle of $C(a)$ in $S(\tau_1,\cdots,\tau_q,P_0)$ and is $q-1$. This yields a contradiction, because the anticanonical line bundle of $M_0$ is $L^{n+1}$ and $L|C(a)$ has Chern class 1, forcing the Chern class of the normal bundle of $C(a)$ in $M_0$ to be $n-1$.

(10.3.1) The zero-set of $\frac{1}{\rho}Y(\tau_1)\wedge\cdots\wedge Y(\tau_q)$ in $S(\tau_1,\cdots,\tau_q,P_0)$ must be a hypersurface in $S(\tau_1,\cdots,\tau_q,P_0)$ (which is biholomorphic to $\mathbb{P}_q$), because it is the section of a line bundle over $S(\tau_1,\cdots,\tau_q,P_0)$. Let $m>0$ be the degree of this hypersurface. Then the zero-set of $X(\tau_1)\wedge\cdots\wedge X(\tau_q)$ in $S(\tau_1,\cdots,\tau_q,P_0)$ is a hypersurface of degree $q+r+m$ in $S(\tau_1,\cdots,\tau_q,P_0)$. On the other hand the restriction of $X(\tau_1)\wedge\cdots\wedge X(\tau_q)$ to $S(\tau_1,\cdots,\tau_q,P_0)$ is a holomorphic section of the anticanonical line bundle of $S(\tau_1,\cdots,\tau_q,P_0)$ and its zero set in $S(\tau_1,\cdots,\tau_q,P_0)$ must be a hypersurface of degree $q+1$, contradicting $r\geq 1$ and $m\geq 1$.

In handling Case 2 of (10.3) I followed a suggestion of Jun–Muk Hwang in doing the ounting of Chern classes in (10.3.1) which simplified my original lengthier argument.

*References*


B] E. Brieskorn, Über holomorphe $\mathbb{P}_n$–Bündel über $\mathbb{P}_1$, Math. Ann. 157 (1965), 343–357.

C–V] E. Calabi and E. Vesentini, On compact locally symmetric Kähler manifolds, Ann. f Math. 71 (1960), 472–507.

F–N] A. Frölicher and A. Nijenhuis, A theorem on stability of complex structures, Proc. Nat. Acad. Sci., U.S.A. 43 (1957), 239–241.

F] A. Fujiki, On the Douady space of a compact complex space in the category $\mathscr{C}$, Nagoya Math. J. 85 (1982); II, Publ. Res. Inst. Math. Sci. 20 (1984), no.3, 461–489.

G] H. Grauert, Ein Theorem der analytischen Garbentheorie und die Modulräume complexer Strukturen, Publ. Math. I.H.E.S. No.5 (1960), 5–60.

G–H] Ph. Griffiths and J. Harris, Principles of Algebraic Geometry, John Wiley & Sons, New York, 1978.

H] F. Hirzebruch, Über eine Klasse von einfach–zusammenhängenden komplexen Mannifaltigkeiten, Math. Ann. 124 (1951), 77–86.

K] K. Kodaira, On the structure of compact complex analytic surfaces I, Amer. J. Math. 86 (1964), 751–798.

K–S] K. Kodaira and D.C. Spencer, On deformation of complex analytic structures II, Ann. of Math. 67 (1958), 403–466.

Ku] M. Kuranishi, On the locally complete families of complex analytic structures, Ann. of Math. 75 (1962), 536–577.

M] J. Morrow, Minimal normal compactification of $\mathbb{C}^2$, *Complex Analysis*, Vol. 1, Geometry of Singularities, ed. H. L. Resnikoff and R. O. Wells, Jr., Rice University Studies Vol. 59 (1973), pp. 97–112.

N1] I. Nakamura, Characterizations of $\mathbb{P}^3$ and hyperquadrics $Q^3$ in $\mathbb{P}^4$, Proc. Japan Acad, 62, Ser. A (1986), 230–233.

N2] I. Nakamura, Moishezon threefolds homeomorphic to $\mathbb{P}^3$, J. math. Soc. Japan 39 1987), 521–534.

P1] T. Peternell, A rigidity theorem for $\mathbb{P}_3(\mathbb{C})$, Manuscripta Math. 50 (1985), 397–428.

P2] T. Peternell, Algebraic structures on certain 3–folds, Math. Ann. 274 (1986), 133–156.

R] O. Riemenschneider, Über die Anwendung algebraischer Methoden in der Deformationstheorie komplexer Räume, Math. Ann. 187 (1970), 40–55.

[R] H. Rossi, Vector fields on analytic spaces, Ann. of Math. 78 (1963), 455–467.

[S] Y.-T. Siu, Nondeformability of the complex projective space, J. reine angew. Math. 399 (1989), 280–219.

[Sp] M. Spivakovsky, Sandwiched surface singularities and the Nash resolution for surfaces, Ph.D. thesis, Harvard University, 1985.

[T] H. Tsuji, Deformation of complex projective spaces.

[U] K. Ueno, Classification theory of algebraic varieties and compact complex spaces, Lecture Notes in Math. Vol. 439, Springer Verlag 1975.

Author's address: Department of Mathematics, Harvard University, Cambridge, MA 02138, U.S.A.

# SOME ASPECTS OF HODGE THEORY
# ON NON-COMPLETE ALGEBRAIC
# MANIFOLDS

by

INGRID BAUER

AND

SIEGMUND KOSAREW

Introduction. This paper continues the study of the Dolbeault and de Rham cohomology groups on smooth algebraic varieties $U$ which admit a compactification $X$ of a certain type. These types are divided into two classes, the so-called "concave" or "convex" situation, notions originating from complex analytic geometry. In the first case, $Y := X \backslash U$ is of "high" codimension in $X$ whereas in the second one, $Y$ is a divisor, $X$ is smooth and the normal bundle $N_{Y|X}$ of $Y$ in $X$ satisfies a suitable positivity condition (as for instance "k-ample"). In $[B\text{-}K]_1$ we obtained the degeneration domain of the Hodge to de Rham spectral sequence and in $[B\text{-}K]_2$ some vanishing theorems of Kodaira-Akizuki-Nakano type (including weak Lefschetz results) are established. Following a suggestion of T. Ohsawa, the vanishing theorem in the convex case can be in fact strengthened where $N_{Y|X}$ is only assumed to be globally generated.

The Hodge symmetry in our context is proved essentially by methods of algebraic geometry. In the convex case, the condition on $Y$ of being a divisor is very crucial for all the results to hold, since an example, taken from Grauert-Riemenschneider ($[G\text{-}R]_2$; Abschnitt 3, C, D) disproves almost everything in the higher codimensional case (even for $N_{Y|X}$ being ample and globally generated).

If $U$ is 1-convex as a complex manifold (which corresponds to the case where $N_{Y|X}$ is ample), then the situation is quite well understood (for instance by results of Flenner, Navarro Aznar, Steenbrink/van Straten).

In the last section of this paper, we wrote up a proof of Steenbrink's vanishing theorem in [St], using the method of cyclic coverings. We were told that E. Viehweg obtained this result by the same approach.

We remark that a parallel development of this theory, but in the analytic context, has been established by Ohsawa and Takegoshi in a series of papers [Oh]., [Oh-T].. Their tools are $L^2$-techniques and estimates of $\bar{\partial}$. Despite of this, our results are independent of theirs, as we stick to more algebraic notions.

We are very grateful to several people for their interest in our work, especially to T. Ohsawa with whom we had many enlightening discussions and to V. Navarro Aznar for some extremely useful hints. Moreover, our thanks go to H. Flenner and J. Steenbrink.

After finishing this paper, we received a preprint of D. Arapura [Ar] in which some of our results (essentially the concave case) are proved by using a suitable extension of Steenbrink's vanishing theorem.

## 1. Statement of the results

We start with fixing some notations. Let $U$ be a smooth algebraic variety over a field $K$ (of characteristic zero) and

$$E_1^{i,j}(U) = H^j(U,\Omega^i_{U/K}) \Longrightarrow H^n_{DR}(U/K) = E^n(U)$$

the Hodge to de Rham spectral sequence. Moreover, let $F^\cdot = F^\cdot(U)$ denote the Hodge filtration on the de Rham cohomology of $U$. We put

$$h^{i,j}(U) := \dim_K H^j(U,\Omega^i_{U/K}) ,$$

$$h^n(U) := \dim_K H^n_{DR}(U/K) .$$

If $A \subset U$ is a divisor with normal crossings, then

$$\Omega^\cdot_{U/K}\langle A\rangle$$

is as usual the de Rham complex with logarithmic poles along $A$.

Let $X$ be an algebraic variety over $\mathbb{C}$ (not necessarily smooth), then

$$IH^\cdot(X,L)$$

denotes the intersection cohomology of $X$ with respect to the *middle* perversity for a locally constant sheaf $L$ on the regular part of the analytification $X^{an}$ of $X$.

Now we are going to state our results.

**Theorem I.** *Let* $X$ *be an integral proper* $K$*-scheme,* $Y \subset X$ *a closed subscheme and* $U := X\setminus Y$ *. Assume that* $U$ *is smooth over* $K$*. Then the following assertions hold*

(1) $h^{i,j}(U)$ , $h^n(U)$ *are finite for* $n,j < \operatorname{codim}(Y,X)-1$ *and arbitrary* $i$ ,

(2) $h^n(U) = \sum_{i+j=n} h^{i,j}(U)$ for $n < \mathrm{codim}(Y,X)-1$ (*$E_1$-degeneration*) ,

(3) *for* $i+j = \mathrm{codim}(Y,X)-1$ , *there is a natural injective map*

$$gr_i H^{i+j}_{DR}(U/K) \hookrightarrow H^j(U,\Omega^i_{U/K})$$

(*where* gr *is taken with respect to the filtration* $F^\cdot(U)$) ,

(4) *if* X *is projective, then*

$$h^{i,j}(U) = h^{j,i}(U) \ , \qquad i+j < \mathrm{codim}(Y,X)-1$$

(*Hodge-symmetry*) ,

(5) *for* $K = \mathbb{C}$ , *the natural map*

$$IH^n(X,\mathbb{C}) \longrightarrow H^n_{DR}(U/\mathbb{C})$$

*is bijective for* $n \leq \mathrm{codim}(Y,X)-1$ *and injective for* $n = \mathrm{codim}(Y,X)$ ,

(6) *let* $\pi : \tilde{X} \longrightarrow X$ *be a proper modification with center contained in* Y , *such that* $\tilde{X}$ *is smooth and* $\tilde{Y} := \pi^{-1}(Y)_{red}$ *is a normal crossing divisor. Then the natural map*

$$H^j(\tilde{X},\Omega^i_{\tilde{X}/K}\langle\tilde{Y}\rangle) \longrightarrow H^j(U,\Omega^i_{U/K})$$

*is bijective for* $i+j < \mathrm{codim}(Y,X)-1$ .

In the convex situation, we have analogously

<u>Theorem II.</u> *Let* X *be an integral scheme, proper and smooth over* Spec(K) *and* $Y \subset X$ *a divisor,* $U := X \setminus Y$ . *We assume that the normal bundle* $N_{Y|X}$ *of* Y *in* X *is* k-*ample (on* Y). *Then the following holds*

(1) $h^{i,j}(U)$ , $h^n(U)$ *are finite for* $j \geq k+1$ *and* $n \geq \dim X+k+1$ ,

(2) $h^n(U) = \sum_{i+j=n} h^{i,j}(U)$ *for* $n \geq \dim X+k+1$ *($E_1$-degeneration)* ,

(3) *for* $i+j = \dim X+k$ , *there is a natural surjective map*

$$H^j(U,\Omega^i_{U/K}) \twoheadrightarrow gr_i H^{i+j}_{DR}(U/K) ,$$

(4) *if* X *is projective and each section of* $N_{Y|X}$ *extends to some open Zariski neighborhood of* Y *in* X , *then*

$$h^{i,j}(U) = h^{j,i}(U) \qquad \textit{for} \quad i+j \geq \dim X+k+1 \qquad \textit{(Hodge symmetry)}$$

(5) *if* Y *is a reduced normal crossing divisor in* X *with smooth components and the normal bundle* $N_{Y|X}$ *is* k*-ample, then the natural map*

$$H^j(X,\Omega^i_{X/K}\langle Y\rangle) \longrightarrow H^j(U,\Omega^i_{U/K})$$

*is bijective for* $i+j \geq \dim X+k+1$ . —

The following result has been conjectured by T. Ohsawa (and in fact, he proved an analytic analogue of it):

<u>Vanishing theorem.</u> *Let* X,Y *be as in theorem* II *and* Y *a reduced normal crossing divisor with smooth components. Assume that* $N_{Y|X}$ *is globally generated (on* Y*). We fix* $L \in \mathrm{Pic}(X)$ . *Then*

(1) *if* $L$ *is* $\ell$*-ample on* X

$$H^j(U,\Omega^i_{U/K} \otimes L) = 0$$

*for* $i+j \geq \dim X+\ell+1$ ,

(2) *if* $L|Y$ *is* $\ell$*-ample on* $Y$

$$\dim_K H^j(U,\Omega^i_{U/K} \otimes L) < \infty$$

*for* $i + j \geq \dim X + \ell + 1$ . —

Remarks. This last theorem strengthens significantly the corresponding result in $[B\text{-}K]_2$.

If we assume in theorem (II) that $Y$ is not a divisor, but $N_{Y|X}$ is still ample (and globally generated), then the degeneration and symmetry assertion is *not* true in general. This will be shown by an example in section 7 . The expected range in the higher codimensional case would be

$$n \geq \dim X + k + \operatorname{rank}(N_{Y|X}) .$$

In this context one should mention a result of M . Schneider in [Sch] that says (for $K = \mathbb{C}$ and $Y$ is a divisor) if $N_{Y|X}$ is q-concave (which corresponds to (q-1)-ample) and $X$ is Kähler, then $U$ is even *hyper*-q-*convex* in the sense of Grauert-Riemenschneider. A similar result does not hold for $\operatorname{codim}(Y,X) > 1$ (compare $[G\text{-}R]_2$) although $U$ is still $(q + \operatorname{rank}(N_{Y|X}) - 1)$-convex.

We should mention that in theorems (I), (II), the Hodge structure on $H^n_{DR}(U)$ is pure of weight $n$ (for $K = \mathbb{C}$) in the appropriate range of $n$ . This follows from our proof below.

## 2. The first critical domain of the spectral sequence

Here we are going to show the assertions (3) of theorem (I) and (II). In fact, they are immediate consequences of (1), (2) and the following lemma

(2.1) Lemma. *Let*

$$E_1^{i,j} \Rightarrow E^n , \quad d_r^{i,j} : E_r^{i,j} \longrightarrow E_r^{i+r,j-r+1}$$

*be a biregular spectral sequence of abelian groups and* $(i_o,j_o,r_o) \in \mathbb{Z}\times\mathbb{Z}\times\mathbb{N}_{>o}$. *Then*

(1) *if* $d_r^{i_o,j_o} = 0$ *for all* $r \geq r_o$ , *there is a natural surjection*

$$E_{r_o}^{i_o,j_o} \twoheadrightarrow gr_{i_o}(E^{i_o+j_o})$$

(2) *if* $d_r^{i_o-r,j_o+r-1} = 0$ *for all* $r \geq r_o$ , *there is a natural injection*

$$E_{r_o}^{i_o,j_o} \hookrightarrow gr_{i_o}(E^{i_o+j_o}) . \text{ —}$$

The proof is trivial.

## 3. The concave case

This section is devoted to the proof of theorem (I). The parts (1) and (2) have been already established in $[B\text{-}K]_1$ and (3) follows from section 2 .

### 3.1 *Connection with intersection cohomology; weak Lefschetz*

Let here $X$ be a projective pure dimensional variety over the field $K = \mathbb{C}$ and denote by $S$ the intersection cohomology complex of the constant sheaf $\mathbb{C}$ on $X$ with respect to a fixed perversity $p$ . Adopting the notations as in the book [Bo], we take a stratification $X_\bullet = X_2 \supset X_3 \supset \ldots$ of $X$ , such that

$$U_k := X \setminus X_k ,$$

and

$$S_{m-k} := U_{k+1} \setminus U_k = X_k \setminus X_{k+1} , \quad m := \dim_{\mathbb{R}} X ,$$

is a pure $(m-k)$-dimensional manifold (or empty).
We denote by $j_k : U_k \hookrightarrow U_{k+1}$ , $i_k : S_{m-k} \hookrightarrow U_{k+1}$ the natural inclusions.

(3.1.1) Lemma. *The restriction map*

$$H^\nu(U_{k+1},S) \longrightarrow H^\nu(U_k,S)$$

*is bijective for* $\nu \leq p(k)$ *and injective for* $\nu = p(k) + 1$ .

*Proof*. Since we have $H^\nu(U_k,S) \cong H^\nu(U_{k+1},R(j_k)_*(j_k)^*S)$ for all $\nu$ , we consider the distinguished triangle

$$\begin{array}{ccc} S & \longrightarrow & R(j_k)_*(j_k)^*S \\ & \nwarrow \quad \swarrow {\scriptstyle [1]} & \\ & (i_k)_!(i_k)^!S & . \end{array}$$

Now by [Bo] p.87 (1"c), we have

$$H^{\nu}(i_k^! S)_x = 0 \ , \quad \text{for} \quad x \in S_{m-k} \quad \text{and} \quad \nu \leq p(k) + 1 \ .$$

This gives the assertion by using the cohomological spectral sequence

$$E_2^{p,q} = H^p(U_{k+1}, H^q((i_k)_!(i_k)^! S)) \Longrightarrow H^{p+q}(U_{k+1}, (i_k)_!(i_k)^! S) \ .$$

Q.E.D.

<u>(3.1.2) Lemma.</u> *Let* $n_o \geq 2$ *be an integer,* $Y := X \setminus U_{n_o}$ *and assume that*

$$U_2 = U_3 = \ldots = U_{n_o} \subset U_{n_o+1} \subset \ldots \ .$$

*Then the natural map*

$$\beta_p^{\nu} : I_p H^{\nu}(X, \mathbb{C}) \longrightarrow H^{\nu}(U_{n_o}, \mathbb{C})$$

*is bijective for* $\nu \leq p(n_o)$ *and injective for* $\nu = p(n_o) + 1$ .

*Proof:* Follows immediately from (3.1.1) . Q.E.D.

<u>(3.1.3) Remark.</u> If $X_\bullet$ is an algebraic Whitney stratification in (3.1.2), and $p = m$ is the middle perversity, then the map

$$\beta^{\nu} = IH^{\nu}(X, \mathbb{C}) \longrightarrow H^{\nu}(X \setminus Y, \mathbb{C})$$

is bijective for $\nu \leq \operatorname{codim}_{\mathbb{C}}(Y,X) - 1$ and injective for $\nu = \operatorname{codim}_{\mathbb{C}}(Y,X)$ . In particular, the assertion (5) of theorem (I) follows from this.

<u>(3.1.4) Proposition.</u> *Let* $D \subset X$ *be a hyperplane section which is transversal to each stratum. Then, in the situation of* (3.1.3) , *the restriction map*

$$H^{\nu}(X \setminus Y, \mathbb{C}) \longrightarrow H^{\nu}((X \setminus Y) \cap D, \mathbb{C})$$

*is bijective for* $\nu \leq \min(\mathrm{codim}_{\mathbb{C}}(Y,X)-1, \dim_{\mathbb{C}} X-2)$ .

The *proof* follows at once from the weak Lefschetz theorem for the intersection homology in [G-M] §7 and the remark above.

Note that (3.1.4) is a slight improvement of theorem (III) in $[B\text{-}K]_2$ .

## 3.2 *Hodge symmetry*

The idea how to prove the Hodge symmetry in the concave case has been communicated to us by V. Navarro Aznar to whom we want to express our thanks for that. The situation is that of theorem (I). We may assume the field K to be algebraically closed.

<u>(3.2.1) Remark.</u> Let $D \subset X$ be an ample divisor such that $D_U := U \cap D$ is smooth. Then the proof of the weak Lefschetz theorem in $[B\text{-}K]_2$ shows that the restriction map

$$H^j(U,\Omega^i_{U/K}) \longrightarrow H^j(D_U,\Omega^i_{D_U/K})$$

is bijective for $i + j < \mathrm{codim}(Y,X) - 2$ and injective for $i + j =$ $= \mathrm{codim}(Y,X) - 2$ .

<u>(3.2.2) Lemma.</u> *Let* X *and* Y *be as above. Then there exist ample divisors* $D_1,\dots,D_r \subset X$ *such that the following holds*

(1) $D_{\leq s} := D_1 \cap \dots \cap D_s$ *are integral for* $1 \leq s \leq r$ ,

(2) $Y \cap D_{\leq r} = \emptyset$ , $Y \cap D_{\leq s} \neq \emptyset$ *for* $s < r$ ,

(3) $\mathrm{codim}(Y \cap D_{\leq s}, X \cap D_{\leq s}) = \mathrm{codim}(Y,X)$ , *for* $1 \leq s \leq r$ *) ,

---

*) we use the convention that $\mathrm{codim}(\emptyset,Z) = \dim Z + 1$

(4) $U \cap D_{\leq s}$ *is smooth for* $1 \leq s \leq r$ .

*Proof.* The assertion follows inductively from the following fact:

Let $X, Y$ be as above. Then there exists an ample divisor $D \subset X$ (compare [F] p.110) such that

(1) $D$ is integral,

(2) $\operatorname{codim}(Y \cap D, D) = \dim D - \dim(Y \cap D) = \dim X - \dim Y = \operatorname{codim}(Y,X)$ ,

(3) $U \cap D$ is smooth/K . Q.E.D

(3.2.3) Corollary. *In the situation of* (3.2.2) , *the map*

$$H^j(U,\Omega^i_{U/K}) \longrightarrow H^j(U \cap D_{\leq r}, \Omega^i_{U \cap D_{\leq r}/K})$$

*is bijective for* $i + j < \operatorname{codim}(Y,X) - 2$ *and injective for* $i + j = \operatorname{codim}(Y,X) - 2$ . —

(3.2.4) Proposition. *Let* $X$ *and* $Y$ *be as in theorem* (I) . *Then*

$$h^{i,j}(U) = h^{j,i}(U)$$

*for* $i + j < \operatorname{codim}(Y,X) - 1$ .

*Proof.* We use (3.2.1), (3.2.2) and put $Z := U \cap D_{\leq r} = D_{\leq r}$ . Then $Z$ is smooth and complete. The assertion is now clear for $i + j < \operatorname{codim}(Y,X) - 2$ . If $i + j = \operatorname{codim}(Y,X) - 2 =: n_o$ , then $h^{i,j}(U) \leq h^{i,j}(Z)$ . By $E_1$-degeneration and the Lefschetz theorem (3.1.4), we have for the Betti-numbers

$$b_{n_o}(U) = \sum_{i+j=n_o} h^{i,j}(U) \leq \sum_{i+j=n_o} h^{i,j}(Z) = b_{n_o}(Z) ,$$

$$b_{n_o}(U) = b_{n_o}(Z) .$$

This shows the proposition. Q.E.D.

(3.2.5) Remark. The proof of (3.2.4) shows especially that the mixed Hodge structure on

$$H^n(U^{an},\mathbb{C}) , \qquad n < \operatorname{codim}(Y,X) - 1$$

is in fact pure (since $Z$ above is smooth and projective).

## 3.3 *Log - poles*

Let $X$, $Y$, $\tilde{X}$, $\tilde{Y}$, be as in theorem (I), part (6), and $D_1,\dots,D_r$ as in (3.2.2). Again we put

$$Z := D_1 \cap \dots \cap D_r \subset U .$$

Then $Z$ is smooth and complete.

Since we may extend all these data in a standard way over a base scheme Spec(A) of finite type over $\mathbf{Z}$ such that also appropriate base change results hold for relative (log)-Dolbeault and (log)-de Rham cohomology, we assume now that $K$ is a perfect field of characteristic $p$ large enough. In this case one has a fairly general splitting result of Deligne/Illusie, see [D-I] and, as a consequence, a commutative diagram

$$\begin{array}{ccc}
\bigoplus\limits_{i+j=n} H^j(\tilde{X},\Omega^i_{\tilde{X}/K}\langle\tilde{Y}\rangle) & \xrightarrow[\oplus\beta_{i,j}]{\text{restr.}} & \bigoplus\limits_{i+j=n} H^j(U,\Omega^i_{U/K}) \\
\Big\downarrow \phi\ \wr & & \Big\downarrow \text{restr.} \\
H^n(\tilde{X},\Omega^{\bullet}_{\tilde{X}/K}\langle\tilde{Y}\rangle) & & \bigoplus\limits_{i+j=n} H^j(Z,\Omega^i_{Z/K}) \\
\Big\downarrow \varepsilon_n & & \wr\ \Big\downarrow \phi' \\
H^n_{DR}(U/K) & \xrightarrow[\beta_n]{\text{restriction}} & H^n_{DR}(Z/K)
\end{array}$$

(observe that here all data lift simultaneously mod $p^2$ ).

We suppose that $n < \operatorname{codim}(Y,X) - 1$ (in order to guarantee base change for de Rham cohomology on $U$ ) and obtain the injectivity of all $\beta_{i,j}$ with $i + j = n$ . Indeed, by base change theory and (3.1.4) , the maps $\varepsilon_n$ and $\beta_n$ are bijective for the above $n$ .

Lifting to characteristic zero, we may conclude that our original $\beta^{i,j}$ is injective for $i + j < \operatorname{codim}(Y,X) - 1$ . By a dimension counting argument as in (3.2) together with $E_1$-degeneration and weak Lefschetz, we see that $\beta^{i,j}$ is in fact bijective. Q.E.D.

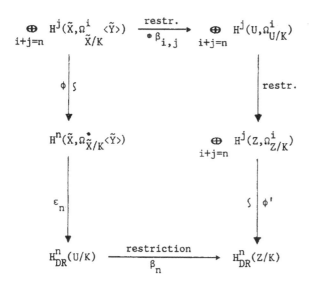

## 4. Log - poles (convex case)

We consider the situation where $X$ is a smooth algebraic variety over a field $K$ of characteristic zero and $Y := \bigcup_{a \in I} Y_a$ is a reduced normal crossing divisor in $X$ with smooth components $Y_a$. We assume the index set $I$ to be totally ordered. Then there is an exact sequence of de Rham complexes (compare for instance [D-I] p.267)

$$(4.1) \qquad 0 \longrightarrow \Omega_X^\bullet\langle Y\rangle(-Y) \longrightarrow \Omega_X^\bullet \longrightarrow \bigoplus_{a \in I} \Omega_{Y_a}^\bullet \xrightarrow{\delta^1} \bigoplus_{a<b} \Omega_{Y_a \cap Y_b}^\bullet \xrightarrow{\delta^2} \cdots .$$

We define

$$\tilde{\Omega}_Y^\bullet := \operatorname{Ker}(\delta^1),$$

(4.2)

$$K^i := \operatorname{Ker}(\Omega_X^i \otimes_{\mathcal{O}_X} \mathcal{O}_Y \longrightarrow \tilde{\Omega}_Y^i).$$

Then there is the following vanishing result

(4.3) Proposition. *We assume that $Y$ is proper over $K$. Let $L \in \operatorname{Pic}(Y)$ be $k$-ample and $m \geq 1$ an integer. Then*

(1) $H^j(Y, \tilde{\Omega}_Y^i \otimes L^m) = 0$ *for* $i + j \geq \dim Y + k + 1$.

*If the normal bundle $N = N_{Y|X}$ is in addition generated by global sections, then*

(2) $H^j(Y, K^i \otimes N \otimes L^m) = 0$ *for* $i + j \geq \dim X + k + 1$,

(3) $H^j(Y, \Omega_X^i \otimes_{\mathcal{O}_X} (N \otimes L^m)) = 0$ *for* $i + j \geq \dim X + k + 1$.

*Proof.* The first part follows easily from the sequence (4.1) and the standard vanishing theorem for k-ample line bundles on complete smooth algebraic varieties.

Clearly (3) is a consequence of (1) and (2). So it remains to prove (2). We denote by $I = \mathcal{O}_X(-Y)$ the idealsheaf of $Y$. By tensoring the exact sequence

$$0 \longrightarrow \Omega_X^i\langle Y\rangle(-Y) \longrightarrow \Omega_X^i \longrightarrow \tilde{\Omega}_Y^i \longrightarrow 0$$

with $\mathcal{O}_Y$, we obtain

$$0 \longrightarrow \mathcal{T}or_1^X(\tilde{\Omega}_Y^i, \mathcal{O}_Y) \longrightarrow \Omega_X^i\langle Y\rangle \otimes I/I^2 \longrightarrow \Omega_X^i \otimes \mathcal{O}_Y \longrightarrow \tilde{\Omega}_Y^i \longrightarrow 0 .$$

Since

$$\mathcal{T}or_1^X(\tilde{\Omega}_Y^i, \mathcal{O}_Y) = \tilde{\Omega}_Y^i \otimes_{\mathcal{O}_X} I$$

(using the sequence $0 \longrightarrow I \longrightarrow \mathcal{O}_X \longrightarrow \mathcal{O}_Y \longrightarrow 0$ ), we conclude

$$K^i = \operatorname{Coker}(\tilde{\Omega}_Y^i \otimes I \longrightarrow \Omega_X^i\langle Y\rangle \otimes I/I^2)$$

$$= \operatorname{Coker}(\Omega_X^i \otimes I/I^2 \longrightarrow \Omega_X^i\langle Y\rangle \otimes I/I^2) .$$

Now we consider the standard weight filtration $W_\bullet$ on $\Omega_X^i\langle Y\rangle$, so

$$\Omega_X^i = W_o \subset W_1 \subset \ldots \subset W_i = \Omega_X^i\langle Y\rangle ,$$

$$W_n/W_{n-1} \cong \bigoplus_{a_1<\ldots<a_n} \Omega^{i-n}_{Y_{a_1}\cap\ldots\cap Y_{a_n}} .$$

If we put

$$\overline{W}_n := W_n/W_o$$

then

$$W_o = 0 ,$$

$$\overline{W}_n/\overline{W}_{n-1} \cong W_n/W_{n-1} ,$$

$$K^i \cong \overline{W}_i \otimes N^{-1} \quad .$$

As we know

$$H^j(Y,(W_n/W_{n-1}) \otimes L^m) \;=\; 0$$

for $i + j \geq \dim X + k + 1$ , we get

$$H^j(Y,\overline{W}_i \otimes L^m) \;= 0$$

in the same range by using the short exact sequences

$$0 \longrightarrow \overline{W}_{n-1} \otimes L^m \longrightarrow \overline{W}_n \otimes L^m \longrightarrow (W_n/W_{n-1}) \otimes L^m \longrightarrow 0 \; .$$

Q.E.D.

Next we take the short exact sequence

$$(4.4) \qquad 0 \longrightarrow \Omega_X^i(mY) \longrightarrow \Omega_X^i((m+1)Y) \longrightarrow \Omega_X^i \otimes N_{Y|X}^{m+1} \longrightarrow 0$$

and consider the natural commutative diagram

$$(4.5) \qquad \begin{array}{ccc} H^j(X,\Omega_X^i) & \longrightarrow & H^j(X,\Omega_X^i\langle Y\rangle) \\ \downarrow & & \downarrow \\ \varinjlim_{m\geq 0} H^j(X,\Omega_X^i(mY)) & \xrightarrow{\;\sim\;} & H^j(U,\Omega_X^i) \end{array}$$

where $U := X \backslash Y$ .

<u>(4.6) Lemma.</u> *Let* Y *be as in* (4.3) *with* $N_{Y|X}$ *k-ample. Then the natural map*

(1) $H^j(X,\Omega_X^i\langle Y\rangle) \longrightarrow H^j(X,\Omega_X^i(Y))$

*is bijective for* $i + j \geq \dim X + k + 1$ *and surjective for* $i + j \geq \dim X + k$ ,

(2) $H^j(X, \Omega^i_X(mY)) \longrightarrow H^j(X, \Omega^i_X((m+1)Y))$ , $m \geq 1$

*is bijective for* $i + j > \dim X + k + 1$ *and surjective for* $i + j \geq \dim X + k + 1$ .

*Proof.* The first part follows from (4.1) , (4.2) and the vanishing result (4.3) (1) . For the second part use (4.4) and (4.3) (3) . Q.E.D.

<u>(4.7) Proposition.</u> *Let* $Y$ *be as in* (4.6) . *Then the following holds*

(1) *The map*

$$H^j(X, \Omega^i_X(Y)) \longrightarrow H^j(U, \Omega^i_X)$$

*is bijective for* $i + j > \dim X + k + 1$ *and surjective for* $i + j \geq \dim X + k + 1$ .

(2) *If* $X$ *is in addition proper over* $K$ , *then the natural maps*

$$H^j(X, \Omega^i_X\langle Y\rangle) \longrightarrow H^j(X, \Omega^i_X(Y)) \longrightarrow H^j(U, \Omega^i_X)$$

*are all bijective for* $i + j \geq \dim X + k + 1$ .

*Proof.* The first part is an easy consequence of (4.5) , (4.6) (2) .

In order to show (2) , we know that

$$h^{i,j}(X\langle Y\rangle) := \dim_K H^j(X, \Omega^i_X\langle Y\rangle) = \dim_K H^j(X, \Omega^i_X(Y)) \geq h^{i,j}(U)$$

for $i + j \geq \dim X + k + 1$ . But by the $E_1$-degeneration theorems ([D-I], $[B\text{-}K]_1$), we get for the Betti-numbers in dimensions $n \geq \dim X + k + 1$

$$b_n(U) = \sum_{i+j=n} h^{i,j}(X\langle Y\rangle) \geq \sum_{i+j=n} h^{i,j}(U) = b_n(U)$$

and the desired equality holds. Q.E.D.

<u>(4.8) Remark.</u> The second part of the vanishing theorem in section 1 is a consequence of (4.3) (3) together with the sequence (4.4) and the isomorphism

$$H^j(U,\Omega^i_{U/K} \otimes \mathcal{L}) \cong \varinjlim_{m\geq 0} H^j(X,\Omega^i_{X/K}(mY) \otimes \mathcal{L}) \quad .$$

. Hodge symmetry (convex case)

irst we treat an easy case, namely where $\mathcal{O}_X(Y)$ is k-ample on X .

5.1) Proposition. *Let X be proper and smooth over K and $Y \subset X$ a divi-or such that $\mathcal{O}_X(Y)$ is k-ample on X . Then*

$$H^j(X \setminus Y, \Omega^i_{X/K}) = 0$$

*r* $i+j \geq \dim X + k + 1$.

*roof.* By the vanishing result [So] Proposition (1.12), we have

$$H^j(X, \Omega^i_{X/K}(nY)) = 0$$

or $i+j \geq \dim X + k + 1$ and all $n > 0$ . Consequently

$$H^j(X \setminus Y, \Omega^i_{X/K}) = \varinjlim_n H^j(X, \Omega^i_{X/K}(nY))$$

$$= 0$$

or $i+j \geq \dim X + k + 1$ . Q.E.D.

order to prove the general case where only the conditions in theorem II (4) re satisfied, we need some preparations.

5.2) Proposition. *Let X be a projective algebraic manifold over $\mathbb{C}$ and $\subset X$ a divisor with k-ample normal bundle, $U := X \setminus Y$ . Moreover, let $L \in \mathrm{Pic}(X)$ with $L|Y \cong N_{Y|X}$ and such that L is globally generated.*

*en the restriction map*

$$H^n(X, \mathbb{C}) \longrightarrow H^n(U, \mathbb{C})$$

*surjective for* $n \geq \dim X + k + 1$ .

*Proof.* We proceed by induction on $k$. So let us assume $k = 0$, i.e. $N_{Y|X}$ is ample. In this case $U$ is a strongly pseudoconvex manifold with a compact exceptional analytic subset $E \subset U$. If we fix $n \geq \dim X+1$, then in the natural commutative diagram

$$\begin{array}{ccc} H^n(X) & \longrightarrow & H^n(U) \\ \uparrow & & \downarrow \beta \\ H^n_E(X) & \xrightarrow[\alpha]{} & H^n(E) \end{array}$$

the map $\alpha$ is surjective by [N] Prop. (5.1)(iii) and $\beta$ is bijective (comparing $\Omega^{\cdot}_U$ and $(\Omega^{\cdot}_U)^{\wedge E}$). This shows the assertion for $k = 0$.

In the induction step "k-1 => k", we may equally show the dual statement:

$$H^n_c(U) \longrightarrow H^n(X) \quad \text{injective for} \quad n \leq \dim X - k - 1.$$

The proper map given by global sections of $L$

$$\phi : X \longrightarrow P := \mathbb{P}(\Gamma(X, L))$$

has, restricted to $Y$, at most k-dimensional fibres. Then by a result of Hironaka [Hi] Theorem (2.6) (see also [So] proof of (1.12)), there is an ample divisor $D$ on $X$ such that $\phi \mid Y \cap D$ has at most (k-1)-dimensional fibres. In the commutative diagram

$$\begin{array}{ccc} H^n_c(U) & \xrightarrow{\gamma} & H^n(X) \\ \beta \downarrow & & \downarrow \\ H^n_c(U \cap D) & \xrightarrow[\gamma_D]{} & H^n(D) \end{array}$$

$\beta$ is injective for $n \leq \dim X - k - 1$ by $[B\text{-}K]_2$ Theorem (IV). Now, since $L|D$ is again globally generated and

$$L|Y \cap D \cong N_{Y|X} \mid Y \cap D \cong N_{Y \cap D|D}$$

is (k-1)-ample, $\gamma_D$ is also injective by induction hypothesis for $n \leq \dim D - (k-1) - 1 = \dim X - k - 1$ which proves the assertion.

Q.E.D.

(5.3) Corollary. *In the situation of (5.2), the mixed Hodge structure on $H^n(U,\mathbb{C})$ is pure of weight $n$ for $n \geq \dim X + k + 1$.*

*Proof.* Since $H^n(X)$ is pure of weight $n$ and $H^n(U)$ is a quotient of $H^n(X)$ for $n \geq \dim X + k + 1$, it follows that $H^n(U)$ is pure of weight $n$ too (see $[D]_2$ "Hodge II" Théorème (2.3.5)(iii)).

Q.E.D.

(5.4) Proposition. *Let X,Y be as in (5.2), then*

$$h^{i,j}(U) = h^{j,i}(U)$$

*for* $i+j \geq \dim X + k + 1$.

*Proof.* We consider a proper birational map

$$\pi : \tilde{X} \longrightarrow X$$

such that $\tilde{X}$ is smooth, $\pi^{-1}(Y)_{red} =: \tilde{Y}$ is a normal crossing divisor on $\tilde{X}$ and $\pi$ is an isomorphism outside $Y$. Then $H^n(\tilde{X}\backslash\tilde{Y}) = H^n(U)$ is pure of weight $n$ for $n \geq \dim X + k + 1$ and therefore

$$h^{i,j}(\tilde{X}\langle\tilde{Y}\rangle) = h^{j,i}(\tilde{X}\langle\tilde{Y}\rangle), \qquad i+j \geq \dim X + k + 1$$

(this fact follows from "Hodge II", $[D]_2$ Théorème (1.2.10)(iv) and (2.3.7) together with the $E_1$-degeneration of the log-Hodge => de Rham spectral sequence).

Using the "characteristic p" argument in (3.3), we may derive

$$h^{i,j}(\tilde{X}\langle\tilde{Y}\rangle) = h^{i,j}(U), \qquad i+j \geq \dim X + k + 1$$

and thus the assertion.

Q.E.D.

(5.5) Remark. If we assume $Y$ to be smooth, then the map

$$H^n(X,\mathbb{C}) \longrightarrow H^n(Y,\mathbb{C})$$

is surjective for $n \geq \dim X + k$ (see [So] Prop. (2.6)). Consequently

$$H^m_Y(X,\mathbb{C}) \longrightarrow H^m(X,\mathbb{C})$$

is injective for $m \leq \dim X - k$ and the Hodge structure on $H^m(U,\mathbb{C})$ is pure of weight $m$ for $m \leq \dim X - k - 1$.

---

In order to treat the case where $\mathcal{O}_X(Y)$ is only globally generated in some open Zariski-neighborhood of $Y \subset X$, we need the following lemma.

(5.6) Lemma. *Let* $\pi : \tilde{U} \longrightarrow U$ *be the blowing up of a smooth separated* $\mathbb{C}$-*scheme* $U$ *in a smooth complete center* $Z \subset U$. *Put* $\tilde{Z} := \pi^{-1}(Z)$.

*Then the natural sequence*

$$0 \longrightarrow H^j(U,\Omega^i_U) \longrightarrow H^j(\tilde{U},\Omega^i_{\tilde{U}}) \longrightarrow H^j(\tilde{Z},\Omega^i_{\tilde{Z}}) \longrightarrow 0$$

*is exact for all* $i,j \in \mathbb{N}$.

*Proof.* Take a smooth compactification $X$ of $U$. Then $Z \cap Y = \emptyset$, where $Y := X \backslash U$. Let $\pi : \tilde{X} \longrightarrow X$ be the blowing up of $Z$ in $X$. Replacing $U$ by $X$ and $\tilde{U}$ by $\tilde{X}$, the result follows from $[D]_1$ Prop. (4.3) and [M]. Since all the maps

$$H^j_Y(X,\Omega^i_X) \longrightarrow H^j_Y(\tilde{X},\Omega^i_{\tilde{X}})$$

are bijective, the assertion is obtained from a short diagram chasing.

Q.E.D.

(5.7) Remark. In (5.6) we have obviously the equivalence

$$h^{i,j}(U) = h^{j,i}(U) \iff h^{i,j}(\tilde{U}) = h^{j,i}(\tilde{U}) .$$

*Proof of part (4) in theorem* (II).

We may assume $K = \mathbb{C}$ . Let $W$ be an open Zariski neighborhood of $Y$ in $X$ where $\mathcal{O}_X(Y)$ is generated by global sections $s_o,\ldots,s_r \in \Gamma(W,\mathcal{O}_X(Y))$ . Denote by $V \subset \Gamma(W,\mathcal{O}_X(Y))$ the subvectorspace generated by the $s_\nu$ . Obviously the map

$$\phi : W \longrightarrow P := \mathbb{P}(V)$$

restricted to $Y$ , has at most k-dimensional fibres.

Now by results of Hironaka, we may find a birational morphism

$$\pi : \tilde{X} \longrightarrow X$$

which is a sequence of blowing up's with smooth centers such that $\pi$ is an isomorphism over $W$ and $\phi$ can be extended to a map

$$\tilde{\phi} : \tilde{X} \longrightarrow P .$$

Setting $L := \tilde{\phi}^{-1}(\mathcal{O}_P(1))$ , we have $L|Y \cong N_{Y|X}$ and $L$ is generated by global sections.

Now by (5.2), the restriction map $(\tilde{U} := \tilde{X}\setminus Y)$

$$H^n(\tilde{X},\mathbb{C}) \longrightarrow H^n(\tilde{U},\mathbb{C})$$

is surjective for $n \geq \dim \tilde{X} + k + 1$ and furthermore

$$h^{i,j}(\tilde{U}) = h^{j,i}(\tilde{U}) , \qquad i+j \geq \dim \tilde{X} + k + 1 .$$

Using (5.7) we get the Hodge symmetry on $U$ . Q.E.D.

<u>(5.8). Remark.</u> In the situation of theorem (II), part (4) the Hodge structure on $H^n(U)$ (for $K = \mathbb{C}$) is pure of weight $n$ for $n \geq \dim X + k + 1$ . This follows from the injectivity of $H^n(U) \longrightarrow H^n(\tilde{U})$ in the proof above (compare also (5.6)).

## 6. Kodaira vanishing with weakly semi-positive normal bundle

In this section we give a proof of the vanishing result mentioned at the beginning. The crucial steps for that were already established in section 4.

Let $X$ be a complete smooth K-variety, $Y \subset X$ a reduced normal crossing divisor with smooth components such that the normal bundle $N = N_{Y|X}$ is generated by global sections (i.e. "weakly semi-positive"). Moreover, we fix an $\ell$-ample line bundle $L \in \mathrm{Pic}(X)$ .

First we note that the natural map

$$H^j(X,\Omega_X^i(mY) \otimes L) \longrightarrow H^j(X,\Omega_X^i((m+1)Y) \otimes L), \qquad m \geq 0$$

is bijective for $i+j > \dim X + \ell + 1$ and surjective for $i+j \geq \dim X + \ell + 1$ . This follows from (4.3)(3). As a consequence

$$H^j(X,\Omega_X^i \otimes L) \longrightarrow H^j(U,\Omega_X^i \otimes L)$$

is surjective for $i+j \geq \dim X + \ell + 1$ . So we are reduced to the compact situation where the result is known (see [So] (1.12)).

---

We want to mention that there is also a proof of the vanishing theorem (at least in the case where $Y$ is smooth) with characteristic $p$ methods. We are going to explain shortly our argument.

We need the following

<u>(6.1) Lemma.</u> *Let $Y$ be a proper scheme over a field $K$ and $N,L \in \mathrm{Pic}(Y)$ such that $N$ is globally generated and $L$ is $\ell$-ample.*

*Then for any coherent locally free sheaf $F$ on $Y$ there exists a natural number $\mu_o = \mu_o(F)$ such that*

$$H^j(Y,F \otimes N^n \otimes L^\mu) = 0$$

*for all* $n \in \mathbb{N}$ , $\mu \geq \mu_o$ *and* $j \geq \ell+1$ .

*Proof* (sketch). The result will be shown by induction on $(n, \dim Y)$ . The cases $n = 0$ or $\dim Y = 0$ are clear. Now, for arbitrary $n$ and $\dim Y$ , we choose $s \in \Gamma(Y,N)$ such that the map $\mathcal{O}_Y \longrightarrow N$, $1 \longmapsto s$ is injective (this may be seen by some tricky application of the theorem of "generic flatness"). The long exact cohomology sequence associated to $(Z := \{s=0\})$

$$0 \longrightarrow F \otimes N^{n-1} \otimes L^{\mu} \longrightarrow F \otimes N^{n} \otimes L^{\mu} \longrightarrow F \otimes (N^{n}|Z) \otimes L^{\mu} \longrightarrow 0$$

gives the induction step. Q.E.D.

Imitating the proof of $[B\text{-}K]_2(3.1)$, we conclude with the help of (6.1)

<u>(6.2) Proposition.</u> *Let* X *be a proper smooth scheme over an arbitrary field* K *and* $Y \subset X$ *a divisor with globally generated normal bundle,* $U := X\backslash Y$ . *Furthermore, let* $L \in \mathrm{Pic}(X)$ *be* $\ell$*-ample and* $F \in \underline{\mathrm{Coh}}(X)$ *locally free.*

*Then there exists a number* $\mu_o = \mu_o(F)$ *such that*

$$H^j(U, F \otimes L^{\mu}) = 0$$

*for* $\mu \geq \mu_o$ *and* $j \geq \ell+1$ .

The next step is the reduction of the powers of $L$ via characteristic $p$ by applying $[B\text{-}K]_2$ Lemma (1.1). We get the following statement

<u>(6.3) Theorem.</u> *Let* k *be a perfect field of characteristic* $p \gg 0$ *and* X *be proper and smooth over* Spec(k) . *Moreover, let* $Y \subset X$ *be a divisor with globally generated normal bundle and* $L \in \mathrm{Pic}(X)$ $\ell$*-ample.*

*Then, if* $U := X\backslash Y$ *is liftable to the Witt-vectors* $W_2(k)$ *of length two,*

$$H^j(U, \Omega^i_{U/k} \otimes L) = 0$$

*for* $i+j \geq \dim X + \ell + 1$ .

In order to obtain part one of our vanishing theorem in section one, we proceed in a standard manner. First of all we extend the given data over an appropriate subring of the base field, then we need results from base change theory in order to conclude from (6.3) to the characteristic zero case. This means that we have to prove the following finiteness assertion.

(6.4) Proposition. *Let* $f : \mathfrak{X} \longrightarrow S$ *be a proper and smooth morphism of noetherian schemes,* $\mathfrak{Y} \subset \mathfrak{X}$ *a divisor with relatively globally generated normal bundle (with respect to* $f|\mathfrak{Y}$*) and* $L \in \mathrm{Pic}(\mathfrak{X})$ *$\ell$-ample for* $\mathfrak{Y}/S$*. Moreover, let* $f|\mathfrak{Y} : \mathfrak{Y} \longrightarrow S$ *be smooth,* $\mathfrak{U} := \mathfrak{X} \backslash \mathfrak{Y}$*.*

*Then*

$$R^j(f|\mathfrak{U})_*(\Omega^i_{\mathfrak{U}/S} \otimes L)$$

*is coherent for* $i+j \geq \dim(\mathfrak{X}/S) + \ell + 1$ .

*Proof.* Obviously one has

$$R^j(f|\mathfrak{U})_*(\Omega^i_{\mathfrak{U}/S} \otimes L) \cong \varinjlim_n R^j f_*(\Omega^i_{\mathfrak{X}/S} \otimes L \otimes \mathcal{O}(n\mathfrak{Y})) .$$

By the exact sequence

$$0 \longrightarrow \mathcal{O}(n\mathfrak{Y}) \longrightarrow \mathcal{O}((n+1)\mathfrak{Y}) \longrightarrow N^{n+1}_{\mathfrak{Y}|\mathfrak{X}} \longrightarrow 0$$

we notice that it is sufficient to show

$$R^j f_*(\Omega^i_{\mathfrak{X}/S} \otimes L \otimes N^{n+1}_{\mathfrak{Y}|\mathfrak{X}}) = 0$$

for $i+j \geq \dim(\mathfrak{X}/S) + \ell+1$ . For this we use the short exact sequence (where $I$ is the ideal sheaf of $\mathfrak{Y}$)

$$0 \longrightarrow I/I^2 \otimes \Omega^{i-1}_{\mathfrak{Y}/S} \longrightarrow \Omega^i_{\mathfrak{X}/S}|\mathfrak{Y} \longrightarrow \Omega^i_{\mathfrak{Y}/S} \longrightarrow 0$$

and the relative version of the Kodaira vanishing theorem for $\ell$-ample line bundles. Q.E.D.

## 7. An example.

In this paragraph we want to give an example in order to show that the result of our theorem (II) is sharp in some sense.

If $X$ is as in (II) and $Y \subset X$ is a locally complete intersection with k-ample normal bundle $N_{Y|X}$ in $X$, then it follows by [Ok] p. 19, that $U := X\setminus Y$ is $(\mathrm{rg}\, N_{Y|X}+k)$-convex. This implies that the $h^{i,j}(U)$ are finite for $j \geq k + \mathrm{rg}\, N_{Y|X}$. Because of this, one might expect the following generalization of theorem (II):

(7.1) Let $X$ be a compact algebraic manifold, $Y \subset X$ a locally complete intersection and $U := X\setminus Y$. We assume that the normal bundle $N_{Y|X}$ of $Y$ in $X$ is k-ample (on Y). Then

(1) $h^{i,j}(U)$, $h^n(U)$ are finite for $j \geq \mathrm{rg}\, N_{Y|X}+k$ and $n \geq \dim X+k+\mathrm{rg}\, N_{Y|X}$, $i$ arbitrary,

(2) $h^n(U) = \sum\limits_{i+j=n} h^{i,j}(U)$ for $n \geq \dim X+k+\mathrm{rg}\, N_{Y|X}$.

In the case where $Y$ is a divisor, this is exactly theorem (II).

In the following we want to show that for $\mathrm{codim}(Y,X) = 2$ the statement (7.1) is not true. For this we follow a construction due to Grauert and Riemenschneider in $[G\text{-}R]_2$.

(7.2) Example. Let $Y$ be a two dimensional complex projective manifold and $\mathcal{O}_Y(1)$ a very ample line bundle on $Y$. Then one has a surjection

$$\mathcal{O}_Y \oplus \mathcal{O}_Y \oplus \mathcal{O}_Y \longrightarrow\!\!\!\!\!\to \mathcal{O}_Y(1)$$

and this gives rise to an exact sequence

$$0 \longrightarrow \mathcal{E} := \text{kernel}(\phi) \longrightarrow 3\mathcal{O}_Y(-1) \xrightarrow{\phi} \mathcal{O}_Y \longrightarrow 0 \quad .$$

One sees easily that $\mathcal{E}$ is locally free of rank 2 and the dual $\mathcal{E}^\vee$ is globally generated and ample.

If $E$ denotes the corresponding geometric vector bundle of $\mathcal{E}$ (we always identify a vector bundle with its sheaf of sections), then we consider $Y \subset X := \overline{E^\vee} = \text{Proj}(S^\cdot(\mathcal{E} \oplus \mathcal{O}_Y))$ , $U := X \setminus Y$ . The normal bundle $N_{Y|X}$ of $Y$ in $X$ is just $E^\vee$ , so that the assumptions of (7.1) are fulfilled for $k = 0$ . We compare now $h^n(U)$ with the sum of the $h^{i,j}(U)$ for $n = \dim X + k + \text{rg}\, N_{Y|X} = 4 + 0 + 2 = 6$ .

Then one gets

$$H^6(U,\mathbb{C}) \cong H^6(\mathbb{P}(E^\vee),\mathbb{C}) \cong \mathbb{C} \ ,$$

because $U$ is a line bundle over $\mathbb{P}(E^\vee)$ which is an oriented six dimensional topological manifold.

Furthermore

$$\bigoplus_{i+j=6} H^j(U,\Omega^i_U) = H^4(U,\Omega^2_U) \oplus H^3(U,\Omega^3_U) \oplus H^2(U,\Omega^4_U)$$

$$= H^3(U,\Omega^3_U) \oplus H^2(U,\Omega^4_U) \ .$$

For the last equality, we use Siu's vanishing theorem. By duality theory we have

$$H^2(U,\Omega^4_U) = H^2_c(U,\mathcal{O}_U)^\vee$$

and by $[G\text{-}R]_2$, it follows that these cohomology groups are not zero.

The canonical projection of $U$ onto $\mathbb{P}(E^\vee)$ induces an injective morphism

$$H^3(\mathbb{P}(E^\vee),\Omega^3_{\mathbb{P}(E^\vee)}) \longrightarrow H^3(U,\Omega^3_U)$$

and so it follows that $h^{3,3}(U) \geq 1$ , because $H^3(\mathbb{P}(E^\vee),\Omega^3_{\mathbb{P}(E^\vee)})$ is isomorphic to $\mathbb{C}$ via the trace map.

So this calculation implies

$$1 = h^6(U) \neq \sum_{i+j=6} h^{i,j}(U) \geq 2 .$$

We remark in addition that the Hodge symmetry doesn't hold either, since

$$H^2(U,\Omega^4_U) \neq 0 ,$$

$$H^4(U,\Omega^2_U) = 0 .$$

## 8. Another proof of Steenbrink's vanishing theorem.

First we state the result.

(8.1) Theorem ([St], Thm. 2). *Let* $X'$ *be a* n*-dimensional projective variety over* $\mathbb{C}$, $Y' \subset X'$ *a closed subscheme such that* $X' \setminus Y'$ *is smooth,* $L$ *an ample line bundle on* $X'$ and $\pi : X \longrightarrow X'$ *a proper birational morphism such that* $X$ *is nonsingular,* $Y := \pi^{-1}(Y')_{red}$ *is a divisor with simple normal crossings on* $X$ *and* $\pi : X \setminus Y \longrightarrow X' \setminus Y'$ *is an isomorphism.*

*Then the following holds*

(1) $H^j(X, I_Y\Omega_X^i\langle Y\rangle \otimes \pi^* L) = 0$ *for* $i+j > n$,

(2) $H^j(X, \Omega_X^i\langle Y\rangle \otimes \pi^* L^{-1}) = 0$ *for* $i+j < n$,

(3) $R^j\pi_*(I_Y\Omega_X^i\langle Y\rangle) = 0$ *for* $i+j > n$.

*Here* $I_Y$ *is the reduced ideal sheaf of the divisor* $Y$.

(8.2) Remark. The statements (1), (2), and (3) of (8.1) are all equivalent.

*Proof.* (1) is equivalent to (2) by Serre-duality.

(1) => (3) : Let $k \gg 0$ such that $R^j\pi_*(I_Y\Omega_X^i\langle Y\rangle) \otimes L^k$ is globally generated for all $i,j$ and

$$H^\nu(X', R^j\pi_*(I_Y\Omega_X^i\langle Y\rangle) \otimes L^k) = 0 \quad \text{for} \quad \nu \geq 1,$$

$i,j$ arbitrary. Then it follows by the spectral sequence

$$E_1^{\nu,j} := H^\nu(X', R^j\pi_*(I_Y\Omega_X^i\langle Y\rangle) \otimes L^k) \Rightarrow H^{\nu+j}(X, I_Y\Omega_X^i\langle Y\rangle \otimes \pi^* L^k)$$

and (1) that

$$H^o(X',R^j\pi_*(I_Y\Omega^i_X\langle Y\rangle) \otimes L^k) = H^j(X,I_Y\Omega^i_X\langle Y\rangle \otimes \pi^*L^k) = 0 \quad \text{for } i+j>n$$

and hence

$$R^j\pi_*(I_Y\Omega^i_X\langle Y\rangle) = 0 \quad \text{for } i+j > n ,$$

because $R^j\pi_*(I_Y\Omega^i_X\langle Y\rangle) \otimes L^k$ is generated by global sections.

(3) => (1) : We choose again $k \gg 0$ as above and conclude in the same way that

$$H^j(X,I_Y\Omega^i_X\langle Y\rangle \otimes \pi^*L^k) = H^o(X',R^j\pi_*(I_Y\Omega^i_X\langle Y\rangle) \otimes L^k) = 0 \quad \text{for } i+j>n .$$

The statement (1) is now obtained by reducing the power of $L$ to one via char.p-methods using a modified version (namely for differential forms with log-poles) of Lemma (1.1) in $[B\text{-}K]_2$ , compare also [D-I](2.9), (4.2).

Q.E.D.

(8.3) *Proof of the vanishing theorem.* Because of (8.2) it suffices to prove (2). For this let $Y = \bigcup_{\ell=1}^{r} Y_\ell$ resp. $I_Y = I_1 \cdot \ldots \cdot I_r$ be the decomposition of $Y$ in its irreducible smooth components.

It exists an ideal $K \subset O_{X'}$ such that $\pi$ is the blowing up of $X'$ in $K$, which implies that $J := K \cdot O_X = I_1^{\nu_1} \cdot \ldots \cdot I_r^{\nu_r}$ for certain $\nu_\ell \geq 1$ . We choose $m_1,\ldots,m_r \in \mathbb{N} \setminus \{0\}$ such that $m_1\nu_1 = \ldots = m_r\nu_r =: m$ . By [An] Thm. 1.3, [K] Thm. 17, there is a manifold $\tilde{X}$ and a finite morphism $\sigma : \tilde{X} \longrightarrow X$ with the properties

(a) $\sigma^*Y_\ell = m_\ell\tilde{Y}_\ell$ , $\tilde{Y}_\ell$ prime divisor ,

(b) $\tilde{Y} := \sum_{\ell=1}^{r} \tilde{Y}_\ell$ is a divisor with normal crossings on $\tilde{X}$ with smooth components $\tilde{Y}_\ell$ .

Because $\sigma^*(J) = (I_{\tilde{Y}_1} \dots I_{\tilde{Y}_r})^m$ and $J$ is relatively ample with respect to $\pi$, we see that $I_{\tilde{Y}} := I_{\tilde{Y}_1} \cdot \ldots \cdot I_{\tilde{Y}_r}$ is relatively ample with respect to $\pi \circ \sigma$. Then for $k \gg 0$, $I_{\tilde{Y}} \otimes (\pi \circ \sigma)^* L^k$ is ample on $\tilde{X}$, which implies by "Kodaira vanishing theorem" that

$$H^j(\tilde{X}, \Omega^i_{\tilde{X}}\langle\tilde{Y}\rangle \otimes I_{\tilde{Y}} \otimes (\pi \circ \sigma)^* L^k) = 0 \quad \text{for} \quad i+j > n$$

or equivalently (with Serre-duality)

$$H^j(\tilde{X}, \Omega^i_{\tilde{X}}\langle\tilde{Y}\rangle \otimes (\pi \circ \sigma)^* L^{-k}) = 0 \quad \text{for} \quad i+j < n .$$

By an analogous char. p-argument as in the remark above we may assume $k=1$.

Since $\sigma$ is finite, we get:

$$H^j(X, \sigma_*(\Omega^i_{\tilde{X}}\langle\tilde{Y}\rangle) \otimes \pi^* L^{-1}) = 0 \quad \text{for} \quad i+j < n ,$$

and finally by (8.4) it follows that $\Omega^i_X\langle Y\rangle$ is a direct summand in $\sigma_*\Omega^i_{\tilde{X}}\langle\tilde{Y}\rangle$ which implies that

$$H^j(X, \Omega^i_X\langle Y\rangle \otimes \pi^* L^{-1}) = 0 \quad \text{for} \quad i+j < n \qquad \text{Q.E.D.}$$

<u>(8.4) Lemma.</u> *With the notations of above, the following holds:*

(1) $\sigma$ *is flat*,

(2) $\sigma_*(\Omega^i_{\tilde{X}}\langle\tilde{Y}\rangle) = \Omega^i_X\langle Y\rangle \otimes \sigma_* O_{\tilde{X}}$ .

*Proof.* (1) Finite morphisms between manifolds are always flat.

(2) One has a canonical injective map

$$\sigma^*\Omega^i_X\langle Y\rangle \hookrightarrow \Omega^i_{\tilde{X}}\langle\tilde{Y}\rangle$$

which is an isomorphism outside $\sigma^{-1}(\mathrm{Sing}(Y))$. This induces

$$\sigma_*\sigma^*\Omega^i_X\langle Y\rangle = \Omega^i_X\langle Y\rangle \otimes \sigma_*\mathcal{O}_{\tilde{X}} \hookrightarrow \sigma_*\Omega^i_{\tilde{X}}\langle\tilde{Y}\rangle$$

which is bijective outside a subvariety of codimension at least two. Since the sheaf on the left side is locally free and $\sigma_*\Omega^i_{\tilde{X}}\langle\tilde{Y}\rangle$ is torsionfree, it is in fact an isomorphism (this argument is essentially due to E. Viehweg).

Q.E.D.

This vanishing result has many interesting consequences, as beautifully explained by Steenbrink in [St] . Especially Grauert-Riemenschneider vanishing theorem ($[G\text{-}R]_1$) can be immediately obtained from it and, moreover, also the singular version of Kodaira vanishing due to V. Navarro Aznar in [G-N-P-P], p. 149.

At the end we want to formulate a generalization of (8.1) (which can be shown by an induction argument):

(8.5) Theorem. *Let* $f : X \longrightarrow X'$ *be a proper morphism between algebraic varieties over* $\mathbb{C}$ *where* $X$ *is smooth and let* $A \subset X$ *be a complete normal crossing divisor such that*

(1) $A = f^{-1}f(A)$ *as a set ,*

(2) $f : X\backslash A \longrightarrow X'\backslash f(A)$ *is smooth of relative dimension* $k$ .

*Then*

$$R^jf_*(I_A\Omega^i_X\langle A\rangle) = 0$$

*for all* $i,j$ *with* $i+j \geq \dim X + k + 1$ . *Here* $I_A$ *denotes the reduced ideal sheaf of* $A$ .

References.

[An] V. Ancona: Vanishing and Nonvanishing Theorems for Numerically Effective Line Bundles on Complex Spaces. Ann. Mat. pura e app. 149, 153-164 (1987)

[Ar] D. Arapura: Local Cohomology of Sheaves Differential Forms and Hodge Theory. Preprint, Purdue University 1989

$[B\text{-}K]_1$ I. Bauer, S. Kosarew: On the Hodge spectral sequence for some classes of non-complete algebraic manifolds. Math. Ann. 284, 577-593 (1989)

$[B\text{-}K]_2$ I. Bauer, S. Kosarew: Kodaira vanishing theorems on non-complete algebraic manifolds. To appear in Math. Z.

[Bo] A. Borel et al.: Intersection cohomology. Progress in Math. vol. 50, Birkhäuser Verlag, Boston - Basel - Stuttgart 1984

$[D]_1$ P. Deligne: Théorèmes de Lefschetz et critères de dégénérescence de suites spectrales. Publ. Math. I.H.E.S. 35 (1968), 107-126

$[D]_2$ P. Deligne: Théorie de Hodge II, III. Publ. Math. I.H.E.S. 40 (1971), 5-57; 44 (1974), 5-77

[D-I] P. Deligne, L. Illusie: Relèvements modulo $p^2$ et décomposition du complexe de deRham. Invent. math. 89, 247-270 (1987)

[F] H. Flenner: Die Sätze von Bertini für lokale Ringe. Math. Ann. 229, 97-111 (1977)

[G-M] M. Goresky, R. MacPherson: Intersection Homology II. Invent. math. 71, 77-129 (1983)

[G-N-P-P] F. Guillén, V. Navarro Aznar, P. Pascual-Gainza, F. Puerta: Hyperrésolutions cubiques et descente cohomologique. Lect. Notes in Math. 1335, Springer Verlag, Berlin Heidelberg New York London Paris Tokyo 1988

$[G\text{-}R]_1$ H. Grauert, O. Riemenschneider: Verschwindungssätze für analytische Kohomologiegruppen auf komplexen Räumen. Invent. math. 11, 263-292 (1970)

[G-R]$_2$ H. Grauert, O. Riemenschneider: Kählersche Mannigfaltigkeiten mit hyper-q-konvexem Rand. Problems in Analysis, Symp. in Honor of S. Bochner, p.61-79, Princeton University Press 1970

[Hi] H. Hironaka: Smoothing of algebraic cycles of small dimensions. Amer. J. Math. 90, 1-54 (1968)

[K] Y. Kawamata: Characterization of abelian varieties. Compositio Math. 43, 253-276 (1981)

[M] Y. Manin: Correspondences, motifs and monoidal transformations. Math. USSR Sb. 6 (1968), 439-469 (engl. translation of Mat. Sb. 77 (1968))

[N] V. Navarro Aznar: Sur la théorie de Hodge des variétés algébriques à singularités isolées. Astér. 130, 272-307 (1985)

[Oh]$_1$ T. Ohsawa: A Reduction Theorem for Cohomology Groups of Very Strongly q-Convex Kähler Manifolds. Invent. math. 63, 335-354 (1981)

[Oh]$_2$ T. Ohsawa: Hodge Spectral Sequence on Compact Kähler Spaces. Publ. RIMS, Kyoto Univ. 23, 265-274 (1987)

[Oh]$_3$ T. Ohsawa: Hodge Spectral Sequence and Symmetry on Compact Kähler Spaces. Publ. RIMS, Kyoto Univ. 23, 613-625 (1987)

[Oh]$_4$ T. Ohsawa: Hodge Spectral Sequence on Pseudoconvex Domains II. To appear in : Proc. of Intern. Coll. Compl. Analysis, Bucureşti 1989

[Oh-T]$_1$ T. Ohsawa, K. Takegoshi: A Vanishing Theorem for $H^p(X,\Omega^q(B))$ on Weakly 1-Complete Manifolds. Publ. RIMS, Kyoto Univ. 17, 723-733 (1981)

[Oh-T]$_2$ T. Ohsawa, K. Takegoshi: Hodge spectral sequence on pseudoconvex domains. Math. Z. 197, 1-12 (1988)

[Ok] C. Okonek: Concavity, Convexity and Complements in Complex Spaces. Math. Gott. 27 (1985)

[Sch] M. Schneider: Lefschetzsätze und Hyperkonvexität. Invent. math. 31, 183-192 (1975)

[So] A.J. Sommese: Submanifolds of Abelian Varieties. Math. Ann. 233, 229-256 (1978)

[St] J. Steenbrink: Vanishing theorems on singular spaces. Astér. 130, 330-341 (1985)

SONDERFORSCHUNGSBEREICH 170
"GEOMETRIE UND ANALYSIS"
MATH. INST. DER UNIVERSITÄT
BUNSENSTR. 3/5
D-3400 GÖTTINGEN

FEDERAL REPUBLIC OF GERMANY

# $L^p$-COHOMOLOGY AND SATAKE COMPACTIFICATIONS

Steven Zucker*
Department of Mathematics
The Johns Hopkins University
Baltimore, MD 21218 USA

Let $D$ be a Riemannian symmetric space of non-compact type, and $\Gamma$ an arithmatically-defined subgroup of the automorphism group of $D$. Then $M = \Gamma\backslash D$ is, at worst, the quotient of a manifold by a finite group, and has finite volume. If $D$ is Hermitian, $D$ can be realized as a bounded symmetric complex domain with the Bergman metric, and $M$ is, in fact, a quasiprojective complex algebraic variety; furthermore, such $M$ can be defined over a number field (this is the theory of Shimura varieties).

It goes back more than 30 years to study the automorphic forms and $L^2$ harmonic forms on these spaces (see [19]). The latter are, for general reasons, (since $M$ is complete), isomorphic to the reduced $L^2$-cohomology (see our (3.5)), which in turn is the same as the [homological] $L^2$-cohomology under a closed-range hypothesis that is satisfied when $D$ is equal-rank, e.g. when $D$ is Hermitian (see (5.6)). The $L^2$ harmonic forms on an arithmetic quotient always make up a finite-dimensional space (see (5.5)).

It had become apparent that a topological interpretation of the $L^2$-cohomology would be useful in the study of Shimura varieties. Two cases were fairly "classical": if $M$ is compact, one can simply invoke the Hodge theorem; and if $D$ is the Poincaré upper half-plane, it is contained in a theorem of Eichler and Shimura (see [28: §12] for a rendition closer to the spirit of this article). After previous work of Borel, Garland-Hsiang, and the author (see [24]) on the mapping from $L^2$-cohomology to ordinary de Rham cohomology, we were led to the so-called Zucker Conjecture, which asserted the isomorphism of the $L^2$-cohomology of $M$ in the Hermitian case and the Goresky-MacPherson intersection cohomology of its Baily-Borel Satake compactification $M^*$ (see §6), which is a projective algebraic variety. After low-rank cases had been verified (see [3],[6],[9],[24],[25:II]), the conjecture was finally proved in full by Looijenga [17] and Saper-Stern [21], who made their discoveries known early in 1987.

There was always something unsettling about the way in which the conjecture

* Supported in part by the National Science Foundation through Grant DMS-8800355.

depended on the Hermitian structure, for the methods of [24] do not. This produced our point of view in [25:II] that the topological space $M^*$ is just a well-behaved Satake compactification of $M$ (see our §7). Borel then came up with an extension of our conjecture to the larger setting of equal-rank symmetric spaces (see (8.1)). Though the methods of [21] make use of the special properties of the root system in the Hermitian case, it seems likely that Saper and Stern can adapt their argument to prove Borel's version. In contrast, the proof of (6.1) given in [17] uses the complex, indeed the algebraic, structure of $M^*$ in an essential way; however, this has led to the exciting discovery by Rapoport that the weights that enter so crucially in the local $L^2$-cohomology (the $\alpha$'s in (6.5)) and the weights of Frobenius in local intersection cohomology (in the reduction modulo a prime) correspond, and the consequent simplification of Looijenga's argument.

A bizarre case of Borel's conjecture (see 8.2, d)) is consistent with the wild idea that these conjectures might extend to <u>every</u> Satake compactification of arithmetic quotients of <u>any</u> symmetric space. Because of the failure of closed-range, such a vast generalization must be stated for <u>reduced</u> $L^2$-cohomology, or better for $L^p$-cohomology with $p$ slightly bigger than 2, and this appears here as Conjecture (8.4). It should be remarked that a single cohomology group on $M$ is conjectured to admit a topological interpretation (as intersection cohomology) on all Satake compactifications of $M$, so it would imply that all of these latter spaces have isomorphic intersection cohomology, a startling statement.

The article is organized as follows:

CONTENTS

## §1. DIFFERENTIAL FORMS AND COHOMOLOGY.

Let $M$ be a $C^\infty$ manifold, $A^i(M)$ the space of $C^\infty$ i-forms on $M$ with $\mathbb{C}$-coefficients (one could work with $\mathbb{R}$-coefficients as well), and

(1.1) $$A^\bullet(M) = \bigoplus_i A^i(m).$$

The exterior derivative $d : A^i(M) \to A^{i+1}(M)$ satisfies $d^2 = 0$, making (1.1) into a cochain complex. The de Rham cohomology of $M$ (with $\mathbb{C}$-coefficients) is the cohomology of $A^\bullet(M)$:

(1.2) $$H^i_{dR}(M) = (\ker d) \cap A^i(M)/dA^{i-1}(M).$$

Note that $H^\bullet_{dR}(M) = \bigoplus_i H^i_{dR}(M)$ is a graded ring, with exterior product as multiplication.

(1.3) Theorem (de Rham). $H^\bullet_{dR}(M)$ is isomorphic, as a ring, to the topological cohomology $H^\bullet(M,\mathbb{C})$ with its cup-product.

(1.4) Remark. The de Rham theorem is actually more explicit. If $\phi \in A^i(M)$ and $d\phi = 0$, then it defines a functional on i-dimensional geometric cycles on $M$:

$$T_\phi(Z) = \int_Z \phi\,;$$

the value of this is unchanged if a boundary is added to $Z$, or if $\phi$ is changed by an exact form.

(1.5) Example. For a basic example, take $M$ to be contractible (e.g., a coordinate disc). Then (1.3) asserts

(1.5.1) $$H^i_{dR}(M) = \begin{cases} \mathbb{C} & \text{if } i = 0, \\ 0 & \text{if } i > 0. \end{cases}$$

The direct verification of this — i.e., that closed forms of positive degree are exact — is the Poincaré lemma. One actually deduces (1.3) for general $M$ from the Poincaré lemma by standard patching arguments (i.e., sheaf theory).

## §2. $L^2$-COHOMOLOGY.

Let $M$ now denote a Riemannian manifold. There is an associated notion of $L^2$ differential forms on $M$: for $\phi \in A^\bullet(M)$, the Riemannian structure gives a (pointwise) squared-length function $|\phi|^2$; integrate this against the Riemannian volume density to define the $L^2$ seminorm

$$(2.1)\qquad ||\phi||^2 = \int_M |\phi|^2 \, dV_M .$$

One says that $\phi$ is $L^2$ when $||\phi|| < \infty$.

We consider the following very elementary, yet important, example.

(2.2) Example. Take $M = \mathbb{R}^+$, the Euclidean half-line $(0,\infty)$, with coordinate $r$. The $L^2$ semi-norms for $A^i(\mathbb{R}^+)$ are simply

$$(2.2.1)\qquad \begin{aligned} i = 0:&\quad ||f||^2 = \int_0^\infty |f(r)|^2 dr, \\ i = 1:&\quad ||gdr||^2 = \int_0^\infty |g(r)|^2 dr. \end{aligned}$$

The de Rham theorem for $\mathbb{R}^+$ ((1.3) or (1.5)) can be expressed in a simple way: for

$$d: A^0(\mathbb{R}^+) \to A^1(\mathbb{R}^+), \qquad f \mapsto g = f',$$

$d$ is surjective, and $\ker d$ consists of the constant functions. If we now impose $L^2$ conditions, we get

$$d: L^2A^0(\mathbb{R}^+) \dashrightarrow L^2A^1(\mathbb{R}^+),$$

i.e., for $L^2$ forms, $d$ is only densely defined, $\ker d = 0$, and one can verify in several ways that the image (range) of $d$ is a proper, dense subspace, hence is of infinite (algebraic) codimension.

(2.3) Definition. i) On any Riemannian manifold $M$, the $L^2$ de Rham complex is the $L^2$ domain of $d$, and is denoted $A^\bullet_{(2)}(M)$.

ii) The $L^2$-cohomology of $M$ is the cohomology of $A^\bullet_{(2)}(M)$, viz.,

$$H^i_{(2)}(M) = (\ker d) \cap A^i_{(2)}(M)/dA^{i-1}_{(2)}(M).$$

It is worth emphasizing that the $L^2$ de Rham complex, hence also the $L^2$-cohomology, is independent of the Riemannian metric only when $M$ is compact. Note that the inclusion of $A^\bullet_{(2)}(M)$ in $A^\bullet(M)$ induces a linear mapping

$$(2.4)\qquad H^\bullet_{(2)}(M) \longrightarrow H^\bullet(M,\mathbb{C}),$$

whose image can be described as those de Rham cohomology classes that have $L^2$ representatives. In general, $H^\bullet_{(2)}(M)$ is not a ring, however, for the exterior product of two $L^2$ differential forms is only, a priori, $L^1$. (On the other hand, his accounts, in part, for the self-duality of $L^2$-cohomology, under reasonable ypotheses.)

2.5) Variant. i) Let $\mathbb{E}$ be a local system on the $C^\infty$ manifold M, i.e., the heaf of germs of horizontal sections of a flat complex vector bundle on M. Then $A^\bullet(M,\mathbb{E})$, the $C^\infty$ de Rham complex with coefficients in $\mathbb{E}$, is defined, and its co-omology is $H^\bullet(M,\mathbb{E})$.

ii) If M is Riemannian, and $\mathbb{E}$ is metrized (i.e., the vector undle is given some Hermitian metric, not recessarily flat), then one can define he $L^2$ de Rham complex $A^\bullet_{(2)}(M,\mathbb{E})$, and thus $L^2$-cohomology with coefficients in $\mathbb{E}$. s in (2.4), there is

$$H^\bullet_{(2)}(M,\mathbb{E}) \to H^\bullet(M,\mathbb{E}).$$

iii) Let w be a positive $C^\infty$ function on M, and denote by $\mathbb{C}_w$ he constant one-dimensional sheaf, metrized with

$$|1|^2 = w.$$

Write $\mathbb{C}$ for $\mathbb{C}_1$.) Put

$$H^\bullet_{(2)}(M,\mathbb{E};w) = H^\bullet_{(2)}(M,\mathbb{E}\otimes\mathbb{C}_w),$$

he $L^2$-cohomology of $\mathbb{E}$ with weight w (and use analogous notation for the $L^2$ e Rham complexes).

2.6) Example (2.2) cont'd. In the notation of $L^2$-cohomology, we can express 2.2) as

$$H^0_{(2)}(\mathbb{R}^+,\mathbb{C}) = 0,$$

2.6.1) $$H^1_{(2)}(\mathbb{R}^+,\mathbb{C}) \text{ is infinite-dimensional},$$

$$[H^i_{(2)}(\mathbb{R}^+,\mathbb{C}) = 0, \text{ for reasons of dimension, if } i > 1].$$

he unpleasantness about $H^1_{(2)}$ can be eliminated by imposing a weight. For future urposes, we take exponential weight functions

$$w(r) = e^{-kr} \qquad (k \neq 0),$$

or which one has instead of (2.2.1):

2.6.2) $$||f||^2 = \int_0^\infty |f(r)|^2 e^{-kr}dr,$$

etc. One gets (see [24:(2.40)]):

$$H^0_{(2)}(\mathbb{R}^+,\mathbb{C};e^{-kr}) = \begin{cases} \mathbb{C} & \text{if} \quad k > 0, \\ 0 & \text{if} \quad k < 0; \end{cases}$$

(2.6.3)

$$H^1_{(2)}(\mathbb{R}^+,\mathbb{C};e^{-kr}) = 0 \quad \text{for all} \quad k \neq 0.$$

## §3. $L^p$-COHOMOLOGY.

Why not? Let $\mathbb{E}$ be a metrized local system on the Riemannian manifold $M$. Changing the 2's in §2 to p's, one defines the $L^p$ seminorm on $\mathbb{E}$-valued differential forms

$$||\phi||^p_{(p)} = \int_M |\phi|^p dV_M,$$

the $L^p$ de Rham complex $A^\bullet_{(p)}(M,\mathbb{E})$, the $L^p$-cohomology $H^\bullet_{(p)}(M,\mathbb{E})$, and weighted $L^p$-cohomology.

The reason for singling out the case $p = 2$ is, of course, that the $L^2$ spaces are Hilbert spaces, hence are also self-dual. From this, one derives the Hodge theorem (see (3.6); see [24:§1] or [25:I] for a very general version), which expresses $L^2$-cohomology in terms of $L^2$ harmonic forms. For $p \neq 2$, the $L^p$ spaces are only Banach spaces; they come in dual pairs $(L^p,L^q)$, where

(3.1) $$p^{-1} + q^{-1} = 1.$$

One can hope for the corresponding duality of $H^\bullet_{(p)}(M,\mathbb{E})$ and $H^\bullet_{(q)}(M,\mathbb{E}^*)$.

(3.2) Example. We show that the $L^p$-cohomology of $\mathbb{R}^+$ has the same description as its $L^2$-cohomology, viz.

$$H^0_{(p)}(\mathbb{R}^+,\mathbb{C};e^{-kr}) = \begin{cases} \mathbb{C} & \text{if} \quad k > 0, \\ 0 & \text{if} \quad k \leq 0; \end{cases}$$

(3.2.1)

$$H^1_{(p)}(\mathbb{R}^+,\mathbb{C};e^{-kr}) \begin{cases} = 0 & \text{if} \quad k \neq 0, \\ \text{is infinite-dimensional if} & k = 0. \end{cases}$$

Indeed, we can adapt the discussion in [24:(2.40)]. The statement about $H^0_{(p)}$ is trivial. The vanishing of $H^1_{(p)}$ comes down to the elementary question: does an $L^p$ function (in the weighted sense) have an $L^p$ anti-derivative? If

$$g dr \in A^1_{(p)}(\mathbb{R}^+,\mathbb{C};w),$$

hen

$$f(r) = \int_0^r g(\rho)d\rho \quad \varepsilon\ A^0_{(p)}(\mathbb{R}^+,\mathbb{C};w)$$

f there is a positive function $\lambda$ such that

$$[w(\rho)\lambda(\rho)]^{-1} \int_\rho^\infty w(r)\left( \int_0^r \lambda(s)^{q-1}\, ds \right)^{p-1} dr$$

s a bounded function on $\mathbb{R}^+$. When $w(r) = e^{-kr}$ with $k > 0$, one can choose $\lambda(\rho) = {}^{\varepsilon\rho}$ with $0 < \varepsilon << 1$. There is an analogous criterion for getting $-\int_r^\infty g(\rho)d\rho$ to e in $A^0_{(p)}(\mathbb{R}^+,\mathbb{C};w)$, that handles the case $k < 0$. This leaves $k = 0$. The infi-ite-dimensionality of $H^1_{(p)}(\mathbb{R}^+,\mathbb{C})$ follows (again) from the open mapping theorem f functional analysis, for we can see that the range of $d$ is not closed. Here, losed range is equivalent to a uniform estimate for $f \in A^0_{(p)}$

$$||f||_{(p)} \leq \text{const.}||df||_{(p)} \quad ;$$

ut $f(r) = e^{-\alpha r}$ ($\alpha > 0$ arbitrary) gives $|f'(r)| = \alpha|f(r)|$.

3.3) <u>Remark</u>. At the expense of destroying the homological nature of $L^p$-coho-ology (which, for our purposes, is too high a cost), one can eliminate the effect f non-closed range by defining the <u>reduced</u> $L^p$-cohomology:

$$\overline{H}^i_{(p)}(M,\mathbb{E}) = (\ker d) \cap A^i_{(p)}(M,\mathbb{E})/\text{closure of } dA^{i-1}_{(p)}(M,\mathbb{E}),$$

hich is automatically a Banach space.

3.4) <u>Acknowledgment</u>. The upgrading in my mind of $L^p$-cohomology for $p \neq 2$ from he status of a curiosity is the result of input from W. Casselman and R. MacPherson. xtant treatment of the topic occurs in the work of V. Gol'dshtein, V. Kuz'minov nd I. Shvedov [11], R. Lockhart [16], and P. Pansu [20:Ch.3].

We return to the case of $p = 2$. Without further explanation, we state a form f the Hodge theorem for $L^2$-cohomology that is sufficient for many purposes.

3.5) <u>Theorem</u>. Let $M$ be a complete Riemannian manifold, and $\mathbb{E}$ a metrized ocal system on $\mathbb{E}$. Let $h^i_{(2)}(M,\mathbb{E})$ denote the space of $L^2$ $\mathbb{E}$-valued harmonic ifferential i-forms on $M$. Then

i) $h^i_{(2)}(M,\mathbb{E})$ consists of closed forms;

ii) the canonical mapping of $h^i_{(2)}(M,\mathbb{E})$ into the reduced $L^2$-cohomology group $\overline{H}^i_{(2)}(M,\mathbb{E})$ is an isomorphism.

(3.6) Corollary. Under the hypothesis of (3.5), suppose that $H^i_{(2)}(M,\mathbb{E})$ is finite-dimensional. Then the canonical mapping

$$h^i_{(2)}(M,\mathbb{E}) \longrightarrow H^i_{(2)}(M,\mathbb{E})$$

is an isomorphism.

(3.7) Remark. If $m = \dim M$, there is an isometry

$$* : L^2A^i(M,\mathbb{E}) \longrightarrow L^2A^{m-i}(M,\mathbb{E}^*),$$

which takes $h^i_{(2)}(M,\mathbb{E})$ onto $h^{m-i}_{(2)}(M,\mathbb{E}^*)$.

## §4. SYMMETRIC SPACES AND THEIR DISCRETE QUOTIENTS.

Let $G$ be a real semi-simple Lie group with finite center, and $K$ a maximal compact subgroup. Let $\underline{g}$ and $\underline{k}$ denote their Lie algebras. The homogeneous space

$$D = G/K \tag{4.1}$$

for $G$ is a Riemannian symmetric space of non-compact type, when $D$ is equipped with any G-invariant metric — we shall always understand $D$ to be so metrized — and all symmetric spaces of non-compact type are of the form (4.1). The tangent space to $D$ at the basepoint is canonically identified with $\underline{g}/\underline{k}$.

We distinguish two important classes of such symmetric spaces.

(4.2) Definition. i) One says that $D$ (or $G$) is Hermitian if there is a G-invariant complex structure on $D$.

ii) One says that $D$ (or $G$) is equal-rank if the $\mathbb{C}$-ranks of $G$ and $K$ are equal, i.e., if $G$ admits compact Cartan subgroups.

(4.3) Remarks. i) Hermitian symmetric spaces are automatically Kählerian. Whether a symmetric space is Hermitian is determined by the center of $K$; if $G$ is simple (and non-compact), it is Hermitian precisely when the center of $K$ is not discrete. (See [14:Ch.VIII,§4].)

ii) "Hermitian" implies "equal-rank". Moreover, any equal-rank $D$ is of even dimension, for then $\underline{g}/\underline{k}$ is spanned by non-compact root vectors, hich come in $\pm$ pairs. One can define thereby the "complex dimension" of any qual-rank D.

4.4) Examples. i) The following give Hermitian symmetric spaces:

a) $G = SU(n,1)$, $K = U(n)$, $D \simeq$ unit ball in $\mathbb{C}^n$,

b) $G = Sp(2r,\mathbb{R})$, $K = U(r)$, $D \simeq$ Siegel upper half-space $\mathcal{H}_r$ of enus r.

ii) For $G = SO(r,s)$, with $r \leq s$, one can take $K = S(O(r) \times O(s))$. ince the $\mathbb{C}$-rank of G is $[\![(r+s)/2]\!]$, and that of K is $[\![r/2]\!] + [\![s/2]\!]$, one sees hat G is equal-rank unless both r and s are odd. It is Hermitian only when $= 2$.

Let $\Gamma$ be a discrete subgroup of G that acts without fixed points on D. hen

4.5) $$M = \Gamma\backslash D$$

s a manifold, with $\pi_1(M) \simeq \Gamma$. It inherits a (complete) Riemannian metric from . Any finite-dimensional representation

$$\rho : G \to GL(E)$$

ive rise to a local system $\mathbb{E}$ on any such M. The corresponding flat vector undle also admits the structure of locally homogeneous vector bundle associated o $\rho|_K$, and thereby $\mathbb{E}$ gets a metrization from a so-called "admissible inner roduct" on E. (See [23:§2].)

By the latter, one also can identify

4.6) $$A^{\bullet}(M,\mathbb{E}) \simeq [A^0(\Gamma\backslash G) \otimes \Lambda^{\bullet}(\underline{g}/\underline{k})^* \otimes E]^K,$$

he invariants under K, acting on the three factors of the tensor product via in order) right-translation, the adjoint representation, and $\rho$. Likewise,

4.7) $$L^2A^{\bullet}(M,\mathbb{E}) \simeq [L^2A^0(\Gamma\backslash G) \otimes \Lambda^{\bullet}(\underline{g}/\underline{k})^* \otimes E]^K.$$

n these terms, the Laplacian is given, by a well-known formula of Kuga, as a imple expression in the Casimir element of the universal enveloping algebra of (see [19:p.385]); in fact, the harmonic forms become certain eigenfunctions of he Casimir.

## §5. ARITHMETIC QUOTIENTS.

An interesting class of large discrete groups $\Gamma$ is comprised of the following:

(5.1) Definition. A subgroup $\Gamma$ of $G$ is said to be arithmetically-definable (or arithmetic) if one can find a $\mathbb{Q}$-algebraic subgroup $\mathfrak{G}$ of some $GL(n)$ such that

i) $\mathfrak{G}(\mathbb{R}) \simeq G$,

ii) $\Gamma$ is within finite index of $G \cap GL(n,\mathbb{Z})$.

(5.2) Remark. One should note that a group $G$ produces the same symmetric space as its finite covering groups, so assuming that $G$ be linear is not serious.

By passing to a subgroup of finite index of any such $\Gamma$, one can eliminate all non-trivial elements of finite order (see [2:§17]), which have fixed points in $D$. For torsion-free arithmetic subgroups of $G$, we obtain a manifold $M$ as in (4.5), whose volume can be shown to be finite. Conversely, by a theorem of Margulis [18], unless $D$ has factors of rank one (in which case there are counterexamples), whenever $M$ has finite volume and is non-compact, $\Gamma$ is arithmetically-definable. We assume henceforth that $\Gamma$ is arithmetic.

(5.3) Example. This is a continuation of (4.4,i,b). Let $\ell$ be a positive integer. Set

$$\Gamma_r(\ell) = \ker\{Sp(2r,\mathbb{Z}) \to Sp(2r,\mathbb{Z}/\ell\mathbb{Z})\}. \tag{5.3.1}$$

This arithmetic group is the principal congruence subgroup of level $\ell$; it is torsion-free for $\ell \geq 3$. The quotient $M = \Gamma_r(\ell)\backslash\mathcal{H}_r$ is the moduli space of principally polarized r-dimensional Abelian varieties over $\mathbb{C}$, with level $\ell$ structure (i.e., with a basis for the points of order $\ell$ specified). As such, $M$ is important in algebraic geometry and number theory; it is, in fact, a quasi-projective complex algebraic variety.

It is actually the case that whenever $D$ is Hermitian, $M$ is a quasi-projective variety. This was proved in [1], where a certain space of automorphic forms for $\Gamma$ was used to construct an embedding of $M$ in projective space. The closure of $M$, which we denote by $M^*$, is the Baily-Borel Satake compactification of $M$. Its topological structure will be treated later, in §7.

From what we have said here and in §4, the spaces $h^i_{(2)}(M,\mathbb{E})$ are interesting spaces coming from the group theory. These were studied in [19] when M is _compact_, where one has

(5.4) $$h^i_{(2)}(M,\mathbb{E}) \simeq H^i_{(2)}(M,\mathbb{E}) = H^i(M,\mathbb{E}).$$

When M is non-compact, there is a partially corresponding assertion given by (3.5,ii). For applications in number theory, a topological interpretation of the $L^2$-cohomology was desired (see [15], for example, for a discussion after the fact), and this has recently been established, as we shall describe in the next section. First, we point out the following useful facts:

(5.5) _Theorem_ [7]. In $L^2A^\bullet(M,\mathbb{E})$, there are only finitely many independent eigenforms for the Laplacian with eigenvalues in any given bounded set. In particular, $h^i_{(2)}(M,\mathbb{E})$ is a finite-dimensional space for all i.

(5.6) _Theorem_ [5]. If M (i.e. D) is equal-rank, then $H^i_{(2)}(M,\mathbb{E})$ is finite-dimensional for all i, hence is isomorphic to $h^i_{(2)}(M,\mathbb{E})$.

§6. ENTER INTERSECTION COHOMOLOGY.

The methods and calculations that appeared in [24] produced the following:

(6.1) _Conjecture_ (1980). Let M be Hermitian. Then

$$H^\bullet_{(2)}(M,\mathbb{E}) \simeq IH^\bullet_{\bar{m}}(M^*,\mathbb{E}).$$

Here, $IH_{\bar{m}}$ refers to intersection (co)homology of (lower-)middle perversity, the relatively new topological invariant of Goresky and MacPherson for stratified spaces [12] that takes into account the complexity of their singularities. We will not reproduce the complete definition here (for which, see [4] or [12]), but instead we state some of its relevant properties. Very fundamental, it is of a local-to-global nature (cf.(1.5)), i.e. determined from a complex of sheaves. It is also a locally finite-dimensional theory, and reduces to ordinary cohomology on manifolds. Though it is not a cohomology theory in the usual sense – indeed, it is not a homotopy invariant, nor is it even functorial for mappings – it does satisfy Poincaré duality on spaces whose strata are of even (real) codimension, e.g. complex analytic varieties. With the above said, one can observe that the Poincaré duality property of $L^2$ harmonic forms (3.7) and the Hermitian case of (5.6) are necessary conditions for (6.1) to hold.

In 1987, Conjecture (6.1) was given two essentially different proofs [17],[21].

An account of these appears in [25:III]. Both are based on the local characterization of intersection cohomology (see [12:II(3.3),(4.1),(6.1)] or [4:V,§4]); we state explicitly the most important (and most difficult) point to be verified:

(6.2) Vanishing property. For all $j > 0$, every point of the complex codimension $j$ stratum of $M^*$ has a fundamental system of neighborhoods $U^*$, such that for $U = U^* \cap M$,

$$H^i_{(2)}(U,\mathbb{E}) = 0 \quad \text{if} \quad i \geq j.$$

In general, the neighborhood structure of a stratified space is not so complicated; one can take $U^*$ to be topologically

$$U^* \simeq V \times \mathrm{cone}(L), \tag{6.3}$$

where $V$ is a disc in the stratum, the vertex of the cone lies on the stratum, and $L$ – the link of the stratum – is itself a stratified space, having one stratum fewer than $U^*$. We then have, for (6.2), a diffeomorphism

$$U \simeq V \times \mathbb{R}^+ \times L^{reg}. \tag{6.4}$$

The most direct approach to (6.2) is to compute the $L^2$-cohomology of $U$ by determining $L^{reg}$, and writing down the metrics of $M$ and $\mathbb{E}$ in terms of (6.4), at least asymptotically. This had been carried out, yielding a formula [25:II(3.19)] (see also [25:III,§2]) of the form:

$$H^{\bullet}_{(2)}(U,\mathbb{E}) \simeq \bigoplus_{\alpha} [H^{\bullet}_{(2)}(\mathbb{R}^+,\mathbb{C};w_1^{\alpha}) \otimes_{\mathbb{C}} H^{\bullet}_{(2)}(Y,\mathbb{H}^{\bullet}_{\alpha};w_2^{\alpha})]. \tag{6.5}$$

Here, $\alpha$ represents a Lie-theoretic weight, which can be construed as an integer; $w_1^{\alpha}$ is an exponential weight as in (2.6); $Y$ is an arithmetic quotient of some symmetric space (usually not Hermitian, and it may have Euclidean factors); $\mathbb{H}^{\bullet}_{\alpha}$ is a suitable local system on $Y$ of the sort we have been considering since §4; and $w_2^{\alpha}$ is a weight function derived from $w_1^{\alpha}$. Behind this is the fact that $L^{reg}$ in (6.4) is a nilmanifold fibration over $Y$ and $V$ can always be ignored. The problem then, resolved in one way or another in [17] and [21], is to see that the exponential weights distribute themselves in such a way that, when (2.6.3) is combined with vanishing properties of the cohomology on $Y$, the desired vanishing (6.2) results. This is very complicated business, so it did not surprise me that it took seven years for Conjecture (6.1) to be settled. We present some very easy examples, which go back to [24], in §9.

## 7. SATAKE COMPACTIFICATIONS.

Since it is really only the underlying topological space of $M^*$ that enters (6.1), we were led to view $M^*$ as one member of a certain class of spaces, namely the Satake compactifications of $M$. These were first defined in [22]. We recall their construction, following the treatment in [26].

Let $\Delta$ be a set of simple $\mathbb{Q}$-roots for the algebraic group $\mathfrak{G}$ of (5.1), relative to a maximal $\mathbb{Q}$-split (multiplicative) torus. In the case that $\mathfrak{G}$ gives the standard rational structure for $G$, $\Delta$ will also be a set of simple roots for the $\mathbb{R}$- (or so-called restricted) root system of $G$. Inside $D$ is the corresponding Euclidean hyperquadrant $A$, with a canonical identification

(7.1) $$A \underset{-}{\sim} (\mathbb{R}^+)^\Delta = (0,\infty)^\Delta .$$

One makes the corner

(7.2) $$\overline{A} = (0,\infty]^\Delta , \text{ then}$$

(7.3) $$D(\Delta) = D \overset{A}{\times} \overline{A} ,$$

which is used in [8] to generate the manifold-with-corners $\overline{D}$,

which contains $D$ as a dense open set. Any discrete group $\Gamma$ that is arithmetic with respect to $\mathfrak{G}$ acts nicely on $\overline{D}$, and then

(7.4) $$\overline{M} = \Gamma\backslash\overline{D}$$

is a compactification of $M$.

We write $\partial\overline{M}$ for the complement ("boundary") of $M$ in $\overline{M}$, and do similarly for other compactifications later. It is clear from (7.3) that $\partial\overline{M}$ decomposes according to the $A$-orbits in $\overline{A}$, where $A$ is taken to act componentwise by extended multiplication; these are the faces of $\overline{A}$:

(7.5) $$\overline{A}'_T = \{a \in \overline{A} : a^\beta = \infty \text{ if and only if } \beta \notin T\}.$$

This gives rise to the decomposition into faces:

(7.6) $$\partial\overline{M} = \coprod_{T \subsetneq \Delta} (\partial\overline{M})_T .$$

The pieces in (7.6) can be described as arithmetic quotients associated to the $\mathbb{Q}$-parabolic subgroups of $G$; for this, we refer the reader to [8:§9] or [25:II,§1].

According to [26:(3.10)], the Satake compactifications of $M$ are certain quotient spaces of $\overline{M}$, formed by making identifications in $\partial\overline{M}$. (In the case of non-standard Q-structures, we have to satisfy two conceivably non-vacuous assumptions, stated in [26:(3.3)], but we will not concern ourselves with that here.) Let $\mathcal{S}$ denote the collection of non-empty subsets of $\Delta$. Thus, $\mathcal{S}$ has $2^{|\Delta|}-1$ elements; $|\Delta|$ is called, by the way, the Q-rank of $\mathfrak{G}$. For each $S \in \mathcal{S}$, we give an A-equivariant quotient $A_S^*$ of $\overline{A}$, such that there is a projection

(7.7) $$A_S^* \to A_T^* \quad \text{whenever} \quad S \supset T,$$

and starting from

$$D \overset{A}{\times} A_S^*$$

one produces the Satake compactification $M_S^*$. To effect this, we recall that the Dynkin diagram of $\Delta$ makes the latter into a graph, so there is a notion of <u>connected</u> subsets.

(7.8) <u>Definition</u>. Fix $S \in \mathcal{S}$.

i) A subset $\Theta$ of $\Delta$ is said to be <u>S-connected</u> if all of its connected components meet $S$.

ii) Given any $T \subset \Delta$, its <u>S-connected component</u> $\kappa(T)$ is the largest S-connected subset of $T$; it is equal to the union of all components of $T$ that meet $S$.

(7.9) <u>Definition</u>. $A_S^* = \coprod_{\Theta\ S\text{-conn.}} A'_\Theta$, with the quotient topology determined from $\overline{A}$ under the projections $A'_T \to A'_{\kappa(T)}$.

We see at once that (7.7) holds. Referring to [26:§3] for details (there are some errata, however; see our appendix), one gets

(7.10) $$M_S^* = \coprod_{\Theta\ S\text{-conn.}} M'_\Theta,$$

topologized as a quotient of $\overline{M}$, where $M'_\Theta$ is a finite union of arithmetic quotients of the symmetric subspace of $D$ whose Q-root system is determined by $\Theta$. From the construction, one sees (cf. (7.7)):

(7.11) <u>Proposition</u> [26:(3.9)]. Whenever $S \supset T$, the identity mapping of $M$ extends to a continuous surjection $M_S^* \to M_T^*$.

(7.12) Remark. Note that for $S = \Delta$, we have $A_\Delta^* = \overline{A}$. However, $M_\Delta^*$ is a proper quotient of $\overline{M}$; it is the maximal Satake compactification of M, mapping onto all other ones.

In the Hermitian case, the compactification $M^*$ of Baily-Borel is one of the minimal Satake compactifications. Specifically, – we may assume that $\Delta$ is connected – it is known that in the Hermitian case, the classification type of the $\mathbb{Q}$-root system is either $BC_r$ or $C_r$ $(r = |\Delta|)$. The Dynkin diagram is therefore linear, with a distinguished end root $\beta_r$ that is respectively shorter or longer than the other simple roots. Taking $S = \{\beta_r\}$ gives $M^* = M_S^*$.

The Satake compactifications $M_S^*$ are all stratified spaces, with (7.10) providing the decomposition into strata. By a mild abuse of language, we will refer to the $M'_\Theta$ in (7.10) also as the boundary components of M in $M_S^*$. The description of the latter as a quotient of $\overline{M}$ is useful for calculating the links.

## §8. FURTHER CONJECTURES.

Conjectures that identify $L^2$-cohomology and intersection cohomology for algebraic varieties have a ten-year history (see [10]). However, it should be apparent by now that we find it unnatural that our conjecture (6.1) be stated in the context of arithmetic quotients only for the case where M is an algebraic variety (i.e., the Hermitian case).

With the results of [5], and in particular Theorem (5.6), in hand, Borel proposed the following extension of (6.1) (see also [27:(2.5)]).

(8.1) Conjecture (1983). Let M be an arithmetic quotient of an equal-rank symmetric space, and let $M_S^*$ be a Satake compactification of M such that all boundary components are equal rank. Then

(8.1.1) $$H^\bullet_{(2)}(M,\mathbb{E}) \simeq IH^\bullet_{\overline{m}}(M_S^*,\mathbb{E}).$$

If one restricts to the case of standard $\mathbb{Q}$-structures, then it is quite easy to list all Satake compactifications that satisfy the hypotheses of (8.1) (see [25:II(A.2)]). For that, one first examines the classification of simple real Lie algebras that give rise to irreducible symmetric spaces of non-compact type (taken from [14:p.518]): AI($G = SL(r+1,\mathbb{R})$), AII($G = SU^*(2r)$), AIII($G = SU(r,s)$, $r \le s$), BDI($G = SO(r,s)$, $r \le s$; B if $r+s$ odd, D otherwise), DIII($G = SO^*(2r)$), CI($G = Sp(2r,\mathbb{R})$), CII($G = Sp(r,s)$, $r \le s$), EI-IX, FI-II, $G_2$. In the preceding, the capital letter refers to the classification type of the complexification; thus, there are seven series of classical type and twelve exceptional cases. Of these, all but AI(r > 1), AII, DI(r and s odd), EI, and EIV are equal-rank. (The Hermitian cases are AI(r = 1), AIII, BDI(r = 2), DIII, CI, EIII, EVII.) However, it is very easy to find non-equal-rank boundary components in most Satake compactifications (recall (7.10)), and we are left with the following very short list:

(8.2) Proposition. Let M be a standard arithmetic quotient of an equal-rank symmetric space, and $M^*_S$ a Satake compactification for which all boundary components are also equal-rank. Then the Dynkin diagram of $\Delta$ is linear, and $M^*_S$ is one of the following:

a) the Baily-Borel compactification in an Hermitian case,

b) the unique Satake compactification in case FII (where r = 1),

c) the analogue of the Baily-Borel compactification for BI(r > 2) and CII,

d) any Satake compactification in cases BI(r = 2), CI(r = 2), and $G_2$.

It is (8.2,d) that is very striking, and we focus a little on that. Since the left-hand side of (8.1.1) is independent of S, we are asserting that all three Satake compactifications have isomorphic intersection cohomology. Recall that by (7.11), we have a diagram

(8.3)
$$\begin{array}{ccc} & M^*_\Delta & \\ {}^{\pi_1}\swarrow & & \searrow{}^{\pi_2} \\ M^*_{\{\beta_1\}} & & M^*_{\{\beta_2\}} \end{array}$$

In [12:II,(6.2)], a criterion (homological smallness) is given for a mapping to induce an isomorphism on intersection cohomology. However, neither $\pi_1$ nor $\pi_2$ is homologically small, so the isomorphism holds for more mysterious global reasons.

We suspect now that this is part of a very general phenomenon. With some coaxing from MacPherson, we have made the following:

(8.4) Conjecture (1989). Let $M^*_S$. be any Satake compactification of any arithmetic quotient M. Then

$$\bar{H}^\bullet_{(2)}(M,\mathbb{E}) \simeq H_{(2+\varepsilon)}(M,\mathbb{E}) \simeq IH^\bullet_{\bar{m}}(M^*_S,\mathbb{E})$$

for all sufficiently small $\varepsilon > 0$.**

**See note at end of text.

(8.5) Remarks. i) This would imply, for $S = \Delta$, the main result in [13]. Indeed, it is our vague hope that the assertion for general $S$ should be deduced from the maximal one by considering direct images.

ii) One will have to impose self-duality assumptions on $\mathbb{E}$ (non-vacuous when $M$ is not equal-rank) in (8.4).

The idea behind the first isomorphism in (8.4) is that it should be possible to prove, without much difficulty, the $L^p$ analogue of (6.5), viz.

(8.6) $$H^{\bullet}_{(p)}(U,\mathbb{E}) \simeq \bigoplus_{\alpha} H^{\bullet}_{(p)}(\mathbb{R}^+,\mathbb{C};w_1^{\alpha}(p)) \otimes_{\mathbb{C}} H^{\bullet}_{(p)}(Y,\mathbb{H}^{\bullet}_{\alpha};w_2^{\alpha}(p)).$$

From the mechanism by which the weights arise, we have

(8.7) $$w_1^{\alpha}(p) = w_1^{\alpha}(2)e^{(p-2)\alpha r} ;$$

hence, the $L^p$ weights are a perturbation of the $L^2$ weights, such that when $\alpha$ is giving the trivial $L^2$ weight ($k = 0$ in (2.6), which implies $\alpha > 0$), $w_1^{\alpha}(p)$ has $k < 0$ when $p > 2$.

## §9. EXAMPLES WITH ISOLATED SINGULARITIES.

It is always enlightening to consider the simplest cases of (6.1) and (8.1) – or of most mathematics, for that matter – to gain some feeling for what is involved. Most of the difficulties disappear when the space $Y$ in (6.5) is smooth, i.e., for the highest-dimensional singular stratum of $M^*$, and everything is simplest in the cases where $M^*$ is a space with only isolated singularities. We present here three examples of the latter, giving the description of deleted neighborhoods of a singular point as Riemannian manifolds. In the following, we write $T^k$ for $(S^1)^k$, the k-dimensional compact torus; and we use the notation $L = F \urcorner B$ to mean 'L is a (certain) fiber bundle with base $B$ and fiber $F$, such that the universal covering space $\tilde{L}$ is the product $\tilde{F} \times \tilde{B}$'.

(9.1) Examples. a) Take $G = SU(n,1)$, as in (4.4,i,a), with the standard $\mathbb{Q}$-structure. Then

$$U = \mathbb{R}^+ \times (T^1 \urcorner T^{2(n-1)}),$$
$$ds^2 = dr^2 + e^{-4r}\,du^2 + e^{-2r}\,dv^2.$$

b) The Hilbert modular varieties come from $G = SL(2,\mathbb{R})^n$, but with a non-standard rational structure of $\mathbb{Q}$-rank one. Here,

$$U = \mathbb{R}^+ \times (T^n \rtimes (T^{n-1}),$$

$$ds^2 = dr^2 + e^{-2r}\, du^2 + dx^2$$

c) Take $G = SO(m,1)$, a case of (4.4,ii) — so that $D$ is m-dimensional real hyperbolic space — with standard $\mathbb{Q}$-structure. Then

$$U = \mathbb{R}^+ \times T^{m-1},$$

$$ds^2 = dr^2 + e^{-2r}\, du^2.$$

The above formulas display that the metric on $U$ (or more precisely, its universal cover) can be described as a (multiply-)warped product metric.

In (a), (b), and the equal-rank case of (c) $m = 2n$, the space $M$ is of complex dimension n, and (6.2) becomes

(9.2) $$H^i_{(2)}(U,\mathbb{E}) = 0 \quad \text{if} \quad i \geq n.$$

The (hitherto unmentioned) rest of the description of intersection cohomology,

(9.3) $$IH^i_{\bar{m}}(U,\mathbb{E}) \cong IH^i_{\bar{m}}(L,\mathbb{E}) \quad \text{if} \quad i < j$$

requires here

(9.4) $$H^i_{(2)}(U,\mathbb{E}) \cong H^i(L,\mathbb{E}) \quad \text{if} \quad i < n.$$

We can give some intuition as to why (9.2) and (9.4) hold, and this is clearest in the case $\mathbb{E} = \mathbb{C}$.

The right-hand factor in the expression for $U$ is the link $L$ of the singular points (cf.(6.4)). We get for the volume density

(9.5) $$e^{-\mu r}\, dr dV_L,$$

where in the three examples, respectively:

(9.6)
a) $\mu = 2n$
b) $\mu = n$
c) $\mu = m-1$

Now, let $\phi$ be an i-form on $L$, and consider its pullback $\tilde{\phi}$ to $\mathbb{R}^+ \times L$. Suppose that $\phi$ is homogeneous with respect to the factors of $U$ as indicated in (9.1). We see that

(9.7) $$|\tilde{\phi}|^2 = e^{2\alpha r}|\phi|^2,$$

where $\alpha$ is a non-negative integer satisfying in each case:

(9.8)
a) $\alpha = i$ or $\alpha = i + 1$,
b) $\max\{0, i-n+1\} \leq \alpha \leq \min\{i,n\}$,
c) $\alpha = i$.

By combining (9.5) and (9.7), we get:

(9.9) $$||\tilde{\phi}||^2 = ||\phi||^2 \int_0^\infty e^{-kr}\, dr,$$

with $k = \mu - 2\alpha$.

If we admit that $H^\bullet(L)$ will be spanned by homogeneous elements, we can decompose:

(9.10) $$H^\bullet(L,\mathbb{C}) \simeq \bigoplus_k H^\bullet(L)_k.$$

Then (9.9) suggests that

(9.11) $$H^\bullet_{(2)}(L,\mathbb{C}) \simeq \bigoplus_{k>0} H^\bullet(L)_k.$$

This is in fact correct, and a proof of it follows from the assertion that the considerations leading to (9.9) describe an embedding:

(9.12) $$\bigoplus_k [A^\bullet_{(2)}(\mathbb{R}^+,\mathbb{C};e^{-kr}) \otimes_{\mathbb{C}} H^\bullet(L)_k] \hookrightarrow A^\bullet_{(2)}(U,\mathbb{C})$$

(choose cohomology representatives on $L$), and one proves:

(9.13) Proposition. The inclusion in (9.12) induces an isomorphism on cohomology:

$$H^\bullet_{(2)}(U,\mathbb{C}) \xleftarrow{\sim} \bigoplus_k [H^\bullet_{(2)}(\mathbb{R}^+,\mathbb{C};e^{-kr}) \otimes_{\mathbb{C}} H^\bullet(L)_k].$$

If we now recall (2.6.3), we see that (9.11) follows, provided we know that $H^\bullet(L)_0 = 0$, for the unweighted $L^2$-cohomology group $H^1_{(2)}(\mathbb{R}^+,\mathbb{C})$ is huge (2.6.1).

We see now that (9.2) and (9.4) become assertions purely about the weights in $H^{\bullet}(L,\mathbb{C})$; — for the general local system $\mathbb{E}$, the analysis is similar, by the way — it must be shown that:

(9.14) <u>Proposition.</u> i) $H^{\bullet}(L)_0 = 0$,

ii) $H^i(L)_k = 0$ for $i \geq n$, $k > 0$,

[ii*) $H^i(L)_k = 0$ for $i < n$, $k < 0$].

If we now look at (9.6) and (9.8) again, we can see that (9.14) is obvious only in case (c). In that case, one can note more: if $m = 2n+1$ (the non-equal-rank case), $H^n(L,\mathbb{C})$ has the trivial weight, from which it follows that $H^{n+1}_{(2)}(U,\mathbb{C})$ is infinite-dimensional, and $H^{n+1}_{(2)}(M,\mathbb{C})$ must be likewise.

For the other two cases, (9.8) offers a choice of weights in most degrees. There, one must invoke the structure of the fiber bundles, in particular their non-triviality, to rule out the damaging possibilities. In fact, one shows (see [24:§6]):

(9.15) <u>Proposition.</u> The actual weight for $H^i(L,\mathbb{C})$ is:

a) $\alpha = i$ if $i < n$, and $\alpha = i + 1$ if $i \geq n$.

b) $\alpha = 0$ if $i < n$, and $\alpha = n$ if $i \geq n$.

We close with some comments about the intersection cohomology for spaces of complex dimension $n$ with isolated singularities, such as the preceding. Let

$$\iota : M \to M^*$$

denote the inclusion: Then

$$\text{(9.16)} \qquad IH^{\bullet}_{\bar{m}}(M^*,\mathbb{E}) = H^{\bullet}(M^*,\tau_{<n}R\iota_*\mathbb{E})$$

(refer to [12:II,§1] for notation). This gives, in the case of constant coefficients, the well-known formula:

$$\text{(9.17)} \qquad IH^i_{\bar{m}}(M^*,\mathbb{C}) \simeq \begin{cases} H^i(M) & \text{if } i < n, \\ H^i(M^*) & \text{if } i > n, \\ \operatorname{im}\{H^n(M^*) \to H^n(M)\} & \text{if } i = n. \end{cases}$$

It is only in the situation with isolated singularities that it is possible to express intersection cohomology wholly in terms of ordinary cohomology. In general,

(9.18) $$IH^i_{\mathfrak{m}}(M^*,\mathbb{E}) \xrightarrow{\sim} H^i(M,\mathbb{E}) \quad \text{if} \quad i < \operatorname{codim}_{\mathbb{C}}(\partial M^*),$$

and there is a fundamental sheaf-theoretic description of intersection cohomology [12:II,(3.1)], due to Deligne, based on an iteration of the construction that appears in the right-hand side of (9.16) (cf.(9.3)).

Appendix. ERRATA TO [26].

There is an annoying "typographical" error on page 333: equation (3.6(3)) should read

$$X_1^*({}_QQ_T) = \coprod_{\substack{\Xi\ \tau\text{-connected} \\ T \subseteq \omega(\Xi)}} e({}_QQ_{\Xi \cup T}).$$

From here, we see that $p_1^*(P)$ in (3.6(4)) maps onto $e({}_QQ_{\Xi \cup T})$. But since we want to identify $X_1^*({}_QQ_T)$ with its open image in $X_1^*({}_QQ_{\kappa(T)})$, via the projections

$$e({}_QQ_{\Xi \cup T}) \to e({}_QQ_{\Xi \cup \kappa(T)}) = e({}_QQ_{\Xi})$$

(as $T \subseteq \omega(\Xi)$ implies $\kappa(T) \subseteq \Xi$), we may as well do it now. With this convention, lines 6-7 on p.334 become correct, whereas lines 8-9 are still false. Similarly, line 1 on p.336 should then assert:

"... are identified homeomorphically along $X_1^*({}_QQ_T)$."

Fortunately, the rest of the paper is not affected.

Thanks go to L. Saper for pointing out the error.

Note added. Regrettably, there are counterexamples to (8.4), even when $\mathbb{E} = \mathbb{C}$. This was first explicitly noted by W. Casselman. Nonetheless, it remains an interesting problem to determine the natural generalization of (8.1).

## REFERENCES


[1] Baily, W., Borel, A.: Compactification of arithmetic quotients of bounded symmetric domains. Ann. of Math. 84(1966), 442-528.

[2] Borel, A.: Introduction aux Groupes Arithmétiques. Paris: Hermann, 1969.

[3] Borel, A.: $L^2$-cohomology and intersection cohomology of certain arithmetic varieties. In: Emmy Noether in Bryn Mawr. Springer-Verlag, 1983, 119-131.

[4] Borel, A., et al.: Intersection Cohomology. Boston: Birkhäuser, 1984.

[5] Borel, A., Casselman, W.: $L^2$-cohomology of locally symmetric manifolds of finite volume. Duke Math. J. 50(1983), 625-647.

[6] Borel, A., Casselman, W.: Cohomologie d'intersection et $L^2$-cohomologie de variétés arithmétiques de rang rationnel 2. C.R. Acad. Sci. Paris 301(1985), 369-373.

[7] Borel, A., Garland, H.: Laplacian and discrete spectrum of an arithmetic group. Amer. J. Math. 105(1983), 309-355.

[8] Borel, A., Serre, J.-P.: Corners and arithmetic groups. Comm. Math. Helv. 48(1973), 436-491.

[9] Casselman, W.: Introduction to the $L^2$-cohomology of arithmetic quotients of bounded symmetric domains. Adv. Stud. Pure Math. 8(1986), 69-93.

[10] Cheeger, J., Goresky, M., MacPherson, R.: $L^2$-cohomology and intersection homology of singular varieties. In: Seminar on Differential Geometry, Annals of Math. Studies 102(1982), 303-340.

[11] Gol'dshtein, V., Kuz'minov, V., Shvedov, I.: $L_p$-cohomology of non-compact Riemannian manifolds. Preprint.

[12] Goresky, M., MacPherson, R.: Intersection homology theory. I, Topology 19 (1980), 135-162; II, Inv. Math. 72(1983), 77-129.

[13] Goresky, M., MacPherson, R.: Weighted cohomology of Satake compactifications. Preprint, 1988.

[14] Helgason, S.: Differential Geometry, Lie Groups, and Symmetric Spaces. Academic Press, New York-San Francisco-London, 1978.

[15] Langlands, R.: Representation theory and arithmetic. AMS Proc. Symp. Pure Math. 48(1988), 25-33.

[16] Lockhart, R.: Fredholm, Hodge and Liouville Theorems on noncompact manifolds. Transactions AMS 301(1987), 1-35.

[17] Looijenga, E.: $L^2$-cohomology of locally symmetric varieties. Comp. Math. 67 (1988), 3-20.

[18] Margulis, G.: Arithmetic properties of discrete subgroups. Russ. Math. Surveys 29(1974), 107-156 (translation of Uspekhi Mat. Nauk. 29(1974), 49-98).

[19] Matsushima, Y., Murakami, S.: On vector bundle valued harmonic forms and automorphic forms on symmetric Riemannian manifolds. Ann. of Math. 78(1963), 365-416.

[20] Pansu, P.: Thèse.

[21] Saper, L., Stern, M.: $L_2$-cohomology of arithmetic varieties. Proc. Nat'l Acad. Sci. USA 84(1987), 5516-5519; Ann. of Math. (to appear).

[22] Satake, I.: On compactifications of the quotient spaces for arithmetically defined discontinuous groups. Ann. of Math. 72(1960), 555-580.

23] Zucker, S.: Locally homogeneous variations of Hodge structure. L'Ens. Math. 27(1981), 243-276.

24] Zucker, S.: $L^2$-cohomology of warped products and arithmetic groups. Inv. Math. 70(1982), 169-218.

25] Zucker, S.: $L^2$-cohomology and intersection homology of locally symmetric varieties. I, AMS Proc. Symp. Pure Math. 40, Part 2 (1983), 675-680; II, Comp. Math. 59(1986), 339-398; III, Proc. Hodge Theory (Luminy, 1987), Astérisque.

26] Zucker, S.: Satake compactifications. Comm. Math. Helv. 58(1983), 312-343.

27] Zucker, S.: $L^2$-cohomology of Shimura varieties. In: Automorphic Forms, Shimura Varieties, and L-functions. Academic Press, 1989, 377-391.

28] Zucker, S.: Hodge theory with degenerating coefficients: $L_2$ cohomology in the Poincaré metric. Ann. of Math. 109(1979), 415-476.

Though this article is mostly expository, it does contain some new material at will not be published elsewhere.

# HARMONIC MAPS AND KÄHLER GEOMETRY

Jürgen Jost*
Math. Inst. RUB,
Postf. 102148,
D-4630 Bochum, West Germany;
Department of Mathematics,
Harvard University,
Cambridge, MA 02138

Shing-Tung Yau
Department of Mathematics,
Harvard University,
Cambridge, MA 02138

## 1 Introduction

It is the purpose of this article to describe the role of harmonic maps in Kähler geometry, not by aiming at a complete survey, but rather by describing some central applications and crucial ideas, often not in their utmost generality. We shall solely treat applications to nonpositively curved Kähler manifolds as here the theory of harmonic maps shows itself to be most successful.

For the reader who wants to get a wider perspective of the analytic background or the geometric setting, we recommend [J1] resp. [J2]. We shall not describe the theory of harmonic maps from or between Riemann surfaces nor its applications in Kähler geometry in detail; some aspects can be found in [J3]. For further information, the survey article [EL] and its bibliography should be useful.

We now summarize the contents of the following paragraphs: In §2, we define harmonic and pluriharmonic maps, the latter being harmonic when restricted to any complex submanifold of their Kähler manifold of definition.

*The first author wants to express his gratitude to J. Noguchi and T. Ohsawa for organizing a very stimulating conference, and to Mr. Taniguchi for his generous support of mathematics. The authors' research described in this article was supported by DFG and Stiftung Volkswagenwerk (first author) and NSF DMS 8711394 (second author).

We also observe that the pullback of a convex function under a harmonic or pluriharmonic map is subharmonic resp. plurisubharmonic.

In §3, existence and uniqueness results for harmonic maps into nonpositively curved Riemannian manifolds are given.

In §4, special properties of harmonic maps from a Kähler manifold into another one of so-called strongly (semi) negative curvature are described. These properties force a harmonic map to be pluriharmonic and sometimes even to be $\pm$ holomorphic. This is applied in §5 where various rigidity results are given. The principle is that one starts with some smooth map, say a homotopy equivalence to be concrete, deforms it into a harmonic map by results from §3 and shows the harmonic map to be $\pm$ holomorphic by the results of §4. Such constructions then yield the rigidity of complex structures. The method can also be viewed as a way to prove the existence of holomorphic maps in given homotopy classes, provided suitable geometric conditions are satisfied.

In §6, we describe some further applications of pluriharmonic maps. We give a criterion that a subvariety of a Kähler manifold cannot be blow-down without creating a singularity. We also construct plurisubharmonic exhaustion functions on certain holomorphic fiber bundles. The result is an application of an existence theorem for a suitable harmonic map given in §3, the pluriharmonicity of this map from §4, and the composition property described in §2.

In §7, we discuss algebraic varieties over function fields. Fibrations by algebraic varieties give rise to harmonic maps into Riemannian symmetric spaces of noncompact type via the period mapping of Griffiths. In the case of curves, one obtains holomorphic maps into the Mumford-Deligne compactification of moduli space or the universal modular curve (neglecting the problem of singularities which turns out to be easily avoided by passing to finite covers). It was first suggested by Grauert- Reckziegel [GR] to use the geometry of these spaces to study fibrations by curves. Noguchi [N] carried this out by using the hyperbolicity of moduli space in the sense of Kobayashi. Noguchi's results have recently been complemented by Imayoshi- Shiga [IS], and new proofs of the theorems of Manin and Arakelov ("Mordell and Shafarevitch problems over function fields") follow. We describe here the approach of [JY6] to these problems. It makes use of the Weil-Petersson geometry of moduli space. It not only improves the results of Manin, Grauert, and Arakelov but also yields various generalizations and complements.

# 2 Definition and Elementary Properties of Harmonic and Pluriharmonic Maps

The concept of harmonic maps is one of Riemann geometry; consequently, for the definition of a harmonic map, we do not need a complex structure.

Thus, let $M$ and $N$ be Riemannian manifolds, $f : M \to N$ a smooth map. The differential $df$ then defines a section of $T^*M \otimes f^*(TN)$. Denoting by $\nabla$ the covariant derivative in this bundle, induced from the Levi-Civita connections of $M$ and $N$, we define $f$ to be *harmonic* if

$$\tau(f) := trace \nabla df = 0 \tag{2.1}$$

where the trace is taken in $T^*M$. $\tau(f)$ is called the tension field of $f$. We can also express this equation in local coordinates: let $m = dimM$, $n = dimN$, $(\gamma_{\alpha\beta})_{\alpha\beta=1,\dots,m}$ be the metric tensor of $M$, $(\gamma^{\alpha\beta}) = (\gamma_{\alpha\beta})^{-1}$, $\gamma := det(\gamma_{\alpha\beta})$, $(g_{ij})_{i,j=1,\dots,n}$ the metric tensor of $N$, $\Gamma^i_{jk} = \frac{1}{2}\sum_{l=1}^n g^{il}, (g_{jl,k} + g_{kl,j} - g_{jk,l})$, where an index after a comma denotes a partial derivative in the resp. coordinate direction. Then (2.1) is equivalent to

$$\frac{1}{\sqrt{\gamma}} \sum_{\alpha,\beta=1}^{m} (\frac{\partial}{\partial x^\alpha}(\sqrt{\gamma}\gamma^{\alpha\beta}\frac{\partial}{\partial x^\beta}f^i) + \gamma^{\alpha\beta}\sum_{j,k=1}^{n}\Gamma^i_{jk}\frac{\partial f^j}{\partial x^\alpha}\frac{\partial f^k}{\partial x^\beta}) = 0 \tag{2.2}$$

for $i = 1,\dots,n$. Of course, $\Gamma^i_{jk}$ is evaluated at $f(x)$. Thus, we obtain a nonlinear elliptic system of partial differential equations, where the leading term is the Laplace- Beltrami operator of $M$ and the nonlinearity arises from the geometry of $N$. Actually, the harmonic map system constitutes the Euler-Lagrange equations for a variational problem. We define the energy of $f$ as

$$E(f) := \frac{1}{2}\int_M |df|^2 dvol(M), \tag{2.3}$$

where the energy density $|df|^2$ can be written in local coordinates as

$$|df|^2 = \sum_{\alpha,\beta=1}^{m}\sum_{i,j=1}^{n}\gamma^{\alpha\beta}(x)g_{ij}(f(x))\frac{\partial f^i}{\partial x^\alpha}\frac{\partial f^j}{\partial x^\beta}. \tag{2.4}$$

If $M$ is compact, so that every smooth map has finite energy, then the harmonic maps are precisely the smooth critical points of the energy functional $E$. It turns that if $N$ has nonpositive curvature, the only case we shall be concerned with, then every critical point of $E$ is automatically smooth, so that we do not have to address the regularity problem here. If $M$ is noncompact, then of course there may exist harmonic maps of infinite energy.

We want to describe some elementary properties of harmonic maps. If $f : M \to N$ is harmonic, then

$$\Delta_M |df|^2 = |\nabla df|^2 + \sum_{\alpha=1}^{m} \langle df(\mathrm{R^M ic} \cdot e_\alpha), df(e_\alpha)\rangle$$

$$- \sum_{\alpha,\beta=1}^{n} \langle R^N(df(e_\alpha), df(e_\beta))df(e_\beta), df(e_\alpha))\rangle, \qquad (2.5)$$

where $\Delta_M$ is the Laplace-Beltrami operator of $M$, described in (2.2) in local coordinates, $< -, - >$ denotes the scalar product in $T^*M \otimes f^*(TN)$ induced by the metrics of $M$ and $N$, $(e_\alpha)_{\alpha=1,\ldots,m}$ is a local orthonormal frame on $M$, $\mathrm{R^M ic}$ is the Ricci tensor of $M$, and $R^N$ is the curvature tensor of $N$, with the sign convention of [KN]. We observe that in our case, where $N$ has nonpositive curvature, the first and third term on the right hand side of (2.5) are nonnegative.

We also need the chain rule. Let $f = M \to N$, $h : N \to \mathbf{R}$ be smooth. Then

$$\Delta_M(h \circ f) = \sum_{\alpha=1}^{m} D^2h(df(e_\alpha), df(e_\alpha)) + \langle(\mathrm{grad} h) \circ f, \tau(f)\rangle, \qquad (2.6)$$

where $D^2h$ is the tensor of the second covariant derivatives of $h$ and we otherwise use the notations defined above. In particular, if $f$ is harmonic,

$$\Delta_M(h \circ f) = \sum_{\alpha=1}^{n} D^2h(df(e_\alpha), df(e_\alpha)). \qquad (2.7)$$

We now look at the case where $M$ is a Kähler manifold. The following slight change of notation will be convenient: we now put $m = dim_{\mathbf{C}} M$. We denote the Kähler metric of $M$ in local coordinates by $(\gamma_{\alpha\overline{\beta}})_{\alpha,\beta=1,\ldots,m}$. The important point is that for a Kähler metric, the Laplace-Beltrami operator becomes

$$\Delta_M = \sum_{\alpha,\beta=1}^{m} \gamma^{\alpha\overline{\beta}} \frac{\partial^2}{\partial z^\alpha \partial z^{\overline{\beta}}}.$$

Thus, (2.2) becomes

$$\sum_{\alpha,\beta=1}^{m} \gamma^{\alpha\overline{\beta}} \left(\frac{\partial^2 f^i}{\partial z^\alpha \partial z^{\overline{\beta}}} + \sum_{j,k=1}^{n} \Gamma^i_{jk} \frac{\partial f^j}{\partial z^\alpha} \frac{\partial f^k}{\partial z^{\overline{\beta}}}\right) = 0 \qquad (2.8)$$

for $i = 1, ..., n$. We define $f : M \to N$ to be pluriharmonic, if

$$\gamma^{\alpha\overline{\beta}}\left(\frac{\partial^2 f^i}{\partial z^\alpha \partial z^{\overline{\beta}}} + \sum_{j,k=1}^{n} \Gamma^i_{jk} \frac{\partial f^j}{\partial z^\alpha} \frac{\partial f^k}{\partial z^{\overline{\beta}}}\right) = 0 \tag{2.9}$$

for $i = 1, ..., n$ and for all $\alpha, \beta = 1, ..., m$. Of course, a pluriharmonic map is harmonic. Eq. (2.9) means that the restriction of $f$ to every complex curve or, more generally, every complex submanifold (of course, not necessarily compact) is harmonic. This might explain the terminology.

The following generalization of (2.6) will be useful ($M$ Kähler, $N$ Riemannian, $f : M \to N$, $h : N \to \mathbf{R}$ smooth):

$$\gamma^{\alpha\overline{\beta}} \frac{\partial^2 (h \circ f)}{\partial z^\alpha \partial z^{\overline{\beta}}} = \gamma^{\alpha\overline{\beta}} D^2 h\left(\frac{\partial f}{\partial z^\alpha}, \frac{\partial f}{\partial z^{\overline{\beta}}}\right) + \left\langle (\operatorname{grad} h) \circ f, \gamma^{\alpha\overline{\beta}} \frac{\partial^2 f}{\partial z^\alpha \partial z^{\overline{\beta}}}\right\rangle \tag{2.10}$$

for $\alpha, \beta = 1, .., m$. Thus, if $f$ is pluriharmonic,

$$\gamma^{\alpha\overline{\beta}} \frac{\partial^2 (h \circ f)}{\partial z^\alpha \partial z^{\overline{\beta}}} = \gamma^{\alpha\overline{\beta}} D^2 h\left(\frac{\partial f}{\partial z^\alpha}, \frac{\partial f}{\partial z^{\overline{\beta}}}\right) \tag{2.11}$$

for $\alpha, \beta = 1, ..., m$.

We now assume that $N$ is also a Kähler manifold, and put now $n := dim_{\mathbf{C}} N$. Since for a Kähler manifold all Christoffel symbols with mixed indices vanish ($\Gamma^i_{j\overline{k}} = 0$, etc.), the harmonic map equation has the same form as (2.8), where the coordinates in $N$ are now complex. Likewise, the equations for a pluriharmonic map, namely (2.9), remain the same in complex coordinates. We observe from (2.9) that a holomorphic map between Kähler manifolds is pluriharmonic and hence in particular harmonic. As holomorphic maps are defined in terms of the complex structure whereas harmonic maps are defined in terms of the Riemann structure, this fact expresses the compatibility of the complex and the Riemannian structure embodied in the definition of a Kähler manifold.

We want to record a consequence of the chain rules (2.7) and (2.11) for harmonic resp. pluriharmonic maps:

**Lemma 2.1.** *(i) Let $M, N$ be Riemannian manifolds, $f : M \to N$ harmonic, and $h : N \to \mathbf{R}$ convex. Then $f \circ h$ is subharmonic.*

*(ii) Let $M$ be Kähler, $N$ Riemannian, $f : M \to N$ pluriharmonic, and $h : N \to \mathbf{R}$ convex. Then $f \circ h$ is plurisubharmonic.*

*(iii) Let $M, N$ be Kähler, $f : M \to N$ holomorphic, $h : N \to \mathbf{R}$ plurisubharmonic. Then $f \circ h$ is plurisubharmonic.*

# 3 Existence and Uniqueness Results for Harmonic Maps

In this section, $M$ and $N$ will be Riemannian manifolds. We shall continue to use the notation established in §2. In general, one cannot expect general existence theorems for holomorphic or pluriharmonic maps as the corresponding elliptic systems are overdetermined. (We shall see in §4, however, that under certain circumstances harmonic maps are automatically pluriharmonic or even holomorphic.) This is different for harmonic maps. Here, we have one elliptic, second order equation for each dependent variable. Technical difficulties, however, arise from the fact that the nonlinear terms couple the individual equations.

The existence problem for compact $M$ and $N$ was first solved by Al'ber [Al1], [Al2] and Eells-Sampson [ES] under the assumption that $N$ has nonpositive sectional curvature. In this case, (2.5) yields

$$\Delta_M |df|^2 \geq -\text{const.} |df|^2, \tag{3.1}$$

as observed above, and this inequality plays a crucial role in the analysis. Actually, the method to solve (2.1) consisted in studying the associated parabolic system (the so-called 'heat flow'): Let $g : M \to N$ be smooth (in fact, continuity of $g$ suffices): One then seeks a map

$$f : M \times [0, \infty) \to N$$

solving

$$f(x, 0) = g(x)$$

for all $x \in M$ (initial values) and

$$\tau(f(x,t)) = \frac{\partial}{\partial t} f(x,t) \tag{3.2}$$

for all $x \in M$, $t > 0$ (the tension field $\tau$ of course involves only the variable $x \in M$, but not $t$). The aim is to show that a solution of this problem exists for all 'time' $t > 0$, and, as $t \to \infty$, $f(x,t)$ converges to a harmonic map $f(x)$ homotopic to $g(x)$. Under the assumption that $N$ has nonpositive sectional curvature, this can be achieved; the result is

**Theorem 3.1.** *Let $M$,$N$ be compact Riemannian manifolds. Suppose $N$ has nonpositive sectional curvature. Then every continuous map $g : M \to N$ is homotopic to a harmonic map $f$, and $f$ can be obtained by the parabolic method just described.*

**Remarks:** As counterexamples show, the assumptions on $N$–compactness and nonpositive sectional curvature–are necessary for Theorem 3.1 to hold.

One then asks whether the harmonic map of Theorem 3.1 is unique in its homotopy class. Again, this problem was first solved by Al'ber [Al1], [Al2]; a little later, the result was also discovered by Hartman [Ha]:

**Theorem 3.2.** *Let $M$ be compact, and $N$ complete, with nonpositive sectional curvature. Let $f_1, f_2 : M \to N$ be homotopic harmonic maps. Then they are homotopic through a parallel family of harmonic maps; more precisely, there exists a smooth map*

$$F : M \times [0,1] \to N$$

*with*

$$F(x,0) = f_1(x), F(x,1) = f_2(x)$$

*for all $x \in M$, for which each map $f(\cdot, s)$ ($g \in [0,1]$) is harmonic. All these maps have the same energy. The curves $F(x,\cdot)$ are geodesics parametrized proportionally to arclength, with length independent of $x \in M$, and $\frac{\partial}{\partial s}F(x,s)$ is a parallel vector field. In particular, if $N$ has negative sectional curvature, a harmonic map is unique in its homotopy class unless it is constant or maps $M$ onto a closed geodesic of $N$.*

We now want to discuss a more general existence problem. We let $X$ and $Y$ be simply connected Riemannian manifolds, with Y complete and of nonpositive sectional curvature. We denote the isometry groups of $X$ and $Y$ by $I(X)$ and $I(Y)$ resp.

We let $\Gamma$ be a discrete subgroup of $I(X)$, with compact quotient $X/\Gamma$ ($\Gamma$ may have fix points), and we suppose

$$\rho : \Gamma \to I(Y)$$

is a homomorphism. We point out that we do not assume that $\rho(\Gamma)$ is discrete.

If $M$ and $N$ are compact as before, $f : M \to N$ continuous, with universal covers $X = \tilde{M}, Y = \tilde{N}$, and $\Gamma = \pi_1(M)$, then $f$ defines a homomorphism

$$f_* : \pi_1(M) \to \pi_1(N).$$

Since $\pi_1(M)$ and $\pi_1(N)$ operate by deck transformations on $X$ and $Y$, we see that the previously defined situation is a special case of the present setting.

One then seeks a $\rho$-equivariant harmonic map $f : X \to Y$, i.e., a harmonic map satisfying

$$f(\gamma x) = \rho(\gamma) f(x)$$

for all $x \in X$, $\gamma \in \Gamma$. This question was first studied in special cases (other than those covered by Theorem 3.1) by Diederich-Ohsawa [DO] and Donaldson [Dn]. The general problem was then treated by Corlette [C], Jost-Yau [JY5], and Labourie [Lb].

In order to describe the result, we need some terminology. A subgroup of $I(Y)$ is called parabolic if all elements have a common fix point on $Y^\infty$. $Y^\infty$ here is a boundary of $Y$ introduced to compactify $Y$ and defined as follows: Two geodesic rays

$$\gamma_1, \gamma_2 : [0, \infty) \to N$$

parametrized by arclength are called asymptotic

$$dist(\gamma_1(t), \gamma_2(t)) \leq const.$$

independent of $t$. $Y^\infty$ then is defined as the set of all equivalence classes of geodesic rays. The action of $I(Y)$ obviously extends to $Y^\infty$.

We also need the following notion introduced in [Lb]: A subgroup $G$ of $I(Y)$ is called reductive, if there exists a closed convex subset $C$ of $Y$, left invariant by $G$, isometric to $C_1 \times E$, with $E$ Euclidean, and if $G = G_1 \times G_2$, where $G_1$ and $G_2$ act isometrically on $C_1$ resp. $E_1$ and where $G_1$ does not have a fixed point in $C_1^\infty$.

If $Y$ is a symmetric space of noncompact type, this notion agrees with the one of a reductive subgroup $G$ of the automorphism group of $Y$, reductive being defined here by the property that the Zariski closure of $G$ (the smallest algebraic subgroup of $Aut(Y)$ containing $G$) does not have a nontrivial nilpotent normal subgroup.

The existence theorem then is:

**Theorem 3.3.** *Let $X$,$Y$ be simply connected Riemannian manifolds, and $Y$ complete with nonpositive sectional curvature. Let $\Gamma$ be a discrete subgroup of $I(X)$, $X/\Gamma$ compact, and let*

$$\rho : \Gamma \to I(Y)$$

*be a homomorphism. If $\rho(\Gamma)$ is reductive, then there exists a $\rho$-equivariant harmonic map $f : X \to Y$. In particular, this is the case, if $\rho(\Gamma)$ is not contained in a parabolic subgroup of $Y$, or if there exists a totally geodesic Euclidean subspace $E$ of $Y$ left invariant by $\rho(\Gamma)$.*

For the proof, one chooses any continuous $\rho$-equivariant map $g : X \to Y$ (since $Y$, complete, nonpositively curved, simply connected, is contractible, there are no topological obstructions for the existence of such a map.) One studies the parabolic system (3.2), (3.3). Uniqueness of solutions implies that $f(\cdot, t)$ stays $\rho$-equivariant for every $t > 0$. A similar analysis as in the proof of Theorem 3.2 yields uniform $C^1$ estimates for the solution $f$. We choose $x_o \in X$. If $f(x_o, t_n)$ stays in a bounded subset of $Y$ for some sequence $t_n \to \infty$, then after selection of a subsequence, $f(\cdot, t_n)$ is shown to converge to a $\rho$- equivalent harmonic map. If $f(x_o, t)$ leaves every bounded subset as $t \to \infty$, we select $y_o \in Y$ and observe that for any $\gamma_1 \in \Gamma$,

$$dist(\rho(\gamma) y_o, y_o)$$

and

$$dist(\rho(\gamma) f(x_o, t), f(x_o, t)) = dist(f(\gamma x_o, t), f(x_o, t))$$

are both bounded independent of $t$, the latter because of the uniform $C^1$-estimates for $f(x, t)$. Thus, for a suitable sequence $t_n \to \infty$, the geodesic arcs from $\rho(\gamma) y_o$ to $\rho(\gamma) f(x_o, t)$ and from $y_o$ to $f(x_o, t_n)$ converge to geodesic rays, and since the distance between two geodesics in a nonpositively curved simply connected manifold is a convex function of the arclength parameter, these geodesic rays are asymptotic, hence define the same point in $Y^\infty$. This means that $\rho(\gamma)$ fixes this point in $Y^\infty$. Since this holds for every $\gamma \in \Gamma$, $\rho(\Gamma)$ is contained in a parabolic subgroup of $I(Y)$.

Nevertheless, there is still a chance that a solution exists. If there exist elements $\sigma_n \in I(Y)$ with the property that $\sigma_n f(x_o, t_n)$ is contained in a fixed compact subset of $Y$, then $\sigma_n f(x, t_n)$ is $\sigma_n \rho \sigma_n^{-1}$ equivariant and converges for $t_n \to \infty$ to a harmonic map $f_\sigma$. If the limit of the conjugations $\sigma_n \rho \sigma_n^{-1}$ is again a conjugate of $\rho$, say $\sigma \rho \sigma^{-1}$, then $f_\sigma$ is $\sigma \rho \sigma^{-1}$ equivariant and $f := \sigma^{-1} f_\sigma$ then is a $\rho$-equivariant harmonic map. Arguments of this kind establish Theorem 3.3.

# 4 Special Properties of Harmonic Maps from Kähl Manifolds into Negatively Curved Manifolds

We start by examining formula (2.5). If $f : M \to N$ is harmonic and $N$ has nonpositive sectional curvature, then the first and third terms on the right hand side of the formula for $\Delta|df|^2$ are nonnegative:

$$\Delta|df|^2 = |\nabla df|^2 + \sum_{\alpha=1}^{m}\langle df(\mathrm{R}^{\mathrm{M}}\mathrm{ic}e_\alpha), df(e_\alpha)\rangle$$

$$- \sum_{\alpha,\beta=1}^{m} \langle R^N(df(e_\alpha), df(e_\beta))df(e_\beta), df(e_\alpha)\rangle. \tag{4.1}$$

If the second term would also be nonnegative, then for a compact $M$, this would force $|df|$ to be constant and all terms on the right hand side to vanish; in particular $\nabla df \equiv 0$, meaning that $f$ is totally geodesic. Of course, the second term is nonnegative, if $M$ has nonnegative Ricci curvature. While this allows to conclude that compact Riemannian manifolds of negative sectional curvature have quite different topological properties from those of nonnegative Ricci curvature (by Theorem 3.1 any continuous map $g : M \to N$ is homotopic to a harmonic one which by the above formula would have to be constant or map $M$ onto a closed geodesic under the present curvature assumptions), this type of argument is obviously insufficient to draw any conclusion in case both $M$ and $N$ have negative curvature.

In this section, we want to discuss identities which are similar to (4.1) but avoid this shortcoming. More precisely, we shall discuss the following three basic results on which our applications in §5 shall be based:

- a Bochner type identity for harmonic maps between Kähler manifolds, discovered by Siu [S1];
- a similar identity for maps from a Kähler into a Riemannian manifold, discovered by Sampson [Sa];
- a harmonic foliation induced by a harmonic map from a Kähler manifold into a hyperbolic Riemann surface, discovered by Jost-Yau [JY1].

In order to establish the Bochner type identities, we use the method of Sampson [Sa]. Looking at the derivation of (2.5), we realize that the Ricci curvature term arises from the contraction by the inverse domain metric involved in the definition of

$$|df|^2 = \sum_{\alpha,\beta=1}^{m}\sum_{i,j=1}^{n} \gamma^{\alpha\beta}(x)g_{ij}(f(x))\frac{\partial f^i}{\partial x^\alpha}\frac{\partial f^j}{\partial x^\beta}.$$

Therefore, it is natural to avoid this contraction and work with a differential form instead of a function; we use, assuming that $M$ and $N$ are Kähler

$$\sum_{\alpha,\beta=1}^{m} \varphi_{\alpha\beta} dz^\alpha \otimes dz^\beta := \sum_{\substack{\alpha,\beta \\ i,j}}^{m} \varphi_{i\bar{j}} \frac{\partial f^i}{\partial z^\alpha} \frac{\partial f^{\bar{j}}}{\partial z^\beta} dz^\alpha \otimes dz^\beta . \tag{4.2}$$

This is the $(2,0)$ part of the pull-back of the image metric under $f$. Differentiating and contracting, we obtain the $(1,0)$ form

$$\sum_{\alpha=1}^{m} \xi_\alpha dz^\alpha := \sum_{\alpha=1}^{m} \sum_{\beta,\delta=n}^{m} \gamma^{\beta\bar{\delta}} \nabla_{\partial\backslash\partial z^{\bar{\delta}}} \varphi_{\alpha\beta} dz^\alpha =$$

$$\sum_{\alpha=1}^{m} \sum_{\beta,\delta=1}^{m} \sum_{i,j=1}^{n} \gamma^{\beta\bar{\delta}} \varphi_{i\bar{j}} (\nabla_{\partial\backslash\partial z^{\bar{\delta}}} \frac{\partial f^i}{\partial z^\alpha}) \frac{\partial f^{\bar{j}}}{\partial z^\beta} dz^\alpha , \tag{4.3}$$

where the last identity already assumes that $f$ is harmonic. One then computes, using again the harmonicity of $f$,

$$\operatorname{div}(\sum_{\alpha=1}^{m} \xi_\alpha dz^\alpha) = \sum_{\alpha,\beta,\delta,\eta=1}^{m} \sum_{i,j=1}^{n} \gamma^{\alpha\bar{\eta}} \gamma^{\beta\bar{\delta}} \varphi_{i\bar{j}} (\nabla_{\partial\backslash\partial z^{\bar{\delta}}} \frac{\partial f^i}{\partial z^\alpha})(\nabla_{\partial\backslash\partial z^{\bar{\eta}}} \frac{\partial f^{\bar{j}}}{\partial z^\beta})$$

$$+\frac{1}{2} \sum_{\alpha,\beta,\delta,\eta=1}^{m} \sum_{i,j,k,l=1}^{n} \gamma^{\alpha\bar{\eta}} \gamma^{\beta\bar{\delta}} R_{i\bar{j}k\bar{l}}$$

$$\left(\frac{\partial f^i}{\partial z^\alpha} \frac{\partial f^{\bar{j}}}{\partial z^\beta} - \frac{\partial f^i}{\partial z^\beta} \frac{\partial f^{\bar{j}}}{\partial z^\alpha}\right) \cdot \left(\frac{\partial f^{\bar{k}}}{\partial z^\delta} \frac{\partial f^l}{\partial z^\eta} - \frac{\partial f^{\bar{k}}}{\partial z^\eta} \frac{\partial f^l}{\partial z^\delta}\right). \tag{4.4}$$

Here, $R_{i\bar{j}k\bar{l}}$ denotes the curvature tensor of $N$. The first term on the right hand side of (4.4) is nonnegative. For the discussion of the second one, we need the following definition of Siu [S1]:

**Definition 4.1:** The curvature tensor of $N$ is called strongly (semi)negative if

$$\sum_{i,j,k,l=1}^{n} R_{i\bar{j}k\bar{l}} (A^i B^{\bar{j}} - C^i D^{\bar{j}}) \overline{(A^l B^{\bar{k}} - C^l D^{\bar{k}})} > 0$$

$(\geq 0)$ for any $A^i, B^i, C^i, D^i \in C$, provided

$$A^i B^{\bar{j}} - C^i D^{\bar{j}} \neq 0$$

for at least one pair $(i,j)$.

**Remark:** Strongly (semi) negative curvature implies negative (resp. nonpositive) sectional curvature.

We conclude from the divergence theorem

**Lemma 4.1.** *Let $M$,$N$ be Kähler manifolds, $M$ compact, $N$ strongly seminegatively curved, and $f : M \to N$ harmonic. Then $f$ is pluriharmonic, and the curvature term in (4.4) also vanishes.*

We observe that in this case the $(2,0)$ tensor

$$\sum_{\alpha,\beta=1}^{m} \varphi_{\alpha\beta} dz^\alpha \otimes dz^\beta$$

is holomorphic. If $m = 1$, i.e., $M$ is a Riemann surface, this holds without the curvature restriction on $N$ and the compactness of $M$, as is seen from (4.3).

For applications in the next section, we already note that all Hermitian locally symmetric spaces of noncompact or Euclidean type have strongly seminegative curvature, and those of rank 1 even have strongly negative curvature, see [S1].

Similarly, for harmonic maps from a Kähler manifold $M$ into a Riemannian manifold $N$, one considers

$$\sum_{\alpha,\beta=1}^{m} \varphi_{\alpha\beta} dz^\alpha \otimes dz^\beta := \sum_{\alpha,\beta=1}^{m} \sum_{i,j=1}^{n} \varphi_{ij} \frac{\partial f^i}{\partial z^\alpha} \frac{\partial f^j}{\partial z^\beta} dz^\alpha \otimes dz^\beta \tag{4.5}$$

and differentiates and computes to get

$$\sum_{\alpha=1}^{m} \xi_\alpha dz^\alpha := \sum_{\alpha,\beta,\delta=1}^{m} \sum_{i,j=l}^{n} \gamma^{\beta\bar{\delta}} \varphi_{ij} (\nabla_{\partial\backslash\partial z^{\bar{\delta}}} \frac{\partial f^i}{\partial z^\alpha}) \frac{\partial f^j}{\partial z^\beta} dz^\alpha \tag{4.6}$$

and

$$div(\sum_{\alpha=1}^{m} \xi_\alpha dz^\alpha) = \sum_{\alpha,\beta,\delta,\eta=1}^{m} \sum_{i,j=1}^{n} \gamma^{\alpha\bar{\eta}} \gamma^{\beta\bar{\delta}} g_{ij} (\nabla_{\partial\backslash\partial z^{\bar{\delta}}} \frac{\partial f^i}{\partial z^\alpha})(\nabla_{\partial\backslash\partial z^{\bar{\eta}}} \frac{\partial f^j}{\partial z^\beta})$$

$$- (\sum_{\alpha,\beta,\delta,\eta=1}^{m} \gamma^{\alpha\bar{\eta}} \gamma^{\beta\bar{\delta}} R_{ijkl} \frac{\partial f^i}{\partial z^\beta} \frac{\partial f^j}{\partial z^\alpha} \frac{\partial f^k}{\partial z^{\bar{\delta}}} \frac{\partial f^l}{\partial z^{\bar{\eta}}}). \tag{4.7}$$

Again, the first term on the right hand side is nonnegative, and for the definition of the second one, we employ the following definition of Sampson [Sa],

**Definition 4.2:** The curvature of $N$ is called Hermitian seminegative if

$$\sum_{i,j,k,l=1}^{n} R_{ijkl} A^i B^j \overline{A^k} \overline{B^l} \leq 0$$

for arbitrary complex $A^i, B^i$.

As before, we conclude

**Lemma 4.2.** *Let $M$ be a compact Kähler manifold, $N$ a Riemannian manifold of Hermition seminegative curvature, and $f : M \to N$ harmonic. Then $f$ is pluriharmonic, and also the curvature term in (4.7) vanishes identically.*

Again, under the conditions of the lemma, $\sum_{\alpha,\beta=1}^{m} \varphi_{\alpha\beta} dz^\alpha \otimes dz^\beta$ is a holomorphic $(2,0)$ form, and again for $m = 1$, this holds without a curvature condition on $N$ and for not necessarily compact $M$.

The curvature tensor of Riemannian locally symmetric spaces of non-compact or Euclidean type is Hermitian seminegative, and so Lemma 4.2 applies in this case. Let $N = G/K$ be such a locally symmetric space, with corresponding Cartan decomposition

$$g = k + p$$

of the Lie algebras. We realize $g$ as an algebra of real matrices and identify the tangent space of $G/K$ with $p$ in the standard manner. Sampson then concludes [Sa1] from an analysis of the vanishing of the curvature expression in (4.7)

**Corollary 4.1.** *Under the conditions of Lemma 4.2, $f$ maps each holomorphic tangent space $T_x^{(1,0)}M$ $(z \in M)$ onto an abelian subalgebra of $p^{\mathbf{C}}$, the complexification of $p$ identified with $T_{f(z)}N$.*

We now come to the third point, namely the holomorphic foliation of [JY1].

**Lemma 4.3.** *Let $M$ be a compact Kähler manifold, $\Sigma$ a hyperbolic Riemann surface, and $f : M \to \Sigma$ harmonic. If $z_o \in M$ and $df(z_o) \neq 0$, then there exists a neighborhood $U$ of $z_o$ for which the local level sets*

$$\{f \equiv c\} \cap U, (c \in \Sigma)$$

*are unions of complex analytic hypersurfaces, and the corresponding foliation is holomorphic.*

# 5 Rigidity Results

In this section, we discuss strategies to show that under certain negativity conditions on the image curvature harmonic maps from Kähler manifolds are holomorphic. Applications will be rigidity results for complex structures. We start with the following result of Siu [S1]:

**Theorem 5.1.** *Let $N$ be a compact quotient (without singularities) of an irreducible Hermitian symmetric space of noncompact type, with $n := dim_{\mathbb{C}} N \geq 2$. Let $M$ be a compact Kähler manifold homotopy equivalent to $N$. Then $M$ is $\pm$ biholomorphically equivalent to $N$.*

The proof proceeds by applying Theorem 3.1 to obtain a harmonic homotopy equivalence $f : M \to N$, using Lemma 4.1 and examining consequences of the vanishing of the curvature term in (4.4). It turns out that the vanishing of this term together with the fact that $f$ as homotopy equivalence has $rank_{\mathbb{R}} df = 2n$ somewhere implies by considerations from linear algebra that $f$ has to be $\pm$ holomorphic and then, as a homotopy equivalence, also bijective.

Theorem 5.1 extends Mostow's rigidity theorem [Mo] in the Hermitian symmetric case by achieving rigidity not only in the class of locally symmetric spaces but in the wider class of Kähler manifolds. Since Mostow's theorem was extended to locally symmetric spaces of finite volume, not necessarily compact, by Prasad [Pa] in the case of *rank* 1 and by Margulis [Ma1] to higher rank, it is natural to ask whether Theorem 5.1 also holds in the finite volume case. This was studied in [JY2,3]; the results are

**Theorem 5.2.** *Let $N$ be a locally irreducible Hermitian locally symmetric space of noncompact type, rank 1 (i.e., a quotient of the unit ball in $C^n$), $dim_{\mathbb{C}} N \geq 2$, and finite volume. Let $M$ be a Kähler manifold, properly homotopically equivalent to $N$. Assume that there exists a compact Kähler manifold $\overline{M}$ containing $M$, with $\overline{M} \backslash M$ being a hypersurface which may have normal crossings (like a union of complex hyperplanes in $\mathbb{C}^n$), but smooth otherwise. Then $M$ is $\pm$ biholomorphic to $N$.*

The assumption on the existence of a compactifying divisor is necessary because $N$ can be compactified in such a way and one can change the complex structure of the complement of a complex hypersurface by deforming this hypersurface topologically. The above theorem then implies that this hypersurface cannot be deformed as a complex hypersurface. Whether the smoothness assumption is really needed is not clear, but if $M$ is quasiprojective it can always be satisfied by Hironaka's theorem.

Of course, the result fails to hold for $dim_{\mathbb{C}} N = 1$, as then $N$ is a hyperbolic Riemann surface and can consequently be deformed. The possible deformations are described by Teichmüller theory.

**Theorem 5.3.** *Let $N$ be a locally irreducible Hermitian locally symmetric manifold of noncompact type, rank $> 1$, and finite volume. Let $M$ be a quasiprojective manifold, with a projective compactification $\overline{M}$ with $\overline{M} \backslash M$ a subvariety of $codim_{\mathbb{C}} \geq 3$ in $\overline{M}$. If $M$ is properly homotopy equivalent to $N$, then they are $\pm$ biholomorphically equivalent.*

Again, the requirement of a suitable compactification of $M$ is necessary. Whether the codimension requirement on the compactifying variety is necessary, is not known, however. In any case, Theorem 5.3 covers all infinitely many cases where Margulis' theorem applies, except one, namely quotients of the Siegel upper half plane of degree 2. This space has complex dimension 3 and is compactified by a one dimensional variety, so that here the codimension is only 2.

The codimension requirement arises in the proof for the following reason: In order to carry over the constructions of §4 to the noncompact setting, one needs a harmonic map of finite energy. In order to produce a harmonic homotopy equivalence of finite energy, one first has to construct some - not necessarily harmonic - homotopy equivalence of finite energy. This would be easy if there would exist smooth compactifications $\overline{M}$ and $\overline{N}$ and a continuous homotopy equivalence between them. A priori, however, the compactifications of $M$ and $N$ may be topologically quite different, and this may lead to the following geometric problem:

We call a nontrivial element of $\pi_1(M)$ or $\pi_1(N)$ geometrically short, if it represents the trivial element in $\pi_1(\overline{M})$ resp. $\pi_1(\overline{N})$. If $g : M \to N$ is a homotopy equivalence mapping a geometrically short element of $\pi_1(M)$ into an element of $\pi_1(N)$ which is not geometrically short, then no map homotopic to $g$ can have bounded energy density. While this phenomenon cannot happen in the case where $N$ has rank 1 as $N$ can then be compactified by adding a finite number of points, in the higher rank case the assumption on the codimension of the compactifying variety implies that $M$ carries a metric the volume form of which decays so fast towards $\overline{M} \backslash M$ that one can still achieve finite global energy.

**Remarks:** (1) The preceding rigidity results hold in somewhat more generality than described. For example, they hold for $N$ of strongly negative curvature, instead of just being Hermitian symmetries of rank 1.

(2) One can also prove a rigidity result for the moduli space of conformal structures on a given surface of *genus* $\geq 2$. The result is similar to Theorem 5.3. In particular, one needs again an assumption on the codimension of a compactifying variety. This is unfortunate, because as most components of the compactifying divisor of the Baily-Borel compactification have codimension 3, there is one component of codimension 2.

As previously mentioned, if $N$ is a Riemann surface or a product of Riemann surfaces, one cannot expect that harmonic maps are holomorphic. Nevertheless, one has, as a consequence of Lemma 4.3 ([JY1]:

**Theorem 5.4.** *Let $M$ be a compact Kähler manifold, $\overline{\Sigma}$ a compact hyperbolic Riemann surface, and $g : M \to \Sigma$ continuous such that*

$$g_* = H_2(M, \mathbf{R}) \to H_2(\Sigma, \mathbf{R})$$

*is nontrivial. Then there exists a compact hyperbolic Riemann surface $\Sigma'$, a holomorphic map $h : M \to \Sigma'$, and a harmonic map $\varphi : \Sigma' \to \Sigma$ so that the harmonic map $f := \varphi \circ h$ is homotopic to $\varphi$. In particular, letting $\triangle$ the hyperbolic unit disk, the lift*

$$\tilde{h} : \tilde{N} \to \triangle$$

*to universal covers is a nonconstant bounded holomorphic function on $\tilde{N}$.*

Studying the properties of the holomorphic foliation of Lemma 4.3 in more detail, one also shows [JY1] for $n = 2$, [Mk 1,3] for the general case, with a simplified proof in [S3]:

**Theorem 5.5.** *Let $N$ be a compact quotient of $\triangle^n$ ($\triangle$ = hyperbolic unit disk), i.e., $N = \triangle^n / \Gamma$, $\Gamma$ discrete, cocompact, fixpoint free. If the Kähler manifold $M$ is homotopically equivalent to $N$, then $M$ has the complex structure of $\triangle^n / \Gamma'$, with $\Gamma'$ isomorphic to $\Gamma$ (as abstract groups). If $\Gamma$ is irreducible, i.e., no finite cover of $\triangle^n / \Gamma$ is a nontrivial product, then $\Gamma$ and $\Gamma'$ are isomorphic as subgroups of $\text{Aut}(\triangle^n)$ (possibly after a change of orientation) and $M$ and $N$ are biholomorphically equivalent.*

In particular, a Kähler manifold homotopically equivalent to a product of hyperbolic compact Riemann surfaces has the complex structure of such a product itself. Again, this result can be extended to the finite volume case, cf. [JY3].

We next state an example of the results of Sampson [Sa1], a consequence of Corollary 4.1.

**Theorem 5.6.** *Let $f$ be a harmonic map from a compact Kähler manifold into a compact quotient of real hyperbolic space. Then*

$$rank_{\mathbf{R}} df \leq 2$$

*everywhere.*

**Corollary 5.1.** *A compact Kähler manifold $M$ with $dim_{\mathbf{C}} M > 1$ cannot be minimally immersed into a space of constant negative curvature.*

We now turn to applications of Theorem 3.3. The next result is taken from [JY5]; similar results were obtained by Carlson and Toledo [CT] by essentially the same method.

**Theorem 5.7.** *Let $\Gamma$ be a discrete subgroup of the isometry group of the simply-connected Kähler manifold, with compact quotient $X/\Gamma$. Let*

$$\rho : \Gamma \rightarrow SU(n, 1)$$

*be a homomorphism, and assume that $\rho(\Gamma)$ has reductive Zariski closure. Then one of the following three possibilities holds:*

1. *There exists a nontrivial $\pm$ holomorphic $\rho$- equivariant map*

$$f : X \rightarrow B^n$$

*where $B^n := \{z \in \mathbf{C}^n : |z| < 1\}$.*

2. *There exists a homomorphism*

$$\sigma : \Gamma \rightarrow Aut(R),$$

*with $R = \mathbf{C}$ or $\Delta$, the unit disk, a $\pm$ holomorphic $\sigma$-equivariant map*

$$\pi : M \rightarrow R$$

*and a harmonic map $h : R \rightarrow B^n$, with*

$$f = h \circ \pi.$$

3. *$\rho(\Gamma)$ is compact.*

*If $\rho(\Gamma)$ is discrete, then also $\sigma(\Gamma)$ is discrete, and there exists a nontrivial homomorphism*

$$g = \Gamma \to \mathbf{Z}.$$

*If $\rho(\Gamma)$ is discrete and compact, it is finite.*

In the proof one studies the $\rho$-equivariant harmonic map

$$f : X \to B^n$$

of Theorem 3.3. If $f$ is constant, case 3) holds. If $rank_{\mathbf{R}} df \leq 2$, one constructs a suitable holomorphic foliation similar to the one of Lemma 4.3, and deduces 2). If $rank_{\mathbf{R}} df > 2$ at some point, $f$ is holomorphic as in Lemma 4.1.

Similarly, using Theorem 5.6 one shows

**Theorem 5.8** *Let $X$ and $\Gamma$ be as before. Let*

$$\rho : \Gamma \to SO(n, 1)$$

*be a homomorphism, with reductive Zariski closure of $\rho(\Gamma)$. Then one of the following two possibilities holds:*

1. *There exists a homomorphism*

$$\sigma : \Gamma \to Aut(R),$$

*with $R = \Delta$ or $\mathbf{C}$, a $\sigma$-equivariant $\pm$-holomorphic map*

$$\pi : M \to R,$$

*a harmonic map*

$$h : R \to H^n := \text{ real hyperbolic } n\text{-space},$$

*and*

$$f = h \circ \pi$$

*is harmonic $\rho$-equivariant.*

2. *$\rho(\Gamma)$ is compact. If $\rho(\Gamma)$ is discrete, again similar conclusions as in Theorem 5.7 hold.*

Combining Theorems 5.7 and 5.8 with a result of Mok [Mk1], one obtains ([JY5]):

**Corollary 5.2.** *Let $\Gamma$ be a discrete cocompact subgroup of the automorphism group of a Hermitian symmetric space $G/K$ of noncompact type. Assume that*

$$\Gamma\backslash G/K$$

*is irreducible and of rank $\geq 2$. Then for every discrete homomorphism*

$$\rho : \Gamma \to SU(n,1) \text{ or } SO(n,1),$$

*$\rho(\Gamma)$ is finite.*

Corollary 5.2 also follows from a recent work of Margulis [Ma2]. Corollary 5.2 has recently been extended by Spazier and Zimmer [SZ] by a different method (not using harmonic maps) in various directions, in particular to the real symmetric case. Again, for Theorems 5.7 and 5.8 and Corollary 5.2, one only needs a suitable negativity requirement on the image curvature instead of the symmetric structure ($B^n$ or $H^n$) of the target.

# 6 Pluriharmonic Maps

In this section, we give some further applications of pluriharmonic maps.

**Theorem 6.1.** *Let $M$ be a compact Kähler manifold, $\Sigma \subset M$ a subvariety. Suppose there exists a continuous map $g : M \to N$ where $N$ is a compact Kähler manifold of strongly seminegative curvature, and suppose $g$ is not homotopic to a map $g'$ with*

$$rank_{\mathbf{R}} dg'(z) < 2dim_{\mathbf{C}}\Sigma \text{ for all } z \in \Sigma$$

*(i.e., $g(\Sigma)$ is not homotopic to something of lower dimension than $\Sigma$). Then $\Sigma$ cannot be blown down to a lower dimensional variety $\Sigma'$ in such a way the resulting space $M'$ is still a smooth manifold.*

**Proof:** Assume there is a smooth manifold $M'$ obtained by blowing $\Sigma$ down to $\Sigma'$. Let $U$ be a neighborhood of $\Sigma'$. By Theorem 3.1, there exists a harmonic map $f : M \to N$ homotopic to $g$, and by Lemma 4.1, $f$ is pluriharmonic. In particular, $f$ is harmonic with respect to any Kähler metric on $U\backslash\Sigma'$ as $M\backslash\Sigma$ is biholomorphic to $M'\backslash\Sigma'$. We are going to show that $f|U\backslash\Sigma'$ extends to a smooth harmonic map

$$f' : U \to N.$$

But then the image of $\Sigma$ under $f$ can have at most the dimension of $\Sigma'$, contradicting the assumption of the theorem.

We shall only show the extension at the smooth points of $\Sigma'$ because then the same argument works for the smooth points of the subvariety $\Sigma'' = sing\Sigma'$, and so on. For every smooth point $z_o$ of $\Sigma'$, there exists $U_o \subset U$, biholomorphic to the unit ball in some $\mathbf{C}^n$ ($n \geq 2$ is the codimension of $\Sigma'$ in $M'$), with $U_o \cap \Sigma' = z_o$. As $f$ is pluriharmonic, $f|U_o\backslash\{z_o\}$ is harmonic, and the following lemma implies that $f\backslash U_o\{z_o\}$ extends to a smooth harmonic map

$$f_o : U_o \to N.$$

Moreover, we can choose $U_o$ to depend smoothly on $z_o$, and since then the values of $f$ on $\partial U_o$ depend smoothly on $z_o$, the extensions $f_o$ likewise do. This is a consequence of *a priori* estimates and uniqueness for the boundary value problem. (These results are similar to the ones presented in §3, and can be found, e.g., in [J2].) Then the extension is harmonic outside $\Sigma'$, and smooth everywhere, hence harmonic everywhere.

In order to complete the proof, it only remains to show

**Lemma 6.1.** *Let $B(z,R)$ be a ball in a Riemannian manifold of real dimension $\geq 3$, with center $z$. Let $N$ be a compact Riemannian manifold of nonpositive sectional curvature, and*

$$f : B(z,R)\backslash\{z\} \to N$$

*be harmonic. Then $f$ extends to a (smooth) harmonic map*

$$f : B(z,R) \to N.$$

**Proof:** Since the dimension of the domain is at least 3, $B(z,R)\backslash\{z\}$ is simply connected, and we can lift to the universal cover $\tilde{N}$ of $N$. We choose some $p \in \tilde{N}$ and put

$$h(q) := d^2(p,q) \text{ for } p \in \tilde{N},$$

where $d(\cdot,\cdot)$ denotes geodesic distance on $\tilde{N}$. Since $N$ has nonpositive sectional curvature, $h$ is strictly convex. By Lemma 2.1, $h \circ f$ is subharmonic, and in fact by (2.7)

$$\Delta(h \circ f) \geq c|\nabla f|^2, \tag{6.1}$$

where $c > 0$ is a lower bound for the Hessian of $h$ on $f(B(z,R)\backslash\{z\})$. (We denote $f$ and its lift to universal cover by the same letter, and we observe that $f(B(z,R)\backslash\{z\})$ is bounded in $\tilde{N}$ because $N$ is compact.) Equation (6.1) implies that $(h \circ f)$ is a bounded subharmonic function, and we may consequently extend it to a bounded subharmonic function on all of $B(z,R)$. Integrating (6.1), we then conclude

$$\int_{B(z,R)} |df|^2 < \infty,$$

i.e., $f$ has finite energy on $B(z,R)$. Since the domain dimension is at least 3, one can consequently extend $f$ to a weak solution (in the sense of PDE) of the harmonic map system on $B(z,R)$. A theorem of Hildebrandt-Kaul-Widman [HKW] then implies that $f$ is a smooth harmonic map on $B(z,R)$, because $\tilde{N}$ has nonpositive curvature. qed.

**Remark:** One can also show removability of isolated singularities for harmonic maps on a two-dimensional domain. $N$ need not have nonpositive curvature; only $\tilde{N}$ has to carry a strictly convex function. Then one concludes as above that $f$ has finite energy, and a theorem of Sacks-Uhlenbeck [SU] implies that $f$ cannot have an isolated singularity.

The following result was proved by Diederich-Ohsawa [DO]:

**Theorem 6.2.** *Let $M$ be a compact Kähler manifold, $\pi : \Omega \to M$ a (locally trivial) holomorphic fiber bundle with fiber the unit disk $D$. Then $\Omega$ is weakly 1-complete, meaning that it carries a smooth plurisubharmonic exhaustion function.*

In order to describe a proof that can be partially generalized, we start with

**Lemma 6.2.** *Let $M$ be a compact Kähler manifold, and $\pi : \Omega \to M$ be a holomorphic fiber bundle with fiber $Y$, a complete simply connected Kähler manifold of strongly seminegative curvature. Suppose that the corresponding homomorphism*

$$\rho : \pi_1(M) \to I(Y)$$

*has reductive image. Then $\Omega$ carries a smooth plurisubharmonic exhaustion function.*

**Proof:** Denote the universal cover of $M$ by $X$. By
Theorem 3.3, there exists a $\rho$-equivariant harmonic map

$$f : X \to Y$$

(we consider $\pi_1(M)$ as a discrete subgroup of $I(X)$ in the standard way). By Lemma 4.1, $f$ is pluriharmonic. Denote the geodesic distance of $Y$ by $d(\cdot,\cdot)$ and define

$$g(z) := d^2(z, h \circ \pi(z)).$$

By Lemma 2.1, $g$ is plurisubharmonic since $d^2$ is strictly convex because $N$ has nonpositive sectional curvature. $g$ obviously is also an exhaustion function. This proves the result. qed.

Having seen Lemma 6.2, it remains to study the case where the image of

$$\rho : \pi_1(M) \to I(Y)$$

is not reductive.

As explained in the proof of Theorem 3.3, in this case the group $\Delta := \rho(\Gamma)$ fixes some point $y_o \in Y^\infty$. Let $P$ be the parabolic subgroup of $I(Y)$ of all isometries fixing $y_o$. Thus $\Delta \subset P$. For each $z \in Y$, we let $\gamma$ be the geodesic ray from $z$ to $y_o$, parametrized by arclength, with $z = \gamma(0)$. We let $b_z$ be the associated Busemann function, i.e.,

$$b_z(y) := \lim_{t\to\infty} (d(h, \gamma(t)) - t)$$

where $(d\cdot,\cdot)$ is the distance function of $y$. The level sets of $b_z$ are called horospheres centered at $y_o$, and any two points on the same horosphere yield the same Busemann function. Moreover, $b_z$ is a convex function, and if $(y_n)_{n\in\infty}$ converges to some $y \in Y^\infty$ along some geodesic ray, then

$$\lim_{n\to\infty} b_z(y_n) = \infty \text{ if } y \neq y_o$$

and

$$\lim_{n\to\infty} b_z(y_n) = -\infty \text{ if } y = y_o.$$

We fix some $z_o \in Y$. We also observe that $P$ maps horospheres into horospheres, and this yields a homomorphism

$$\sigma : P \to \mathbf{R}$$

by putting

$$\sigma(\varphi) := b_{z_o}(\varphi(B)) - b_{z_o}(B)$$

for any horosphere $B$, for $\varphi \in P$. We put

$$\rho' := \sigma \circ \rho.$$

By Theorem 3.3, there exists a $\rho'$-equivariant harmonic map

$$f' : X \to R,$$

$X$ being the universal cover of $M$. $f'$ is pluriharmonic by Lemma 4.2. We interpret this as associating to each $x \in X$ a horosphere $B(f'(x))$. Then

$$g(z) := b_z(B(f'(\pi(z))))$$

is a plurisubharmonic function. The only drawback is that $g$ approaches $-\infty$ instead of $\infty$ when $z$ tends to $y_o$ along a geodesic ray.

In the case where $Y$ is the unit disk $D$, one can map $D$ conformally onto the upper half plane $H$ in $\mathbf{C}$, $y_o$ going to $\infty$, and exploit the fact that the horospheres centered at $\infty$ then are straight lines in the Euclidean geometry of $\mathbf{C}$. One then constructs a $\rho$-equivariant harmonic map

$$f_o : X \to \mathbf{C}$$

leaving those horospheres invariant and puts

$$g_o(z) := |z - f_o(\pi(z))|^2.$$

This then is a plurisubharmonic function which is bounded on $\partial H \backslash \{\infty\}$, and tends to $\infty$ for $z \to \infty$, faster than $g$ tends to $-\infty$. Thus

$$g + g_o$$

yields a plurisubharmonic exhaustion function. For details for this last step, we refer to [DO].

# 7 Harmonic Mappings and Algebraic Varieties over Function Fields

We start by recalling the Mordell and Shafarevitch problems over function fields. In geometric terminology, one is dealing with an algebraic surface ibered by curves with generic fiber of genus $g$. While the cases $g = 0, 1$ were already treated in the work of the Italian school and Kodaira, for $g \geq 2$ the so- called Mordell and Shafarevitch conjectures for curves over function fields were solved by Manin [M] (with another proof by Grauert Ga]), and Arakelov [A], resp. (building on earlier work of Parshin [P1]). Arakelov's theorem says that for a fixed base curve $C$, and $S \subset C$ consisting of finitely many points, and $g \geq 2$, there exist only finitely many nontrivial curves over $C$, of genus $g$ over $C \backslash S$. The theorem of Manin says that such a nontrivial curve over $C$ admits at most finitely many sections, and even a trivial curve ('trivial' meaning constant over a finite cover of $C$) admits at most finitely many nontrivial sections.

Now a fibration as in Arakelov's theorem induces a holomorphic map from $C$ into $\overline{M}_g$, the Deligne- Mumford compactification of the moduli space curves of genus $g$. This map satisfies the technical condition of local liftabilty o the Teichmüller space level near the singularities of $M_g$, because it arises from a topological fiber space. Likewise, a section induces a locally liftable, holomorphic map into a compactification of the universal modular curve $\mathcal{M}_g$. For the sake of simplicity, we shall therefore neglect the question of the singularities of $M_g$ and $\mathcal{M}_g$ in the subsequent discussion and pretend that hey are smooth.

It was first suggested by Grauert-Reckziegel [GR] to use the geometry of $M_g$ and $\mathcal{M}_g$, in order to deduce the above results. In this setting, the proof consists of two steps. The first one is boundedness, namely to show hat there are only finitely many homotopy classes of maps from $C$ into $\overline{M}_g$ or $\overline{\mathcal{M}}_g$ with the image of $C \backslash S$ contained in the interior that contain holomorphic maps. The second one is rigidity, namely to show that each such homotopy class contains only finitely many holomorphic maps, in particular no nontrivial families of holomorphic maps.

A boundedness proof along these lines was obtained by Noguchi [N] who exploited the fact that $M_g$ is a hyperbolic space in the sense of Kobayashi. Therefore, holomorphic maps into $M_g$ are equicontinuous and can hence only be contained in finitely many homotopy classes. A rigidity proof complementing Noguchi's approach was found by Imayoshi-Shiga [IS].

Results of Tromba [T] (negativity of the curvature) and Wolpert [W] existence of convex exhaustion functions) on the geometry of Teichmüller pace equipped with its Weil-Petersson metric make it possible to use harmonic maps to prove the above theorems ([JY6]). Since the holomorphic ectional curvature of the Weil-Petersson metric is bounded from above, the Schwarz lemma (O[Y], [R]) implies equicontinuity of holomorphic maps into

$M_g$ or $\mathcal{M}_g$, yielding boundedness as before. For the rigidity part, one looks at a possible limit of homotopic holomorphic maps from $C$ into $\overline{M}_g$ or $\overline{\mathcal{M}}_g$, with $C \backslash S$ contained in the interior as before. This limit either maps $C \backslash S$ again in the interior, or it maps all of $C$ into the boundary, and in both cases one can use uniqueness considerations from the theory of harmonic maps to derive a contradiction. This implies that no nontrivial sequence of homotopic holomorphic maps can exist in this setting, yielding rigidity.

Of course, this approach works as naturally if one has an arbitrary compact Kähler manifold $N$ and a divisor $A$ instead of a curve $C$ and a finite collection $S$ of points.

One can also combine the preceding constructions with arguments as in §5 in order to obtain existence results for fibrations by holomorphic curves. Namely, one first constructs a suitable harmonic map and then shows that it is holomorphic. This depends on Schumacher's result [Su] that the Weil-Petersson metric is strongly negatively curved in the sense of Definition 4.1. Such constructions can be applied in three settings:

1. $B$ a compact Kähler surface, $C$ a compact curve, $\varphi : B \to C$ smooth, $T$ a divisor with normal crossings on $B$, $S$ a divisor on $C$ with:

   - $B \backslash T = \varphi^{-1}(C/S)$,
   - $\varphi$ has maximal rank on $B \backslash T$,
   - $C \backslash S$ carries a complete hyperbolic metric,
   - $\varphi$ is nontrivial on the second homology.

   One then has a curve $C'$ and a divisor $S'$, with $C' \backslash S'$ homeomorphic to $C \backslash S$ and a holomorphic map

   $$f : B \to C',$$

   homotopic to $\varphi$, with $B \backslash T \subset C' \backslash S'$. $f$ has maximal rank on $B \backslash T$, and the fibers of $f$ on $B \backslash T$ then have the same genus as those of $\varphi$ on $B \backslash T$. Again, this generalizes to higher dimensions.

2. $N$ is a compact Kähler manifold, $dim_{\mathbb{C}} N \geq 2$, $\eta : N \to \overline{M}_g$ is continuous (and locally

   liftable near the singularities of $\overline{M}_g$), with:

   - $\eta(N \backslash A) \in M_g$ for some divisor $A \in N$ with normal crossings,

- $\eta$ is not homotopic to a map into $\overline{M}_g \backslash M_g$ or to one of real $rank \leq 2$ everywhere.

One then obtains a holomorphic map

$$h : N \backslash A \to M_g$$

which is homotopic to $\eta$ and extends to a holomorphic map

$$h : N \to \overline{M}_g.$$

One then obtains a fiber space $M$ over $N$ with fiber over $z \in N$ given by the curve defined by $h(z) \in \overline{M}_g$. The fibers over $N \backslash A$ are then smooth curves of genus $g$.

3. $M$ a compact Kähler manifold, $dim_{\mathbb{C}} M \geq 2$, $\kappa : M \to \mathcal{M}_g$ continuous (and locally liftable as before) with:

   - $\kappa(M \backslash A) \in \mathcal{M}_g$ for some divisors $A \in M$ with normal crossings,
   - $\kappa$ is not homotopic to a map into $\overline{\mathcal{M}}_g \backslash \mathcal{M}_g$ or to one of real $rank \leq 2$ everywhere,
   - the homology class of the fibers of the natural projection $\pi : \mathcal{M}_g \to M_g$ is contained in the image of $H_2(M)$ under $\kappa$.

   One then obtains a $\pm$ holomorphic map

   $$k : M \to \overline{\mathcal{M}}_g$$

   with

   $$k(M \backslash A) \in \mathcal{M}_g.$$

   Since $\mathcal{M}_g$ is naturally fibered by curves of genus $g$, then so is $M \backslash A$.

Let us remark that existence results of a different nature have been obtained by Shiga [Sh].

With these techniques, one can also study variations of the singular set. For example, one has

**Theorem 7.1** *Let $C$ be a compact algebraic curve, $S_1, S_2$ linearly equivalent divisors on $C$; assume that $C \backslash S_1$ and $C \backslash S_2$ are not conformally equivalent. Let $f_i : B_i \to C$, $i = 1, 2$, induce fiber spaces over $C \backslash S_i$ with fibers of genus $g \geq 2$. Suppose these fiber spaces are topologically equivalent. Let*

*$L : \mathbf{CP} \to Div(C)$ be a linear system, $L(z_i) = S_i$, $i = 1,2$. Then there exists a holomorphic family or fiber spaces including the two given ones, described by a holomorphic map*

$$H : C \times \mathbf{CP}^m \to \overline{M}_g$$

*with*

$$H(C\backslash L(z) \times \{z\}) \in M_g$$

*for all $z \in \mathbf{CP}^m$ and*

$$H_{|C\backslash L(z_i)} = h_i,$$

*where $h_i : C\backslash S_i \to M_g$ is the map induced by $f_i$, $i = 1,2$.*

Again, this can be generalized to higher dimensions.

The preceding constructions also work for fibrations by algebraic varieties other than curves. The case of fibrations by Abelian varieties was solved by Faltings [F]. While boundedness holds as before, for rigidity one needs an additional condition, even for nontrivial families. Namely, any family of Abelian varieties can be multiplied by a constant Abelian variety, and as the latter can be deformed, also the product can be deformed with necessarily being trivial itself. A more sophisticated example is presented in [F].

The above techniques then apply to reprove these results (the boundedness part was also obtained in [N]). More generally, if one has a fibration of a Kähler manifold $M$ by algebraic varieties, one obtains the period mapping of Griffiths [Gi]

$$\varphi : N\backslash A \to H\backslash G/\Gamma,$$

where $N$ is the base of the fibration, $A$ a divisor, the homogeneous complex manifold is the period domain, and the discrete group $\Gamma$ is the global monodromy group. In general, $H\backslash G$ does not have nonpositive curvature so that the harmonic map techniques do not apply directly, but there is a natural projection

$$\pi : H\backslash G \to K\backslash G$$

onto a Riemannian symmetric space of noncompact type. The composition $\pi \circ \varphi$, while in general not holomorphic, is still harmonic, and thus harmonic map techniques can be applied after all as $K\backslash G$ has nonpositive sectional curvature.

# References


[Al 1] Al'ber, S. I., On n-dimensional problems in the calculus of variation in the large, *Sov. Math. Dokl. 5*, 700-794 (1964).

[Al 2] Al'ber, S. I., Spaces of mappings into a manifold with negative curvature, *Sov. Math. Dokl. 9*, 6-9 (1967).

[A] Arakelov, S. J., Families of algebraic curves with fixed degeneracies, *Math. USSR Izv. 5*, 1277-1302 (1971).

[CT] Carlson, J., and D. Toledo, Harmonic mappings of Kähler manifolds to locally symmetric spaces, *Publ. Math. IHES*, to appear.

[C] Corlette, K., Flat G-bundles with canonical metrics, *J. Diff. Geom. 28*, 361-382 (1988).

[DO] Diederich, K., and T. Ohsawa, Harmonic mappings and disk bundles over compact Kähler manifolds,

[Dn] Donaldson, S., Twisted harmonic maps and the self-duality equations, *Proc. London Math. Soc. 55*, 127- 131 (1987).

[EL] Eells, J., and L. Lemaire, Another report on harmonic maps, *Bull. London Math. Soc. 20*, 385-524 (1988).

[ES] Eells, J., and J. Sampson, Harmonic mappings of Riemannian manifolds, *Amer. J. Math. 85*, 109-160 (1964).

[F] Faltings, G., Arakelov's theorem for Abelian varieties, *Inv. Math. 73*, 337-347 (1983).

[Ga] Grauert, H., Mordells Vermutung über rationale Punkte auf algebraischen Kurven und Funktionenkörper, *Publ. Math. IHES 25*, 363-381 (1965).

[GaR] Grauert, H., and H. Reckziegel, Hermitesche Metriken und normale Familien holomorpher Abbildungen, *Math. Z. 89*, 108-125 (1965).

[Gi] Griffiths, P., periods of integrals on algebraic manifolds: Summary of main results and discussion of open problems, *Bull. AMS 76*, 228-296 (1970).

[Ha] Hartman, P., On homotopic harmonic maps, *Can. J. Math. 19*, 673-687 (1967).

[HKW] Hildebrandt, S., H. Kaul, and K.-O. Widman, An existence theorem for harmonic mappings of Riemannian manifolds, *Acta Math. 138*, 1-16 (1977).

[IS] Imayoshi, Y., and H. Shiga, A finiteness theorem for holomorphic families of Riemann surfaces, in: *Holomorphic functions and moduli II*, pp. 207-220, ed. by D. Drasin, *et al.*, Springer, 1988.

[J1] Jost, J., Harmonic mappings between Riemannian manifolds, *Proc. CMA Vol. 4*, ANU-Press, Canberra, 1984.

[J2] Jost, J., Nonlinear methods in Riemannian and Kählerian geometry, Birkhauser, Basel, Boston, 1988.

[J3] Jost, J., Two dimensional geometric variational problems, Wiley-Interscience, to appear.

[JY1] Jost, J., and S.-T. Yau, Harmonic mappings and Kähler manifolds, *Math. Ann. 262*, 145-166 (1983).

[JY2] Jost, J., and S.-T. Yau, A strong rigidity theorem for a certain class of compact analytic surfaces, *Math. Ann. 271*, 143-152 (1985).

[JY3] Jost, J., and S.-T. Yau, The strong rigidity of locally symmetric complex manifolds of rank one and finite volume, *Math. Ann. 275*, 291-304 (1986).

[JY4] Jost, J., and S.-T. Yau, On the rigidity of certain discrete groups and algebraic varieties, *Math. Ann. 278*, 481-496 (1987).

[JY5] Jost, J., and S.-T. Yau, Harmonic maps and group representations, to appear in volume in honor of M. doCarmo (B. Lawson, ed.).

[JY6] Jost, J., and S.-T. Yau, Harmonic mappings and algebraic varieties over function fields, preprint.

[KN] Kobayashi, S., and K. Nomizu, Foundations of differential geometry, Vol. 1, Wiley, New York, 1963.

[Lb] Labourie, F., Existence d'applications harmoniques tordues à valeurs dans les variéteś à courbure négative, preprint.

[M] Manin, Ju. I., Rational points of algebraic curves over function fields, *Amer. Math. Soc. Transl. 50*, 189-234 (1966).

[Ma] Margulis, G. A., Discrete groups of motion of manifolds of nonpositive curvature, *AMS Transl. 190*, 33-45 (1977).

[Ma2] Margulis, G. A., Discrete subgroups of Lie groups, monograph, to appear.

[Mk1] Mok, N., The holomorphic or antiholomorphic character of harmonic maps into irreducible compact quotients of polydiscs, *Math. Ann. 272*, 197-216 (1985).

[Mk2] Mok, N., Uniqueness theorems of Hermitian metrics of seminegative curvature on quotients of bounded symmetric domains, *Ann. Math. 125*, 105-152 (1987).

[Mk3] Mok, N., Strong rigidity of irreducible quotients of polydiscs of finite volume, *Math. Ann. 282*, 555-578 (1988).

[Mo] Mostow, G., Strong rigidity of locally symmetric spaces, *Ann. Math. Studies 78*, Princeton, 1973.

[N] Noguchi, J., Moduli spaces of holomorphic mappings into hyperbolically imbedded complex spaces and locally symmetric spaces, *Inv. math. 93*, 15-34 (1988).

[P] Parshin, A. N., Algebraic curves over function fields, *I., Math. USSR Izv. 2*, 1145-1170 (1968).

[Pr] Prasad, G., Strong rigidity of Q-rank 1 lattices, *Inv. math. 21*, 255-286 (1973).

[R] Royden, H., The Ahlfors-Schwarz lemma in several complex variables, *Comment. Math. Helv. 55*, 547-558 (1980).

[SU] Sacks, J., and K. Uhlenbeck, The existence of minimal immersions of 2-spheres, *Ann. Math. 113*, 1-24 (1981).

[Sa] Sampson, J., Applications of harmonic maps to Kähler geometry, *Contemp. Math. 49*, 125-134 (1986).

[Su] Schumacher, G., Harmonic maps of the moduli space of compact Riemann surfaces, *Math. Ann. 275*, 455-466 (1986).

[Sh] Shiga, Lecture at the Taniguchi Symposium, Katata, August 1989.

[S1] Siu, Y.-T., The complex analyticity of harmonic maps and the strong rigidity of compact Kähler manifolds, *Ann. Math. 112*, 73-111 (1980).

[S2] Siu, Y.-T., Complex analyticity of harmonic maps, vanishing, and Lefschetz theorems, *J. Diff. Geom. 17*, 555-138 (1982).

[S3] Siu, Y.-T., Strong rigidity for Kähler manifolds and the construction of bounded holomorphic functions, pp. 124-151, in: R. Howe (ed.), Discrete groups in geometry and analysis, Birkhauser, Boston, Basel, 1987.

[SZ] Spatzier, and R. Zimmer, Fundamental groups of negatively curved manifolds and actions of semisimple groups, Preprint.

[T] Tromba, A., On a natural affine connection on the space of almost complex structures and the curvature of Teichmüller space with respect to its Weil-Petersson metric, *Man. Math. 56*, 475-497 (1986).

[W1] Wolpert, S., Chern forms and the Riemann tensor for the moduli space of curves, *Inv. Math. 85*, 119-145 (1986).

[W2] Wolpert, S., Geodesic length functions and the Nielsen problem, *J. Diff. Geom. 25*, 275-296 (1987).

[Y] Yau, S.-T., A general Schwarz lemma for Kähler manifolds, *Amer. J. Math. 100*, 197-203 (1978).

# Complex-Analyticity of Pluriharmonic Maps and their Constructions

Yoshihiro Ohnita *and* Seiichi Udagawa

Department of Mathematics, Tokyo Metropolitan University,
Fukasawa, Setagaya, Tokyo 158, Japan
*and*
Nihon University, Department of Mathematics,
School of Medicine, Itabashi, Tokyo 173, Japan

## Introduction.

The theory of harmonic maps is one of the most interesting and important topics in differential geometry. Let $M$ and $N$ be compact Riemannian manifolds. Consider harmonic maps between $M$ and $N$. If $N$ is nonpositively curved, we already know the general existence theorem due to Eells and Sampson [E-S] and the rigidity and classification theorems and their remarkable applications (cf. [E-L1,2]). If $N$ is nonnegatively curved, in general it is very hard to expect such results. So, in particular, when $N$ is a projective space, a compact Lie group and a symmetric space of compact type, it is a good way to find a nice class of harmonic maps into such manifolds and study its properties and structures.

Let $\varphi : M \longrightarrow N$ be a smooth map from a complex manifold into a Riemannian manifold. Let $\varphi^{-1}TN$ be the pull-back of the tangent bundle of $N$ over $M$ by $\varphi$. The differential of $\varphi$, denoted by $d\varphi$, may be interpreted as a smooth section of $T^*M \otimes \varphi^{-1}TN$. By the complex structure of $M$, the complexification of the tangent bundle of $M$, denoted by $TM^{\mathbf{C}}$, is decomposed as

$$TM^{\mathbf{C}} = TM^{1,0} \oplus TM^{0,1}, \tag{0.1}$$

where $TM^{1,0}$(resp. $TM^{0,1}$) is the $\sqrt{-1}$(resp. $-\sqrt{-1}$)-eigenspace of the complex structure tensor of $M$. By (0.1), $d\varphi$ is decomposed as

$$d\varphi = \partial\varphi + \bar{\partial}\varphi. \tag{0.2}$$

By the $\bar{\partial}$-operator of $M$ and the pull-back connection $\nabla^{\varphi}$ on $\varphi^{-1}TN$, we define the $(0,1)$-exterior covariant derivative of $\partial\varphi$, denoted by $D''\partial\varphi$, as

$$(D''\partial\varphi)(\bar{Z}, W) = \nabla^{\varphi}_{\bar{Z}}(\partial\varphi(W)) - \partial\varphi(\bar{\partial}_{\bar{Z}} W) \tag{0.3}$$

for $Z, W \in C^{\infty}(TM^{1,0})$. Then $\varphi$ is called *pluriharmonic* if

$$D''\partial\varphi \equiv 0. \tag{0.4}$$

If $\varphi$ is a pluriharmonic map, then $\varphi$ is a harmonic map with respect to any Kähler metric on $M$. Note that any holomorphic or antiholomorphic map from a complex manifold into a Kähler manifold is pluriharmonic. If $M$ is a Riemann surface, the pluriharmonicity of $\varphi$

coincides with the usual harmonicity of $\varphi$. Moreover, we observe that $\varphi$ is pluriharmonic if and only if for any holomorphic curve $\iota : C \longrightarrow M$, $\varphi \circ \iota : C \longrightarrow N$ is always harmonic. Thus the pluriharmonic maps from complex manifolds can be considered as a natural extension of harmonic maps from Riemann surfaces.

The purpose of the present lecture is to discuss the relationship of complex-analyticity, constancy and stability of pluriharmonic maps of compact Kähler manifolds into positively curved Riemannian manifolds and Hermitian symmetric spaces of compact type, and the construction and classification of pluriharmonic maps into compact Lie groups and symmetric spaces, in particular unitary groups and complex Grassmann manifolds.

In 1980, Siu [Si1] proved the strong rigidity of compact irreducible symmetric bounded domains. In his work, the existence theorem of harmonic maps into manifolds of nonpositive sectional curvatures due to [E-S] and the complex-analyticity of harmonic maps from compact Kähler manifolds into compact quotient of irreducible symmetric bounded domains played essential roles. As a consequence of his analysis, it is observed that any harmonic map from a compact Kähler manifold into a Kähler manifold of strongly nonpositive curvature tensor is always pluriharmonic. Hermitian symmetric spaces of noncompact type are typical examples of Kähler manifolds of strongly nonpositive curvature tensor. Moreover, Siu [Si4] proved that if $\varphi : M \longrightarrow N$ is a harmonic map from a compact Kähler manifold into an irreducible Hermitian symmetric space of noncompact type satisfying the condition $\max_M \operatorname{rank}_{\mathbf{R}} d\varphi \geq 2p(N) + 1$, then $\varphi$ is holomorphic or anti-holomorphic ($\pm$-holomorphic), where $p(N)$ is an integer defined as the degree of strong nondegeneracy of the holomorphic bisectional curvature of $N$, which was computated by Siu [Si4] for classical type and by Zhong [Zh] for exceptional type.

Our main interest is the case where the target manifold is of nonnegative curvature. In this case, the situation is considerably different from that for the case of nonpositive curvature. It seems to be difficult to expect general theorems on existence and complex-analyticity of harmonic maps. In fact, there exists so many non$\pm$-holomorphic but pluriharmonic maps into irreducible Hermitian symmetric spaces of compact type (cf. [O-U2],[O-V]). And we can find a lot of non-pluriharmonic but harmonic maps from some specific compact Kähler manifolds into complex projective spaces (cf. [Oh2]).

On the other hand, it is important to study the stability of harmonic maps in our case. For the case where the target manifold is of nonpositive sectional curvature, it is known that every harmonic map from a compact Riemannian manifold is stable, that is, the second variation of the energy functional is nonnegative. Recall that any $\pm$-holomorphic map from a compact Kähler manifold into a Kähler manifold is energy-minimizing in its homotopy class. Our first interest is in investigating the stability, complex-analyticity and constancy of pluriharmonic maps from Kähler manifolds and the second is in constructing non$\pm$-holomorphic but pluriharmonic maps. We can also find many non-pluriharmonic but harmonic maps into complex projective spaces ([Oh2], [B-G-T]).

In Section 6, we prove that any stable harmonic map from a complex projective space with Fubini-Study metric into a Kähler manifold with positive bisectional curvature is $\pm$-holomorphic.

In Section 7, we prove that any pluriharmonic map from a compact complex manifold with positive first Chern class (outside the singularity set of $\operatorname{codim}_{\mathbf{C}} \geq 2$) into some complex Grassmann manifolds of low dimension may be constructed by certain explicit method from a holomorphic (or a rational) map into a complex Grassmann manifold. However, the problem for the explicit construction of any pluriharmonic map into a general complex Grassmann manifold is still open.

Table of Contents.



## 1. Pluriharmonic maps.

Let $M$ be a connected complex manifold of complex dimension $m$. Denote by $J$ the complex structure tensor of $M$. Let $N$ be a Riemannian manifold with a metric $h$. The pluriharmonicitiy of a map depends only on the complex structure of a domain complex manifold. From the definition we see

**(1.1) Proposition.** *A smooth map $\varphi : M \longrightarrow N$ is pluriharmonic if and only if, for any holomorphic map $\nu : S \longrightarrow M$ from a complex manifold $S$ to $M$, $\varphi \circ \nu : S \longrightarrow N$ is also pluriharmonic.*

**(1.2) Proposition.** *Let $(M, g, J)$ be a Hermitian manifold of fundamental 2-form $\omega$, $\omega(X,Y) = g(JX,Y)$. Then $(M,g,J)$ is cosymplectic Hermitian or semi-Kähler, $d\omega^{m-1} = 0$, (resp. Kähler, $d\omega = 0$) if and only if any pluriharmonic map $\varphi : (U,J) \longrightarrow N$, where $U$ is an arbitrary open subset of $M$, is always harmonic (resp. (1,1)-geodesic* (cf. [Rw])*) with respect to $g$.*

*Proof.* We observe that

$$(D''\partial\varphi)(\bar{Z}, W) = (\nabla d\varphi)(\bar{Z}, W) + d\varphi((\nabla^g - \bar{\partial})_{\bar{Z}} W),$$

where $\nabla d\varphi$ denote the second fundamental form of the map $\varphi$. Hence

$$\mathrm{tr}_g D''\partial\varphi = \frac{1}{2}\tau_\varphi + d\varphi(\sum_\alpha (\nabla^g - \bar{\partial})_{\bar{u}_\alpha} u_\alpha),$$

where $\tau_\varphi$ is the tension field of $\varphi$ and $\{u_\alpha\}$ is a local unitary frame field of $TM^{1,0}$ relative to $g$. Note that $\sum_\alpha (\nabla^g - \bar{\partial})_{\bar{u}_\alpha} u_\alpha \doteqdot 0$ (resp. $\nabla^{0,1} = \bar{\partial}$ ) if and only if $d\omega^{m-1} = 0$ (resp. $d\omega = 0$). (1.2) follows from them. Q.E.D.

**(1.3) Lemma.** *If $\varphi : M \longrightarrow N$ is pluriharmonic, then*

$$R^N(d\varphi(Z), d\varphi(V))d\varphi(\bar{W}) = 0$$

*for each $Z, V, W \in T_xM^{1,0}$ and each $x \in M$.*

This follows from $(D'' \circ D'')\partial\varphi = 0$ and $\bar{\partial} \circ \bar{\partial} = 0$.

**(1.4) Proposition.** *Let $\mathcal{R}^N : \wedge^2 TN \longrightarrow \wedge^2 TN$ denote the curvature operator of $N$. Assume that $\mathcal{R}^N \geq 0$. If $\varphi : M \longrightarrow N$ is pluriharmonic, then the curvature form $R^\varphi$ of the induced connection $\nabla^\varphi$ is of type $(1,1)$. Hence $\varphi^{-1}TN^{\mathbf{C}}$ has a unique holomorphic vector bundle structure whose $\bar{\partial}$-operator coincides with the (0,1)-part of $\nabla^\varphi$. Relative to the holomorphic vector bundle structure and the induced fibre metric, the connection $\nabla^\varphi$ is a Hermitian connection.*

*Remark.* Conversely, if $\varphi : M \longrightarrow N$ is a harmonic map from a compact Kähler manifold and $R^\varphi$ is of type (1,1), then $\varphi$ is pluriharmonic.

For a smooth map $\varphi : M \longrightarrow N$, we define the *energy form* $\varepsilon(\varphi)$ of $\varphi$ as the (1,1)-part of $\varphi^* h$. $\varepsilon(\varphi)$ is a nonnegative (1,1)-form on $M$. The *energy* of $\varphi$ relative to a Hermitian metric $g$ on $M$ is

$$E_\omega(\varphi) = \frac{1}{(m-1)!}\int_M \varepsilon(\varphi) \wedge \omega^{m-1}.$$

Here $\omega$ is a fundamental 2-form of $g$. By Ricci identity a simple computation shows

$$\begin{aligned} &m(m-1)\sqrt{-1}(\partial\bar{\partial}\varepsilon(\varphi)) \wedge \omega^{m-2} \\ &= \{|D''\partial\varphi|^2 - |\mathrm{tr}_g D''\partial\varphi|^2 - \sum_{i,j=1}^{m} h(\mathcal{R}^N(\partial_i\varphi \wedge \partial_j\varphi), \overline{\partial_i\varphi \wedge \partial_j\varphi})\}\omega^m, \end{aligned}$$

where $\{\partial_i\varphi\}$ are the components of $\partial\varphi$ with respect to a local unitary frame field of $TM^{1,0}$ relative to $g$. This is a slight extension of Bochner type identity of [Si1].

A smooth map $\varphi : M \longrightarrow N$ is called *pluriconformal* (cf. [O-V]) if the (2,0)-part $(\varphi^* h)^{(2,0)} = \overline{(\varphi^* h)^{(0,2)}}$ of $\varphi^* h$ vanishes identically. $\varphi$ is pluriconformal if and only if, for each holomorphic curve $i : C \longrightarrow M$, $\varphi \circ i : C \longrightarrow N$ is always (weakly) conformal. If $\varphi$ is pluriharmonic, then $(\varphi^* h)^{(2,0)}$ is a holomorphic section of $\otimes^2 T^* M^{1,0}$. Therefore, we see the following

**(1.5) Proposition** (cf. [O-U2]). *If $M$ is a compact complex manifold with positive first Chern class $c_1(M) > 0$, then any pluriharmonic map $\varphi : M \longrightarrow N$ is always pluriconformal.*

Here we present some examples of pluriharmonic maps.

(1) When $N$ is a Kähler manifold, any holomorphic map $\varphi : M \longrightarrow N$ is pluriharmonic and pluriconformal.

(2) When $M$ is a Kähler manifold, any totally geodesic map $\varphi : M \longrightarrow N$ is pluriharmonic. Particularly, totally geodesic immersions of Hermitian symmetric spaces into symmetric spaces are such nice examples (cf. [C-N], [O-U2], [Te]).

(3) When $M$ is a Kähler manifold and $N$ is a quaternionic Kähler manifold, any totally complex immersion $\varphi : M \longrightarrow N$ is pluriharmonic (cf. [Ts]).

(4) When $M$ is a compact Kähler manifold and $N$ is a Kähler manifold with strongly nonpositive curvature tensor or a Riemannian manifold with nonpositive curvature operator, any harmonic map $\varphi : M \longrightarrow N$ is always pluriharmonic (cf. [Si3], [O-U2]).

(5) When $M$ is a complex projective space $\mathbf{C}P^m$ with the Fubini-Study metric and $N$ is an arbitrary Riemannian manifold, any stable harmonic map $\varphi : M \longrightarrow N$ is always pluriharmonic ([Oh1]).

## 2. Constancy of pluriharmonic maps from Kähler manifolds into positively curved Riemannian manifolds.

Let $\varphi : (M, g, J) \longrightarrow (N, h)$ be a pluriharmonic map from a compact Kähler manifold of complex dimension $m$ into a Riemannian manifold. Let $\varepsilon(\varphi)$ denote the energy form of $\varphi$. If $\partial\varphi$ is not injective everywhere on $M$, $(\varepsilon(\varphi))^m = \varepsilon(\varphi) \wedge \ldots \wedge \varepsilon(\varphi) = 0$ on $M$, which implies $[\varepsilon(\varphi)]^m = 0$ as an element of $H^*(M, \mathbf{R})$. If $\mathrm{b}_2(M) = \dim H^2(M, \mathbf{R}) = 1$, then $[\varepsilon(\varphi)] = 0$ as an element of $H^2(M, \mathbf{R})$, hence $\varepsilon(\varphi) = 0$ since $M$ is compact Kähler and $\varepsilon(\varphi)$ is nonnegative. Thus we obtain

**(2.1) Proposition** ([O-U2]). *Let $M$ be a compact Kähler manifold with $\mathrm{b}_2(M) = 1$.*
*(1) If $\mathrm{rank}_{\mathbf{C}}\partial\varphi < m$ on $M$, $\varphi$ is constant.*
*(2) If $c_1(M) > 0$ and $\varphi$ is not constant, then there exists an open subset $U$ of $M$ such that (i) $\varphi : U \longrightarrow N$ is an immersion, (ii) $(U, \varphi^*h, J)$ is a Kähler manifold, and (iii) $\varphi : (U, \varphi^*h, J) \longrightarrow (N, h)$ is a pluriharmonic isometric immersion.*

$(\varphi^*h)^{(2,0)}$ is an obstruction for the pluriconformality of $\varphi$ (cf. Section 1) and it is given by $\sqrt{-1}h(\partial\varphi, \partial\varphi)$. By the Ricci identity we get

**(2.2) Proposition** ([O-U2]). *Let $\varphi : (M, g) \longrightarrow (N, h)$ be a harmonic map from Kähler manifold to a Riemannian manifold. Then,*

$$\sum_{i,j} g^{k\bar{i}} g^{l\bar{j}} \nabla_{\bar{j}} \nabla_{\bar{i}} h(\partial_k\varphi, \partial_l\varphi) \tag{2.3}$$
$$= \sum h(\nabla_{\bar{j}}\partial_k\varphi, \nabla_l\partial_{\bar{i}}\varphi) g^{k\bar{i}} g^{l\bar{j}} - \sum h(\partial_k\varphi, R^N(\partial_{\bar{i}}\varphi, \partial_{\bar{j}}\varphi)\partial_l\varphi) g^{k\bar{i}} g^{l\bar{j}},$$

*where $R^N$ denotes the curvature tensor of $N$.*

The term involving the curvature of $N$ in (2.3) is closely related to the condition of positivity on totally isotropic 2-planes due to Micallef and Moore [M-M]. By (2.3) this curvature term is zero if $\varphi$ is pluriharmonic.

*Definition.* Let ( , ) be the complex extension of the real inner product of the tensor bundles on $N$ induced by $h$. Then a complex subspace $\sigma \subset T_pN^{\mathbf{C}}$, $p \in N$ is called *totally isotropic* if $(z, z) = 0$ for all $z \in \sigma$. The *complex sectional curvature* for $\sigma$, denoted by $\tilde{K}(\sigma)$, defined by

$$\tilde{K}(\sigma) = (R^N(Z, W)\bar{W}, \bar{Z}) \tag{2.4}$$

where $\{Z,W\}$ is a unitary basis for $\sigma$. We say that $N$ has *positive curvature on totally isotropic 2-planes* if $\tilde{K}(\sigma) > 0$ for any totally isotropic 2-plane $\sigma$ of $T_pN^{\mathbf{C}}$ and $p \in N$.

Note that this curvature condition is defined only for $\dim N \geq 4$. If the sectional curvature $K$ of $N$ is pointwise 1/4-pinched, namely $(1/4)\delta < K \leq \delta$, where $\delta$ is a positive function on $N$, then $N$ has positive curvature on totally isotropic 2-planes ([M-M]). And if $N$ has positive curvature operator, then $N$ has positive curvature on totally isotropic 2-planes. They [M-M] proved that any simply connected compact $n$-dimensional Riemannian manifold which has positive curvature on totally isotropic 2-planes with $n \geq 4$ is homeomorphic to a sphere.

Now assume that $c_1(M) > 0$ and $N$ has positive curvature on totally isotropic 2-planes. Then we have $(\varphi^*h)^{(2,0)} = \overline{(\varphi^*h)^{(0,2)}} = 0$, hence $\mathrm{span}_{\mathbf{C}}\{\partial_i\varphi, \partial_j\varphi; 1 \leq i \neq j \leq m\}$ is a totally isotropic 2-planes. Using the fact that the curvature term in (2.3) is zero and the assumption that $N$ has positive curvature on totally isotropic 2-planes, we get that $\mathrm{rank}_{\mathbf{C}}\partial\varphi \leq 1$ on $M$. This, together with (2.1), yields

**(2.5) Theorem** ([O-U2]). *Let $\varphi : M \longrightarrow N$ be a pluriharmonic map from a compact Kähler manifold with $c_1(M) > 0$ and $b_2(M) = 1$ into a Riemannian manifold which has positive curvature on totally isotropic 2-planes. Then, one of the following cases occurs*
*(1) $\varphi$ is a constant map.*
*(2) $\dim_{\mathbf{C}} M = 1$ and $\varphi$ is a branched minimal immersion.*

## 3. Complex-analyticity of pluriharmonic maps between Kähler manifolds

Let $N$ be a Kähler manifold and let $Q^{(1,1)}$ be the curvature operator acting on (1,1)-forms of $N$. We say that $N$ has *strongly positive* (resp. *nonnegative, negative, nonpositive*) *curvature tensor* if

$$(3.1) \qquad (Q^{(1,1)}(\xi), \bar{\xi}) > 0 \quad (\text{resp. } \geq 0, < 0, \leq 0)$$

for any nonzero $\xi = Z_1 \wedge \bar{W}_1 + Z_2 \wedge \bar{W}_2$, $Z_1, Z_2, W_1, W_2 \in T_xN^{1,0}$. If $N$ has strongly positive curvature tensor, then the sectional curvature of $N$ is positive. Indeed, the positivity of the sectional curvature is equivalent to the condition

$$(Q^{(1,1)}(\xi), \bar{\xi}) > 0$$

for any nonzero $\xi = Z \wedge \bar{W} + \bar{Z} \wedge W$, $Z, W \in T_xN^{1,0}$.

Now we state Siu's result.

**(3.2) Theorem** ([Si1],[Si4]). *Let $\varphi : M \longrightarrow N$ be a harmonic map from a compact Kähler manifold. Then, $\varphi$ is $\pm$-holomorphic provided (i) $N$ has strongly negative curvature tensor* *and* $\max_M \operatorname{rank}_{\mathbf{R}} d\varphi \geq 4$ *or (ii)* $N$ *is a compact quotient of an irreducible symmetric bounded domain and* $\max_M \operatorname{rank}_{\mathbf{R}} d\varphi \geq 2p(N)+1$.

In fact, by (2.3) and Stokes' theorem, we have

$$\begin{aligned}(3.3)\qquad &\sum h(\partial_k\varphi, R^N(\partial_{\bar i}\varphi, \partial_{\bar j}\varphi)\partial_l\varphi)g^{k\bar i}g^{l\bar j}\\ &=\sum h(Q^N(\partial_{\bar i}\varphi^{1,0}\wedge\partial_{\bar j}\varphi^{0,1}-\partial_{\bar j}\varphi^{1,0}\wedge\partial_{\bar i}\varphi^{0,1}),\\ &\qquad \partial_k\varphi^{1,0}\wedge\partial_l\varphi^{0,1}-\partial_l\varphi^{1,0}\wedge\partial_k\varphi^{0,1})g^{k\bar i}g^{l\bar j}=0.\end{aligned}$$

Because the 1st equality of (3.3) and the strong nonpositivity of the curvature tensor of $N$, which is the case for both (i) and (ii), imply that the right hand side of (2.3) is nonnegative. Then one obtains (3.2) by analyzing the 2nd equality of (3.3). The proof of (ii) in (3.2) is more technical. Set $\xi = \bar{*}\wedge^p(\bar\partial\varphi^{1,0})$, which is a $(m, m-p)$-form on $M$ with values in $\varphi^*\wedge^p TN^{1,0}$, where $\bar{*}$ is the generalized Hodge star operator composed with the complex conjugation. Let $\Delta_c$ be the complex Laplacian defined by $\Delta_c = \bar D^*\bar D + \bar D\bar D^*$, where $\bar D$ is the covariant exterior derivative which sends $\varphi^*\wedge^p TN^{1,0}$-valued (p,q)-forms to $\varphi^*\wedge^p TN^{1,0}$-valued (p,q+1)-forms. Then we have

$$\begin{aligned}(3.4)\qquad &(\Delta_c\xi,\bar\xi)\\ &=-\sum(\nabla_i\nabla_{\bar j}\xi,\bar\xi)g^{i\bar j}\\ &\quad-(1/(p+1)!(p-1)!)\\ &\quad\sum R^N_{\bar\beta_1\alpha_1\gamma\bar\delta}\varphi^{\bar\beta_1\ldots\bar\beta_p\gamma}_{i_1\ldots i_{p+1}}\varphi^{\alpha_1\ldots\alpha_p\bar\delta}_{\bar j_1\ldots\bar j_{p+1}}h_{\alpha_2\bar\beta_2}\cdots h_{\alpha_p\bar\beta_p}g^{i_1\bar j_1}\cdots g^{i_{p+1}\bar j_{p+1}},\end{aligned}$$

where

$$\begin{aligned}(3.5)\qquad &\partial\varphi^{\bar\beta_1}\wedge\ldots\wedge\partial\varphi^{\bar\beta_p}\wedge\partial\varphi^{\gamma}\\ &=(1/(p+1)!)\sum\varphi^{\bar\beta_1\ldots\bar\beta_p\gamma}_{i_1\ldots i_p j}dz^{i_1}\wedge\ldots\wedge dz^{i_p}\wedge dz^{j}.\end{aligned}$$

Choose a point $x\in N$ such that $\operatorname{rank}_{\mathbf{R}}(d\varphi)_x \geq 2p(N)+1$. Set $p=p(N)$ and $r=\operatorname{rank}_{\mathbf{C}}(\bar\partial\varphi^{1,0})_x$. If $\varphi$ is harmonic, then the left hand side of (3.4) vanishes. However, the right hand side of (3.4) is nonnegative after integrating it over $M$, because the curvature term is negative for nonzero $\varphi^{\bar\beta_1\ldots\bar\beta_p\gamma}_{i_1\ldots i_{p+1}}$, which is directly shown by Siu and Zhong [Si4, Zh] using the properties of the curvature tensor of $N$. Thus, we have $\varphi^{\bar\beta_1\ldots\bar\beta_p\gamma}_{i_1\ldots i_{p+1}}=0$. If $r\geq p+1$, then $\varphi$ is anti-holomorphic at $x$. If $r=0$, $\varphi$ is holomorphic at $x$. If $0<r\leq p$, this and the strong nondegeneracy of the bisectional cuvature of $N$ contradict the assumption $\operatorname{rank}_{\mathbf{R}}(d\varphi)_x\geq 2p+1$. We omit the details. Therefore, $\varphi$ is $\pm$-holomorphic at $x$. However, there exists an open subset $U$ of $N$ such that $\operatorname{rank}_{\mathbf{R}} d\varphi \geq 2p+1$ on $U$, hence $\varphi$ is $\pm$-holomorphic on $U$. Then, the result follows from the following

**(3.6) Theorem** (Unique continuation theorem [Si1]). *Let $\varphi : M \longrightarrow N$ be a harmonic map between Kähler manifolds. If $\varphi$ is $\pm$-holomorphic on some open subset of $M$, then $\varphi$ is $\pm$-holomorphic on $M$.*

In the following, we assume that $\varphi : (M, g) \longrightarrow (N, h)$ *be a pluriharmonic map between Kähler manifolds* unless otherwise stated.

Modifying the proof of (3.2), we have

**(3.7) Theorem** ([Ud3]). *Assume that $N$ is an irreducible Hermitian locally symmetric space of compact or noncompact type. Then, $\varphi$ is $\pm$-holomorphic, provided* $\max_M \operatorname{rank}_{\mathbf{R}} d\varphi \geq 2p(N) + 1$.

**(3.8) Proposition** ([O-U2]). *Assume that $N$ has strongly positive curvature tensor. Then, one of the following cases occurs*

*(i) $\varphi$ is $\pm$-holomorphic,*

*(ii)* $\operatorname{rank}_{\mathbf{C}} \partial\varphi \leq 1$ *on $M$.*

As a consequence of (2.1) and (3.7), we have

**(3.9) Theorem** ([O-U2]). *Assume that $M$ is compact with $c_1(M) > 0$ and $b_2(M) = 1$, and $N$ is an irreducible Hermitian symmetric space of compact type. If $\dim_{\mathbf{C}} M \geq p(N)+1$, then $\varphi$ is $\pm$-holomorphic.*

As a consequence of (2.1) and (3.8), we have

**(3.10) Theorem** ([O-U2]). *Assume that $M$ is an $m(\geq 2)$-dimensional compact Kähler manifold with $b_2(M) = 1$ and $N$ has strongly positive curvature tensor. Then $\varphi$ is $\pm$-holomorphic.*

*Remark.* If $N$ has strongly negative curvature tensor, the assumption of pluriharmonicity of $\varphi$ in (3.10) may be relaxed to that of harmonicity.

**(3.11) Lemma** ([O-U2]). *Assume that $M$ is a compact Kähler manifold with $c_1(M) > 0$. Then*

$$\sum h_{\alpha\bar{\beta}} \partial_i \varphi^\alpha \partial_j \varphi^{\bar{\beta}} dz^i \otimes dz^j = 0. \tag{3.12}$$

In fact, the left hand side of (3.12) is a holomorphic section of $\otimes^2 T^* M^{1,0}$, hence it is identically zero because $c_1(M) > 0$.

**(3.13) Corollary** ([O-U2]). *Assume that $M$ is a compact Kähler manifold with $c_1(M) > 0$ and $N$ is a Riemann surface. Then $\varphi$ is $\pm$-holomorphic.*

This is a direct consequence of (3.12), one dimensionality of $N$ and (3.6).

*Remark.* For the case of $\dim_{\mathbf{C}} M = 1$, (3.13) is due to Eells-Wood [E-W1]. For the case where $M$ has nonpositive Ricci curvature, there is a result of [Na], see also [On].

Consider the (1,1)-forms $\varepsilon'(\varphi)$ and $\varepsilon''(\varphi)$ on $M$ defined by

$$\varepsilon'(\varphi) = \sqrt{-1}\sum h_{\alpha\bar\beta}\partial_i\varphi^\alpha\partial_{\bar j}\varphi^{\bar\beta} dz^i \wedge dz^{\bar j},$$
$$\varepsilon''(\varphi) = \sqrt{-1}\sum h_{\alpha\bar\beta}\partial_i\varphi^{\bar\beta}\partial_{\bar j}\varphi^{\alpha} dz^i \wedge dz^{\bar j}.$$

Note that $\varepsilon(\varphi) = \varepsilon'(\varphi) + \varepsilon''(\varphi)$ and $(\varphi^*\omega^N)^{(1,1)} = \varepsilon'(\varphi) - \varepsilon''(\varphi)$, where $\varepsilon(\varphi)$ is the energy form of $\varphi$ defined in Section 1 and $(\varphi^*\omega^N)^{(1,1)}$ is a (1,1)-part of the pull-back of the Kähler form $\omega^N$ of $N$. $\varepsilon'(\varphi)$ and $\varepsilon''(\varphi)$ are real nonnegative (1,1)-forms on $M$ and they are also $d$-closed by the pluriharmonicity of $\varphi$. Therefore by the argument similar to (2.1) we obtain

**(3.14) Proposition** ([O-U2]). *Assume that $M$ is an $m$-dimensional compact Kähler manifold with $b_2(M) = 1$. If $\mathrm{rank}_{\mathbf{C}}(\bar\partial\varphi^{1,0}) < m$ (resp. $\mathrm{rank}_{\mathbf{C}}(\partial\varphi^{1,0}) < m$) on $M$, then $\varphi$ is holomorphic (resp. anti-holomorphic).*

**(3.15) Corollary** ([O-U2]). *Assume that $M$ is a compact Kähler manifold with $b_2(M) = 1$. If $\dim_{\mathbf{C}} N < \dim_{\mathbf{C}} M$, then $\varphi$ is a constant map.*

As a consequnce of (3.11), (3.14) and (3.6), we obtain

**(3.16) Theorem** ([O-U2]). *Assume that $M$ is a compact Kähler manifold with $c_1(M) > 0$ and $b_2(M) = 1$. If $\dim_{\mathbf{C}} N = \dim_{\mathbf{C}} M$, then $\varphi$ is $\pm$-holomorphic.*

## 4. Stability of harmonic maps from compact Kähler manifolds

The fundamental idea of the use of *stability* is due to Lawson and Simons [L-S], who considered the variation of minimal subvarieties along holomorphic vector fields on the ambient complex projective space. Their method is also applicable to the harmonic map theory in the situation where the domain or the target manifold is a Hermitian symmetric space of compact type.

Let $\varphi : (M, g) \longrightarrow (N, h)$ be a harmonic map between compact Riemannian manifolds. Consider the index form

$$I_\varphi(V, V) = \int_M \langle \mathcal{J}_\varphi(V), V\rangle dv_g \tag{4.1}$$

for $V \in C^\infty(\varphi^{-1}TN)$, where $\mathcal{J}_\varphi$ is the Jacobi operator for $\varphi$. Assume that $M$ is a Hermitian symmetric space of compact type. Set $V = d\varphi(J_M X)$, where $J_M$ is the complex structure tensor of $M$ and $X$ is a Killing vector field on $M$.

**(4.2) Proposition** ([O-U1]). *Let $\varphi : M \longrightarrow N$ be a harmonic map from a Hermitian symmetric space of compact type into a Riemannian manifold. If $\varphi$ is stable, then*

$$\mathcal{R}^M_{(1,1)}(D''\partial\varphi) = 0, \tag{4.3}$$

*where $\mathcal{R}^M_{(1,1)}$ is the curvature operator acting on the space of (1,1)-forms on $M$, which is extended to the operator for vector bundle valued forms on $M$.*

**(4.4) Theorem** ([Oh1]). *Every stable harmonic map from a complex projective space $CP^m$ with the Fubini-Study metric to any Riemannian manifold is pluriharmonic.*

Next we assume that $(M, g)$ is a Riemannian manifold and $N$ is a Hermitian symmetric space of compact type. Set $V = J_N X$, where $J_N$ is the complex structure tensor of $N$ and $X$ is a Killing vector field on $N$. Then we obtain

$$\sum_{i=1}^{m}\{J_N R^N(X, d\varphi(e_i))d\varphi(e_i) - R^N(J_N X, d\varphi(e_i))d\varphi(e_i)\} = 0 \tag{4.5}$$

for any $X \in C^\infty(TN)$, where $\{e_i\}_{i=1}^m$ is a local orthonormal frame field of $M$. Denote by $\mathcal{Q}^N$ the curvature operator acting on $TN^{1,0} \cdot TN^{1,0}$, the symmetric square of $TN^{1,0}$. Then (4.5) implies

**(4.6) Proposition** ([O-U1]). *If $\varphi : M \longrightarrow N$ is a stable harmonic map from a compact Riemannian manifold into a Hermitian symmetric space of compact type. Then*

$$\mathcal{Q}^N(\sum_i d\varphi^{1,0}(e_i) \cdot d\varphi^{1,0}(e_i)) = 0. \tag{4.7}$$

If $N$ is irreducible, it is known that Ker $\mathcal{Q}^N = 0$ (cf. [Bo],[C-V],[It]), which implies that

$$\sum_i d\varphi^{1,0}(e_i) \cdot d\varphi^{1,0}(e_i) = 0. \tag{4.8}$$

Therefore, we obtain

**(4.9) Theorem** ([B-B-B-R], [O-U1]). *Any stable harmonic map from a compact Riemann surface into an irreducible Hermitian symmetric space of compact type is $\pm$-holomorphic.*

Another conclusions of the equation (4.8) due to [B-B-B-R] are as follows : The image of $d\varphi$ is $J_N$-invariant and hence of even dimension. If $d\varphi$ is injective on some open subset $U$ of $M$, then $\varphi$ is a local diffeomorphism of $U$ onto a complex submanifold of $N$. Thus $U$ has the complex structure tensor $J$ defined by $J = d\varphi^{-1} \circ J_N \circ d\varphi$ with respect to which $\varphi$ is holomorphic. Moreover, $J$ is Hermitian with respect to the given metric on $M$. Differentiating covariantly the equation $d\varphi \circ J = J_N \circ d\varphi$ and using the parallelism of $J_N$ and the harmonicity of $\varphi$, we see that tr $\nabla J = 0$, that is, the Hermitian structure $(U, g, J)$ has coclosed fundamental 2-form. Thus one obtain

**(4.10) Theorem** ([B-B-B-R]). *If $\varphi : M \longrightarrow N$ is a stable harmonic immersion of a compact Riemannian manifold into an irreducible Hermitian symmetric space of compact type, then $M$ is of even dimension and $M$ admits a complex structure which is cosymplectic Hermitian relative to the given metric $g$, and $\varphi$ is holomorphic relative to this complex structure.*

If $\dim U = 4$, then a fundamental 2-form is closed if and only if it is coclosed. Thus, in this case $(U, g, J)$ is a Kähler manifold.

**(4.11) Theorem** ([B-B-B-R]). *Let $\varphi : M \longrightarrow N$ be a stable harmonic map from a real analytic 4-dimensional Riemannain manifold into an irreducible Hermitian symmetric space of compact type. Assume that $\max_M \operatorname{rank}_{\mathbf{R}} d\varphi \geq 3$. Then $M$ has a unique Kähler structure $J$ such that $\varphi$ is holomorphic with respect to $J$.*

The following formula (4.12) will be used in Section 5. When $M$ is a Riemann surface, (4.13) is the same as a formula of [M-M].

**(4.12) Proposition.** *Let $\varphi : (M, g) \longrightarrow (N, h)$ be a harmonic map from a compact Kähler manifold to a Riemannian manifold. Then, the index form $I_\varphi$ for $\varphi$ is given by*

$$\begin{aligned}(4.13)\qquad & I_\varphi(V, \bar{V}) \\ &= 2\int_M \sum_\alpha \{|\nabla_{\bar{u}_\alpha} V|^2 - \langle R^N(d\varphi(u_\alpha), V)\bar{V}, d\varphi(\bar{u}_\alpha)\rangle\} dv_g\end{aligned}$$

*for any $V \in C^\infty(\varphi^{-1}TN^{\mathbf{C}})$, where $\{u_\alpha\}$ is a unitary basis of $T_pM^{1,0}$, $p \in M$. In particular, if $(N, h)$ is a Kähler manifold, (4.13) becomes*

$$\begin{aligned}(4.14)\qquad & I_\varphi(V, \bar{V}) \\ &= 2\int_M \sum_\alpha \{|\nabla_{\bar{u}_\alpha} V|^2 - \langle R^N(\partial\varphi^{0,1}(u_\alpha), V)\bar{V}, \overline{\partial\varphi^{0,1}(u_\alpha)}\rangle\} dv_g\end{aligned}$$

*for any $V \in C^\infty(\varphi^{-1}TN^{1,0})$.*

*Proof.* The usual second variational formula is given by (cf. [E-L2])

$$I_\varphi(V, \bar{V}) = \int_M \sum_i \{\langle \nabla_{e_i} V, \nabla_{e_i} \bar{V}\rangle - \langle R^N(V, d\varphi(e_i))d\varphi(e_i), \bar{V}\rangle\} dv_g,$$

where $\{e_i\}$ is an orthonormal basis of $T_xM$, $x \in M$. Choose a local unitary frame field $\{u_\alpha\}$ of $TM^{1,0}$ such that $(\nabla u_\alpha)_x = 0$. We compute

$$\begin{aligned}&\sum_i \langle R^N(V, d\varphi(e_i))d\varphi(e_i), \bar{V}\rangle \\ &= \sum_\alpha \langle R^N(V, d\varphi(u_\alpha))d\varphi(\bar{u}_\alpha), \bar{V}\rangle + \sum_\alpha \langle R^N(V, d\varphi(\bar{u}_\alpha))d\varphi(u_\alpha), \bar{V}\rangle\end{aligned}$$

$$
\begin{aligned}
&= \sum_\alpha \langle R^N(V, d\varphi(u_\alpha))d\varphi(\bar u_\alpha), \bar V\rangle - \sum_\alpha \langle R^N(d\varphi(\bar u_\alpha), d\varphi(u_\alpha))V, \bar V\rangle \\
&\quad - \sum_\alpha \langle R^N(d\varphi(u_\alpha), V)d\varphi(\bar u_\alpha), \bar V\rangle \\
&= 2\sum_\alpha \langle R^N(d\varphi(u_\alpha), V)\bar V, d\varphi(\bar u_\alpha)\rangle + \sum_\alpha \langle R^N(d\varphi(u_\alpha), d\varphi(\bar u_\alpha))V, \bar V\rangle,
\end{aligned}
$$

and

$$
\begin{aligned}
&\sum_i \langle \nabla_{e_i} V, \nabla_{e_i} \bar V\rangle = \sum_\alpha \langle \nabla_{u_\alpha} V, \nabla_{\bar u_\alpha} \bar V\rangle + \sum_\alpha \langle \nabla_{\bar u_\alpha} V, \nabla_{u_\alpha} \bar V\rangle \\
&= \sum_\alpha \nabla^M_{\bar u_\alpha} \langle \nabla_{u_\alpha} V, \bar V\rangle - \sum_\alpha \langle \nabla_{\bar u_\alpha} \nabla_{u_\alpha} V, \bar V\rangle + \sum_\alpha \langle \nabla_{\bar u_\alpha} V, \nabla_{u_\alpha} \bar V\rangle \\
&= \sum_\alpha \nabla^M_{\bar u_\alpha} \langle \nabla_{u_\alpha} V, \bar V\rangle - \sum_\alpha \langle \nabla_{u_\alpha} \nabla_{\bar u_\alpha} V, \bar V\rangle \\
&\quad + \sum_\alpha \langle R^N(d\varphi(u_\alpha), d\varphi(\bar u_\alpha))V, \bar V\rangle + \sum_\alpha \langle \nabla_{\bar u_\alpha} V, \nabla_{u_\alpha} \bar V\rangle \\
&= (\sum_\alpha \nabla^M_{\bar u_\alpha} \langle \nabla_{u_\alpha} V, \bar V\rangle - \sum_\alpha \nabla^M_{u_\alpha} \langle \nabla_{\bar u_\alpha} V, \bar V\rangle) + 2\sum_\alpha \langle \nabla_{\bar u_\alpha} V, \nabla_{u_\alpha} \bar V\rangle \\
&\quad + \sum_\alpha \langle R^N(d\varphi(u_\alpha), d\varphi(\bar u_\alpha))V, \bar V\rangle.
\end{aligned}
$$

We show

$$
\int_M \sum_\alpha \nabla^M_{\bar u_\alpha} \langle \nabla_{u_\alpha} V, \bar V\rangle dv_g = \int_M \sum_\alpha \nabla^M_{u_\alpha} \langle \nabla_{\bar u_\alpha} V, \bar V\rangle dv_g = 0.
$$

In general, for any $(1,0)$-form $\xi = (\xi_\alpha)$ on $M$, we have

$$
d(\xi \wedge \omega_M^{m-1}) = d\xi \wedge \omega_M^{m-1} = \bar\partial \xi \wedge \omega_M^{m-1},
$$

which implies that

$$
\int_M (\sum_\alpha \nabla^M_{\bar u_\alpha} \xi_\alpha) dv_g = 0.
$$

Similarly, for any $(0,1)$-form $\bar\xi = (\bar\xi_\alpha)$ we have

$$
\int_M (\sum_\alpha \nabla^M_{u_\alpha} \bar\xi_\alpha) dv_g = 0.
$$

Thus, we get (4.13). Q.E.D.

*Remark.* Burstall and Rawnsley also got the formula (4.13).

## 5. Stability of pluriharmonic maps from Kähler manifolds into irreducible Hermitian symmetric spaces of compact type

In this section we assume that $\varphi$ is a *stable pluriharmonic map from a compact Kähler manifold into an irreducible Hermitian symmetric space of compact type* unless otherwise stated. Here a stable pluriharmonic map means a pluriharmonic and stable map as a harmonic map. Let $R^N = (R^N_{\bar\alpha\beta\gamma\bar\delta})$ be the curvature tensor of $N$, where $R^N_{\bar 1 1 1 \bar 1} > 0$ for Riemann sphere $N = \mathbf{C}P^1$. By (4.5), the stability of $\varphi$ implies

$$\sum_{\beta,\gamma,i} R^N_{\bar\alpha\beta\gamma\bar\delta}\varphi^\beta_{\bar i}\varphi^\gamma_{\bar i} = 0, \tag{5.1}$$

for any $\alpha, \delta = 1, \ldots, n = \dim_{\mathbf{C}} N$, where $\varphi^\alpha_{\bar i} = h(d\varphi^{1,0}(e_i), \bar u_\alpha)$ and $\{e_i\}$, $\{u_\alpha\}$ are unitary bases for $TM^{1,0}$ and $TN^{1,0}$, respectively. On the other hand, by (2.3), the pluriharmonicity of $\varphi$ implies

$$\sum_{\alpha,\beta,\gamma,\delta} R^N_{\bar\alpha\beta\gamma\bar\delta}(\varphi^{\bar\alpha}_i\varphi^\beta_{\bar j} - \varphi^{\bar\alpha}_j\varphi^\beta_{\bar i})(\varphi^\gamma_{\bar i}\varphi^{\bar\delta}_j - \varphi^\gamma_{\bar j}\varphi^{\bar\delta}_i) = 0, \tag{5.2}$$

for any $i, j = 1, \cdots, m = \dim_{\mathbf{C}} M$. It follows from (5.1), (5.2) and the nonnegativity of the bisectional curvature of $N$ that

$$\sum_{\alpha,\beta,\gamma,\delta} R^N_{\bar\alpha\beta\gamma\bar\delta}\varphi^{\bar\alpha}_i\varphi^\beta_{\bar i}\varphi^\gamma_{\bar j}\varphi^{\bar\delta}_j = 0 \tag{5.3}$$

for any $i, j = 1, \ldots, m$. Thus, by Siu and Zhong's calculation of the degree of strong non-degeneracy of the bisectional curvature of $N$, we obtain

**(5.4) Theorem ([Ud3]).** *$\varphi$ is $\pm$-holomorphic provided (i) $N = \mathbf{C}P^n$ or (ii) $\max_M\{\mathrm{rank}_{\mathbf{C}}(\partial\varphi^{1,0}) + \mathrm{rank}_{\mathbf{C}}(\bar\partial\varphi^{1,0})\} \geq p(N) + 1$.*

*Remark.* The condition $\max_M \mathrm{rank}_{\mathbf{R}} d\varphi \geq 2p(N) + 1$ implies the condition $\max_M\{\mathrm{rank}_{\mathbf{C}}(\partial\varphi^{1,0}) + \mathrm{rank}_{\mathbf{C}}(\bar\partial\varphi^{1,0})\} \geq p(N) + 1$. In fact, using (5.4) (ii) and (3.14), we may show the following

**(5.5) Theorem ([O-U2]).** *Assume that $b_2(M) = 1$. If $\dim_{\mathbf{C}} M \geq p(N)$, then $\varphi$ is $\pm$-holomorphic.*

On the other hand, (5.1) and the irreducibility of $N$ implies

$$\sum_i \{\varphi^\alpha_{\bar i}\varphi^\beta_{\bar i} + \varphi^\beta_{\bar i}\varphi^\alpha_{\bar i}\} = 0 \tag{5.6}$$

for any $\alpha, \beta = 1, \ldots, n$. This, together with (3.11) and (3.14), implies

**(5.7) Theorem ([O-U2]).** *Assume that $c_1(M) > 0$ and $b_2(M) = 1$. Then, $\varphi$ is $\pm$-holomorphic.*

As an application of (5.7), we obtain

**(5.8) Theorem ([O-U2]).** *Any stable totally geodesic isometric immersion between irreducible Hermitian symmetric spaces of compact type is ±-holomorphic.*

This follows from (5.7) and the following proposition

**(5.9) Proposition ([O-U2]).** *Let $\varphi : M \longrightarrow N$ be a totally geodesic isometric immersion between compact Riemannian manifolds. Then, $\varphi$ is stable as a harmonic map if and only if $\varphi$ is stable as a minimal immersion and the identity map of $M$ is stable as a harmonic map.*

Note that the identity map of compact Kähler manifold is holomorphic, hence stable.

*Remark.* If $\varphi$ is a totally geodesic isometric immersion, the complex structure tensor defined by $J = d\varphi^{-1} \circ J_N \circ d\varphi$ is a Kähler complex structure, hence it follows from (4.10) that if $\varphi : M \longrightarrow N$ is a stable totally geodesic isometric immersion from compact Riemannian manifold into irreducible Hermitian symmetric space of compact type, then there exists a unique Kähler structure on $M$ such that $\varphi$ is holomorphic with respect to this Kähler structure.

**(5.10) Proposition.** *Let $\varphi : M \longrightarrow N$ be a stable pluriharmonic map of a compact homogeneous Kähler manifold into a Kähler manifold with positive bisectional curvature. Then, $\varphi$ is ±-holomorphic.*

*Proof.* In (4.14) set $V = d\varphi^{1,0}(v)$, where $v$ is a holomorphic vector field on $M$. Then, since $\varphi$ is pluriharmonic, we have

$$\nabla_{\bar{Z}} V = (\nabla d\varphi)^{1,0}(v, \bar{Z}) + d\varphi(\nabla^M_{\bar{Z}} v) = 0$$

for any $Z \in C^\infty(TM^{1,0})$. By the stability of $\varphi$ and the positivity of the bisectional curvature of $N$, we obtain $\partial\varphi^{0,1}(Z) = 0$ or $\partial\varphi^{1,0}(W) = 0$ for each $Z, W \in T_pM^{1,0}$, $p \in M$. Therefore, $\varphi$ is ±-holomorphic. Q.E.D.

## 6. Stable harmonic maps from complex projective spaces

As consequences of (2.5), (3.15) and (4.4), we have

**(6.1) Theorem ([O-U2]).** *Let $\varphi : \mathbf{C}P^m \longrightarrow N$ be a stable harmonic map into a Riemannian manifold $N$. Then, $\varphi$ is constant, if one of the following conditions is assumed :*
*(i) $N$ has positive curvature on totally isotropic 2-planes and $\dim N \geq 4$.*
*(ii) $N$ has positive Ricci curvature and $\dim N = 3$.*
*(iii) $N$ is a Kähler manifold and $\dim_{\mathbf{C}} N < m$.*

(6.1) (i) is a partial answer to harmonic map version of Lawson-Simons' conjecture [L-S]. See also [Ho], [Ok] for other partial answers to this conjecture.
As consequences of (3.10), (3.16), (5.7) and (4.4), we have

**(6.2) Theorem** ([O-U2]). *Let $\varphi : \mathbf{C}P^m \longrightarrow N$ be a stable harmonic map into a Kähler manifold N. Then, $\varphi$ is ±-holomorphic, if one of the following conditions is assumed :*
*(i) N has strongly positive curvature tensor.*
*(ii)* $\dim_{\mathbf{C}} N = m$.
*(iii) N is an irreducible Hermitian symmetric space of compact type.*

*Remark.* (6.1) (i), (ii) for $m = 1$ is due to [M-M]. (6.2) (iii) was first obtained by [Oh1] in the case when a target manifold is a complex projective space and by [B-B-B-R] in a different way in the case when a target manifold is an irreducible Hermitian symmetric space of compact type. In (6.2) (i), the case of $m = 1$ is due to [S-Y], who proved that any stable harmonic map from a Riemann sphere to a Kähler manifold with positive bisectional curvature is ±-holomorphic. We prove the following generalization of (6.2) (i).

**(6.3) Theorem.** *Let $\varphi : \mathbf{C}P^m \longrightarrow N$ be a stable harmonic map into a Kähler manifold with positive bisectional curvature. Then, $\varphi$ is ±-holomorphic.*

*Proof of* (6.3). It follows from (4.4) and (5.10). Q.E.D.

## 7. Construction of non±-holomorphic but pluriharmonic maps into complex Grassmann manifolds

In [O-U2], the authors gave a construction of non±-holomorphic but pluriharmonic maps of complex manifold into complex Grassmann manifold from holomorphic maps into complex projective space using the twister fibrations of generalized complex flag manifolds over a complex Grassmann manifold. In this section, we study the alternative method of using the second fundamental form of holomorphic Hermitian vector bundles. In 7.C, we prove a factorization theorem for any pluriharmonic map into certain complex Grassmann manifold.

7.A. **Fundamental tools.** Let $E$ be a unitary vector bundle over a complex manifold $M$ with an Hermitian fibre metric $h$ and a connection $\nabla^E$ compatible with $h$. Let $F$ be a complex subbundle of $E$. Then $F$ also becomes a unitary vector bundle over $M$ with respect to the Hermitian metric induced from $E$. We denote by $H$ the orthogonal complement of $F$ in $E$ with respect to $h$. Similarly, $H$ is also a unitary vector bundle over $M$. The Gauss formulas for subbundles are given by

$$\nabla^E_X v = \nabla^F_X v + A^{F,H}_X(v) \tag{7.1}$$

for any $v \in C^\infty(F)$ and any $X \in C^\infty(TM)$,

$$\nabla^E_X w = \nabla^H_X w + A^{H,F}_X(w) \tag{7.2}$$

for any $w \in C^\infty(H)$ and any $X \in C^\infty(TM)$, where $\nabla^E$, $\nabla^F$ and $\nabla^H$ are the Hermitian connections of $E$,$F$ and $H$, respectively, and $A^{F,H}$ (resp. $A^{H,F}$) is Hom$(F,H)$ (resp. Hom$(F,H)$)-valued 1-form on $M$. Then we have

$$A^{F,H} = -(A^{H,F})^*, \tag{7.3}$$

where $(\ )^*$ denotes the adjoint of $(\ )$ with respect to $h$. According to the decomposition $T^*M^{\mathbf{C}} = T^*M^{1,0} \oplus T^*M^{0,1}$, we decompose

$$A^{F,H} = A^{F,H}_{(1,0)} + A^{F,H}_{(0,1)}. \tag{7.4}$$

Let $D$ be the exterior covariant diffrentiation with respect to the induced connection $\nabla^{F^*\otimes H}$, and $D'$ and $D''$ denote the (1,0)- and (0,1)-parts of $D$, respectively, $D = D' + D''$. The (0,1)-exterior covariant derivative $D''A^{F,H}_{(1,0)}$ of $A^{F,H}_{(1,0)}$ is defined by

$$(D''A^{F,H}_{(1,0)})(\bar{Z}, W) = \nabla_{\bar{Z}} A^{F,H}_{W} - A^{F,H}_{\bar{\partial}_Z W} \quad \text{and} \tag{7.5}$$
$$\nabla_{\bar{Z}} A^{F,H}_{W} = \nabla^{H}_{\bar{Z}} \circ A^{F,H}_{W} - A^{F,H}_{W} \circ \nabla^{F}_{\bar{Z}}$$

for $Z, W \in C^\infty(TM^{1,0})$. Similarly, $D'A^{F,H}_{(0,1)}$ is defined. Assume that $E$ is a Hermitian holomorphic vector bundle and $F$ is a holomorphic subbundle of $E$. We endow $H$ with a holomorphic vector bundle structure by the isomorphism $H \cong E/F$. Then, $\mathrm{Hom}(F, H)$ is a Hermitian holomorphic vector bundle over $M$. $A \in C^\infty(T^*M^{1,0} \otimes \mathrm{Hom}(F, H))$ is called a *holomorphic section* if $D''A \equiv 0$.

Let $\varphi : M \longrightarrow G_k(\mathbf{C}^n)$ be a smooth map from a complex manifold into a complex Grassmann manifold of $k$-dimensional complex subspaces in $\mathbf{C}^n$. Let $\underline{\mathbf{C}}^n = M \times \mathbf{C}^n$ be a trivial bundle over $M$. We equip $\underline{\mathbf{C}}^n$ with the trivial Hermitian fibre metric, holomorphic vector bundle structure and Hermitian connection. We denote by $\underline{\varphi}$ the subbundle of $\underline{\mathbf{C}}^n$ whose fibre $\underline{\varphi}_x$ at $x \in M$ is given by $\underline{\varphi}_x = \varphi(x)$ for any $x \in M$. Let $T$ be a universal bundle over $G_k(\mathbf{C}^n)$, which is a subbundle of the trivial bundle $G_k(\mathbf{C}^n) \times \mathbf{C}^n$. Then, $\underline{\varphi} = \varphi^{-1}T$. Conversely, given a smooth complex subbundle $S$ of $\underline{\mathbf{C}}^n$ with rank $S = k$, we can make a smooth map $\varphi : M \longrightarrow G_k(\mathbf{C}^n)$ such that $S = \varphi^{-1}T$. Thus, we may identify a map $\varphi : M \longrightarrow G_k(\mathbf{C}^n)$ with a subbundle $\underline{\varphi}$ of $\underline{\mathbf{C}}^n$ with rank $\underline{\varphi} = k$. We know that the holomorphic tangent bundle $TG_k(\mathbf{C}^n)^{1,0}$ of $G_k(\mathbf{C}^n)$ is isomorphic to $\mathrm{Hom}(T, T^\perp)$ in the natural way, where $T^\perp$ is the orthogonal complement of $T$ in $G_k(\mathbf{C}^n) \times \mathbf{C}^n$ with respect to the Hermitian metric on $G_k(\mathbf{C}^n) \times \mathbf{C}^n$. Moreover, there is a connection preserving holomorphic isometry $j : TG_k(\mathbf{C}^n)^{1,0} \longrightarrow \mathrm{Hom}(T, T^\perp)$ with respect to the Levi-Civita connection on $TG_k(\mathbf{C}^n)^{1,0}$ and the Hermitian connection on $\mathrm{Hom}(T, T^\perp)$ induced from the flat connection on $G_k(\mathbf{C}^n) \times \mathbf{C}^n$ (cf. [B-W], [Wol]). In fact, $j$ is given by

$$j(Z)\sigma = A^{T,T^\perp}_{Z}(\sigma) \tag{7.6}$$

where $Z \in C^\infty(TG_k(\mathbf{C}^n)^{1,0})$ and $\sigma \in C^\infty(T)$. The pull-back of $j$ by $\varphi$ is also denoted by $j$, that is, $j : \varphi^{-1}TG_k(\mathbf{C}^n)^{1,0} \longrightarrow \mathrm{Hom}(\underline{\varphi}, \underline{\varphi}^\perp)$, where $\underline{\varphi}^\perp = \varphi^{-1}T^\perp$ is the orthogonal complement of $\underline{\varphi}$ in $\underline{\mathbf{C}}^n$. Set

$$A^{\varphi} = A^{\underline{\varphi}, \underline{\varphi}^\perp}, \quad A^{\varphi^\perp} = A^{\underline{\varphi}^\perp, \underline{\varphi}}. \tag{7.7}$$

Set $A^{\varphi} = A^{\varphi}_{(1,0)} + A^{\varphi}_{(0,1)}$ and $A^{\varphi^\perp} = A^{\varphi^\perp}_{(1,0)} + A^{\varphi^\perp}_{(0,1)}$. Then, by (7.3) we have

$$A^{\varphi}_{(1,0)} = -(A^{\varphi^\perp}_{(0,1)})^*, \qquad A^{\varphi}_{(0,1)} = -(A^{\varphi^\perp}_{(1,0)})^*. \tag{7.8}$$

Note that $\underline{\varphi}$ (resp. $\underline{\varphi}^{\perp}$) is a holomorphic (resp. an anti-holomorphic) subbundle of $\underline{C}^n$ if and only if $A^{\varphi}_{(0,1)} = 0$ (resp. $A^{\varphi}_{(1,0)} = 0$), that is, $\nabla^{\underline{C}^n}_{\bar{Z}} C^{\infty}(\underline{\varphi}) \subset C^{\infty}(\underline{\varphi})$ (resp. $\nabla^{\underline{C}^n}_{Z} C^{\infty}(\underline{\varphi}) \subset C^{\infty}(\underline{\varphi})$) for any $Z \in C^{\infty}(TM^{1,0})$. By (7.6) we see that

$$j(\partial\varphi^{1,0}) = A^{\varphi}_{(1,0)}, \qquad j(\bar{\partial}\varphi^{1,0}) = A^{\varphi}_{(0,1)}. \tag{7.9}$$

Thus, we may say that $\varphi$ is a holomorphic (resp. an anti-holomorphic) map if and only if $A^{\varphi}_{(0,1)} = 0$ (resp. $A^{\varphi}_{(1,0)} = 0$), if and only if $\underline{\varphi}$ is a holomorphic (resp. an anti-holomorphic) subbundle of $\underline{C}^n$. Write $A^{\varphi}_{(1,0)} = \sum_{i=1}^{m} A_{\partial/\partial z^i} \otimes dz^i$ and $A^{\varphi}_{(0,1)} = \sum_{i=1}^{m} A_{\partial/\partial \bar{z}^i} \otimes d\bar{z}^i$ etc. , where $\{z^i\}$ is a local complex coordinate system of $M$ and $m = \dim_{\mathbf{C}} M$. We denote by $\nabla^{\varphi}$ the pull-back connection on $\varphi^{-1}TG_k(\mathbf{C}^n)^{1,0}$ and $\mathrm{Hom}(\underline{\varphi}, \underline{\varphi}^{\perp})$. Then $A^{\varphi}$ satisfies the identity

$$(DA^{\varphi})(X,Y) = \nabla^{\varphi}_X A^{\varphi}_Y - \nabla^{\varphi}_Y A^{\varphi}_X - A^{\varphi}_{[X,Y]} = 0$$

for each $X, Y \in C^{\infty}(TM)$. Hence we have

$$0 = DA^{\varphi} = D' A^{\varphi}_{(0,1)} + D'' A^{\varphi}_{(1,0)}. \tag{7.10}$$

Then $\varphi$ is pluriharmonic if and only if

$$D'' A^{\varphi}_{(1,0)} \equiv 0, \text{ equivalently } \quad D' A^{\varphi}_{(0,1)} \equiv 0.$$

**7.B. Osculating space methods of constructing pluriharmonic maps.** First of all, we establish the following

**(7.11) Proposition.** *Let $\varphi : M \longrightarrow G_k(\mathbf{C}^n)$ be a pluriharmonic map from a complex manifold. Then, each of $\underline{\varphi}$ and $\underline{\varphi}^{\perp}$ admits a unique holomorphic vector bundle structure such that the Hermitian connection with respect to this holomorphic structure and the induced Hermitian fibre metric coincides with the induced connection.*

*Proof of* (7.11). Since $\varphi^{-1}TG_k(\mathbf{C}^n)^{1,0} \cong \mathrm{Hom}(\underline{\varphi}, \underline{\varphi}^{\perp}) \cong \underline{\varphi}^* \otimes \underline{\varphi}^{\perp}$, by (1.4) we have

$$R^{\underline{\varphi}^* \otimes \underline{\varphi}^{\perp}}(Z, W) = R^{\underline{\varphi}^* \otimes \underline{\varphi}^{\perp}}(\bar{Z}, \bar{W}) = 0 \tag{7.12}$$

for any $Z, W \in TM^{1,0}$. Note that $R^{\underline{\varphi}^* \otimes \underline{\varphi}^{\perp}} = R^{\underline{\varphi}^*} \otimes I_{\underline{\varphi}^{\perp}} + I_{\underline{\varphi}^*} \otimes R^{\underline{\varphi}^{\perp}}$, where $R^E$ and $I_E$ denote the curvature form and the identity endomorphism of a unitary vector bundle $E$, respectively. Hence, it follows from (7.12) that

$$\begin{aligned} R^{\underline{\varphi}}(Z, W) &= F(Z, W) I_{\underline{\varphi}}, \\ R^{\underline{\varphi}^{\perp}}(Z, W) &= F(Z, W) I_{\underline{\varphi}^{\perp}} \end{aligned} \tag{7.13}$$

for some bilinear form $F : TM^{1,0} \otimes TM^{1,0} \longrightarrow \mathbf{C}$. By (7.1), (7.2), (7.10) and the flatness of $\underline{C}^n$, a direct calculation yields

$$R^{\underline{\varphi}}(\partial/\partial z^i, \partial/\partial z^j) = A^{\varphi^{\perp}}_j \circ A^{\varphi}_i - A^{\varphi^{\perp}}_i \circ A^{\varphi}_j, \tag{7.14}$$

where $A_i = A_{\partial/\partial z^i}$ and $i, j = 1, \cdots, m$. It follows from (7.14) that $(\mathrm{tr}R^{\underline{\varphi}})^{(2,0)} = -(\mathrm{tr}R^{\underline{\varphi}^\perp})^{(2,0)}$ (cf. (7.8)).

Thus, by (7.13) we see that

$$kF = (\mathrm{tr}R^{\underline{\varphi}})^{(2,0)} = -(\mathrm{tr}R^{\underline{\varphi}^\perp})^{(2,0)} = -(n-k)F,$$

hence $F = 0$, which implies $(R^{\underline{\varphi}})^{(2,0)} = (R^{\underline{\varphi}^\perp})^{(2,0)} = 0$. Similarly, $(R^{\underline{\varphi}})^{(0,2)} = (R^{\underline{\varphi}^\perp})^{(0,2)} = 0$. Then, each of $R^{\underline{\varphi}}$ and $R^{\underline{\varphi}^\perp}$ is of type $(1,1)$ . This, together with a theorem of [K-M] (cf. [A-H-S], [Ko]), yields the conclusions. Q.E.D.

Let $E$ be a subbundle of the trivial bundle $\underline{\mathbf{C}}^n$. For any subbundle $F$ of a unitary bundle $E$, the orthogonal complement of $F$ in $E$ is denoted by $E \ominus F$. On the other hand, by $F^\perp$ we always mean the Hermitian orthogonal complement of $F$ in $\underline{\mathbf{C}}^n$, namely, $E \ominus F = F^\perp \cap E$.

**(7.15) Proposition.** *Let $\varphi : M \longrightarrow G_k(\mathbf{C}^n)$ be a pluriharmonic map from a complex manifold. Define $\tilde{\varphi}$ by*

$$\underline{\tilde{\varphi}} = (\underline{\varphi} \ominus \underline{\alpha}) \oplus \underline{\beta}, \tag{7.16}$$

*where $\underline{\alpha}$ and $\underline{\beta}$ satisfy the following conditions*
*(i) $\underline{\alpha}$ is a holomorphic subbundle of $\underline{\varphi}$,*
*(ii) $\underline{\beta}$ is a holomorphic subbundle of $\underline{\varphi}^\perp$,*
*(iii) $A^{\varphi}_{(1,0)}(\underline{\alpha}) \subseteq T^*M^{1,0} \otimes \underline{\beta}$, $\quad A^{\varphi^\perp}_{(1,0)}(\underline{\beta}) \subseteq T^*M^{1,0} \otimes \underline{\alpha}$.*
*Then, $\tilde{\varphi}$ is a pluriharmonic map from $M$ into $G_r(\mathbf{C}^n)$ for some $r$.*

*Proof.* Set $\underline{\gamma} = \underline{\alpha}^\perp \cap \underline{\varphi}$, $\underline{\delta} = \underline{\beta}^\perp \cap \underline{\varphi}^\perp$. Then, we have $\underline{\varphi} = \underline{\alpha} \oplus \underline{\gamma}$, $\underline{\varphi}^\perp = \underline{\beta} \oplus \underline{\delta}$, $\underline{\tilde{\varphi}} = \underline{\gamma} \oplus \underline{\beta}$, $\underline{\tilde{\varphi}}^\perp = \underline{\alpha} \oplus \underline{\delta}$. By the assumptions, we obtain

$$A^{\alpha,\gamma}_{(0,1)} = 0, \quad A^{\alpha,\delta}_{(1,0)} = 0, \quad A^{\beta,\delta}_{(0,1)} = 0, \quad A^{\beta,\gamma}_{(1,0)} = 0, \tag{7.17}$$

where, for simplicity of the notation, we write $A^{\alpha,\beta}$ in place of $A^{\underline{\alpha},\underline{\beta}}$ etc. . The pluriharmonicity of $\varphi$, (7.8) and (7.10) imply that

$$\begin{aligned}(D'A^{\varphi}_{(0,1)})(Z,\bar{W}) &= (D''A^{\varphi}_{(1,0)})(\bar{W},Z) = 0,\\ (D'A^{\varphi^\perp}_{(0,1)})(Z,\bar{W}) &= (D''A^{\varphi^\perp}_{(1,0)})(\bar{W},Z) = 0\end{aligned} \tag{7.18}$$

for any $Z, W \in C^\infty(TM^{1,0})$. By (7.17), we get

$$\begin{aligned}A^{\varphi}_{(1,0)} &= A^{\alpha,\beta}_{(1,0)} + A^{\gamma,\beta}_{(1,0)} + A^{\gamma,\delta}_{(1,0)},\\ A^{\varphi}_{(0,1)} &= A^{\alpha,\beta}_{(0,1)} + A^{\alpha,\delta}_{(0,1)} + A^{\gamma,\delta}_{(0,1)},\end{aligned} \tag{7.19}$$

and

$$A^{\widetilde{\varphi}}_{(1,0)} = A^{\gamma,\delta}_{(1,0)} + A^{\beta,\alpha}_{(1,0)} + A^{\beta,\delta}_{(1,0)}, \tag{7.20}$$
$$A^{\widetilde{\varphi}}_{(0,1)} = A^{\gamma,\alpha}_{(0,1)} + A^{\beta,\alpha}_{(0,1)} + A^{\gamma,\delta}_{(0,1)}.$$

By (7.17) and (7.20) direct computations yield

$$\begin{aligned}(D'A^{\tilde{\varphi}}_{(0,1)})(Z,\bar{W}) =&(D'A^{\gamma,\alpha}_{(0,1)})(Z,\bar{W}) + (D'A^{\beta,\alpha}_{(0,1)})(Z,\bar{W}) \\ &+ (D'A^{\gamma,\delta}_{(0,1)})(Z,\bar{W}) + A^{\delta,\alpha}_{Z} \circ A^{\gamma,\delta}_{\bar{W}} - A^{\beta,\alpha}_{\bar{W}} \circ A^{\gamma,\beta}_{Z},\end{aligned} \tag{7.21}$$

and

$$\begin{aligned}(D''A^{\tilde{\varphi}}_{(1,0)})(\bar{W},Z) =&(D''A^{\gamma,\delta}_{(1,0)})(\bar{W},Z) + (D''A^{\beta,\alpha}_{(1,0)})(\bar{W},Z) \\ &+ (D''A^{\beta,\delta}_{(1,0)})(\bar{W},Z) + A^{\alpha,\delta}_{\bar{W}} \circ A^{\beta,\alpha}_{Z} - A^{\gamma,\delta}_{Z} \circ A^{\beta,\gamma}_{\bar{W}}.\end{aligned} \tag{7.22}$$

It follows from (7.10), (7.21) and (7.22) that

$$\begin{aligned}(D'A^{\tilde{\varphi}}_{(0,1)})(Z,\bar{W}) &= (D''A^{\tilde{\varphi}}_{(1,0)})(\bar{W},Z) \\ &= (D''A^{\gamma,\delta}_{(1,0)})(\bar{W},Z) + (D''A^{\beta,\alpha}_{(1,0)})(\bar{W},Z).\end{aligned} \tag{7.23}$$

On the other hand, by (7.18) and (7.19) we have

$$\begin{aligned}(D''A^{\varphi}_{(1,0)})(\bar{W},Z) &= (D''A^{\alpha,\beta}_{(1,0)})(\bar{W},Z) + (D''A^{\gamma,\delta}_{(1,0)})(\bar{W},Z) \\ &\quad + (D''A^{\gamma,\beta}_{(1,0)})(\bar{W},Z) + A^{\delta,\beta}_{\bar{W}} \circ A^{\gamma,\delta}_{Z} - A^{\alpha,\beta}_{Z} \circ A^{\gamma,\alpha}_{\bar{W}} \\ &= 0,\end{aligned} \tag{7.24}$$

and

$$\begin{aligned}(D''A^{\varphi^\perp}_{(1,0)})(\bar{W},Z) &= (D''A^{\beta,\alpha}_{(1,0)})(\bar{W},Z) + (D''A^{\delta,\gamma}_{(1,0)})(\bar{W},Z) \\ &\quad + (D''A^{\delta,\alpha}_{(1,0)})(\bar{W},Z) + A^{\gamma,\alpha}_{\bar{W}} \circ A^{\delta,\gamma}_{Z} - A^{\beta,\alpha}_{Z} \circ A^{\delta,\beta}_{\bar{W}} \\ &= 0.\end{aligned} \tag{7.25}$$

In particular, we have $(D''A^{\gamma,\delta}_{(1,0)})(\bar{W},Z) = (D''A^{\beta,\alpha}_{(1,0)})(\bar{W},Z) = 0$. Thus, we get that $\tilde{\varphi}$ is pluriharmonic. Q.E.D.

*Remark.* We may use $A^{\varphi}_{(0,1)}$ in place of $A^{\varphi}_{(1,0)}$. In this case, we take $\underline{\alpha}$ and $\underline{\beta}$ as the anti-holomorphic subbundles of $\underline{\varphi}$ and $\underline{\varphi}^\perp$, respectively.

We consider $A^{\varphi}_{(1,0)}$ as a bundle homomorphism $A^{\varphi}_{(1,0)} : TM^{1,0} \otimes \underline{\varphi} \longrightarrow \underline{\varphi}^\perp$. Set

$$\mathrm{Im}A^{\varphi}_{(1,0)} = \bigcup_{x\in M} \mathrm{Im}(A^{\varphi}_{(1,0)})_x.$$

Let $t = \max\{\dim \mathrm{Im}(A^{\varphi}_{(1,0)})_x; x \in M\}$. It induces a bundle homomorphism $\wedge^t A^{\varphi}_{(1,0)}$ : $\wedge^t(TM^{1,0} \otimes \underline{\varphi}) \longrightarrow \wedge^t \underline{\varphi}^{\perp}$. By the pluriharmonicity of $\varphi$, $A^{\varphi}_{(1,0)}$ and $\wedge^t A^{\varphi}_{(1,0)}$ are holomorphic bundle homomorphisms. Set

$$V = \{x \in M; (\wedge^t A^{\varphi}_{(1,0)})_x = 0\} = \{x \in M; \dim(\mathrm{Im} A^{\varphi}_{(1,0)})_x < t\},$$

which is an analytic subset of $M$. Then, $\mathrm{Im}A^{\varphi}_{(1,0)}$ is a holomorphic subbundle of $\underline{\varphi}^{\perp}$ over $M \setminus V$. Let $V = \cup_i D_i \cup W$ be a decomposition of $V$ into irreducible components $D_i$ of codimension 1 and the union of components of codimension at least 2. Let $n_i$ be the multiplicity of $\wedge^t A^{\varphi}_{(1,0)}$ at $D_i$. Let $x \in D_i \setminus (\cup_{j \neq i} D_j \cup W)$. $D_i$ is defined by the equation $w = 0$ near $x$, where $w$ is a locally defined holomorphic function. Then we have, near $x$,

$$\wedge^t A^{\varphi}_{(1,0)} = w^{n_i} \sigma,$$

where $\sigma$ is a local holomorphic section of $\mathrm{Hom}(\wedge^t(TM^{1,0} \otimes \underline{\varphi}), \wedge^t \underline{\varphi}^{\perp})$. Thus, the image of $\sigma$ defines a holomorphic subbundle of $\underline{\varphi}^{\perp}$ of rank $t$ over $D_i$ around $x$. In this way, $\mathrm{Im}A^{\varphi}_{(1,0)}$ extends to a holomorphic subbundle of $\underline{\varphi}^{\perp}$ over $M \setminus W$. We denote by $\underline{\mathrm{Im}}A^{\varphi}_{(1,0)}$ this subbundle.

Similarly, we can consider $A^{\varphi}_{(1,0)}$ also as another holomorphic bundle homomorphism $A^{\varphi}_{(1,0)} : \underline{\varphi} \longrightarrow T^*M^{1,0} \otimes \underline{\varphi}^{\perp}$. Set

$$\mathrm{Ker}A^{\varphi}_{(1,0)} = \bigcup_{x \in M} \mathrm{Ker}(A^{\varphi}_{(1,0)})_x.$$

In the same way, $\mathrm{Ker}A^{\varphi}_{(1,0)}$ extends to a holomorphic subbundle of $\underline{\varphi}$ over $M \setminus W'$, which is denoted by $\underline{\mathrm{Ker}}A^{\varphi}_{(1,0)}$, where $W'$ is an analytic subset of $M$. For the examples satisfying the conditions in (7.15), take $\underline{\widetilde{\varphi}} = \underline{\mathrm{Im}}A^{\varphi}_{(1,0)}$ or $\underline{\widetilde{\varphi}} = \underline{\varphi} \ominus \underline{\mathrm{Ker}}A^{\varphi}_{(1,0)}$.

Then, by (7.15) we obtain

**(7.26) Proposition.** *Let $\varphi : M \longrightarrow G_k(\mathbf{C}^n)$ be a pluriharmonic map from a complex manifold. Then, the following map $\widetilde{\varphi}$ becomes a pluriharmonic map from $M \setminus W_{\widetilde{\varphi}}$ into $G_r(\mathbf{C}^n)$ for some $r$ , where $W_{\widetilde{\varphi}}$ is an analytic subset of $M$ with $\mathrm{codim}_{\mathbf{C}} W_{\widetilde{\varphi}} \geq 2$ :*
*(1) $\underline{\widetilde{\varphi}} = \underline{\mathrm{Im}}A^{\varphi}_{(1,0)}$. (2) $\underline{\widetilde{\varphi}} = \underline{\mathrm{Im}}A^{\varphi}_{(0,1)}$.*
*(3) $\underline{\widetilde{\varphi}} = \underline{\varphi} \ominus \underline{\mathrm{Ker}}A^{\varphi}_{(1,0)}$. (4) $\underline{\widetilde{\varphi}} = \underline{\varphi} \ominus \underline{\mathrm{Ker}}A^{\varphi}_{(0,1)}$.*
*(5) $\underline{\widetilde{\varphi}} = (\underline{\varphi} \ominus \underline{\alpha}) \oplus \underline{\mathrm{Im}}(A^{\varphi}_{(1,0)} \mid \underline{\alpha})$, where $\underline{\alpha} \subset \mathrm{Ker}(A^{\varphi^{\perp}}_{(1,0)} \circ A^{\varphi}_{(1,0)})$ is a holomorphic subbundle of $\underline{\varphi}$.*
*(6) $\underline{\widetilde{\varphi}} = (\underline{\varphi} \ominus \underline{\alpha}) \oplus \underline{\mathrm{Im}}(A^{\varphi}_{(0,1)} \mid \underline{\alpha})$, where $\underline{\alpha} \subset \mathrm{Ker}(A^{\varphi^{\perp}}_{(0,1)} \circ A^{\varphi}_{(0,1)})$ is an anti-holomorphic subbundle of $\underline{\varphi}$.*

In general, we can not expect that $W_{\widetilde{\varphi}} = \emptyset$, when $\dim_{\mathbf{C}} M \geq 2$. In the case when $M$ is a Riemann sphere, we know many factorization theorems of harmonic maps for $G_k(\mathbf{C}^n)$ (cf. [E-W2], [B-W], [B-S], [C-Wol], [Wol], [Wd1]). One of them is the following result of Wood

(7.27) **Theorem** ([Wd1]). *Let $\varphi : S^2 \longrightarrow G_k(\mathbf{C}^n)$ be a harmonic map from a Riemann sphere. Set $\varphi_0 = \varphi$ and define a sequence $\{\varphi_m\}$ of harmonic maps as follows : $\underline{\varphi}_m$ = either of the following (1) ~ (4)*

*(1)* $\underline{\mathrm{Im}}A^{\varphi_{m-1}}_{(1,0)}$ *if $\varphi_{m-1}$ is irreducible and* $\deg(\varphi_{m-1}) < 0$ ,

*(2)* $\underline{\mathrm{Im}}A^{\varphi_{m-1}}_{(0,1)}$ *if $\varphi_{m-1}$ is irreducible and* $\deg(\varphi_{m-1}) \geq 0$,

*(3)* $\underline{\varphi}_{m-1} \ominus \underline{\mathrm{Ker}}A^{\varphi_{m-1}}_{(1,0)}$ *if $\varphi_{m-1}$ is $\partial'$-reducible,*

*(4)* $\underline{\varphi}_{m-1} \ominus \underline{\mathrm{Ker}}A^{\varphi_{m-1}}_{(0,1)}$ *if $\varphi_{m-1}$ is $\partial''$-reducible.*

*Then, there exists an integer $s \in \mathbf{N}$ such that $\underline{\varphi}_s = \underline{0}$, that is, $\varphi_s$ is a constant map. If we set $\varphi^i = \varphi_{s-i}$, a sequence $\{\varphi^i\}$ gives a factorization of $\varphi = \varphi^s$.*

Here, we say that $\varphi$ is *$\partial'$-reducible* (resp. *$\partial''$-reducible*) if $\underline{\mathrm{Ker}}A^{\varphi}_{(1,0)} \neq \underline{0}$ (resp. $\underline{\mathrm{Ker}}A^{\varphi}_{(0,1)} \neq \underline{0}$). Moreover, we say that $\varphi$ is *irreducible* if it is neither $\partial'$-reducible nor $\partial''$-reducible. In fact, if $\varphi$ is irreducible, $\wedge^r A^{\varphi}_{(1,0)}$ is a non-zero holomorphic section of the holomorphic line bundle $F = \mathrm{Hom}(\otimes^r T(S^2)^{1,0} \otimes \wedge^r \underline{\varphi}, \wedge^r \underline{\mathrm{Im}}A^{\varphi}_{(1,0)})$ over $S^2$, where $r = \mathrm{rank}\,\underline{\varphi} = \mathrm{rank}\,(\underline{\mathrm{Im}}A^{\varphi}_{(1,0)})$. Therefore, by $c_1(S^2) > 0$ and $c_1(F) \geq 0$ we obtain $c_1(\underline{\mathrm{Im}}A^{\varphi}_{(1,0)}) \geq c_1(\underline{\varphi}) + rc_1(S^2)$. Thus, if $\deg(\varphi_{m-1}) < 0$ then $c_1(\underline{\varphi}_{m-1}) > 0$, hence we obtain $E(\varphi_m) \leq E(\varphi_{m-1}) - C$ for some positive constant $C$, where $E(\psi)$ denotes the energy of a map $\psi$. If $\deg(\varphi_{m-1}) \geq 0$, we also have $E(\varphi_m) \leq E(\varphi_{m-1}) - C$. Since the energy is nonnegative, there exists an integer $i \in \mathbf{N}$ such that $\varphi_i$ is $\partial'$-reducible or $\partial''$-reducible. Then, rank $\underline{\varphi}_{i+1}$ < rank $\underline{\varphi}_i$. Repeating this process, we see that there exists an integer $s \in \mathbf{N}$ such that $\underline{\varphi}_s = \underline{0}$, that is, $\varphi_s$ is a constant map.

The method using the energy reduction is not available for the case where the domain is of higher dimension because the case of rank $\underline{\varphi} < \mathrm{rank}(\underline{\mathrm{Im}}A^{\varphi}_{(1,0)})$ may occur. However, the method used in [B-W] seems to be useful. Now, we prepare the fundamental lemma, which is proved by Ramanathan [Ra] for the case of a Riemann sphere.

(7.28) **Lemma.** *Let $E$ be a Hermitian holomorphic vector bundle over $M \setminus S$ and $A \in H^0(M \setminus S, \otimes^l T^*M^{1,0} \otimes \mathrm{Hom}(E,E))$, $l \in \mathbf{N}$, where $S = \bigcup_{j=1}^k S_j$ and each $S_i$ $(i = 1, \cdots, k)$ is an analytic subset of $M \setminus \bigcup_{j=1}^{i-1} S_j$ with $\mathrm{codim}_{\mathbf{C}} S_i \geq 2$. Assume that $M$ is compact and $c_1(M) > 0$. Then, $A$ is nilpotent, that is, $A^m \equiv 0$ for some positive integer $m \leq \mathrm{rank}\,E$.*

*Proof.* Set

$$\det(A + \lambda I) = \lambda^k + c_1(A)\lambda^{k-1} + \cdots + c_k(A),$$

where $\lambda \in \mathbf{R}$, $k = \mathrm{rank}\,E$ and $I$ is an identity endomorphism of $E$. Then, $c_i(A)(i = 1, \cdots, k)$ is a holomorphic differential on $M \setminus S$. Using Hartogs' extension theorem successively, we see that each $c_i(A)$ extends to a holomorphic differential $\tilde{c}_i(A)$ on $M$ such that $\tilde{c}_i(A)\,|_{M \setminus S} = c_i(A)$. If $c_1(M) > 0$, using the Kähler metric of positive Ricci curvature in the same Kähler class as that of the original Kähler metric ([Ya]) and the vanishing theorem of Bochner type, we get $\tilde{c}_i(A) \equiv 0$, hence $c_i(A) \equiv 0$ on $M \setminus S$. Q.E.D.

The following is due to Burstall and Wood

**(7.29) Lemma ([B-W]).** *Let* $A \in C^\infty(\mathrm{Hom}(\underline{\varphi}, \underline{\psi}))$.
*(1) If* $A$ *is holomorphic and* $\underline{\alpha}$ *is a holomorphic subbundle of* $\underline{\varphi}$, *then* $A \mid \underline{\alpha} : \underline{\alpha} \longrightarrow \underline{\psi}$ *is holomorphic.*
*(2) If* $A$ *is holomorphic and* $\underline{\beta}$ *is an anti-holomorphic subbundle of* $\underline{\psi}$, *then* $\pi \circ A : \underline{\varphi} \longrightarrow \underline{\beta}$ *is holomorphic, where* $\pi : \underline{\psi} \longrightarrow \underline{\beta}$ *is an orthogonal projection.*
*(3) Let* $\underline{\gamma}$ *be a subbundle of* $\underline{\varphi}$ *with* $\underline{\gamma}^\perp \subset \mathrm{Ker}A$. *Then,* $A$ *is holomorphic if and only if* $A \mid \underline{\gamma} : \underline{\gamma} \longrightarrow \underline{\psi}$ *is holomorphic.*
*(4) Let* $\underline{\delta}$ *be a subbundle of* $\underline{\psi}$ *containing the image of* $A$. *Then, if* $A$ *is holomorphic,* $A : \underline{\varphi} \longrightarrow \underline{\delta}$ *is holomorphic.*

7.C. **Explicit construction of pluriharmonic maps from a compact complex manifold with positive first Chern class into a certain complex Grassmann manifold.** Throughout this section, we assume that $M$ is *a compact complex manifold with* $c_1(M) > 0$, unless otherwise stated.

First, we establish the following

**(7.30) Theorem.** *Let* $\varphi : M \setminus S_\varphi \longrightarrow CP^n$ *be a pluriharmonic map, where* $S_\varphi$ *is as in (7.28). Inductively, define a sequence* $\{\varphi_i\}$ *of pluriharmonic maps* $\varphi_i : M \setminus S_{\varphi_i} \longrightarrow \mathbf{C}P^n$ *by* $\underline{\varphi_i} = \underline{\mathrm{Im}}A^{\varphi_{i-1}}_{(1,0)}$ $(i = 1, 2, \cdots)$ *with* $\varphi_0 = \varphi$, *where* $S_{\varphi_i}$ *is as in (7.28). Then, there exists an integer* $s \in \mathbf{N}$ *such that* $\varphi_s$ *is an anti-holomorphic map from* $M \setminus S_{\varphi_s}$ *into* $\mathbf{C}P^n$, *where* $s \leq n-1$.

*Proof.* First, observe that each $A^{\varphi_{i-1},\varphi_i}_{(1,0)}$ is a holomorphic section of $\mathrm{Hom}(\underline{\varphi}_{i-1}, \underline{\varphi}_i)$ by (7.29). If $\varphi_{i-1}$ is non $\pm$-holomorphic map from $M \setminus S_{\varphi_{i-1}}$ into $CP^n$, then, by (3.8), (7.9) and rank $\underline{\varphi}_{i-1} = 1$, we see that rank $(\underline{\mathrm{Im}}A^{\varphi_{i-1}}_{(1,0)}) = 1$ on $M \setminus S_{\varphi_i}$. Thus, $\underline{\varphi}_i$ also defines a pluriharmonic map from $M \setminus S_{\varphi_i}$ into $\mathbf{C}P^n$. Note that $\underline{\varphi}_i$ is orthogonal to $\underline{\varphi}_j$ with $j \neq i$ while $\underline{\varphi}_i$ and $\underline{\varphi}_j$ are contained in $\underline{\varphi}^\perp$. Thus, it is enough to prove that each $A^{\varphi_{i-1}}_{(1,0)}$ has its image in $\underline{\varphi}^\perp$. Suppose that there exists $k \in \mathbf{N}$ such that each $\varphi_j$ with $j \leq k$ is non $\pm$-holomorphic and $A^{\varphi_k,\varphi}_{(1,0)} \neq 0$. Then, we have a non-zero holomorphic endomorphism $A^{\varphi_k,\varphi}_{(1,0)} \circ A^{\varphi_{k-1},\varphi_k}_{(1,0)} \cdots \circ A^{\varphi,\varphi_1}_{(1,0)}$ of $\underline{\varphi}$ over $M \setminus S_{\varphi_k}$. However, this endomorphism must vanish identically by (7.28) because of rank $\underline{\varphi} = 1$. Therefore, each $A^{\varphi_{i-1}}_{(1,0)}$ has its image in $\underline{\varphi}^\perp$ while $\varphi_{i-1}$ is non $\pm$-holomorphic. Lastly, note that if $\varphi_r$ is holomorphic for some $r$ then $\varphi_{r-1}$ is already anti-holomorphic. Q.E.D.

(7.30) may be used to establish the factorization theorem for pluriharmonic maps into complex Grassmann manifolds. For example, let $\varphi : M \longrightarrow G_2(\mathbf{C}^4)$ be a pluriharmonic map. Then, rank $\underline{\varphi} =$ rank $\underline{\varphi}^\perp = 2$. Assume that $\varphi$ is non $\pm$-holomorphic. If $\underline{\mathrm{Ker}}A^{\varphi}_{(0,1)}$ is non-trivial, by (7.26) the replacement of $\underline{\widetilde{\varphi}} = \underline{\varphi} \ominus \underline{\mathrm{Ker}}A^{\varphi}_{(0,1)}$ gives a pluriharmonic map $\widetilde{\varphi}$ from $M \setminus S_{\widetilde{\varphi}}$ into $\mathbf{C}P^3$, hence reduced to (7.30). Suppose that $\underline{\mathrm{Ker}}A^{\varphi}_{(0,1)} = \underline{0}$. Then, by (7.8) we see that $A^{\varphi^\perp}_{(1,0)} : \underline{\varphi}^\perp \longrightarrow \underline{\varphi}$ is surjective. Then, since $A^{\varphi^\perp}_{(1,0)} \circ A^{\varphi}_{(1,0)}$ is nilpotent by (7.28), we have $\mathrm{rank}(\underline{\mathrm{Im}}A^{\varphi}_{(1,0)}) = 1$, hence reduced to (7.30).

For the other complex Grassmann manifolds, some generalizations of the results due to Burstall and Wood [B-W] are necessary.

*Definition* (cf. [B-W]). Let $\varphi : M \longrightarrow G_k(\mathbf{C}^n)$ be a pluriharmonic map from a complex manifold. Denote by $G^{(r)}(\varphi)$, the $r$-th $\partial'$-*Gauss bundle* of $\varphi$ defined by

$$G'(\varphi) = G^{(1)}(\varphi) = \underline{\mathrm{Im}}A^{\varphi}_{(1,0)}, \quad G^{(r+1)}(\varphi) = G'(G^{(r)}(\varphi)).$$

Similarly, define the $r$-th $\partial''$-*Gauss bundle* $G^{(-r)}(\varphi)$ by

$$G''(\varphi) = G^{(-1)}(\varphi) = \underline{\mathrm{Im}}A^{\varphi}_{(0,1)}, \quad G^{(-r-1)}(\varphi) = G''(G^{(-r)}(\varphi)).$$

In particular, set $G'_{\varphi}(\alpha) = \underline{\mathrm{Im}}(A^{\varphi}_{(1,0)}|\underline{\alpha})$ and $G''_{\varphi}(\alpha) = \underline{\mathrm{Im}}(A^{\varphi}_{(0,1)}|\underline{\alpha})$ for a subbundle $\underline{\alpha}$ of $\underline{\varphi}$.

*Definition* (cf. [B-W]). We say that $\varphi$ has *isotropy order* $\geq r (\geq 1)$ if $\underline{\varphi} \perp G^{(i)}(\varphi)$ for $1 \leq i \leq r$.

From the definitions, the induction argument yields

**(7.31) Lemma** (cf. [B-W]). *If $\varphi : M \longrightarrow G_k(\mathbf{C}^n)$ has isotropy order $\geq r(\geq 1)$, then $G^{(i)}(\varphi) \perp G^{(j)}(\varphi)$ for any $i, j$ such that $0 < | i - j | \leq r$.*

We have the following propositions, which are modifications of the results in [B-W].

**(7.32) Proposition.** *Let $\varphi : M \setminus S_{\varphi} \longrightarrow G_2(\mathbf{C}^n)$ be a pluriharmonic map, where $S_{\varphi}$ is as in (7.28). Assume that $\varphi$ has isotropy order $\geq r(\geq 1)$ and all $\underline{\varphi}, G'(\varphi), \cdots, G^{(r)}(\varphi)$ have the $\partial'$-Gauss bundle of rank $\geq 2$ (otherwise, reduced to a pluriharmonic map into $\mathbf{C}P^{n-1}$). Then, the following statements hold*
*(i) $A^{G^{(r)}(\varphi),\varphi}_{(1,0)}$ is holomorphic.*
*(ii) Set $\underline{\alpha} = \underline{\mathrm{Im}}A^{G^{(r)}(\varphi),\varphi}_{(1,0)}$. Then, rank $\underline{\alpha} \leq 1$.*
*(iii) If $\widetilde{\varphi}$ is obtained from $\varphi$ by*

$$\underline{\widetilde{\varphi}} = \underline{\varphi} \ominus \underline{\alpha} \oplus G'_{\varphi}(\alpha)$$

*then $\widetilde{\varphi} : M \setminus S_{\widetilde{\varphi}} \longrightarrow G_k(\mathbf{C}^n)$ is a pluriharmonic map of isotropy order $\geq r + 1$ for some $k$.*

*Remark.* If $\underline{\alpha} = \underline{0}$, $\varphi$ has already isotropy order $\geq r + 1$.

**(7.33) Proposition.** *Let $\varphi : M \setminus S_{\varphi} \longrightarrow G_3(\mathbf{C}^n)$ be a pluriharmonic map, where $S_{\varphi}$ is as in (7.28). Assume that $\varphi$ has isotropy order $\geq r(\geq 1)$ and all $\underline{\varphi}, G'(\varphi), \cdots, G^{(r)}(\varphi)$ have the $\partial'$-Gauss bundle of rank $\geq 3$ (otherwise, reduced to a pluriharmonic map of isotropy order $\geq r$ into $G_2(\mathbf{C}^n)$ or a pluriharmonic map into $\mathbf{C}P^{n-1}$). Then, the following statements hold :*
*(i) $A^{G^{(r)}(\varphi),\varphi}_{(1,0)}$ is holomorphic.*
*(ii) Set $\underline{\gamma} = \underline{\mathrm{Im}}A^{G^{(r)}(\varphi),\varphi}_{(1,0)}$. Then, rank $\underline{\gamma} \leq 2$.*
*(iii) If rank $\underline{\gamma} \leq 1$, then there exists a holomorphic subbundle $\underline{\delta}$ of $\underline{\varphi}$ contained in* $\mathrm{Ker}(A^{\varphi^{\perp}}_{(1,0)}A^{\varphi}_{(1,0)})$ *such that $\widetilde{\varphi}$ obtained by*

$$\underline{\widetilde{\varphi}} = \underline{\varphi} \ominus \underline{\delta} \oplus G'_{\varphi}(\delta)$$

*is a pluriharmonic map from $M\backslash S_{\tilde\varphi}$ into $G_k(\mathbf{C}^n)$ for some $k$ and has isotropy order $\geq r+1$.*
*(iv) If rank $\underline{\gamma} = 2$, there exists a holomorphic subbundle $\underline{\delta}$ of $\underline{\varphi}$ contained in* $\mathrm{Ker}(A^{\varphi^\perp}_{(1,0)}A^{\varphi}_{(1,0)})$ *such that $\tilde\varphi$ obtained by*

$$\underline{\tilde\varphi} = \underline{\varphi} \ominus \underline{\delta} \oplus G'_{\varphi}(\delta)$$

*is a pluriharmonic map from $M \setminus S_{\tilde\varphi}$ into $G_k(\mathbf{C}^n)$ for some $k$ and has isotropy order $\geq r$* and rank $(\underline{\mathrm{Im}}A^{G^{(r)}(\tilde\varphi),\tilde\varphi}_{(1,0)}) \leq 1$.

Since the proof of (7.32) is essentially contained in the proof of the case (iii) in (7.33), we prove only (7.33).

*Proof of* (7.33). Since each $A^{G^{(i)}(\varphi)}_{(1,0)}(i = 0,1,\cdots,r)$ is holomorphic, by (7.29) we see that $A^{G^{(r)}(\varphi),\varphi}_{(1,0)}$ is holomorphic. Thus, we have a holomorphic endomorphism $A^{G^{(r)}(\varphi),\varphi}_{(1,0)} \circ A^{G^{(r-1)}}_{(1,0)} \cdots A^{\varphi}_{(1,0)} : \underline{\varphi} \longrightarrow \underline{\varphi}$, which is nilpotent by (7.28). Therefore, we have $\mathrm{rank}(\underline{\mathrm{Im}}A^{G^{(r)}(\varphi),\varphi}_{(1,0)}) \leq 2(< \mathrm{rank}\underline{\varphi})$. Thus, we obtain (i) and (ii). Set $\underline{R} = (\underline{\varphi} \oplus G'(\varphi) \oplus \cdots \oplus G^{(r)}(\varphi))^{\perp}$. Note that $\underline{R} \neq \underline{0}$ because $G^{(r)}(\varphi)$ has the $\partial'$-Gauss bundle of rank $\geq 3$. Set $\underline{\gamma} = \underline{\mathrm{Im}}A^{G^{(r)}(\varphi),\varphi}_{(1,0)}$. If $\underline{\gamma} = \underline{0}$, then $\varphi$ has already isotropy order $\geq r+1$. Assume that rank $\underline{\gamma} = 1$. By (7.29), $A^{\gamma,G'(\varphi)}_{(1,0)}$ is holomorphic, hence set $\underline{\gamma}_1 = \underline{\mathrm{Im}}A^{\gamma,G'(\varphi)}_{(1,0)}$ and define $\underline{\gamma}_i$ $(2 \leq i \leq r)$ by $\underline{\gamma}_i = \underline{\mathrm{Im}}A^{\gamma_{i-1},G^{(i)}(\varphi)}_{(1,0)}$. Thus, each $\underline{\gamma}_i$ is a holomorphic subbundle of $G^{(i)}(\varphi)$ and each $A^{\gamma_{i-1},\gamma_i}_{(1,0)}$ is holomorphic by (7.29). Let $\underline{\delta} = \underline{\varphi} \cap \underline{\gamma}^{\perp}$ and let $\underline{\delta}_i = G^{(i)}(\varphi) \cap \underline{\gamma}_i^{\perp}$ $(i = 1,\cdots,r)$. Suppose that each $\underline{\gamma}_i$ is non-zero. We have a holomorphic endomorphism $A^{\gamma_r,\gamma}_{(1,0)} \circ A^{\gamma_{r-1},\gamma_r}_{(1,0)} \cdots \circ A^{\gamma,\gamma_1}_{(1,0)} : \underline{\gamma} \longrightarrow \underline{\gamma}$, which must vanish by (7.28). Thus, we have $A^{\gamma_r,\gamma}_{(1,0)} \equiv 0$. Note that each $\underline{\delta}_i$ cannot be zero because $A^{\delta_r,\gamma}_{(1,0)} \neq 0$. If $\underline{\gamma}_i = \underline{0}$ for some $i$, we also have $A^{\gamma_r,\gamma}_{(1,0)} \equiv 0$. If $A^{\gamma_r,R}_{(1,0)} \neq 0$, set $\underline{\gamma}_{r+1} = \underline{\mathrm{Im}}A^{\gamma_r,R}_{(1,0)}$ and $\underline{\delta}_{r+1} = \underline{\gamma}_{r+1}^{\perp} \cap \underline{R}$. Note that all $A^{\gamma_i,\gamma_{i+1}}_{(1,0)}, A^{\delta_i,\delta_{i+1}}_{(1,0)}, A^{\delta_r,\gamma}_{(1,0)}, A^{\gamma_r,R}_{(1,0)}$ and $A^{R,\delta}_{(1,0)}$ are holomorphic (cf. Propositions 1.5 and 4.3 in [B-W]). We have a holomorphic endomorphism $A^{\delta_r,\gamma}_{(1,0)} \cdots A^{\gamma,\gamma_1}_{(1,0)} : \underline{\gamma} \rightarrow \underline{\gamma}_1 \rightarrow \cdots \rightarrow \underline{\gamma}_{r+1} \rightarrow \underline{\delta} \rightarrow \cdots \rightarrow \underline{\delta}_r \rightarrow \underline{\gamma}$, which must vanish by (7.28). Thus, we see that either $A^{\gamma_{r+1},\delta}_{(1,0)} \equiv 0$ or $\mathrm{Im}A^{\gamma_{r+1},\delta}_{(1,0)} \subset \mathrm{Ker}\{A^{\delta_r,\gamma}_{(1,0)} \cdots A^{\delta,\delta_1}_{(1,0)}\}$. If $A^{\gamma_{r+1},\delta}_{(1,0)} \equiv 0$, $\tilde\varphi$ defined by $\underline{\tilde\varphi} = \underline{\varphi} \ominus \underline{\gamma} \oplus \underline{\gamma}_1$ is a pluriharmonic map with isotropy order $\geq r+1$ because $\underline{\gamma} \subset \mathrm{Ker}(A^{\varphi^\perp}_{(1,0)}A^{\varphi}_{(1,0)})$, $G^{(i)}(\tilde\varphi) = \underline{\delta}_i \oplus \underline{\gamma}_{i+1}(i = 1,2,\cdots,r)$ and $G^{(r+1)}(\tilde\varphi) \subset \underline{\gamma} \oplus \underline{R}$. Otherwise, set $\underline{\eta} = \underline{\mathrm{Im}}A^{\gamma_{r+1},\delta}_{(1,0)}$, which is a holomorphic subbundle of $\underline{\delta}$. Note that rank $\underline{\eta} = 1$. Let $\underline{\eta}_1 = \underline{\mathrm{Im}}A^{\eta,\delta_1}_{(1,0)}$ and define $\underline{\eta}_i$ by $\underline{\eta}_i = \underline{\mathrm{Im}}A^{\eta_{i-1},\delta_i}_{(1,0)}(i = 2,\cdots,r+1)$. Set $\underline{\delta}' = \underline{\eta}^{\perp} \cap \underline{\delta}$ and $\underline{\delta}'_i = \underline{\eta}_i^{\perp} \cap \underline{\delta}_i (i = 1,\cdots,r+1)$. By the definition of $\underline{\eta}$, we have $A^{\eta_r,\gamma}_{(1,0)} \equiv 0$, which also holds for the case where $\underline{\eta}_j = \underline{0}$ for some $1 \leq j \leq r$. As before, we have a holomorphic endomorphism

$$A^{\delta'_r,\gamma}_{(1,0)} \cdots A^{\gamma,\gamma_1}_{(1,0)} : \underline{\gamma} \rightarrow \cdots \rightarrow \underline{\gamma}_{r+1} \rightarrow \underline{\eta} \rightarrow \underline{\eta}_1 \rightarrow \cdots \rightarrow \underline{\eta}_{r+1} \rightarrow \underline{\delta}' \rightarrow \underline{\delta}'_1 \rightarrow \cdots \rightarrow \underline{\delta}'_r \rightarrow \underline{\gamma},$$

which must vanish by (7.28). Since $\underline{\gamma} = \underline{\mathrm{Im}}A^{\delta'_r,\gamma}_{(1,0)}$, $\underline{\delta}'_i$ cannot be zero for any $1 \leq i \leq r$.

Therefore, by rank $\underline{\delta}' = 1$ we obtain $A^{\eta_{r+1},\delta'}_{(1,0)} \equiv 0$, which also holds for the case where $\underline{\eta}_j = \underline{0}$ for some $1 \le j \le r+1$. Note that $\underline{\eta} \oplus \underline{\gamma} \subset \mathrm{Ker}(A^{\varphi^\perp}_{(1,0)} A^{\varphi}_{(1,0)})$ because $A^{\eta_r,\gamma}_{(1,0)} = A^{\gamma_r,\gamma}_{(1,0)} = 0$. Set $\underline{\widetilde{\varphi}} = \underline{\delta}' \oplus \underline{\eta}_1 \oplus \underline{\gamma}_1$. Then, we have $G^{(i)}(\widetilde{\varphi}) = \underline{\delta}'_i \oplus \underline{\eta}_{i+1} \oplus \underline{\gamma}_{i+1}$ for $1 \le i \le r$ and $G^{(r+1)}(\widetilde{\varphi}) \subset \underline{\gamma} \oplus \underline{\eta} \oplus \underline{\delta}'_{r+1}$, which imply that $\widetilde{\varphi}$ has isotropy order $\ge r+1$. This gives (iii). Finally, we prove (iv). Assume that rank $\underline{\gamma} = 2$. Define $\underline{\gamma}_i$ and $\underline{\delta}_i$ as before. Since $A^{\gamma_r,\gamma}_{(1,0)} \cdots A^{\gamma,\gamma_1}_{(1,0)} : \underline{\gamma} \to \underline{\gamma}_1 \to \cdots \to \underline{\gamma}_r \to \underline{\gamma}$ is holomorphic, by (7.28) we have $\mathrm{rank}(\underline{\mathrm{Im}} A^{\gamma_r,\gamma}_{(1,0)}) \le 1$. If $\underline{\mathrm{Im}} A^{\gamma_r,\gamma}_{(1,0)} = \underline{0}$, we have a holomorphic endomorphism $A^{\delta_r,\gamma}_{(1,0)} \cdots A^{\gamma,\gamma_1}_{(1,0)} : \underline{\gamma} \to \cdots \to \underline{\gamma}_r \to \underline{\gamma}_{r+1} \to \underline{\delta} \to \underline{\delta}_1 \to \cdots \to \underline{\delta}_r \to \underline{\gamma}$, which must vanish by (7.28). Since $\underline{\gamma} = \underline{\mathrm{Im}} A^{\delta_r,\gamma}_{(1,0)}$, $\underline{\delta}_i$ cannot be zero for any $1 \le i \le r$. Therefore, by rank $\underline{\delta} = 1$ we obtain $A^{\gamma_{r+1},\delta}_{(1,0)} \equiv 0$, which also holds for the case where $\underline{\gamma}_i = \underline{0}$ for some $1 \le i \le r+1$. Thus, $\widetilde{\varphi}$ defined by $\underline{\widetilde{\varphi}} = \underline{\delta} \oplus \underline{\gamma}_1$ is a pluriharmonic map with isotropy order $\ge r+1$ as before. If $\mathrm{rank}(\underline{\mathrm{Im}} A^{\gamma_r,\gamma}_{(1,0)}) = 1$, set $\underline{\alpha} = \underline{\mathrm{Im}} A^{\gamma_r,\gamma}_{(1,0)}$, which is a holomorphic subbundle of $\underline{\gamma}$, and $\underline{\beta} = \underline{\gamma} \cap \underline{\alpha}^\perp$. Set $\underline{\alpha}_1 = \underline{\mathrm{Im}} A^{\alpha,\gamma_1}_{(1,0)}, \underline{\alpha}_i = \underline{\mathrm{Im}} A^{\alpha_{i-1},\gamma_i}_{(1,0)} (i = 2,\cdots,r)$ and $\underline{\beta}_i = \underline{\alpha}_i^\perp \cap \underline{\gamma}_i (i = 1,2,\cdots,r)$. Since $A^{\alpha_r,\alpha}_{(1,0)} \cdots A^{\alpha,\alpha_1}_{(1,0)} : \underline{\alpha} \to \underline{\alpha}_1 \to \cdots \to \underline{\alpha}_r \to \underline{\alpha}$ is holomorphic, we have $A^{\alpha_r,\alpha}_{(1,0)} \equiv 0$ by (7.28). Note that all $A^{\delta_i,\delta_{i+1}}_{(1,0)}, A^{\alpha_i,\alpha_{i+1}}_{(1,0)}, A^{\beta_i,\beta_{i+1}}_{(1,0)}, A^{\delta_r,\beta}_{(1,0)}, A^{\beta_r,\alpha}_{(1,0)}, A^{\alpha_r,R}_{(1,0)}$ and $A^{R,\delta}_{(1,0)}$ are holomorphic (cf. [B-W]). We have a holomorphic endomorphism $A^{\beta_r,\alpha}_{(1,0)} \cdots A^{\alpha,\alpha_1}_{(1,0)} : \underline{\alpha} \to \underline{\alpha}_1 \to \cdots \to \underline{\alpha}_r \to \underline{R} \to \underline{\delta} \to \underline{\delta}_1 \to \cdots \to \underline{\delta}_r \to \underline{\beta} \to \underline{\beta}_1 \to \cdots \to \underline{\beta}_r \to \underline{\alpha}$, which must vanish by (7.28). Since $A^{\delta_r,\beta}_{(1,0)}$ and $A^{\beta_r,\alpha}_{(1,0)}$ are surjective, all $\underline{\beta}_i$ and $\underline{\delta}_i$ cannot be zero. Thus, $\underline{\alpha}_s = \underline{0}$ for some $1 \le s \le r$ or $\mathrm{Im} A^{\alpha_r,R}_{(1,0)} \subset \mathrm{Ker} A^{R,\delta}_{(1,0)}$. Set $\underline{\widetilde{\varphi}} = \underline{\delta} \oplus \underline{\beta} \oplus \underline{\alpha}_1$. Then, we see that $G^{(i)}(\widetilde{\varphi}) = \underline{\delta}_i \oplus \underline{\beta}_i \oplus \underline{\alpha}_{i+1} (i = 1,\cdots,r-1), G^{(r)}(\widetilde{\varphi}) = \underline{\delta}_r \oplus \underline{\beta}_r \oplus \underline{\mathrm{Im}} A^{\alpha_r,R}_{(1,0)}$ and $\underline{\mathrm{Im}} A^{G^{(r)}(\widetilde{\varphi}),\widetilde{\varphi}}_{(1,0)} = \underline{\beta}$, which is of rank 1.

Q.E.D.

In the case (iv) of (7.33), since $\widetilde{\varphi}$ does not always define a pluriharmonic map into $G_k(\mathbf{C}^n)$ with $k \le 3$, we are not able to reduce this case to the case (iii). However, we may prove the following

**(7.34) Proposition.** *Let $\varphi : M \setminus S_\varphi \longrightarrow G_k(\mathbf{C}^n)$ be a pluriharmonic map, where $S_\varphi$ is as in (7.28). Assume that $\varphi$ has isotropy order $\ge r(\ge 1)$ and* $\mathrm{rank}(\underline{\mathrm{Im}} A^{G^{(r)}(\varphi),\varphi}_{(1,0)}) = 1$. *Then, there exists a holomorphic subbundle $\underline{\delta}$ of $\underline{\varphi}$ such that $\widetilde{\varphi}$ obtained by*

$$\underline{\widetilde{\varphi}} = \underline{\varphi} \ominus \underline{\delta} \oplus G'_\varphi(\delta)$$

*is a pluriharmonic map from $M \setminus S_{\widetilde{\varphi}}$ into $G_l(\mathbf{C}^n)$ for some $l$ and has isotropy order $\ge r+1$.*

*Proof.* In the same way as the proof of (7.33) (i), we may show that $A^{G^{(r)}(\varphi),\varphi}_{(1,0)}$ is holomorphic. Set $\underline{\gamma} = \underline{\mathrm{Im}} A^{G^{(r)}(\varphi),\varphi}_{(1,0)}$, which is a holomorphic subbundle of $\underline{\varphi}$. Define $\underline{\gamma}_i$ $(i = 1,\cdots,r+1)$ as in the proof of (7.33). Set $\underline{\delta}_i = \underline{\gamma}_i^\perp \cap G^{(i)}(\varphi)$ $(i = 1,\cdots,r+1)$. Since rank $\underline{\gamma} = 1$ we have $A^{\gamma_r,\gamma}_{(1,0)} \equiv 0$, which also holds for the case where $\underline{\gamma}_i = \underline{0}$ for some $1 \le i \le r$. If $\underline{\gamma}_{r+1} = \underline{0}$ or $A^{\gamma_{r+1},\delta}_{(1,0)} \equiv 0$, $\widetilde{\varphi}$ defined by $\underline{\widetilde{\varphi}} = \underline{\varphi} \ominus \underline{\gamma} \oplus \underline{\gamma}_1$ is a

pluriharmonic map with isotropy order $\geq r+1$. Suppose that $A_{(1,0)}^{\gamma_{r+1},\delta} \neq 0$. Since we have a holomorphic endomorphism $A_{(1,0)}^{\delta_r,\gamma} \cdots A_{(1,0)}^{\gamma,\gamma_1} : \underline{\gamma} \to \underline{\gamma}_1 \to \cdots \to \underline{\gamma}_{r+1} \to \underline{\delta} \to \underline{\delta}_1 \to \cdots \to \underline{\delta}_r \to \underline{\gamma}$, we see that $\underline{\mathrm{Im}}A_{(1,0)}^{\gamma_{r+1},\delta} \subset \mathrm{Ker}(A_{(1,0)}^{\delta_r,\gamma} \cdots A_{(1,0)}^{\delta,\delta_1})$. Set $\underline{\gamma}^1 = \underline{\mathrm{Im}}A_{(1,0)}^{\gamma_{r+1},\delta}$. Note that rank $\underline{\gamma}^1 <$ rank $\underline{\delta}$. Set $\underline{\delta}^1 = (\underline{\gamma}^1)^\perp \cap \underline{\delta}$. Define $\underline{\gamma}_i^1$ by $\underline{\gamma}_1^1 = \underline{\mathrm{Im}}A_{(1,0)}^{\gamma^1,\delta_1}$, $\underline{\gamma}_i^1 = \underline{\mathrm{Im}}A_{(1,0)}^{\gamma_{i-1}^1,\delta_i}$ $(i = 2,\cdots,r+1)$ and set $\underline{\delta}_i^1 = (\underline{\gamma}_i^1)^\perp \cap \underline{\delta}_i (i = 1,\cdots,r+1)$. By the definition of $\underline{\gamma}^1$, we see that $A_{(1,0)}^{\gamma_r^1,\gamma} \equiv 0$. If $\underline{\gamma}_{r+1}^1 = \underline{0}$ or $A_{(1,0)}^{\gamma_{r+1}^1,\delta^1} \equiv 0$, $\widetilde{\varphi}^1$ defined by $\underline{\widetilde{\varphi}}^1 = \underline{\delta}^1 \oplus \underline{\gamma}_1 \oplus \underline{\gamma}_1^1$ is a pluriharmonic map with isotropy order $\geq r+1$. Inductively if $A_{(1,0)}^{\gamma_{r+1}^s,\delta^s} \neq 0$, set $\underline{\gamma}^{s+1} = \underline{\mathrm{Im}}A_{(1,0)}^{\gamma_{r+1}^s,\delta^s}$, $\underline{\delta}^{s+1} = (\underline{\gamma}^{s+1})^\perp \cap \underline{\delta}^s$ and $\underline{\gamma}_1^{s+1} = \underline{\mathrm{Im}}A_{(1,0)}^{\gamma^{s+1},\delta_1^s}$, $\underline{\gamma}_i^{s+1} = \underline{\mathrm{Im}}A_{(1,0)}^{\gamma_{i-1}^{s+1},\delta_i^s}$ $(i = 2,\cdots,r+1)$, $\underline{\delta}_i^{s+1} = (\underline{\gamma}_i^{s+1})^\perp \cap \underline{\delta}_i^s$ $(i = 1,\cdots,r+1)$ for $s = 1,2,\cdots$, where $A_{(1,0)}^{\gamma_r^{s+1},\gamma} = A_{(1,0)}^{\gamma_r^s,\gamma} = 0$ and rank $\underline{\gamma}^{s+1} <$ rank $\underline{\delta}^s$. Therefore, there exists an integer $t \in \mathrm{N}$ such that $\underline{\gamma}_{r+1}^t = \underline{0}$ or $A_{(1,0)}^{\gamma_{r+1}^t,\delta^t} \equiv 0$, and $A_{(1,0)}^{\gamma_r^i,\gamma} \equiv 0$ for any $1 \leq i \leq t$. Note that $\underline{\delta}^t \neq \underline{0}$. Thus, $\widetilde{\varphi}$ defined by $\underline{\widetilde{\varphi}} = \underline{\delta}^t \oplus (\underline{\gamma}_1 \oplus \underline{\gamma}_1^1 \oplus \cdots \oplus \underline{\gamma}_1^t)$ is a pluriharmopnic map with isotropy order $\geq r+1$.
Q.E.D.

Now, we may prove the following

**(7.35) Theorem.** *Let $\varphi : M \setminus S_\varphi \longrightarrow G_k(\mathrm{C}^n)$ be a pluriharmonic map from a compact complex manifold with $c_1(M) > 0$, where $S_\varphi$ is as in (7.28). Assume that $k = 2,3$ and $n \leq 12$. Then, by the successive procedures of the type (1) or (5) in (7.26), $\varphi$ is reduced to a $\pm$-holomorphic map $f : M \setminus S_f$ into $G_r(\mathrm{C}^n)$ for some $r$, namely, a rational map $f : M \longrightarrow G_r(\mathrm{C}^n)$.*

*Proof.* We prove only the case of $k = 3$. The case of $k = 2$ is similar. First, by (7.33) and (7.34) we have a pluriharmonic map $\widetilde{\varphi}$ with isotropy order $\geq 2$. Suppose that $n \leq 11$. Then, one of the bundles $\underline{\widetilde{\varphi}}, G'(\widetilde{\varphi}), G^{(2)}(\widetilde{\varphi})$ has rank $\leq 3$. Moreover, by (7.31) each of $G'(\widetilde{\varphi})$ and $G^{(2)}(\widetilde{\varphi})$ also has isotropy order $\geq 2$. Choosing a bundle which has rank $\leq 3$ among them, we, again, obtain a pluriharmonic map $\widetilde{\varphi}'$ with isotropy order $\geq 3$ by (7.32), (7.33) and (7.34). Since $n \leq 11$, one of the bundles $\widetilde{\varphi}', G^{(i)}(\widetilde{\varphi}')(i = 1,2,3)$ has rank $\leq 2$. Repeating this process, we see that $\varphi$ is reduced to a $\pm$-holomorphic map into $G_r(\mathrm{C}^n)$ for some $r$. If $n = 12$, either one of the bundles $\underline{\widetilde{\varphi}}, G^{(i)}(\widetilde{\varphi})(i = 1,2)$ has rank $\leq 3$ or all of them have rank 4. The former case may be treated in the same way as the case of $n \leq 11$. The latter case implies that rank $G^{(3)}(\widetilde{\varphi}) \leq 3$ because $G^{(3)}(\widetilde{\varphi}) \subset \underline{\widetilde{\varphi}}$. Thus, it may be treated in the same way as the case of $n \leq 11$.
Q.E.D.

Now, we will give the explicit construction of pluriharmonic maps using (7.35). First, we need the following

**(7.36) Proposition** (cf. [B-W]). *Let $\varphi : M \longrightarrow G_k(\mathbf{C}^n)$ be a pluriharmonic map from a complex manifold. Let $\underline{\alpha} \subset \mathrm{Ker}(A^{\varphi^\perp}_{(1,0)} A^{\varphi}_{(1,0)})$ be a holomorphic subbundle of $\underline{\varphi}$ and let $\underline{\widetilde{\varphi}} = \underline{\varphi} \ominus \underline{\alpha} \oplus G'_\varphi(\alpha)$. Then, $G'_\varphi(\alpha)$ is an anti-holomorphic subbundle of $\underline{\widetilde{\varphi}}$, $G'_\varphi(\alpha) \subset \mathrm{Ker}(A^{\widetilde{\varphi}^\perp}_{(0,1)} A^{\widetilde{\varphi}}_{(0,1)})$* and, *if $\underline{\mathrm{Ker}} A^{\varphi}_{(1,0)} = \underline{0}$, then $\varphi$ is obtained from $\underline{\widetilde{\varphi}}$ by*

$$\underline{\varphi} = \underline{\widetilde{\varphi}} \ominus G'_\varphi(\alpha) \oplus G''_{\widetilde{\varphi}}(G'_\varphi(\alpha)).$$

*Proof.* Since $A^{\varphi^\perp}_{(1,0)}(G'_\varphi(\alpha)) = 0$, $G'_\varphi(\alpha)$ is an anti-holomorphic subbundle of $\underline{\widetilde{\varphi}}$. Moreover, we have $A^{\widetilde{\varphi}}_{(0,1)} |_{G'_\varphi(\alpha)} = A^{G'_\varphi(\alpha),\alpha}_{(0,1)} \subset \underline{\alpha}$ because $G'_\varphi(\alpha)$ is a holomorphic subbundle of $\varphi^\perp$. On the other hand, we have

$$A^{\widetilde{\varphi}^\perp}_{(0,1)} |_{\underline{\alpha}} = A^{\alpha,\widetilde{\varphi}}_{(0,1)} = A^{\alpha,\varphi\cap\alpha^\perp}_{(0,1)} + A^{\alpha,G'_\varphi(\alpha)}_{(0,1)} = 0$$

because $\underline{\alpha}$ is a holomorphic subbundle of $\underline{\varphi}$ and $A^{G'_\varphi(\alpha),\alpha}_{(1,0)} = 0$. Thus,

$$G'_\varphi(\alpha) \subset \mathrm{Ker}(A^{\widetilde{\varphi}^\perp}_{(0,1)} A^{\widetilde{\varphi}}_{(0,1)}).$$

If $\underline{\mathrm{Ker}} A^{\varphi}_{(1,0)} = \underline{0}$, then $\underline{\mathrm{Ker}} A^{\alpha,G'_\varphi(\alpha)}_{(1,0)} = \underline{0}$, hence $A^{G'_\varphi(\alpha),\alpha}_{(0,1)} : G'_\varphi(\alpha) \longrightarrow \underline{\alpha}$ is surjective. Therefore, we see that $G''_{\widetilde{\varphi}}(G'_\varphi(\alpha)) = \underline{\alpha}$ and $\underline{\varphi} = \underline{\widetilde{\varphi}} \ominus G'_\varphi(\alpha) \oplus G''_{\widetilde{\varphi}}(G'_\varphi(\alpha))$. Q.E.D.

For the case of $\underline{\mathrm{Ker}} A^{\varphi}_{(1,0)} \neq \underline{0}$, we have

**(7.37) Proposition** (cf. [B-W]). *Let $\varphi : M \longrightarrow G_k(\mathbf{C}^n)$ be a pluriharmonic map from a complex manifold. Assume that $\underline{\mathrm{Ker}} A^{\varphi}_{(1,0)} \neq \underline{0}$. Then, there exists a pluriharmonic map $\psi : M \setminus S_\psi \longrightarrow G_t(\mathbf{C}^n)$ for some $0 \leq t \leq k-1$ and a non-zero anti-holomorphic subbundle $\beta$ of $(\underline{\psi} \oplus G'(\psi))^\perp$ such that $\underline{\varphi} = \underline{\psi} \oplus \underline{\beta}$ over $M \setminus S_\psi$, where $S_\psi$ is as in (7.28). Conversely, given $\psi : M \longrightarrow G_t(\mathbf{C}^n)$* a *pluriharmonic map and a non-zero anti-holomorphic subbundle $\beta$ of $(\underline{\psi} \oplus G'(\psi))^\perp$ then setting $\underline{\varphi} = \underline{\psi} \oplus \underline{\beta}$ gives a pluriharmonic map $\varphi : M \setminus S_\varphi \longrightarrow G_k(\mathbf{C}^n)$ with $\underline{\mathrm{Ker}} A^{\varphi}_{(1,0)} \neq \underline{0}$, where $k = t +$* rank *$\underline{\beta}$.*

*Proof.* Set $\underline{\beta} = \underline{\mathrm{Ker}} A^{\varphi}_{(1,0)}$ and $\underline{\psi} = \underline{\varphi} \ominus \underline{\beta}$. Then, since $A^{\beta,\varphi^\perp}_{(1,0)} = 0$ and $G'(\psi) = G'(\varphi)$ we see that $\underline{\beta}$ is an anti-holomorphic subbundle of $(\underline{\psi} \oplus G'(\psi))^\perp$, which, together with (7.26), yields the first part of the assertion. For the converse, let $\psi : M \longrightarrow G_t(\mathbf{C}^n)$ be a pluriharmonic map and set $\underline{R} = (\underline{\psi} \oplus G'(\psi))^\perp$. Let $\underline{\beta}$ be a non-zero anti-holomorphic subbundle of $\underline{R}$. Since $A^{\psi,R}_{(1,0)} = 0$, we have $A^{R,\psi}_{(0,1)} = 0$, hence $A^{\psi^\perp}_{(0,1)} |_{\underline{\beta}} = 0$. Since $G'(\psi)$ is a holomorphic subbundle of $\underline{\psi}^\perp$, we have $A^{R,G'(\psi)}_{(1,0)} = 0$, which implies that $\underline{\beta}$ is also an anti-holomorphic subbundle of $\underline{\psi}^\perp$. Thus, $\varphi$ defined by $\underline{\varphi} = \underline{\psi} \oplus \underline{\beta}$ is a pluriharmonic map by (7.15) (if we reverse the orientation of $M$ in (7.15)). Moreover, since $A^{\beta,\varphi^\perp}_{(1,0)} = A^{\beta,R\cap\beta^\perp}_{(1,0)} + A^{\beta,G'(\psi)}_{(1,0)} = 0$, we see that $\underline{\beta} \subset \mathrm{Ker} A^{\varphi}_{(1,0)}$, hence $\underline{\mathrm{Ker}} A^{\varphi}_{(1,0)} \neq \underline{0}$. Q.E.D.

By (7.36) and (7.37), we can reverse the procedures in (7.35). Note that if $\varphi$ has isotropy order $\geq r(\geq 1)$ and $\underline{\mathrm{Ker}}A^{\varphi}_{(1,0)} \neq \underline{0}$ then $\widetilde{\varphi}$ defined by $\underline{\widetilde{\varphi}} = \underline{\varphi} \ominus \underline{\mathrm{Ker}}A^{\varphi}_{(1,0)}$ also has isotropy order $\geq r(\geq 1)$.

Thus, we obtain (if we reverse the orientation of $M$)

**(7.38) Theorem.** *Let $\varphi : M \setminus S_\varphi \longrightarrow G_k(\mathbf{C}^n)$ be any pluriharmonic from a compact complex manifold with $c_1(M) > 0$, where $S_\varphi$ is as in (7.28). Assume that $k = 2, 3$ and $n \leq 12$. Then, there is a sequence of pluriharmonic maps $\{\varphi_i\}_{i=0}^N$, $\varphi_i : M \setminus S_{\varphi_i} \longrightarrow G_{k_i}(\mathbf{C}^n)$, such that*
*(i) $\varphi_0$ is a holomorphic map, (ii) $\varphi = \varphi_N$,*
*(iii) each $\varphi_{i+1}$ is obtained from $\varphi_i$ by either of the following :*
*(1) $\underline{\varphi}_{i+1} = G'(\varphi_i)$,*
*(2) $\underline{\varphi}_{i+1} = \underline{\varphi}_i \oplus \underline{\beta}$, where $\underline{\beta}$ is a non-zero holomorphic subbundle of $(\underline{\varphi}_i \oplus G''(\varphi_i))^\perp$,*
*(3) $\underline{\varphi}_{i+1} = \underline{\varphi}_i \ominus \underline{\alpha} \oplus G'_{\varphi_i}(\alpha)$, where $\underline{\alpha} \subset \mathrm{Ker}(A^{\varphi_i^\perp}_{(1,0)} A^{\varphi_i}_{(1,0)})$ is a holomorphic subbundle of $\underline{\varphi}_i$.*

7.D. **A construction of non $\pm$-holomorphic, pluriharmonic maps from $\pm$-holomorphic maps.** The following method of constructing non $\pm$-holomorphic, pluriharmonic maps from $\pm$-holomorphic maps is a generalization of the result in [O-U2].

**(7.39) Proposition.** *Let $\varphi : M \longrightarrow G_k(\mathbf{C}^n)$ be a holomorphic (resp. an anti-holomorphic) map from a complex manifold. Set $\varphi = \varphi_0$ and define a sequence $\{\varphi_i\}$ of pluriharmonic maps, inductively, by*

$$\underline{\varphi}_i = \underline{\mathrm{Im}}A^{\varphi_{i-1}}_{(1,0)} \ (\text{resp. } \underline{\mathrm{Im}}A^{\varphi_{i-1}}_{(0,1)})$$

*for $i = 1, 2, \cdots$, over $M \setminus S_{\varphi_i}$. Then, there exists a positive integer $r$ such that $\underline{\varphi}_r = \underline{0}$. Moreover, $\varphi_{r-1}$ is an anti-holomorphic (resp. a holomorphic) map and $\varphi_s$ for $s = 1, \cdots, r-2$ is a non $\pm$-holomorphic, pluriharmonic map from $M \setminus S_{\varphi_s}$ into $G_{k_s}(\mathbf{C}^n)$ for some $k_s$.*

*Proof.* We prove only the case of $\varphi$ being holomorphic. The anti-holomorphic case is similar. Since $\varphi$ is holomorphic, we have $A^{\varphi}_{(0,1)} \equiv 0$, which implies that $\underline{\varphi} \perp G^{(i)}(\varphi)$ for any $i \in \mathbf{N}$. Therefore, the conclusion is clear. Q.E.D.

*Example.* Let $\varphi : \mathbf{C}P^2 \ni [1, z_1, z_2] \longrightarrow [1, \sqrt{2}z_1, \sqrt{2}z_2, z_1^2, \sqrt{2}z_1z_2, z_2^2] \in \mathbf{C}P^5$ be the second Veronese embedding, where $\{z_1, z_2\}$ is an inhomogeneous coordinate system for $\mathbf{C}P^2$. Then, by an easy computation we see that $\varphi_1$ is a non $\pm$-holomorphic, pluriharmonic map into $G_2(\mathbf{C}^6)$ and $\varphi_2$ is already an anti-holomorphic map into $G_3(\mathbf{C}^6)$.

From this example, it is convenient to make the following definition

*Definition.* Let $\varphi : M \longrightarrow G_k(\mathbf{C}^n)$ be a holomorphic map. Inductively, define $\underline{\varphi}_i = \underline{\mathrm{Im}}A^{\varphi_{i-1}}_{(1,0)}$ for $i = 1, 2, \cdots$, where $\varphi = \varphi_0$. If $\varphi_s$ is anti-holomorphic, we call such a least integer $s$ the *osculating degree* of a holomorphic map $\varphi$.

Then, by a direct computation and a result of Nakagawa, Takagi and Takeuchi [N-T], [T-T] we have the following

**(7.40) Proposition.** *Let $\varphi : M \longrightarrow CP^N$ be a full p-th canonical Kähler embedding of a compact irreducible Hermitian symmetric space of rank $M = r$* (cf. [Na-T], [T-T]). *Then, the osculating degree of $\varphi$ is not more than $pr$.*

## 8. Pluriharmonic maps into compact Lie groups.

Let $G$ be a compact connected Lie group with Lie algebra $\mathbf{g}$. We equip $G$ with a biinvariant Riemannian metric $g_G$ induced from an Ad$G$-invariant Riemannian metric $( \, , \, )$ of $\mathbf{g}$. Denote by $\mu_G$ the Maurer-Cartan form of $G$. For a smooth map $\varphi : M \longrightarrow G$, set $\alpha = \varphi^* \mu_G = \alpha' + \alpha''$. Here $\alpha'$ and $\alpha''$ are the (1,0)-part and the (0,1)-part of $\alpha$. From the Maurer-Cartan equation, we have

$$d\alpha + \frac{1}{2}[\alpha \wedge \alpha] = 0. \tag{8.1}$$

Set $A = \frac{1}{2}\alpha$, $A' = \frac{1}{2}\alpha'$ and $A'' = \frac{1}{2}\alpha''$. Consider a $G$-connection $d_A = d + A$ on $M$. We define $\partial_A = \partial + A'$, $\bar{\partial}_A = \bar{\partial} + A''$. Note that the connection $d_A$ coincides with the connection induced from the Levi-Civita connection of $(G, g_G)$. By (0.4) and (1.4) we have

**(8.2) Lemma** ([O-V]). *(1) $\varphi$ is pluriharmonic if and only if*

$$\bar{\partial}\alpha' + \frac{1}{2}[\alpha' \wedge \alpha''] = 0. \tag{8.3}$$

*(2) If $\varphi$ is pluriharmonic, then*

$$[\alpha' \wedge \alpha'] = [\alpha'' \wedge \alpha''] = 0. \tag{8.4}$$

Therefore, we see that $\varphi : M \longrightarrow G$ is pluriharmonic if and only if $\bar{\partial}_A \circ \bar{\partial}_A = 0$ and $\bar{\partial}_A A' = 0$. Thus, for any $G$-module $V$, trivial bundle $\underline{V} = M \times V$ has a Hermitian holomorphic vector bundle structure such that $\bar{\partial}$-operator coincides with $\bar{\partial}_A$-operator, and $A'$ is a holomorphic $(1,0)$-form with values in End$(\underline{V}, \bar{\partial}_A)$. Moreover, $\{(\underline{V}, \bar{\partial}_A), A'\}$ is a Higgs bundle in the sense of [Hi1,2,Sp] and satisfies the equations

$$\bar{\partial}_A A' = 0, \quad F_A = [A' \wedge (A')^*],$$

where $F_A$ denotes the curvature form of the connection $A$.

In [Uh], the concept of extended solutions for a harmonic map from a Riemann surface into Lie group played a central role. We will estabilish the construction of extended solutions for a pluriharmonic map from a complex manifold into a compact Lie group.

Let $\varphi : M \longrightarrow G$ be a smooth map from complex manifold into compact Lie group. For each $\lambda \in \mathbf{C}^* = \mathbf{C} - \{0\}$, define a $\mathbf{g}^{\mathbf{C}}$-valued 1-form $\alpha_\lambda$ on $M$ by

$$\alpha_\lambda = \frac{1}{2}(1 - \lambda^{-1})\alpha' + \frac{1}{2}(1 - \lambda)\alpha''.$$

Let $G^{\mathbf{C}}$ be the complexification of $G$. For each $\lambda \in \mathbf{C}^*$, consider the linear partial diffrential equation of first order for $\Phi_\lambda : M \longrightarrow G^{\mathbf{C}}$ as follows :

$$\Phi_\lambda^* \mu_{G^{\mathbf{C}}} = \alpha_\lambda$$

or

$$\begin{cases} \partial \Phi_\lambda = \frac{1}{2}(1-\lambda^{-1})\Phi_\lambda \alpha', \\ \bar{\partial}\Phi_\lambda = \frac{1}{2}(1-\lambda)\Phi_\lambda \alpha''. \end{cases}$$

The complete integrability condition for this equation is

$$d\alpha_\lambda + \frac{1}{2}[\alpha_\lambda \wedge \alpha_\lambda] = 0, \tag{8.5}$$

where $\lambda \in \mathbf{C}^*$. Using (8.1), (8.3) and (8.4) we can verify that the equation (8.5) is equivalent to the pluriharmonicity of $\varphi : M \longrightarrow G$. Therefore, we have the following

**(8.6) Theorem** ([Uh],[O-V]). *Assume that $M$ is simply connected. Let $x_0$ be a base point of $M$. If $\varphi : M \longrightarrow G$ is pluriharmonic, then, for any $\sigma : \mathbf{C}^* \longrightarrow G^{\mathbf{C}}$, there exists uniquely $\Phi : \mathbf{C}^* \times M \longrightarrow G^{\mathbf{C}}$ such that $\Phi_\lambda(x_0) = \sigma(\lambda)$ and $\Phi_\lambda^*\mu_{G^{\mathbf{C}}} = \alpha_\lambda$ for each $\lambda \in \mathbf{C}^*$.*

We call this $\Phi$ an *extended solution* for the pluriharmonic map $\varphi$. Moreover, $\Phi$ is called *real* if $\Phi_\lambda : M \longrightarrow G$ for each $\lambda \in S^1 \subset \mathbf{C}^*$, and *based* if $\Phi_\lambda(x_0) = e$ for each $\lambda \in \mathbf{C}^*$.

In the following, *we treat only real extended solutions satisfying $\Phi_1 \equiv e$.*

We can regard the real extended solution $\Phi$ as a map into the infinite dimensional Lie group $\Omega G = \{\gamma : S^1 \longrightarrow G, \text{ smooth}, \gamma(1) = e\}$ of based smooth loops in $G$. $\Omega G$ has an infinite dimensional complex manifold structure $J_1$ and becomes a Kähler manifold with respect to $\mathrm{L}^2_{1/2}$-metric. On the other hand, Burstall and Rawnsley introduced the interesting non-integrable almost complex structure $J_2$ on $\Omega G$ (cf. [E-L2]). If we define a projection $\pi : \Omega G \longrightarrow G$ by $\pi(\gamma) = \gamma(-1)$, this plays a role of universal twistor fibration (cf. [E-L2]). Then, we see that real extended solution $\Phi : M \longrightarrow \Omega G$ is holomorphic with respect to both $J_1$ and $J_2$. As applications of holomorphic map $\Phi$ into $\Omega G$ associated with pluriharmonic map $\varphi$, we obtain the following.

**(8.7) Theorem** ([O-V], cf. [E-L2]). *Assume that $M$ is simply connected and let $\varphi : M \longrightarrow G$ be a pluriharmonic map. Then,*
*(i) there exists $J_1$-holomorphic map $\Phi : M \longrightarrow \Omega G$ such that $\varphi = \pi \circ \Phi$,*
*(ii) if $\mathrm{H}^3(G,\mathbf{Z}) \cong \mathrm{H}^2(\Omega G,\mathbf{Z}) \cong \mathbf{Z}$, then the cohomology class $[(|\delta|^2/16\pi)\varepsilon(\varphi)]$ is integral and*

$$E_\omega(\varphi) = \frac{16m\pi}{m!|\delta|^2}\deg_\omega(\Phi),$$

*where $\deg_\omega(\Phi) = \Phi^*\kappa_1 \wedge \omega^{m-1}$, $\delta$ is the highest root of $\mathbf{g}$, $\kappa_1$ represents a positive generator of $\mathrm{H}^2(\Omega G,\mathbf{Z})$ and $\omega$ is a fundamental 2-form of a Hermitian metric on $M$.*
*(iii) If $M$ has a cosymplectic Hermitian metric $g$, then any nonconstant pluriharmonic map*

*$\varphi : (M,g) \longrightarrow G$ is always unstable as a harmonic map. In particular, any nonconstant harmonic map $\varphi$ from $\mathbf{C}P^m$ with the Fubini-Study metric to $G$ is always unstable.*

We assume that $G = U(N)$. Set $Gr(\mathbf{C}^N) = \{a \in U(N); a^2 = I\}$. Each connected component of $Gr(\mathbf{C}^N)$ is a complex Grassmann manifold $G_k(\mathbf{C}^N)$ $(0 \le k \le N)$ and a totally geodesic submanifold of $U(N)$. There is a natural bijective correspondence between a complex subbundle $\eta$ of the trivial complex vector bundle $\underline{\mathbf{C}}^N = M \times \mathbf{C}^N$ with rank $k$ and a smooth map $\Pi_\eta - \Pi_\eta^\perp : M \longrightarrow G_k(\mathbf{C}^N) \subset U(N)$. Here $\Pi_\eta$ denotes the Hermitian projection onto $\eta$. We can introduce the notion of *uniton* also for pluriharmonic maps in the way similar to [Uh].

Let $\varphi : M \longrightarrow U(N)$ be a pluriharmonic map and $\Phi$ be an real extended solution of $\varphi$. For a smooth map $\Pi - \Pi^\perp = \Pi_\eta - \Pi_\eta^\perp : M \longrightarrow G_k(\mathbf{C}^N)$, set $\Psi_\lambda = \Phi_\lambda(\Pi + \lambda\Pi^\perp) : M \longrightarrow GL(N, \mathbf{C})$ for each $\lambda \in \mathbf{C}^*$ and $\psi = \Psi_{-1} : M \longrightarrow U(N)$.

**(8.8) Lemma.** *The follwing conditions are equivalent each other :*
*(1) $\Psi$ is a real extended solution, namely, $\psi$ is pluriharmonic.*
*(2) $\Pi$ satisfies*

$$\Pi^\perp A'\Pi = 0, \qquad \Pi^\perp(\bar\partial\Pi + A'') = 0. \tag{8.9}$$

*(3) The subbundle $\eta$ is a holomorphic subbundle of $(\underline{\mathbf{C}}^N, \bar\partial_A)$ invariant under $A'$.*

The equation (8.9) is called the *uniton equation* of $\varphi$ and its solution $\pi$ (or $\eta, \Pi - \Pi^\perp : M \longrightarrow Gr(\mathbf{C}^N)$) is called *(smooth) uniton* for $\varphi$.

*Definition.* $\eta$ is called a *meromorphic uniton* for $\varphi$ if $\eta$ is a smooth uniton for $\varphi$ defined over $M \setminus S_\eta$. Here $S_\eta$ is an analytic subset of $M$ with $\mathrm{codim}_{\mathbf{C}} S_\eta \ge 2$.

Note that a meromorphic uniton $\eta$ extends to a coherent subsheaf $\mathcal{S}(\eta)$ of $\mathcal{O}(\underline{\mathbf{C}}^N, \bar\partial_A)$.

**(8.10) Theorem ([Va1],[O-V]).** *Assume that $M$ is compact. Let $\eta$ be a meromorphic uniton for $\varphi$. Then, for any real (1,1)-form $\omega$ with $d\omega^{m-1} = 0$, we have*

$$E_\omega(\psi) - E_\omega(\varphi) = -\frac{8m\pi}{m!}\deg_\omega(\mathcal{S}(\eta)),$$

*where, $\deg_\omega(\mathcal{S}(\eta)) = \int_M c_1(\mathcal{S}(\eta)) \wedge \omega^{m-1}$. In particular, the energy $E_\omega(\psi)$ of $\psi : M \setminus S_\eta \longrightarrow U(N)$ is finite.*

*Example.* $A'$ induces the sheaf homomorphisms

$$B_1 : \mathcal{O}(TM^{1,0}) \otimes \mathcal{O}(\underline{\mathbf{C}}^N, \bar\partial_A) \longrightarrow \mathcal{O}(\underline{\mathbf{C}}^N, \bar\partial_A),$$
$$B_2 : \mathcal{O}(\underline{\mathbf{C}}^N, \bar\partial_A) \longrightarrow \mathcal{O}(T^*M^{1,0}) \otimes \mathcal{O}(\underline{\mathbf{C}}^N, \bar\partial_A).$$

The image subsheaf of $B_1$ and the kernel subsheaf of $B_2$ defines meromorphic unitons $\underline{\mathrm{Im}}A'$, $\underline{\mathrm{Ker}}A'$. These are called *basic unitons* for $\varphi$.

*Definition.* $\varphi : M \longrightarrow U(N)$ is called a *meromorphically pluriharmonic map* if the following two conditions are satisfied :

(i) $\varphi$ is a smooth pluriharmonic map into $U(N)$ defined over $M \setminus S_\varphi$, where $S_\varphi$ is an analytic subset of $M$ with $\mathrm{codim}_{\mathbf{C}} S_\varphi \geq 2$.

(ii) There exists a connected complex manifold $\hat{M}$ and a surjective proper holomorphic map $\tau : \hat{M} \longrightarrow M$ such that $\tau : \hat{M} \setminus \tau^{-1} S_\varphi \longrightarrow M \setminus S_\varphi$ is a biholomorphic diffeomorphism and $\varphi \circ \tau$ extends to a pluriharmonic map $\hat{\varphi} : \hat{M} \longrightarrow U(N)$.

*Remark.* Very recently G. Valli informed to the authors that he showed that if $M$ is a simply connected compact complex manifold and $\varphi : M \setminus S_\varphi \longrightarrow U(N)$ is a pluriharmonic map with finite energy defined outside an analytic subset $S_\varphi$ of $M$ with $\mathrm{codim}_{\mathbf{C}} S_\varphi \geq 2$, then $\varphi : M \longrightarrow U(N)$ is meromorphically pluriharmonic.

**(8.11) Theorem** ([Uh], [Val], [O-V]). *Assume that $M$ satisfies the following condition (A) or (B) :*
*(A) $M$ is a simply connected compact complex manifold.*
*(B) $M$ is a projective algebraic manifold with Hodge metric $\omega$ and the Harder-Narasimhan filtration of $TM^{(1,0)}$ relative to $\omega$*

$$0 = \mathcal{T}_0 \subset \mathcal{T}_1 \subset \ldots \subset \mathcal{T}_{r-1} \subset \mathcal{T}_r = \mathcal{O}(TM^{1,0})$$

*satisfies $\deg_\omega(\mathcal{T}_r / \mathcal{T}_{r-1}) > 0$.*
*Then any meromorphically pluriharmonic map $\varphi : M \longrightarrow U(N)$ has a factorization*

$$\varphi = a(\Pi_1 - \Pi_1^\perp)(\Pi_2 - \Pi_2^\perp) \ldots (\Pi_n - \Pi_n^\perp) \qquad (a \in U(N)),$$

*which has the following properties :*
*(1) Each $\varphi^{(i)} = a(\Pi_1 - \Pi_1^\perp)(\Pi_2 - \Pi_2^\perp) \ldots (\Pi_i - \Pi_i^\perp) : M \longrightarrow U(N)$ $(i = 1, \ldots, n)$ is a meromorphically pluriharmonic map.*
*(2) $\Pi_i$ $(i = 1, \ldots, n)$ is a meromorphic uniton for $\varphi^{(i)}$.*
*(3) $\Pi_1 - \Pi_1^\perp : M \longrightarrow Gr(\mathbf{C}^N)$ is a meromorphic map.*
*(4) In case (A), $n < N$ and $n$ is equal to the minimal uniton number* (cf. [Uh], [O-V]) *of $\varphi$.*
*(5) For any Hermitian metric $\omega$ on $M$, each $E_\omega(\varphi^{(i)})$ is finite. If $M$ has cosymplectic Hermitian metric $\omega$, we have $E_\omega(\varphi^{(i)}) > E_\omega(\varphi^{(i-1)})$ $(i = 1, \ldots, n)$. Moreover, if the de Rham cohompology class $[\omega^{m-1}]$ is integral, then $n \leq \frac{m!}{8m\pi} E_\omega(\varphi)$.*

*Remark.* (1) The methods of factorizations in cases (A) and (B) are quite different. In general the number $n$ of factors of unitons are not same in cases (A) and (B). (2) In case (A), the uniqueness of factorizations holds under the condition similar to [Uh]. (3) We can show that if $M$ is a comapct complex manifold with $c_1(M) > 0$, then $M$ satisfies the condition (B) (cf. [O-V]). (4) The method in case (B) enables us to prove the factorization theorem for pluriharmonic maps into any compact Lie group which is an isometry group of a compact irreducible Hermitian symmetric space in the same way as in [B-R].

*Outline of Proof of* (8.11).

In case (A) : We use the method of [Uh]. For $\varphi$, we can find a real extended solution of the form $\Phi_\lambda = \sum_{\alpha=0}^{n} T_\alpha \lambda^\alpha$ ($\lambda \in \mathbf{C}^*$). Then $T_0$ induces a sheaf homomorphism

$$T_0 : \mathcal{O}(\underline{\mathbf{C}}^N, \bar\partial_A) \longrightarrow \mathcal{O}(\underline{\mathbf{C}}^N, \bar\partial),$$

and the kernel subsheaf of $T_0$ define a meromorphic uniton $\eta$ for $\varphi$. We add this meromorphic uniton $\eta$ to $\Phi$ or $\varphi$. By repeating this process, we can factorize $\Phi_\lambda$ into factors of degree 1 relative to $\lambda$, and we get the factorization of $\varphi$.

In case (B) : We use the method of [Va1], [Va2] and Harder-Narasimhan filtration for coherent sheaves over projective algebraic manifolds ([Ma], [Sh], [Ko]). Let

$$0 = \mathcal{E}_0 \subset \mathcal{E}_1 \subset \ldots \subset \mathcal{E}_s = \mathcal{O}(\underline{\mathbf{C}}^N, \bar\partial_A)$$

be the Harder-Narasimhan filtration of $\mathcal{O}(\underline{\mathbf{C}}^N, \bar\partial_A)$. By using the condition (B), we can show that each $\mathcal{E}_i$ is invariant under $A'$ and $\mathcal{O}(\underline{\mathbf{C}}^N, \bar\partial_A)$ is not semistable, i.e., $s > 1$. We transform $\varphi$ into a meromorphically pluriharmonic map $\psi$, by adding $\eta$ defined by $\mathcal{E}_{s-1}$. By (8.10), $(m!/8m\pi)(E_\omega(\psi) - E_\omega(\varphi))$ is a negative integer. Repeating this process, we get the factorization for $\varphi$.

*Question.* Is there a projective algebraic manifold $M$ which is not simply connected but satisfies the condition (B) ?

**(8.12) Corollary.** *For a generic K3 surface $M$, there exists no nonconstant meromorphically pluriharmonic map of $M$ into a compact symmetric space.*

*Question.* Find a non-pluriharmonic, harmonic map from a K3 surface $M$ to a unitary group $U(N)$.

## References.


[A] M. F. Atiyah, Instantons in two and four dimensions, Commun. Math. Phys. **93** (1984), 437–451.

[A-H-S] M. F. Atiyah, N. J. Hitchin and I. M. Singer, Self-duality in four dimensional Riemannian geometry, Proc. R. Soc. London, A. **362** (1978), 425–461.

[B-G-T] D. Barbasch, J. H. Glazebrook and G. Toth, Harmonic maps between complex projective spaces, to appear in Geom. Dedicata.

[Ba-W1] A. Bahy-El-Dien and J. C. Wood, The explicit construction of all harmonic two-spheres in $G_2(\mathbf{R}^n)$, to appear in J. Reine Angew. Math. .

[Ba-W2] A. Bahy-El-Dien and J. C. Wood, The explicit construction of all harmonic two-spheres in quaternionic projective space, preprint, Univ. of Leeds, 1989.

[Bo] A. Borel, .On the curvature tensor of the Hermitian symmetric manifolds, Ann. of Math. **71** (1960), 508–521.

[B-H] A. Borel and F. Hirzebruch, Characteristic classes and homogeneous spaces I, Amer. J. Math. **80** (1958), 458–538.

[B-B-B-R] D. Burns, F. Burstall, P. de Bartolomeis and J. Rawnsley, Stability of harmonic maps of Kähler manifolds, J. Differential Geom. **30** (1989), 579–594.

[B-B] D. Burns and P. de Bartolomeis, Applications harmoniques stables dans $P^n$, Ann. Scient. Ec. Norm. Sup. **21** (1988), 159–177.

[B-R] F. Burstall and J. Rawnsley, Spheres harmoniques dans les groupes de Lie compacts et curbes holomorphes dans les espaces homogenes, C. R. Acad. Sc. Paris **302** (1986), 709–712.

[B-R-S] F. Burstall, J. Rawnsley and S. Salamon, Stable harmonic 2-spheres in symmetric spaces, Bull. Amer. Math. Soc. **16** (1987), 274–278.

[B-S] F. Burstall and S. Salamon, Tournaments, flags and harmonic maps, Math. Ann. **277** (1987), 249–265.

[B-W] F. Burstall and J. C. Wood, The construction of harmonic maps into complex Grassmannians, J. Differential Goem. **23** (1986), 255–297.

[C-V] E. Calabi and E. Vesentini, On compact, locally symmetric Kähler manifolds, Ann. of Math. 71 (1960), 472–507.

[C-N] B. Y. Chen and T. Nagano, Totally geodesic submanifolds of symmetric spces, II, Duke Math. J. **45** (1978), 405–425.

[C-W] S. S. Chern and J. G. Wolfson, Harmonic maps of the 2-sphere into a complex Grassmann manifold, II, Ann. of Math. **125** (1987), 301–335.

[E-L1] J. Eells and L. Lemaire, A report on harmonic maps, Bull. London Math. Soc. **10** (1978), 1–68.

[E-L2] J. Eells and L. Lemaire, Another report on harmonic maps, Bull. London Math. Soc. **20** (1988), 385–524.

[E-S] J. Eells and J. H. Sampson, Harmonic mappings of Riemannian manifolds, Amer. J. Math. **86** (1964), 109–160.

[E-W1] J. Eells and J. C. Wood, Restrictions on harmonic maps of surfaces, Topology **15** (1976), 263–266.

[E-W2] J. Eells and J. C. Wood, Harmonic maps from surfaces to complex projective spaces, Adv. in Math. **49** (1983), 217–263.

[F] D. S. Freed, Flag manifolds and infinite dimensional Kähler geometry, In "Infinite Dimensional Groups with Applications", Math. Sci. Res. Inst. Publ. 4 (1985), 83–121, Springer-Verlag.

[G-H] P. Griffiths and J. Harris, Principles of Algebraic Geometry, Pure and Applied Math. , A Wiley-Interscience series, 1978.

[Gu] M. A. Guest, Geometry of maps between generalized flag manifolds, J. Differential Geom.**25** (1987), 223–247.

[Hi1] N. J. Hitchin, The self-duality equations on a Riemann surface, Proc. London Math. Soc. **55** (1987), 59–126.

[Hi2] N. J. Hitchin, Harmonic maps from a 2-torus to the 3-sphere, a preprint.

[Ho] R. Howard, The nonexistence of stable submanifolds, varifolds and harmonic maps in sufficiently pinched simply connected Riemannian manifolds, Mich. Math. J. **32** (1985), 321–334.

[Is] T. Ishihara, The Gauss map and non holomorphic harmonic maps, Proc. Amer. Math. Soc. **89** (1983), 661–665.

[It] M. Itoh, On curvature properties of Kähler C-spaces, J. Math. Soc. Japan **30** (1978), 39–71.

[J-S] M. Jaques and Y. Saint-Aubin, Infinite dimensional Lie algebras acting on the solution space of various $\sigma$-models, J. Math. Phys. **28** (1987), 2463–2479.

[Ko] S. Kobayashi, Differential Geometry of Complex Vector Bundles, Publ. Math. Soc. Japan **15**, Iwanami Shoten Publ. and Princeton Univ. Press, 1987.

[K-M] J. L. Koszul and B. Malgrange, Sur certaines structures fibrees complexes, Arch. Math. **9** (1958), 102–109.

[L-S] H. B. Lawson and J. Simons, On stable currents and their application to global problems in real and complex geometry, Ann. of Math. **98** (1973), 427–450.

[Li] A. Lichnerowicz, Applications harmoniques et varietes Kählerinnes, Symp. Math. III (Bologna 1970), 341–402.

[Ma] M. Maruyama, Theorem of Grauert Mülich-Spindler, Math. Ann. **255** (1981), 317–333.

[M-M] M. J. Micallef and J. D. Moore, Minimal two-spheres and the topology of manifolds with positive curvature on totally isotropic two planes, Ann. of Math. **127** (1988), 199–227.

[Na] H. Naito, On the holomorphicity of pluriharmonic maps, "Geometry of Manifolds", Proceedings of the 35th symposium on differential geometry, edited by K. Shiohama, Perspectives in Math. Vol. 8, 1989, 419–425.

[N-T] H. Nakagawa and R. Takagi, On locally symmetric Kähler submanifolds in a complex projective space, J. Math. Soc. Japan **28** (1976), 638–667.

[Ni] S. Nishikawa, Gauss map of Kähler immersions, Tohoku Math. J. **27** (1975), 453–460.

[Oh1] Y. Ohnita, On pluriharmonicity of stable harmonic maps, J. London Math. Soc. **35** (1987), 563–568.

[Oh2] Y. Ohnita, Homogeneous harmonic maps into complex projective spaces, to appear in Tokyo J. Math. .

[O-U1] Y. Ohnita and S. Udagawa, Stable harmonic maps from Riemann surfaces to compact Hermitian symmetric spaces, Tokyo J. Math. **10** (1987), 385–390.

[O-U2] Y. Ohnita and S. Udagawa, Stability, complex-analyticity and constancy of pluriharmonic maps from compact Kähler manifolds, to appear in Math. Z..

[O-V] Y. Ohnita and G. Valli, Pluriharmonic maps into compact Lie groups and factorization into unitons, to appear in Proc. London Math. Soc. .

[Ok] T. Okayasu, Pinching and nonexistence of stable harmonic maps, Tohoku Math. J. **40** (1988), 213–220.

[On] K. Ono, On the holomorphicity of harmonic maps from compact Kähler manifolds to hyperbolic Riemann surfaces, Proc. Amer. Math. Soc. **102** (1988), 1071–1076.

[P-S] A. Pressley and G. Segal, "Loop Groups", Oxford mathematical monographs, Clarendon press, Oxford, 1986.

[Ra] J. Ramanathan, Harmonic maps from $S^2$ to $G_{2,4}$, J. Differential Geom. **19** (1984), 207–219.

[R-S] A. Ramanathan and S. Subramanian, Einstein-Hermitian connections on principal bundles and stability, J. Reine Angew. Math. **390** (1988), 21–31.

[Rw] J. Rawnsley, f-structures, f-twistor and harmonic maps, Geom. Seminar "Luigi Bianchi" II (1984) Lecture Notes in Math. **1164**, 85–159, Springer-Verlag, Berlin.

[Re] R. Remmert, Holomorphe und meromorphe Abbildungen Komplexer Raume, Math. Ann. **133** (1957), 328–370.

[Sa] S. Salamon, Harmonic and holomorphic maps, Geom. Seminar "Luigi Bianchi"II (1984), Lecture Notes in Math. **1164**, Springer-Verlag, Berlin.

[Sm] J. H. Sampson, Applications of harmonic maps to Kähler geometry, Contemp. Math. 49 (1986), 125–134.

[St] M. Sato, Soliton equations as dynamical systems on an infinite dimensional Grassmann manifold, RIMS kokyuroku **439** (1981), 30–46.

[Sh] S. Shatz, The decomposition and specialization of algebraic families of vector bundles, Compositio Math. **35** (1977), 163–187.

[Sp] C. T. Simpson, Constructing variations of Hodge structure using Yang-Mills theory and applications to uniformization, J. Amer. Math. Soc. **1** (1988), 867–918.

[Si1] Y.-T. Siu, The complex analyticity of harmonic maps and the strong rigidity of compact Kähler manifolds, Ann. of Math. **112** (1980), 73–111.

[Si2] Y.-T. Siu, Curvature characterization of hyperquadrics, Duke Math. J. **47** (1980), 641–654.

[Si3] Y.-T. Siu, Strong rigidity of compact quotients of exceptional bounded symmetric domains, Duke Math. J. **48** (1981), 857–871.

[Si4] Y.-T. Siu, Complex-analyticity of harmonic maps, vanishing and Lefschetz Theorems, J. Differential Geom. **17** (1982), 55–138.

[S-Y] Y.-T. Siu and S.-T. Yau, Compact Kähler manifolds of positive bisectional curvature, Invent. Math. **59** (1980), 189–204.

[T-T] R. Takagi and M. Takeuchi, Degree of symmetric Kählerian submanifolds of a complex projective space, Osaka J. Math. **14** (1977), 501–518.

[Ta] K. Takasaki, A new approach to the Yang-Mills equations, Commun. Math. Phys. **94** (1984), 35–59.

[Te] M. Takeuchi, Totally complex submanifolds of quaternionic symmetric spaces, Japan J. Math. **12** (1986), 161–189.

[Ts] K. Tsukada, Parallel submanifolds in a quaternionic projective space, Osaka J. Math. **22** (1985), 187–241.

[Ud1] S. Udagawa, Minimal immersions of Kähler manifolds into complex space forms, Tokyo J. Math. **10** (1987), 227–239.

[Ud2] S. Udagawa, Pluriharmonic maps and minimal immersions of Kähler manifolds, J. London Math. Soc. (2) **37** (1988), 375–384.

[Ud3] S. Udagawa, Holomorphicity of certain stable harmonic maps and minimal immersions, Proc. London Math. Soc. (3) **57** (1988), 577–598.

[Uh] K. Uhlenbeck, Harmonic maps into Lie groups (Classical solutions of the chiral model), J. Differential Geom. **30** (1989), 1-50.

[Va1] G. Valli, On the energy spectrum of harmonic 2-spheres in unitary groups, Topology **27** (1988), 129–136.

[Va2] G. Valli, Some remarks on geodesics in gauge groups and harmonic maps, J. Geom. Phys. **4** (1987), 335–359.

[Va3] G. Valli, Harmonic gauges on Riemann surfaces and stable bundles, Ann. Inst. H. Poincaré (analyse non-lineaire) **6** (1989) 233–245.

[We] R. O. Wells, Differential Analysis on Complex Manifolds, Prentice-Hall, Englewood Cliffs, N. J., 1973.

[Wo1] J. G. Wolfson, Harmonic sequences and harmonic maps of surfaces into complex Grassmann manifolds, J. Differential Geom. **27** (1988), 161–178.

[Wo2] J. G. Wolfson, Harmonic maps of the two-sphere into the complex hyperquadric, J. Differential Geom. **24** (1986), 141–152.

[Wd1] J. C. Wood, The explicit construction and parametrization of all harmonic maps from the two-sphere to a complex Grassmannian, J. Reine Angew. Math. **386** (1988), 1–31.

[Wd2] J. C. Wood, Explicit construction and parametrization of harmonic two-spheres in the unitary group, Proc. London Math. Soc. (3) **58** (1989), 608–624 .

[Ya] S.-T. Yau, On Calabi's conjecture and some results in algebraic geometry, Proc. Nat. Acad. Sci. USA **74** (1977), 1798–1799.

[Z-M] V. E. Zakharov and A. V. Mikhailov, Relativistically invariant two dimensional models of field theory which are integrable by means of the inverse scattering problem method, Sov. Phys. J. E. T. P. **47** (1978), 1017–1027.

[Z-S] V. E. Zakharov and A. B. Shabat, Integration of non-linear equations of mathematical physics by the method of inverse scattering II, Transl. Funk. Anal. **13** (1979), 166–174.

[Zh] J.-Q. Zhong, The degree of strong nondegeneracy of the bisectional curvature of exceptional bounded symmetric domains, Several Complex Variables, Proc. 1981 Hangzhou Conf. , 1984, 127–139.

# Higher Eichler Integrals and
# Vector Bundles over the Moduli of Spinned Riemann surfaces

Kyoji Saito

RIMS, Kyoto Univesity

## § 1. Introduction

The importance of infinite dimensional freedom is recognized more and more in the recent development of mathematics. In this note, we study some infinite dimensional representations of the surface group $\Gamma_g$ of genus $g\geq 2$.

Let $\mathbb{H}$ be the complex upper half plane and let $K_{\mathbb{H}}^{1/2}$ be its half canonical bundle, on which $SL(2,\mathbb{R})$ operates naturally. The group $\Gamma_g$ will operate on $\Gamma(\mathbb{H},\mathcal{O}_{\mathbb{H}}(K_{\mathbb{H}}^{-n/2}))$ $(n\geq 0)$ through a representation $\rho:\Gamma_g\longrightarrow SL(2,\mathbb{R})$. This leads to an increasing tower of local systems $F_{X,i}^n$ $(n,i\geq 0)$ (cf §4) over the Riemann surface $X:=\rho(\Gamma_g)\backslash\mathbb{H}$, which generalize the Eichler integrals. The graduations $G_{X,i}^n:=F_{X,i+1}^n/F_{X,i}^n$ $(n,i\geq 0)$ form vector bundles over the moduli of spinned Riemann surfaces such that i) the sum $\mathcal{G}_X:=\oplus_{n,i}G_{X,i}^n$ is an infinitely generated graded algebra over the subalgebra $\mathcal{G}_{X,0}:=\oplus_n G_{X,0}^n$ (naturally isomorphic to (a part of) the canonical ring of X (cf §6)), and ii) the algebra $\mathcal{G}_X$ admits an exact sequence over $\mathcal{G}_{X,0}$(=a resolution of $\mathcal{G}_{X,0}$)

$$0 \longrightarrow \mathcal{G}_{X,0} \xrightarrow{\ i\ } \mathcal{G}_X \xrightarrow{\ \delta^0\ } \mathcal{G}_X\otimes_{\mathbb{Z}}H^1(X,\mathbb{Z}) \xrightarrow{\ \delta^1\ } (\mathcal{G}_X\oplus \mathbb{C}_X)\otimes_{\mathbb{Z}}H^2(X,\mathbb{Z}) \longrightarrow 0.$$

For details, see §7 Theorem.

The motivation of this work was to understand the transcendence of the complex structure on the moduli space of Riemann surfaces. Details will appear in a forthcoming paper[10].

## § 2. Weil's representation space

We illustrate how the infinite dimensional representation (mentioned in the introduction) comes naturally (or unavoidably) in a construction of the complex structure of the Teichmüller space. Let us start with finite dimensional representations of the Fuchsian surface group of genus $g\geq 2$.

$$\Gamma_g := \langle a_1, b_1, \ldots, a_g, b_g \mid \prod_{i=1}^{g} [a_i, b_i] = 1 \rangle$$

into the Lie group $G := SL(2,\mathbb{R})$. According to A. Weil[13], we denote

$$R_0(\Gamma_g, SL(2,\mathbb{R})) := \{\rho : \Gamma_g \longrightarrow SL(2,\mathbb{R}) \mid \text{an injective homomorphism, whose image is discrete and cocompact}\}.$$

The group $\Gamma_g$ (resp. G) acts on $\rho \in R_0(\Gamma_g, SL(2,\mathbb{R}))$ by composition with the inner automorphism action on $\Gamma_g$ (resp. the target G). Thus we introduce the spaces

$$S^1\mathfrak{X}_g := \Gamma_g \backslash R_0(\Gamma_g, SL(2,\mathbb{R}))$$

$$\mathfrak{X}_g := \Gamma_g \backslash R_0(\Gamma_g, SL(2,\mathbb{R}))/SO(2)$$

$$\mathfrak{T}_g := \Gamma_g \backslash R_0(\Gamma_g, SL(2,\mathbb{R}))/SL(2,\mathbb{R}).$$

It is classical that those spaces are real manifolds of dimension $6g-3$, $6g-4$ and $6g-6$ respectively (cf[5] and its references).

An interpretation of these spaces is as follows. Since $SL(2,\mathbb{R})/SO(2) \simeq \mathbb{H} := \{z \in \mathbb{C} : Im(z) > 0\}$, a representation $\rho : \Gamma_g \longrightarrow SL(2,\mathbb{R})$ determines a pair $(X, K_X)$ of a Riemann surface $X := \rho(\Gamma_g)\backslash\mathbb{H}$ and a spin bundle $\mathcal{O}(K_X^{1/2}) := \left(\mathcal{O}_{\mathbb{H}}(K_{\mathbb{H}}^{1/2})\right)^{\Gamma_g}$ on it (cf. §3). Since $(X, K_X)$ depends only on the conjugacy class of $\rho$, the natural map $\mathfrak{X}_g \longrightarrow \mathfrak{T}_g$ can be regarded as a universal family of spinned Riemann surface of genus $g$ and $S^1\mathfrak{X}_g \longrightarrow \mathfrak{X}_g$ is the unit circle bundle, which is the double covering of the unit circle bundle in the relative tangent bundle $T_{\mathfrak{X}_g/\mathfrak{T}_g}$ w.r.t. the Poincaré metric cf[9]. By this, $\mathfrak{T}_g$ is a finite covering of the classical Teichmüller space, which is well known to carry a natural complex structure (Weil, Ahlfors, Bers, Kodaira,..), so that $\mathcal{T}_g$ carries also a complex structure.

Let us define the complex structures on the real spaces $\mathcal{T}_g$ and $\mathfrak{X}_g$ directly by a use of Eichler-Shimura isomorphism. According to A. Weil [14], the tangent spaces of these spaces are described in terms of cocycles of the group $\Gamma_g$ with coefficients in the Lie algebra $sl(2,\mathbb{R})$ as follows. At each point $\rho$ of $S^1\mathfrak{X}_g$,

$$T_\rho(S^1\mathfrak{X}_g) \xrightarrow[e]{\simeq} Z^1(\Gamma_g, sl(2,\mathbb{R}))$$
$$v \longmapsto e(v),$$

where $$e(v)(\gamma) := \rho(\gamma)^{-1}\partial_v\rho(\gamma) \quad \text{for } {}^\forall\gamma\in\Gamma_g.$$

(i.e. a tangent vector v is mapped to a logarithmic differential operator on the matrix valued function $\rho(\gamma)$ at $\rho$, which takes its values in the Lie algebra). This induces the isomorphisms:

$$T_\rho(\mathfrak{X}_g) \xrightarrow{\simeq} Z^1(\Gamma_\rho, sl(2,\mathbb{R}))/\delta(so(2)) ,$$
$$T_\rho(\mathcal{T}_g) \xrightarrow{\simeq} Z^1(\Gamma_\rho, sl(2,\mathbb{R}))/\delta(sl(2)) =: H^1(\Gamma_g, sl(2,\mathbb{R})).$$

Here $\delta$ is the coboundary map defined by $\delta(m)(\gamma) := m\cdot\gamma - m$ for $\gamma\in\Gamma_g$. Of course, they are real tangent spaces of the real moduli space of discrete subgroups in $SL(2,\mathbb{R})$. Denote

$$\Gamma(\mathbb{H},\theta) := \{\text{all holomorphic vector fields on } \mathbb{H}\},$$

which is obviously a complex infinite dimensional Lie algebra. There is a natural inclusion map among the Lie algebras:

$$\iota : sl(2,\mathbb{R}) \longrightarrow \Gamma(\mathbb{H},\theta), \qquad \iota(X) := (-1,z)X\begin{bmatrix} z \\ 1 \end{bmatrix}\frac{d}{dz}.$$

An $A\in SL(2,\mathbb{R})$ acts on $sl(2,\mathbb{R})$ by the adjoint action $X\cdot ad(A) := A^{-1}XA$ and on $\Gamma(\mathbb{H},\theta)$ by the pullback of vector fields $v\cdot ad(A) := dA_*^{-1}(v)$ such that the actions are equivariant with $\iota$. At each fixed $\rho\in R_0(\Gamma_g, SL(2,\mathbb{R}))$, the composition $ad_\rho := ad\circ\rho$ defines the action of $\Gamma_g$ on the $sl(2,\mathbb{R})$ and $\Gamma(\mathbb{H},\theta)$. Then $\iota$ induces the isomorphisms:

$$H^1(\Gamma_g, sl(2,\mathbb{R})) \simeq H^1(\Gamma_g, \Gamma(\mathbb{H},\theta)),$$

called an Eichler-Shimura isomorphism [3,11], which also induces:

$$Z^1(\Gamma_\rho, sl(2,\mathbb{R}))/\delta(so(2)) \simeq Z^1(\Gamma_g, \Gamma(\mathbb{H},\Theta))/\delta(\Gamma(\mathbb{H},\Theta(-[i])))\,.$$

These isomorphisms define *almost complex structures* on the tangent bundles of $\mathcal{T}_g$ and $\mathfrak{X}_g$. The integrability of these almost complex structures is directly shown in [9].

*Note. The complex structure on* $H^1(\Gamma_g, sl(2,\mathbb{R}))$ *is defined through the representation of* $\Gamma_g$ *on the infinite space* $\Gamma(\mathbb{H},\Theta)$. Any known construction of the complex structure on the Teichmüller space uses infinite dimensional spaces as it is the case here (cf Beilinson-Manin-Shechtmann, Earl-Eels, Hitchin, Kodaira, Tromba, Wolpert etc. See references). Since the representation on $\Gamma(\mathbb{H},\Theta)$ is not unitary and not completely reducible, we construct an increasing tower of finite representations in $\Gamma(\mathbb{H},\Theta)$ in §4.

## § 3. Eichler Shimura isomorphism

We recall the Eichler Shimura isomorphisms [3,11] for $\Gamma_g$.

Let $Sym^n(\mathbb{R}^2)$ be the n-the symmetric tensor space of the vector representation space $\mathbb{R}^2$ of $SL(2,\mathbb{R})$. The space $Sym^n(\mathbb{R}^2)$ is identified with the space of real polynomials of one variable z of degree $\leq$ n, on which $A=\begin{bmatrix} a & b \\ c & d \end{bmatrix} \in SL(2,\mathbb{R})$ acts from the right by

$$\varphi(z)\cdot ad^{n/2}(A) := \varphi\left(\frac{az+b}{cz+d}\right)(cz+d)^n \quad .$$

Let us denote by $\Gamma(\mathbb{H},\mathcal{O}_{\mathbb{H}}(K_{\mathbb{H}}^{-n/2}))$ the space of symbols $\varphi(z)\left(\frac{d}{dz}\right)^{n/2}$ for all holomorphic functions $\varphi(z)$ over $\mathbb{H}$, on which $A=\begin{bmatrix} a & b \\ c & d \end{bmatrix}$ acts by

$$\left(\varphi(z)\left(\frac{d}{dz}\right)^{n/2}\right)\cdot ad^{n/2}(A) := \varphi(Az)\left(\frac{d}{dAz}\right)^{n/2} = \varphi\left(\frac{az+b}{cz+d}\right)(cz+d)^n\left(\frac{d}{dz}\right)^{n/2} .$$

The natural inclusion $\iota_n : Sym^n(\mathbb{R}^2) \longrightarrow \Gamma(\mathbb{H},\mathcal{O}_{\mathbb{H}}(K^{-n/2}))$ $(\varphi \mapsto \varphi\left(\frac{d}{dz}\right)^{n/2})$ is equivariant under the $ad^{n/2}$ action of $SL(2,\mathbb{R})$ and hence with the action $ad_\rho^{n/2}$ of $\Gamma_g$ defined by the composition $ad_\rho^{n/2} := ad^{n/2}\circ\rho$. Note that by this action, one has $\mathcal{O}_{\mathbb{H}}(K_{\mathbb{H}}^{-n/2})/\Gamma_g \simeq \mathcal{O}_X(K_X^{-n/2})$.

*Theorem. The* $\iota_n$ *induces an isomorphism of the 1-st cohomologies*

$$H^1(\Gamma_g, \mathrm{Sym}^n(\mathbb{R}^2)) \simeq H^1(\Gamma_g, \Gamma(\mathbb{H}, \mathcal{O}_{\mathbb{H}}(K^{-n/2}))) .$$

This is a particular case of Eichler-Shimura isomorphisms which treat general discrete groups. M. Eichler calculated the $\mathbb{R}$-rank of the left space and the $\mathbb{C}$-rank of the right space. G. Shimura showed the injectivity of the correspondance by a use of Petersson metric. Generalizations of the isomorphism are done by several authors. See S. Murakami [8] and its reference.

The case n=2 of the Theorem is discussed in § 2.

## § 4. Towers $F^n_{\rho,i}$ $(n, i \geq 0)$ of $\mathbb{C}$-$\Gamma_g$-modules

Recall that the inclusion $\mathrm{Sym}^n(\mathbb{R}^2) \longrightarrow \Gamma(\mathbb{H}, \mathcal{O}_{\mathbb{H}}(K^{-n/2}))$ induces the isomorphism of the 1-cohomology as in §3. It may be natural to ask for an intermediate $\mathbb{C}$-$\Gamma_g$-module $\mathcal{F}$, i.e. $\mathrm{Sym}^n(\mathbb{R}^2) \subset \mathcal{F} \subset \Gamma(\mathbb{H}, \mathcal{O}_{\mathbb{H}}(K_{\mathbb{H}}^{-n/2}))$ such that the natural inclusions induces the isomorphism

$$H^1(\Gamma_g, \mathrm{Sym}^n(\mathbb{R}^2)) \simeq H^1(\Gamma_g, \mathcal{F}) \simeq H^1(\Gamma_g, \Gamma(\mathbb{H}, \mathcal{O}_{\mathbb{H}}(K_{\mathbb{H}}^{-n/2})) \tag{4.1}$$

As an answer to the question, we have the following Lemma, which tells that there does not exist such $\mathcal{F}$ of finite rank but it exists a smallest $\mathcal{F}^n_\rho$ which admits an increasing filtration $(F^n_{\rho,i})_{i=0}^{\infty}$ of $\mathbb{C}$-$\Gamma_g$ submodules of finite rank and $\mathcal{F}^n_\rho = \bigcup_{i=0}^{\infty} F^n_{\rho,i}$.

*Lemma. Let the setting be the same as above. Fix* $\rho \in R_0(\Gamma, G)$.

1. *There exists a* $\mathbb{C}$-$\Gamma$ *submodule* $\mathcal{F}^n_\rho$ *of* $\Gamma(\mathbb{H}, \mathcal{O}_{\mathbb{H}}(K^{-n/2}))$ *with an increasing filtration by* $\mathbb{C}$-$\Gamma$ *submodules of finite rank,*

$$\{0\} \subset F^n_{\rho,0} \subset F^n_{\rho,1} \subset \dots \subset \mathcal{F}^n_\rho , \quad \bigcup_{i=0}^{\infty} F^n_{\rho,i} = \mathcal{F}^n_\rho$$

*such that*

i) $\mathbb{C} \otimes \iota_n : \mathbb{C} \otimes_{\mathbb{R}} \mathrm{Sym}^n(\mathbb{R}^2) \simeq F^n_{\rho,0}$ .

ii) *The* $F^n_{\rho,i+1}$ *is the largest* $\mathbb{C}$-$\Gamma$ *subspace of* $\mathcal{F}^n_\rho$ *containing* $F^n_{\rho,i}$ *such that* $F^n_{\rho,i+1}/F^n_{\rho,i}$ *is a* $\Gamma_g$ *trivial module for* $i\geq 0$.

iii) *The isomorphisms* (4.1) *hold for* $\mathcal{F} = \mathcal{F}^n_\rho$ .

2. i) *If there is a* $\mathbb{C}$-$\Gamma_g$ *submodule* $\mathcal{K}$ *of* $\Gamma(\mathbb{H},\mathcal{O}_{\mathbb{H}}(K_{\mathbb{H}}^{-n/2})$ *with* $\mathrm{Sym}^n(\mathbb{R}^2)\subset\mathcal{K}$ *such that* (4.1) *holds for* $\mathcal{F} = \mathcal{K}$. *Then* $\mathcal{F}^n_\rho \subset \mathcal{K}$.

ii) *Any* $\mathbb{C}$-$\Gamma_g$ *endomorphism of* $\mathcal{F}^n_\rho$, *which is the identity on* $\mathrm{Sym}^n(\mathbb{R}^2)$, *is the identity on* $\mathcal{F}^n_\rho$.

The property of Lemma 1.ii) can be reformulated as an exact sequence

$$0 \longrightarrow F^n_{\rho,i+1}/F^n_{\rho,i} \xrightarrow{\ \delta\ } H^1(\Gamma_g,F^n_{\rho,i}) \longrightarrow H^1(\Gamma_g,\mathcal{F}^n_\rho) \longrightarrow 0$$

for $i\geq 0$, where the first inclusion is induced from the coboundary map $\delta$. Particularly the case $i=0$ is important as we can see below.

*Lemma* 3. *Let the setting be as in Lemma* 1. *The image of* $F_1/F_0$ *in* $\mathbb{C}\otimes_{\mathbb{R}}H^1(\Gamma,R_n)$ *is the eigenspace of* J (*=the complex structure on* $H^1(\Gamma_g,R_n)$ *induced from the Eichler-Shimura isomorphism.*) *for the eigenvalue* $-\sqrt{-1}$. *Hence one has a direct sum decomposition:*

$$F^n_{\rho,1}/F^n_{\rho,0} \oplus \overline{F^n_{\rho,1}/F^n_{\rho,0}} \overset{\delta\oplus\delta*}{\simeq} \mathbb{C}\otimes_{\mathbb{R}}H^1(\Gamma_g,\mathrm{Sym}^n(\mathbb{R}^2)).$$

These Lemma's 1-3 are easy general results on the cohomology groups of finitely generated groups.

## § 5. Regularity

To determine more structures on the spaces $F^n_{\rho,i} \subset \Gamma(\mathbb{H},\mathcal{O}_{\mathbb{H}}(K_{\mathbb{H}}^{-n/2}))$ (for $n,i\in\mathbb{Z}_{\geq 0}$ and $\rho\in R_0(\Gamma_g,SL(2,\mathbb{R}))$), we introduce a concept:

*Definition. Let* $\mathcal{F} = (F_i)$ *be a tower for some* n *and* $\rho$ *as in* §4.

1. *$\mathcal{F}$ is i-regular for $i\geq 1$, if the multiplication of $\gamma-1$ on $F_i$ from the right is a surjection onto $F_{i-1}$ for all $\gamma\in\{a_1,b_1,\dots,a_g,b_g\}$.*

2. *$\mathcal{F}$ is 0-regular, if*

i) $\mathrm{rank}_{\mathbb{R}}(\ker(\gamma-1:R\longrightarrow R))\leq 1$ *for* $\forall\gamma\in\{a_1,b_1,\dots,a_g,b_g\}$,

ii) $R=R(a_i-1)+R(b_i-1)$ *for* $i=1,\dots,2g$.

*where $R$ is the real form* $\mathrm{Sym}^n(\mathbb{R}^2)$ *of* $F_0$ (cf §4 Lemma i)).

The following heredity of the regularity are key Lemmas, whose proofs require some detailed properties of the action of Fuchsian groups and SL(2,$\mathbb{R}$) on $\mathbb{H}$.

*Lemma* 4. *If $\mathcal{F}$ is i-regular, then it is i+1-regular.*

*Lemma* 5. i) *The $\mathcal{F}^0_\rho$ for any $\rho$ is 1-regular.*

ii) *The $\mathcal{F}^n_\rho$ for any $\rho$ and $n\geq 1$ is 0-regular.*

*Corollary.*

$$H^2(\Gamma_g,F^n_{\rho,i}) = \begin{cases} F^0_{\rho,0}\otimes H^2(\Gamma_g,\mathbb{Z}) & \text{for } i=0 \text{ and } n=0,\\ 0 & i=0, \text{ and } n\geq 1,\\ F^n_{\rho,i}/F^n_{\rho,i-1}\otimes H^2(\Gamma_g,\mathbb{Z}) & i\geq 1, \text{ and } n\geq 0.\end{cases}$$

As a consequence, the graduation $gr(\mathcal{F})$ is determined inductively as follows. First let us introduce a notation for the graduation.

$$G^n_{\rho,i} := F^n_{\rho,i+1}/F^n_{\rho,i}$$

for $n,i\geq 0$. By Lemma 1.ii) they are trivial $\Gamma_g$-modules. The long exact sequence of cohomology for $0\longrightarrow F^n_{\rho,i}\longrightarrow F^n_{\rho,i+1}\longrightarrow G^n_{\rho,i}\longrightarrow 0$ combined with the above Corollary yields the following.

*Lemma* 6. *For $n,i\geq 0$, one has exact sequences:*

$$0\longrightarrow G^n_{\rho,i+2}\xrightarrow{\delta^0} G^n_{\rho,i+1}\otimes H^1(\Gamma_g,\mathbb{Z})\xrightarrow{\delta^1} G^n_{\rho,i}\otimes H^2(\Gamma_g,\mathbb{Z})\longrightarrow 0.$$

$$0 \longrightarrow G^n_{\rho,1} \xrightarrow{\delta^0} G^n_{\rho,0} \otimes H^1(\Gamma_g,\mathbb{Z}) \xrightarrow{\delta^1} F \otimes H^2(\Gamma_g,\mathbb{Z}) \longrightarrow 0.$$

*where* $F:=F^0_{\rho,0}\simeq\mathbb{C}$ *for* $n=0$, *and* $:=\{0\}$ *for* $n>0$.

Combining this Lemma with the fact $F^n_{\rho,1}/F^n_{\rho,0} \simeq \Gamma(\mathbb{H},\mathcal{O}(K^{n/2+1}))^{\Gamma_g}$, which will be shown in §6 (G. Bol), one obtains the following expression of a generating function of $\left[G^n_{\rho,i}\right]$'s. Here we denote by $[G]$ the class of $G$ in the K-group of $\mathbb{C}$-vector spaces.

*Corollary.*

$$\sum_{i=0}^{\infty} [G^n_{\rho,i}]t^i = \frac{\left[\Gamma(\mathbb{H},\mathcal{O}(K^{n/2+1}))^{\Gamma_g}\right]}{1-\left[H^1(\Gamma_g,\mathbb{C})\right]t+\left[H^2(\Gamma_g,\mathbb{C})\right]t^2} \quad \text{for } n>0,$$

$$\sum_{i=0}^{\infty} [G^0_{\rho,i}]t^i = \frac{\left[\Gamma(\mathbb{H},\mathcal{O}(K))^{\Gamma_g}\right]-\left[F^0_{\rho,0}\right]t}{1-\left[H^1(\Gamma_g,\mathbb{C})\right]t+\left[H^2(\Gamma_g,\mathbb{C})\right]t^2} \quad \text{for } n=0.$$

In particular these give a formula for the rank $g^n_{g,i}$ of $G^n_{\rho,i}$, which, as obviously seen, depends only on $n,i$ and $g$.

## § 6. Local systems $F^n_{X,i}$ and their characterizing equations

Let us introduce a local system $(\mathbb{H}\times F^n_{\rho,i})/\Gamma_g$ over the quotient Riemann surface $X:=\rho(\Gamma_g)\backslash\mathbb{H}$ by letting $\gamma\in\Gamma$ act on $\mathbb{H}\times F^n_{\rho,i}$ by

$$(z,f)\cdot\gamma:=(\rho(\gamma)^{-1}z, f\cdot ad^{n/2}_{\rho}(\gamma)) \text{ of } \gamma\in\Gamma_g.$$

The commutativity of the following diagrams implies that the local system on the surface X depends only on the spin class $(X,K^{1/2}_X)$ and hence on the class $[\rho]\in \mathrm{Aut}(\Gamma_g)\backslash\mathfrak{T}_g)$ but not on $\rho$.

For $A\in PGL^+(2,\mathbb{R})(=\mathrm{Aut}(SL(2,\mathbb{R}))$ and $\alpha\in\mathrm{Aut}(\Gamma_g)$

$$\begin{array}{ccc} F^n_{\rho,i} & \xrightarrow{ad^{n/2}(A)} & F^n_{\rho\cdot Ad(A),i} \\ \downarrow \gamma & & \downarrow \gamma \\ F^n_{\rho,i} & \xrightarrow{ad^{n/2}(A)} & F^n_{\rho\cdot Ad(A),i} \end{array} \quad \text{and} \quad \begin{array}{ccc} F^n_{\rho,i+1} & \xrightarrow{ad^{n/2}(A)} & F^n_{\rho\cdot Ad(A),i+1} \\ \downarrow \delta_\rho & & \downarrow \delta_{\rho\cdot Ad(A)} \\ Z^1(\Gamma,F^n_{\rho,i}) & \xrightarrow{ad^{n/2}(A)} & Z^1(\Gamma,F^n_{\rho\cdot Ad(A),i}) \end{array}$$

$$\begin{array}{ccc} F^n_{\rho,i} & = & F^n_{\alpha\cdot\rho,i} \\ \downarrow \gamma & & \downarrow \alpha(\gamma) \\ F^n_{\rho,i} & = & F^n_{\alpha\cdot\rho,i} \end{array} \quad \textit{and} \quad \begin{array}{ccc} F^n_{\rho,i+1} & = & F^n_{\alpha\cdot\rho,i+1} \\ \downarrow \delta_\rho & & \downarrow \delta_{\alpha\cdot\rho} \\ Z^1(\Gamma,F^n_{\rho,i}) & \xrightarrow{\varphi} & Z^1(\Gamma,F^n_{\alpha\cdot\rho,i}) \end{array}$$

(*where* $\varphi$ *is a bijection defined by* $\varphi(c)(\gamma):= c(\alpha^{-1}(\gamma)))$.

Hence we denote the local system on X by

$$\mathcal{F}^n_{X,i} := (\mathbb{H}\times F^n_{\rho,i})/\Gamma_g \ .$$

In the following, we give a description of the local systems $\mathcal{F}^n_{X,i}$ as the local systems of solutions of certain differential equations on X. The property ii) of $\mathcal{F}^n_p$ of the Lemma 1 implies explicitly: an element $\varphi(z)\left(\frac{d}{dz}\right)^{n/2}\in\Gamma(\mathbb{H},\mathcal{O}(K^{-n/2}))$ belongs to $F^n_{\rho,i+1}$, if and only if

$$\varphi\left(\frac{az+b}{cz+d}\right)(cz+d)^n - \varphi(z) \in F^n_{\rho,i}$$

for ${}^\vee\rho(\gamma)=\begin{bmatrix} a & b \\ c & d\end{bmatrix}$. Let us take (n+1)-th derivative of the relation for i=0, so that the right hand side vanishes. Applying the formula

$$\left(\frac{d}{dz}\right)^{n+1}\left(\varphi\left(\frac{az+b}{cz+d}\right)(cz+d)^n\right)=\left(\left(\frac{d}{dz}\right)^{n+1}\varphi\right)\left(\frac{az+b}{cz+d}\right)(cz+d)^{-n-2},$$

we obtain the relation:

$$\varphi^{(n+1)}\left(\frac{az+b}{cz+d}\right)(cz+d)^{-n-2} - \varphi^{(n+1)}(z)=0.$$

Hence:

*Assertion* (G. Bol[16]). $\varphi(z)\left(\frac{d}{dz}\right)^{n/2}$ *belongs to* $F^n_{\rho,1}$, *iff* $\varphi^{(n+1)}$ *is an automorphic form of weight* n+2. *Particularly*

$$\partial^{n+1}:\quad F^n_{\rho,1}/F^n_{\rho,0} \simeq \Gamma(\mathbb{H},\mathcal{O}_{\mathbb{H}}(K^{n/2+1}))^{\Gamma_g}.$$

*Remark* 1. The above isomorphism implies that $F^n_{\rho,1}$ is the space of (n+1)-th integrals of the forms in $\Gamma(X,\mathcal{O}_X(K^{n/2+1}))$ on the Riemann surface $X:=\rho(\Gamma)\backslash\mathbb{H}$ and $F^n_{\rho,0}$ is considered as the space of integral constants. The integral is intensively studied by M. Eichler[3].

2. Recall that the space $G^n_{\rho,0}:= F^n_{\rho,1}/F^n_{\rho,0}$ is complex conjugate to the space $H^1(\Gamma_g,\Gamma(\mathbb{H},\mathcal{O}(K^{-n/2}))$ (§4. Lemma 3), which is complex dual to $\Gamma(\mathbb{H},\mathcal{O}(K^{n/2+1})^{\Gamma_g}$ by Serre duality and is isomorphic to $G^n_{\rho,0}$ by the inverse of Bol's map $\partial^{n+1}$. Hence $G^n_{\rho,0}$ is the Hermitian dual

of itself. This is the Petersson metric on $G^n_{\rho,0}$.

We denote by $G^n_{X,i}:=F^n_{X,i+1}/F^n_{X,i}$ the trivial sheaf on X, whose fiber is isomorphic to $G^n_{\rho,i}$. The Bol's isomorphism can be generalized to:

*Lemma 7. For each* $n\geq 0$, *there exists a sequence of differential operators* $D^n_{X,i}$ $(i\in\mathbb{N})$ *of degree* $= g^n_{g,i}$ *over a Riemann surface* X

$$\mathcal{O}_X(K_X^{-n/2}) \xrightarrow{\partial^{n+1}} \mathcal{O}_X(K_X^{n/2+1}) \xrightarrow{D^n_{X,1}} \mathcal{O}_X(K_X^{d^n_{g,1}/2}) \xrightarrow{D^n_{X,2}} \mathcal{O}_X(K_X^{d^n_{g,2}/2}) \longrightarrow \cdots$$

*such that*

i) $\quad \mathrm{Ker}(D^n_{X,i}\circ D^n_{X,i-1}\circ\cdots\circ D^n_{X,0}) = F^n_{X,i}$ ,

ii) $\quad D^n_{X,i}\circ D^n_{X,i-1}\circ\cdots\circ D^n_{X,0}(F^n_{X,i+1}) \simeq G^n_{X,i} \subset \Gamma(X,\mathcal{O}(K_X^{d^n_{g,i}/2}))$.

Here $d^n_{g,i}$ is a sequence of integers inductively defined by $d^n_{g,0}:=n+2$ and $d^n_{g,i+1}:=(d^n_{g,i}+g^n_{g,i})(g^n_{g,i}+1)$. The above i) implies that the space $F^n_{\rho,i}$ may be defined by a linear differential equation The operators $D^n_{X,i}$ are easily described by a use of Wronskians but the author does not know how to determine the image $G^n_{X,i}$ of ii).

As another consequence of the Bol's differential operator $\partial^{n+1}$: $\Gamma(\mathbb{H},\mathcal{O}(K_{\mathbb{H}}^{-n/2}))\longrightarrow\Gamma(\mathbb{H},\mathcal{O}(K_X^{n/2\ +1})$, it is not hard to show the following:

*Lemma 8. The product* $K_{\mathbb{H}}^{n/2+1}\times K_{\mathbb{H}}^{m/2+1} \longrightarrow K_{\mathbb{H}}^{(n+m)/2+2}$ *of the powers of the half canonical bundle* $K_{\mathbb{H}}$ *induces a* $\Gamma_g$*-equivariant product:*

$$\partial^{n+1}(\mathcal{F}^n_X) \times \partial^{m+1}(\mathcal{F}^m_X) \longrightarrow \partial^{n+m+3}(\mathcal{F}^{n+m+2}_X)$$

*which preserves the filtration as follows:*

$$\partial^{n+1}(F^n_{X,i+1}) \times \partial^{m+1}(F^m_{X,j+1}) \longrightarrow \partial^{n+m+3}(F^{n+m+2}_{X,i+j+1})$$

Particularly the product induces a product on the graduation: $G^n_{X,i}\times G^m_{X,j} \longrightarrow G^{n+m+2}_{X,i+j}$ for $i,j,n,m\geq 0$. Therefore the sum

$$\mathcal{G}_X := \bigoplus_{n,i=0}^{\infty} G^n_{X,i}$$

forms a doubly graded algebra. Due to Bol's isomorphism, the i=0 graded part $\mathcal{G}_{X,0} := \bigoplus_{n=0}^{\infty} G^n_{X,0} \simeq \bigoplus_{n=0}^{\infty} H^0(X, \mathcal{O}(K_X^{n/2+1}))$ is (a part of) the half canonical ring for the spinned Riemann surface $(X, K_X)$. One checks easily that coboundary maps $\delta^i$ (i=0,1) in the Lemma 6 are derivation maps of the algebra $\mathcal{G}_X$ over $\mathcal{G}_{X,0}$.

## §7. Relative local systems over $\mathfrak{X}_g$ and Vector bundles over $\mathcal{T}_g$

Recall the universal family $\Phi: \mathfrak{X}_g \longrightarrow \mathcal{T}_g$ of spinned Riemann surfaces, on which $Out(\Gamma_g) := Aut(\Gamma_g)/\Gamma_g$ (the outer automorphism group of $\Gamma_g$) acts equivariantly. The local system $F^n_{X,i}$ over X, as defined in §5, depends up to conjugacy by $SL(2,\mathbb{R}) \times Aut(\Gamma_g)$ on the parameter $\rho$ and hence on $Out(\Gamma_g) \backslash \mathcal{T}_g$. Forgetting about the action of $Out(\Gamma_g)$ for the moment, one obtains a relative local system

$$F^n_{\mathfrak{X}_g, i}$$

over $\mathfrak{X}_g$ such that i) its restriction on each fiber $X_t := \Phi^{-1}(t)$ $(t \in \mathcal{T}_g)$ gives the local system $F^n_{X_t, i}$, ii) it admits an action of $Out(\Gamma_g)$ (The relative local system is defined as a $\Phi^{-1}(\mathcal{A}_{\mathcal{T}_g})$-module).

Since the quotient $F^n_{\mathfrak{X}_g, i+1}/F^n_{\mathfrak{X}_g, i}$ is trivial on each fiber of $\Phi$ and of constant rank $g^n_{g,i}$, it is the pull-back of a vector bundle

$$G^n_{\mathcal{T}_g, i}$$

over $\mathcal{T}_g$ of rank $g^n_{g,i}$. The $Out(\Gamma_g)$ acts on the bundles $G^n_{\mathcal{T}_g, i}$.

Summarizing the Lemma's 1-8 and their Corollarys, we have:

*Theorem. For g>1, there are relative local systems* $\mathcal{F}^n_{\mathfrak{X}_g, i}$ *on* $\mathfrak{X}_g$ *relative to* $\mathcal{T}_g$ *and complex vector bundles* $G^n_{\mathcal{T}_g, i}$ *on* $\mathcal{T}_g$ $(n, i \in \mathbb{Z}_{\geq 0})$

*with the following properties:*

i) $F^{n}_{\mathfrak{X}_g, i+1}/F^{n}_{\mathfrak{X}_g, i} \simeq \Phi^{-1}(G^{n}_{\mathcal{T}_g, i})$ *for* n, i≥0.

ii) $\mathbb{R}\Phi_*(F^{n}_{\mathfrak{X}_g, i}) \simeq \mathbb{R}\Phi_*(\mathcal{O}_{\mathfrak{X}_g}(K^{-n/2}_{\mathfrak{X}_g/\mathcal{T}_g}))\oplus G^{n}_{\mathcal{T}_g, i+1}[1]\oplus G^{n}_{\mathcal{T}, i-1}[2]$ *for* n, i≥0

iii) *The sum* $\mathcal{G}_{\mathcal{T}_g} := \bigoplus_{n,i=0}^{\infty} G^{n}_{\mathcal{T}_g, i}$ *forms a graded algebra over (a part of) the half canonical ring* $\bigoplus_{n=0}^{\infty} \Phi_*(\mathcal{O}_{\mathfrak{X}_g}(K^{n/2+1}_{\mathfrak{X}_g/\mathcal{T}_g})$ *such that there exists derivation operators* $\delta^{i}$ *with a short exact sequence:*

$$0 \longrightarrow \mathcal{G}_{\mathcal{T}_g, 0} \longrightarrow \mathcal{G}_{\mathcal{T}_g} \xrightarrow{\delta^0} \mathcal{G}_{\mathcal{T}_g}\otimes \mathbb{R}^1\Phi_*(\mathbb{C}_{\mathfrak{X}}) \xrightarrow{\delta^1} (\mathcal{G}_{\mathcal{T}_g} \oplus \mathbb{C}_{\mathcal{T}_g})\otimes\mathbb{R}^2\Phi_*(\mathbb{C}_{\mathfrak{X}}) \longrightarrow 0.$$

iv) $$\sum_{i=0}^{\infty} [G^{n}_{\mathfrak{X}_g, i}]t^{i} = \frac{\left[\Phi_*\left(\mathcal{O}_{\mathfrak{X}}(K^{n/2+1}_{\mathfrak{X}/\mathcal{T}})\right)\right]}{1 - \left[\mathbb{R}^1\Phi_*(\mathbb{C}_{\mathfrak{X}})\right]t + \left[\mathbb{R}^2\Phi_*(\mathbb{C}_{\mathfrak{X}})\right]t^2} \quad \textit{for } n>0,$$

$$\sum_{i=0}^{\infty} [G^{0}_{\mathfrak{X}_g, i}]t^{i} = \frac{\left[\Phi_*(\mathcal{O}_{\mathfrak{X}})\right] - \left[\mathbb{C}_{\mathcal{T}}\right]t}{1 - \left[\mathbb{R}^1\Phi_*(\mathbb{C}_{\mathfrak{X}})\right]t + \left[\mathbb{R}^2\Phi_*(\mathbb{C}_{\mathfrak{X}})\right]t^2} \quad \textit{for } n=0.$$

## Literature

[1] A.A. Beilinson, Y.I. Manin, and Y.A. Schechtman: Localization of the Virasoro and Neveu-Schwartz Algebra, Lecture Note Math. 1289

[2] C.G. Earle and J. Eells: A fiber bundle description of Teichmüller theory, J. Diff. Geom. 3 (1969) 19-43.

[3] M. Eichler: Eine Verallgemeinerung der Abelschen Integrale, Math. Z., 67 (1957), 267-298.

[4] W. Goldman: The Symplectic Nature of Fundamental groups of Surfaces, Advances in Math. 54, 200-225 (1984).

[5] W.J. Harvey et al.: Discrete Groups and Automorphic Functions, Proc. of an instructional Conf. organized by London Math. Soc. and the Univ. of Cambridge, Academic Press (1977).

[6] N.J. Hitchin: The self-duality equation on a Riemann surface, Proc. London Math. Soc. (1987).

[7] K. Kodaira and D.C. Spencer: Existence of complex structure on a differentiable family of deformations of compact complex manifolds, Annals of Math., **70**(1959), 145-166.

[8] S. Murakami: Vanishing theorems on cohomology associated to Hermitian symmetric spaces, Ann. Inst. Fourier, Grenoble, **37**, 4 (1989), 225-233.

[9] K. Saito: Moduli Space for Fuchsian Groups, Algebraic Analysis, Vol. II., edit. by Kashiwara & Kawai, Academic Press, (1988), 735-787.

[10] K. Saito: A generalization of Eichler integrals and certain local systems over spinned Riemann surfaces. to appear.

[11] G. Shimura: Sur les intégrales attachées aux formes automorphes, J. Math Soc. Japan, **11** (1959), 291-311.

[12] A.J. Tromba and A.E. Fischer: Almost complex principal bundles and complex structure on Teichmüller space, Jr. reine und ang. Math. **352** (1984) 151-160.

[13] A. Weil: On discrete subgroups of Lie groups, I. Ann. of Math. **72** (1960) 369-384, II. Ann. of Math. **75** (1962) 578-602.

[14] A. Weil: Remarks on the cohomology of groups, Ann. of Math. **80** (1964), 149-157.

[15] S. Wolpert: An oral communication on a construction of a complex structure on the Teichmüller space.

[16] G. Bol: Invarianten linearer Differentialgleichungen. Abh. math. Seminar Hamburger Univ. **16**, 1-28 (1949).

Added in Proof.

1. After submitting this note, the author realized that the relative local systems $\mathcal{F}^n_{\mathfrak{X}_g,i}$ (n,i≥0) over $\mathfrak{X}_g$ and the vector bundles $\mathcal{G}^n_{\mathcal{T}_g}$ (n,i≥0) over $\mathcal{T}_g$ are not holomorphic and that it is necessary to add their complex conjugate part as in the Lemma 3

to recover the holomorphic structure. Since it asks some more works to write it down, we leave the present note as it is here.

2. Prof. Takayuki Oda has pointed out to the author a relation between the formula of the Corollary to Lemma 6, Theorem of [a] and (1.4) Theorem of [b]. It seems interesting to clarify the relationship among the approaches.

[a] Labute, J.P.: On the decending central series of groups with a single defining relation, Journal of Algebra **14** (1970), 16-23.

[b] Kohno, T. and Oda, T.: The Lower Central Series of the Pure Braid Group of an Algebraic Curve, Adv. Studies in Pure Math. **12**, 1987, 201-219.

ol. 1371: S. Glaz, Commutative Coherent Rings. XI, 347 pages. 989.

'ol. 1372: J. Azéma, P.A. Meyer, M. Yor (Eds.), Séminaire de robabilités XXIII. Proceedings. IV, 583 pages. 1989.

ol. 1373: G. Benkart, J.M. Osborn (Eds.), Lie Algebras. ladison 1987. Proceedings. V, 145 pages. 1989.

'ol. 1374: R.C. Kirby, The Topology of 4-Manifolds. VI, 108 ages. 1989.

ol. 1375: K. Kawakubo (Ed.), Transformation Groups. roceedings, 1987. VIII, 394 pages, 1989.

ol. 1376: J. Lindenstrauss, V.D. Milman (Eds.), Geometric spects of Functional Analysis. Seminar (GAFA) 1987–88. VII, 88 pages. 1989.

ol. 1377: J.F. Pierce, Singularity Theory, Rod Theory, and ymmetry-Breaking Loads. IV, 177 pages. 1989.

ol. 1378: R.S. Rumely, Capacity Theory on Algebraic Curves. I, 437 pages. 1989.

ol. 1379: H. Heyer (Ed.), Probability Measures on Groups IX. roceedings, 1988. VIII, 437 pages. 1989.

ol. 1380: H.P. Schlickewei, E. Wirsing (Eds.), Number Theory, lm 1987. Proceedings. V, 266 pages. 1989.

ol. 1381: J.-O. Strömberg, A. Torchinsky, Weighted Hardy paces. V, 193 pages. 1989.

ol. 1382: H. Reiter, Metaplectic Groups and Segal Algebras. I, 128 pages. 1989.

ol. 1383: D.V. Chudnovsky, G.V. Chudnovsky, H. Cohn, M.B. athanson (Eds.), Number Theory, New York 1985–88. Semiar. V, 256 pages. 1989.

ol. 1384: J. Garcia-Cuerva (Ed.), Harmonic Analysis and artial Differential Equations. Proceedings, 1987. VII, 213 ages. 1989.

ol. 1385: A.M. Anile, Y. Choquet-Bruhat (Eds.), Relativistic luid Dynamics. Seminar, 1987. V, 308 pages. 1989.

ol. 1386: A. Bellen, C.W. Gear, E. Russo (Eds.), Numerical lethods for Ordinary Differential Equations. Proceedings, 1987. II, 136 pages. 1989.

ol. 1387: M. Petkovi´c, Iterative Methods for Simultaneous clusion of Polynomial Zeros. X, 263 pages. 1989.

ol. 1388: J. Shinoda, T.A. Slaman, T. Tugué (Eds.), lathematical Logic and Applications. Proceedings, 1987. V, 23 pages. 1989.

ol. 1000: Second Edition. H. Hopf, Differential Geometry in e Large. VII, 184 pages. 1989.

ol. 1389: E. Ballico, C. Ciliberto (Eds.), Algebraic Curves and rojective Geometry. Proceedings, 1988. V, 288 pages. 1989.

ol. 1390: G. Da Prato, L. Tubaro (Eds.), Stochastic Partial ifferential Equations and Applications II. Proceedings, 1988. I, 258 pages. 1989.

ol. 1391: S. Cambanis, A. Weron (Eds.), Probability Theory Vector Spaces IV. Proceedings, 1987. VIII, 424 pages. 1989.

ol. 1392: R. Silhol, Real Algebraic Surfaces. X, 215 pages. 989.

ol. 1393: N. Bouleau, D. Feyel, F. Hirsch, G. Mokobodzki ds.), Séminaire de Théorie du Potentiel Paris, No. 9. roceedings. VI, 265 pages. 1989.

ol. 1394: T.L. Gill, W.W. Zachary (Eds.), Nonlinear emigroups, Partial Differential Equations and Attractors. roceedings, 1987. IX, 233 pages. 1989.

ol. 1395: K. Alladi (Ed.), Number Theory, Madras 1987. oceedings. VII, 234 pages. 1989.

Vol. 1396: L. Accardi, W. von Waldenfels (Eds.), Quantum Probability and Applications IV. Proceedings, 1987. VI, 355 pages. 1989.

Vol. 1397: P.R. Turner (Ed.), Numerical Analysis and Parallel Processing. Seminar, 1987. VI, 264 pages. 1989.

Vol. 1398: A.C. Kim, B.H. Neumann (Eds.), Groups – Korea 1988. Proceedings. V, 189 pages. 1989.

Vol. 1399: W.-P. Barth, H. Lange (Eds.), Arithmetic of Complex Manifolds. Proceedings, 1988. V, 171 pages. 1989.

Vol. 1400: U. Jannsen. Mixed Motives and Algebraic K-Theory. XIII, 246 pages. 1990.

Vol. 1401: J. Steprans, S. Watson (Eds.), Set Theory and its Applications. Proceedings, 1987. V, 227 pages. 1989.

Vol. 1402: C. Carasso, P. Charrier, B. Hanouzet, J.-L. Joly (Eds.), Nonlinear Hyperbolic Problems. Proceedings, 1988. V, 249 pages. 1989.

Vol. 1403: B. Simeone (Ed.), Combinatorial Optimization. Seminar, 1986. V, 314 pages. 1989.

Vol. 1404: M.-P. Malliavin (Ed.), Séminaire d´Algèbre Paul Dubreil et Marie-Paul Malliavin. Proceedings, 1987–1988. IV, 410 pages. 1989.

Vol. 1405: S. Dolecki (Ed.), Optimization. Proceedings, 1988. V, 223 pages. 1989. Vol. 1406: L. Jacobsen (Ed.), Analytic Theory of Continued Fractions III. Proceedings, 1988. VI, 142 pages. 1989.

Vol. 1407: W. Pohlers, Proof Theory. VI, 213 pages. 1989.

Vol. 1408: W. Lück, Transformation Groups and Algebraic K-Theory. XII, 443 pages. 1989.

Vol. 1409: E. Hairer, Ch. Lubich, M. Roche. The Numerical Solution of Differential-Algebraic Systems by Runge-Kutta Methods. VII, 139 pages. 1989.

Vol. 1410: F.J. Carreras, O. Gil-Medrano, A.M. Naveira (Eds.), Differential Geometry. Proceedings, 1988. V, 308 pages. 1989.

Vol. 1411: B. Jiang (Ed.), Topological Fixed Point Theory and Applications. Proceedings. 1988. VI, 203 pages. 1989.

Vol. 1412: V.V. Kalashnikov, V.M. Zolotarev (Eds.), Stability Problems for Stochastic Models. Proceedings, 1987. X, 380 pages. 1989.

Vol. 1413: S. Wright, Uniqueness of the Injective $III_1$ Factor. III, 108 pages. 1989.

Vol. 1414: E. Ramirez de Arellano (Ed.), Algebraic Geometry and Complex Analysis. Proceedings, 1987. VI, 180 pages. 1989.

Vol. 1415: M. Langevin, M. Waldschmidt (Eds.), Cinquante Ans de Polynômes. Fifty Years of Polynomials. Proceedings, 1988. IX, 235 pages. 1990.

Vol. 1416: C. Albert (Ed.), Géométrie Symplectique et Mécanique. Proceedings, 1988. V, 289 pages. 1990.

Vol. 1417: A.J. Sommese, A. Biancofiore, E.L. Livorni (Eds.), Algebraic Geometry. Proceedings, 1988. V, 320 pages. 1990.

Vol. 1418: M. Mimura (Ed. ), Homotopy Theory and Related Topics. Proceedings, 1988. V, 241 pages. 1990.

Vol. 1419: P.S. Bullen, P.Y. Lee, J.L. Mawhin, P. Muldowney, W.F. Pfeffer (Eds.), New Integrals. Proceedings, 1988. V, 202 pages. 1990.

Vol. 1420: M. Galbiati, A. Tognoli (Eds.), Real Analytic Geometry. Proceedings, 1988. IV, 366 pages. 1990.

Vol. 1421: H.A. Biagioni, A Nonlinear Theory of Generalized Functions, XII, 214 pages. 1990.

Vol. 1422: V. Villani (Ed.), Complex Geometry and Analysis. Proceedings, 1988. V, 109 pages. 1990.

Vol. 1423: S.O. Kochman, Stable Homotopy Groups of Spheres: A Computer-Assisted Approach. VIII, 330 pages. 1990.

Vol. 1424: F.E. Burstall, J.H. Rawnsley, Twistor Theory for Riemannian Symmetric Spaces. III, 112 pages. 1990.

Vol. 1425: R.A. Piccinini (Ed.), Groups of Self-Equivalences and Related Topics. Proceedings, 1988. V, 214 pages. 1990.

Vol. 1426: J. Azéma, P.A. Meyer, M. Yor (Eds.), Séminaire de Probabilités XXIV, 1988/89. V, 490 pages. 1990.

Vol. 1427: A. Ancona, D. Geman, N. Ikeda, École d'Eté de Probabilités de Saint Flour

XVIII, 1988. Ed.: P.L. Hennequin. VII, 330 pages. 1990.

Vol. 1428: K. Erdmann, Blocks of Tame Representation Type and Related Algebras. XV. 312 pages. 1990.

Vol. 1429: S. Homer, A. Nerode, R.A. Platek, G.E. Sacks, A. Scedrov, Logic and Computer Science. Seminar, 1988. Editor: P. Odifreddi. V, 162 pages. 1990.

Vol. 1430: W. Bruns, A. Simis (Eds.), Commutative Algebra. Proceedings. 1988. V, 160 pages. 1990.

Vol. 1431: J.G. Heywood, K. Masuda, R. Rautmann, V.A. Solonnikov (Eds.), The Navier-Stokes Equations – Theory and Numerical Methods. Proceedings, 1988. VII, 238 pages. 1990.

Vol. 1432: K. Ambos-Spies, G.H. Müller, G.E. Sacks (Eds.), Recursion Theory Week. Proceedings, 1989. VI, 393 pages. 1990.

Vol. 1433: S. Lang, W. Cherry, Topics in Nevanlinna Theory. II, 174 pages.1990.

Vol. 1434: K. Nagasaka, E. Fouvry (Eds.), Analytic Number Theory. Proceedings, 1988. VI, 218 pages. 1990.

Vol. 1435: St. Ruscheweyh, E.B. Saff, L.C. Salinas, R.S. Varga (Eds.), Computational Methods and Function Theory. Proceedings, 1989. VI, 211 pages. 1990.

Vol. 1436: S. Xambó-Descamps (Ed.), Enumerative Geometry. Proceedings, 1987. V, 303 pages. 1990.

Vol. 1437: H. Inassaridze (Ed.), K-theory and Homological Algebra. Seminar, 1987–88. V, 313 pages. 1990.

Vol. 1438: P.G. Lemarié (Ed.) Les Ondelettes en 1989. Seminar. IV, 212 pages. 1990.

Vol. 1439: E. Bujalance, J.J. Etayo, J.M. Gamboa, G. Gromadzki. Automorphism Groups of Compact Bordered Klein Surfaces: A Combinatorial Approach. XIII, 201 pages. 1990.

Vol. 1440: P. Latiolais (Ed.), Topology and Combinatorial Groups Theory. Seminar, 1985–1988. VI, 207 pages. 1990.

Vol. 1441: M. Coornaert, T. Delzant, A. Papadopoulos. Géométrie et théorie des groupes. X, 165 pages. 1990.

Vol. 1442: L. Accardi, M. von Waldenfels (Eds.), Quantum Probability and Applications V. Proceedings, 1988. VI, 413 pages. 1990.

Vol. 1443: K.H. Dovermann, R. Schultz, Equivariant Surgery Theories and Their Periodicity Properties. VI, 227 pages. 1990.

Vol. 1444: H. Korezlioglu, A.S. Ustunel (Eds.), Stochastic Analysis and Related Topics VI. Proceedings, 1988. V, 268 pages. 1990.

Vol. 1445: F. Schulz, Regularity Theory for Quasilinear Elliptic Systems and – Monge Ampère Equations in Two Dimensions. XV, 123 pages. 1990.

Vol. 1446: Methods of Nonconvex Analysis. Seminar, 1989. Editor: A. Cellina. V, 206 pages. 1990.

Vol. 1447: J.-G. Labesse, J. Schwermer (Eds), Cohomology of Arithmetic Groups and Automorphic Forms. Proceedings, 1989. V, 358 pages. 1990.

Vol. 1448: S.K. Jain, S.R. López-Permouth (Eds.), Non-Commutative Ring Theory. Proceedings, 1989. V, 166 pages. 1990.

Vol. 1449: W. Odyniec, G. Lewicki, Minimal Projections in Banach Spaces. VIII, 168 pages. 1990.

Vol. 1450: H. Fujita, T. Ikebe, S.T. Kuroda (Eds.), Functional-Analytic Methods for Partial Differential Equations. Proceedings, 1989. VII, 252 pages. 1990.

Vol. 1451: L. Alvarez-Gaumé, E. Arbarello, C. De Concini, N.J. Hitchin, Global Geometry and Mathematical Physics. Montecatini Terme 1988. Seminar. Editors: M. Francaviglia, F. Gherardelli. IX, 197 pages. 1990.

Vol. 1452: E. Hlawka, R.F. Tichy (Eds.), Number-Theoretic Analysis. Seminar, 1988–89. V, 220 pages. 1990.

Vol. 1453: Yu.G. Borisovich, Yu.E. Gliklikh (Eds.), Global Analysis – Studies and Applications IV. V, 320 pages. 1990.

Vol. 1454: F. Baldassari, S. Bosch, B. Dwork (Eds.), p-adic Analysis. Proceedings, 1989. V, 382 pages. 1990.

Vol. 1455: J.-P. Françoise, R. Roussarie (Eds.), Bifurcations of Planar Vector Fields. Proceedings, 1989. VI, 396 pages. 1990

Vol. 1456: L.G. Kovács (Ed.), Groups – Canberra 1989. Proceedings. XII, 198 pages. 1990.

Vol. 1457: O. Axelsson, L.Yu. Kolotilina (Eds.), Preconditioned Conjugate Gradient Methods. Proceedings, 1989. V, 196 pages. 1990.

Vol. 1458: R. Schaaf, Global Solution Branches of Two Point Boundary Value Problems. XIX, 141 pages. 1990.

Vol. 1459: D. Tiba, Optimal Control of Nonsmooth Distributed Parameter Systems. VII, 159 pages. 1990.

Vol. 1460: G. Toscani, V. Boffi, S. Rionero (Eds.), Mathematical Aspects of Fluid Plasma Dynamics. Proceedings, 1988. V, 22[illegible] pages. 1991.

Vol. 1461: R. Gorenflo, S. Vessella, Abel Integral Equations. VII, 215 pages. 1991.

Vol. 1462: D. Mond, J. Montaldi (Eds.), Singularity Theory and its Applications. Warwick 1989, Part I. VIII, 405 pages. 199[illegible]

Vol. 1463: R. Roberts, I. Stewart (Eds.), Singularity Theory and its Applications. Warwick 1989, Part II. VIII, 322 pages. 199[illegible]

Vol. 1464: D. L. Burkholder, E. Pardoux, A. Sznitman, Ecole d'Eté de Probabilités de Saint- Flour XIX-1989. Edited by P. [illegible] Hennequin. VI, 259 pages. 1991.

Vol. 1465: G. David, Wavelets and Singular Integrals on Curves and Surfaces. X, 107 pages. 1991.

Vol. 1466: W. Banaszczyk, Additive Subgroups of Topological Vector Spaces. VII, 178 pages. 1991.

Vol. 1467: W. M. Schmidt, Diophantine Approximations and Equations. VIII, 217 pages. 1991.

Vol. 1468: J. Noguchi, T. Ohsawa (Eds.), Prospects in Complex Geometry. Proceedings, 1989. VII, 421 pages. 1991.